Lecture Notes in Computer Science 15211

Founding Editors

Gerhard Goos
Juris Hartmanis

The series Lecture Notes in Computer Science (LNCS), including its subseries Lecture Notes in Artificial Intelligence (LNAI) and Lecture Notes in Bioinformatics (LNBI), has established itself as a medium for the publication of new developments in computer science and information technology research, teaching, and education.

LNCS enjoys close cooperation with the computer science R & D community, the series counts many renowned academics among its volume editors and paper authors, and collaborates with prestigious societies. Its mission is to serve this international community by providing an invaluable service, mainly focused on the publication of conference and workshop proceedings and postproceedings. LNCS commenced publication in 1973.

Luca Maria Aiello · Tanmoy Chakraborty ·
Sabrina Gaito

Editors

Social Networks Analysis and Mining

16th International Conference, ASONAM 2024
Rende, Italy, September 2–5, 2024
Proceedings, Part I

 Springer

Editors
Luca Maria Aiello 🆔
IT University of Copenhagen
Copenhagen, Denmark

Tanmoy Chakraborty 🆔
Indian Institute of Technology Delhi
New Delhi, Delhi, India

Sabrina Gaito 🆔
Università degli Studi di Milano
Milan, Italy

ISSN 0302-9743   ISSN 1611-3349 (electronic)
Lecture Notes in Computer Science
ISBN 978-3-031-78540-5   ISBN 978-3-031-78541-2 (eBook)
https://doi.org/10.1007/978-3-031-78541-2

This Springer imprint is published by the registered company Springer Nature Switzerland AG
The registered company address is: Gewerbestrasse 11, 6330 Cham, Switzerland

If disposing of this product, please recycle the paper.

Preface

With ASONAM 2024, the International Conference on Advances in Social Network Analysis and Mining marked its 16th anniversary as the flagship, premier and leading venue in the rapidly growing domain of social network analysis and mining. It has emerged into one of the most well-established and successful conferences. It is a great pleasure that the acceptance rate has stabilized below 20% for full papers since ASONAM was organized in Istanbul in 2012. Indeed, this reflects maturity and stability in terms of number of submissions, acceptance rate and participation, and ASONAM has earned its permanent position among top-tier international conferences.

This year, we moved from ACM/IEEE as sponsors to having Springer as the Sponsor and publisher of the proceedings. Authors of all papers presented at ASONAM and the co-located events are invited to submit expanded versions of their manuscripts to the prestigious SNAM journal, NetMAHIB journal or the LNSN series, which are characterized by their high visibility and fast processing of submissions. Special thanks to Springer Nature for their various publication venues which have been well integrated with ASONAM to the benefit of both parties.

We gathered over four days to witness interesting and exciting research achievements by various authors who present full, short, poster or demo papers covering a wide spectrum of research contributions to the foundations and applications of social networks. This would not have been possible without the dedication of a large team of motivated research leaders working closely together for twelve months to put together the attractive and intensive scientific program. Their great achievements contributed much to the visibility of ASONAM. I would like to heartily thank them all.

We do not forget in particular the generous support received from the operational organizing team who spent considerable time and effort handling daily issues and activities, answering emails, updating the Websites, etc. Special thanks to Min-Yuh Day, Panagiotis Karampelas, Tansel Ozyer, Mehmet Kaya, Deniz Bestepe, Diaylo Steiman and Jalal Kawash, who worked hard to produce the proceedings, communicate with participants/authors and handle the registration; also special thanks go to the local arrangements team from the University of Calabria. Indeed, without their highly appreciated effort it would have been really very hard to maintain the quality of the social program and keep the trend of providing rich meals and breaks during the conference and the excursion trip organized.

Thank you to all organizers including general chairs, program chairs and the chairs of various tracks and workshops, to participants, to authors who submitted papers, to program committee members and to the reviewers who invested their valuable time and effort to provide timely and comprehensive reviews. We encourage researchers and practitioners to submit again next year to get the opportunity and privilege to present their work at ASONAM 2025.

September 2024   Luca M. Aiello
                                                        Tanmoy Chakraborty
                                                           Sabrina Gaito

Organization

Steering Chair

Reda Alhajj   University of Calgary, Canada

Honorary Chair

Frans N. Stokman   University of Groningen, Netherlands

General Chairs

Andrea Tagarelli   University of Calabria, Italy
Roberto Interdonato   CIRAD, UMR Tetis, France
Jon Rokne   University of Calgary, Canada

Program Committee Chairs

Luca M. Aiello   IT University of Copenhagen, Denmark
Tanmoy Chakraborty   IIT Delhi, India
Sabrina Gaito   University of Milan, Italy

Industry Track Chairs

Francesco Gullo   University of L'Aquila, Italy
Gianmarco De Francisci Morales   CENTAI, Italy

Workshops Chairs

I-Hsien Ting   National University of Kaohsiung, Taiwan
Rajesh Sharma   University of Tartu, Estonia
Lucio La Cava   University of Calabria, Italy

Multidisciplinary Track Chairs

Ester Zumpano   University of Calabria, Italy
Shirin Nilizadeh   University of Texas at Arlington, USA
Carmela Comito   ICAR-CNR, Italy

PhD Forum and Posters Track Chairs

Huzefa Rangwala   George Mason University, USA
Alessia Antelmi   University of Turin, Italy
Domenico Mandaglio   University of Calabria, Italy

Demos and Exhibitions Chairs

Elio Masciari   University of Naples, Italy
Tansel Ozyer   Ankara Medipol University, Turkey

Tutorial Chairs

Pasquale De Meo   University of Messina, Italy
Shady Shehata   YOURIKA Labs, Canada
Davide Vega   University of Uppsala, Sweden

Publicity Chairs

Shang Gao   Jilin University, China
Buket Kaya   Firat University, Turkey
Kashfia Sailunaz   University of Calgary, Canada

Publication Chairs

Min-Yuh Day   National Taipei University, Taiwan
Panagiotis Karampelas   Hellenic Air Force Academy, Greece

Registration Chairs

Jalal Kawash   University of Calgary, Canada
Mehmet Kaya   Firat University, Turkey

Web Chair

Deniz Bestepe   Istanbul Medipol University, Turkey

Additional Reviewers

Abdessamad Benlahbib   FSDM, Morocco
Abdessamad Imine   Loria, France
Abiola Akinnubi   COSMOS-UALR, USA
Adnan Hoq   University of Notre Dame, USA
Aisling Third   Open University, UK
Akira Matsui   Yokohama National University, Japan
Alessandro Visintin   University of Padua, Italy
Alexander Rodriguez   Georgia Institute of Technology, USA
Amrit Poudel   University of Notre Dame, USA
Anastasios Giovanidis   Centre National de la Recherche Scientifique,
                                 France
Anatoliy Gruzd   Toronto Metropolitan University, Canada
Anggy Eka Pratiwi   Indian Institute of Technology Jodhpur, India
Ankan Mullick   IIT Kharagpur, India
Anurag Singh   National Institute of Technology Delhi, India
Arlei Silva   Rice University, USA
Ashwin Shreyas Mohan Rao   University of Southern California, USA
B. Aditya Prakash   Georgia Tech, USA
Bailu Jin   Cranfield University, UK
Bijaya Adhikari   University of Iowa, USA
Billy Spann   University of Arkansas at Little Rock, USA
Bing He   Georgia Institute of Technology, USA
Bohan Jiang   Arizona State University, USA
Casey Doyle   Sandia National Laboratories, USA
Charalampos Chelmis   University at Albany State University of New
                                 York, USA
Christine Largeron   Université de Lyon, France
Constantine Dovrolis   Georgia Institute of Technology, USA
Courtland Vandam   Massachusetts Institute of Technology, USA

David Skillicorn	Queen's University, Canada
Debanjan Datta	Virginia Tech, USA
Dong Wang	University of Illinois Urbana-Champaign, USA
Eduard Dragut	Temple University, USA
Ehsan Ul Haq	Hong Kong University of Science and Technology, China
Etienne Gael Tajeuna	Laval University, Canada
Fattane Zarrinkalam	University of Guelph, Canada
Fernando Terroso-Saenz	Catholic University of Murcia, Spain
Frank Liu	Southern Illinois University, USA
Fujio Toriumi	University of Tokyo, Japan
George Panagopoulos	École Polytechnique, France
Gita Sukthankar	University of Central Florida, USA
Hadassa Daltrophe	Shamoon College of Engineering, Israel
Hamid R. Rabiee	Sharif University of Technology, Iran
Hanjia Lyu	University of Rochester, USA
Hasan Davulcu	Arizona State University, USA
Hitkul Jangra	Indraprastha Institute of Information Technology, Delhi, India
Huimin Zeng	University of Illinois at Urbana-Champaign, USA
Humayun Kabir	Microsoft, USA
Isabel Murdock	Carnegie Mellon University, USA
Jiaming Cui	Georgia Institute of Technology, USA
Jiamou Liu	University of Auckland, New Zealand
Jiten Sidhpura	Sardar Patel Institute of Technology, India
Jose Luis Fernandez-Marquez	University of Geneva, Switzerland
Julio Cesar Soares dos Reis	Federal University of Viçosa, Brazil
Keith Burghardt	University of Southern California, USA
Kenji Yokotani	Tokushima University, Japan
Keyan Guo	University at Buffalo, USA
Kijung Shin	Korea Advanced Institute of Science and Technology, South Korea
Kshiteesh Hegde	Western Digital, USA
Lanyu Shang	University of Illinois Urbana-Champaign, USA
Lara Quijano-Sanchez	Universidad Autónoma de Madrid, Spain
Lu-An Tang	NEC Labs America, USA
Mainuddin Shaik	University of Arkansas at Little Rock, USA
Mehrdad Jalali	Karlsruhe Institute of Technology, Germany
Michael Smit	Dalhousie University, Canada
Mirela Riveni	University of Groningen, The Netherlands
Muhammad Abulaish	South Asian University, India
Nayoung Kim	Arizona State University, USA

Young-Woo Kwon   Kyungpook National University, South Korea
Yue Zhang   Amazon, Inc., USA
Yueqing Liang   Illinois Institute of Technology, USA
Zhenming Liu   College of William and Mary, USA
Zhenrui Yue   University of Illinois Urbana-Champaign, USA
Zhihao Hu   Virginia Tech, USA

Contents – Part I

Research

Research

Scalable High-Performance Community Detection Using Label Propagation in Massive Networks

Sharon Boddu[✉] and Maleq Khan

Department of Electrical Engineering and Computer Science, Texas A&M
University-Kingsville, Kingsville, USA
`sharonbdd25@gmail.com`, `maleq.khan@tamuk.edu`

Abstract. Community detection is the problem of finding naturally
forming clusters in networks. It is an important problem in mining and
analyzing social and other complex networks. Community detection can
be used to analyze complex systems in the real world and has applications in many areas, including network science, data mining, and computational biology. *Label propagation* is a community detection method
that is simpler and faster than other methods such as Louvain, InfoMap,
and spectral-based approaches. Some real-world networks can be very
large and have billions of nodes and edges. Sequential algorithms might
not be suitable for dealing with such large networks. This paper presents
distributed-memory and hybrid parallel community detection algorithms
based on the label propagation method. We incorporated novel optimizations and communication schemes, leading to very efficient and scalable algorithms. We also discuss various load-balancing schemes and
present their comparative performances. These algorithms have been
implemented and evaluated using large high-performance computing systems. Our hybrid algorithm is scalable to thousands of processors and has
the capability to process massive networks. This algorithm was able to
detect communities in the Metaclust50 network, a massive network with
282 million nodes and 42 billion edges, in 654 s using 4096 processors.

Keywords: Community detection · parallel graph algorithms ·
network analysis · graph mining

1 Introduction

Graphs can be used to represent many complex real-world systems and networks. A node in a network can be represented by a vertex in a graph, and a
link between nodes by an edge. Therefore, we use the terms graphs and networks,
as well as nodes and vertices, interchangeably. We can find many networks in
the real world. For example, the world-wide web is a network of HTML pages
connected by hyperlinks [3]. Similarly, social networks are made up of individuals represented as vertices and the interpersonal relationships between them

© The Author(s), under exclusive license to Springer Nature Switzerland AG 2025
L. M. Aiello et al. (Eds.): ASONAM 2024, LNCS 15211, pp. 3–19, 2025.
https://doi.org/10.1007/978-3-031-78541-2_1

as edges [7,11,16]. Identifying the organization of vertices and the presence of a modular structure or communities can help to understand these complex systems [16,17]. Real-world networks often exhibit community structures [16]. A community within a network can be formally said to be a group of nodes with more edges connecting to other vertices within the group than vertices outside the group [17,20]. In many cases, communities tend to emerge organically within complex systems, and these communities can represent meaningful subsystems with a specific purpose or function. In the world-wide web, a community might correspond to a set of web pages with similar content or user interests [3]. In a protein-protein interaction network, a community represents a group of proteins with similar functionalities [15]. A community in a social network might correspond to a group of like-minded individuals with common interests [1].

There are some existing algorithms for community detection. Well-known algorithms for community detection include Louvain [2], Infomap [21], label propagation [20], spectral optimization-based [18], hierarchical [14], minimum-cut-based [6], and betweenness-score-based [17] methods. These methods can find communities with high quality, but often require expensive computational work [24]. Serial implementation of these algorithms can take prohibitively long to find communities in large networks. The parallel computing paradigm enables the processing of massive graphs quickly. Multiple processors can be used to distribute the computational load among them and reduce the total runtime. However, it is challenging to parallelize the community detection algorithm due to irregular communication patterns and computational dependencies. Communications among processors can impede the efficiency and scalability of the algorithm. Hence, it is essential to develop efficient communication strategies.

In this paper, we propose label-propagation-based parallel community detection algorithms for distributed-memory systems. Minimizing communication and achieving good load balancing is crucial to ensuring the scalability of a parallel algorithm. To reduce communication and computation, we have employed several novel optimizations, which resulted in efficient and scalable algorithms. We also studied various load-balancing schemes. The presented algorithm is implemented using Message Passing Interface (MPI) primitives. Additionally, we extended our algorithm to hybrid parallel systems using MPI and OpenMP. The hybrid parallel algorithm combines the advantages of distributed and shared-memory systems, improving speedup and scalability even further. Our main contributions are summarized below.

1. We developed a distributed-memory parallel algorithm based on label propagation. We employed efficient communication strategies to improve performance.
2. We devised novel techniques, which eliminate redundancy in computation and reduce the runtime significantly.
3. We experimented with various communication routines and load-balancing schemes and presented experimental results comparing them.

4. Finally, we present a hybrid algorithm that effectively utilizes the advantages of both shared and distributed memory systems and demonstrates significantly improved performance and scalability.

For the community detection problem, there are a few existing distributed-memory parallel algorithms, which are briefly discussed in Sect. 2.3. Our algorithms are significantly faster and more scalable than the state-of-the-art algorithms such as DistLouvain [8] and DDOLP [12]. For instance, on the Twitter graph with 61M (million) nodes and 1.9B (billion) edges, our hybrid algorithm runs in around 120 s using 4000 processors, whereas the DistLouvain algorithm takes 600 s using the same number of processors. Our hybrid algorithm scales to more than 4000 processors, whereas DDOLP scales up to only one hundred processors. Our parallel algorithms are capable of working with very large graphs. Our algorithm was successfully executed on a massive Metaclust graph that comprised 282 million nodes and 42 billion edges in 654 s using 4096 processors. In contrast, the largest graphs that were tested with DistLouvain and DDOLP were UK graphs with 100 million nodes and 3.3 billion edges and a synthetic graph with 20 million nodes and 400 million edges, respectively.

The remainder of the paper is organized as follows. Some preliminary concepts and related work are discussed in Sect. 2. Our parallel algorithms are presented in Sect. 3, and the experimental results are given in Sect. 4.

2 Background

2.1 Preliminaries

Let $G = (V, E)$ be a graph where V is the set of vertices and E is the set of edges. If $(u, v) \in E$, u and v are called neighbors of each other. N_v denotes the set of neighbors of vertex v; that is, $N_v = \{u : (u, v) \in E\}$ and $d_v = |N_v|$ denotes the degree of v. Communities can be defined as groups of vertices having dense connections within each group but relatively sparse connections between the groups. The communities form a partition of the vertex set V: $P = \{C_1, C_2, \ldots, C_k\}$, where the community $C_i \subset V$ and $C_i \cap C_j = \phi$ for $i \neq j$.

We measure and compare the quality of communities using the normalized mutual information (NMI) metric, which is defined as follows. Suppose that we have two partitions, $P = \{P_1, P_2, \ldots, P_k\}$ and $Q = \{Q_1, Q_2, \ldots, Q_k\}$. The NMI between P and Q is defined as:

$$\text{NMI}(P, Q) = \frac{2 \times \text{I}(P, Q)}{H(P) + H(Q)},$$

where $H()$ represents entropy, and $I()$ represents mutual information between the partitions. $\text{I}(P, Q)$ is given by:

$$\text{I}(P, Q) = \sum_{i=1}^{|P|} \sum_{j=1}^{|Q|} \frac{|P_i \cap Q_j|}{n} \log \frac{n|P_i \cap Q_j|}{|P_i||Q_j|},$$

and $H(P)$ is given by:

$$H(P) = -\sum_{i=1}^{|P|} \frac{|P_i|}{n} \log \frac{|P_i|}{n}.$$

NMI is commonly used to evaluate community detection algorithms by comparing detected communities with ground truth communities.

2.2 Sequential Algorithms

Community detection algorithms can be classified as divisive or agglomerative algorithms. Divisive algorithms start with the entire graph as one community and progressively split the communities based on some measures such as internal degrees [5] and statistical inference [11]. Agglomerative algorithms start with singleton communities (communities with only one vertex) and gradually merge communities using similarity measures such as the Jaccard index, cosine similarity, and clustering coefficient [24]. One well known approach to community detection is modularity maximization. Modularity maximization algorithms start with some initial community assignments and repeatedly update the communities in multiple steps. At each step, communities are updated such that the modularity gain is maximized [4,14]. The Louvain algorithm [2] is a state-of-the-art modularity optimization algorithm for community detection. In contrast to the Louvain method, which optimizes a global modularity score, the label propagation method [20] uses only local information (the labels of the immediate neighbors) for each node and does not require optimizing any global quality score. Algorithm 1 provides the pseudocode of the label propagation method [20]. It starts by assigning a unique community label to each vertex. At this point, each vertex is a separate singleton community. In each subsequent iteration, these labels are updated based on some consensus or selection rule. A common consensus rule is that every vertex is assigned a label that is the majority among the labels of its neighbors, with ties broken arbitrarily. Raghavan et al. [20] reported that for the most large real-world networks, after only five iterations, the labels of 95% of the nodes converged to their final values.

Algorithm 1 Label Propagation Algorithm

1: **for all** $v \in V$ **do**	7: $L'(v) \leftarrow$ *majority label*
2: $L(v) \leftarrow v$	8: **for all** $v \in V$ **do**
3: *Converged* $\leftarrow$ *false*	9: **if** $L'(v) \neq L(v)$ **then**
4: **while** not *Converged* **do**	10: *Converged* $\leftarrow$ *false*
5: *Converged* $\leftarrow$ *true*	11: $L(v) \leftarrow L'(v)$
6: **for all** $v \in V$ **do**	12: **end while**

2.3 Related Parallel Algorithms

Most of the previous parallel algorithms for community detection are for shared-memory systems. There are only a few algorithms for distributed-memory systems. Distributed direction optimizing label propagation (DDOLP) [12] is a distributed-memory parallel algorithm based on label propagation. It propagates a selected subset of vertices called seed vertices in two execution modes, *push* operation and *pull* operation. Push operation is a blind propagation step in which the seed vertices propagate their labels without any resistance. The pull operation propagates labels according to a consensus (similar to the label propagation algorithm). By selectively propagating seed labels, DDOLP artificially initializes core communities, interfering with the detection of natural community cores. These drawbacks make DDOLP unsuitable for applications that need to detect naturally-formed communities. DDOLP was implemented in AM++, and hence we are unable to compare it with our algorithm directly. DDOLP was evaluated on synthetic graphs [12], but not on real-world graphs. Real-world graphs present unique challenges for load-balancing [22]; our algorithm uses a suitable graph partitioning scheme to overcome this problem. The scalability of DDOLP is limited to less than 100 processors [12] on large graphs like uk-2007 (with 105M vertices and 3.3B edges). In contrast, our algorithm scales to thousands of processors, as shown in Sect. 4.

DistLouvain [8] is a state-of-the-art distributed-memory algorithm, which is a parallelization of the modularity-maximizing Louvain algorithm [2]. DistLouvain scales to thousands of processors. We extensively compare our algorithm with DistLouvein and show that our algorithm outperforms DistLouvain in both runtime and scalability (see Sect. 4). Distributed Parallel Louvain Algorithm with Load-balancing (DPLAL) [22] is another algorithm for community detection. DPLAL uses efficient METIS graph partitioning [9] to address load imbalances in real-world graphs. DPLAL code is not publicly available; hence we are not able to directly compare with them. DPLAL is shown to scale up to 512 processors on the orkut graph (3M vertices and 117M edges), whereas our algorithm scales up to 4096 processors and is capable of processing much larger graphs like MetaClust50 (282.2M vertices and 42.8B edges).

3   Our Parallel Algorithms

In this section, we present our parallel algorithms for detecting communities using the label propagation method. First, we present our algorithm for distributed-memory systems. We then discuss how this algorithm can be extended for hybrid systems. We present a few novel optimization techniques to reduce the number of messages and local computation (computation done locally by each processor). These techniques significantly improve the efficiency and scalability of the algorithm. The hybrid extension of our algorithm, together with the optimization techniques, makes it highly efficient and scalable to thousands of processors.

We introduce some notation and definitions that are used to describe our algorithm. The graph $G = (V, E)$ is partitioned and distributed among the processors so that the processor P_i stores the vertices V_i and their corresponding set of edges E_i. Each $v \in V_i$ is said to be local to the processor P_i, and P_i is the owner of v. For any edge $(u, v) \in E$, if the same processor owns both u and v,

u is called a *local neighbor* of v; otherwise, u is a *ghost neighbor* of v. Let N_v be the set of all neighbors of v and $L(v)$ the label of v.

For each vertex $v \in V_i$, we use a list of labels of the neighbors of v. Let NL_v be the multiset containing the labels of the vertices in N_v. A label can appear multiple times in NL_v since different neighbors of v can have the same label. Multiset NL_v can be implemented using an array of size d_v. Thus, the total space required for the NL_v multisets for all $v \in V_i$ is $O(|V_i| + |E_i|)$.

Initially, each vertex is assigned its id as its label. The processor P_i initializes $L(v) = v$ and $NL_v = N_v$ for each $v \in V_i$. Then, the algorithm proceeds in synchronized iterations; that is, a processor continues to the next iteration only after all processors complete their current iteration. In each iteration, a processor executes three major steps: a) update the labels of the vertices, b) check for convergence of the labels, and c) exchange the labels with the other processors. The details of each of these steps are described below.

a) **Computing the Majority Label:** Array NL_v contains the labels of the neighbors of v. For each $v \in V_i$, processor P_i finds the most frequent label, called the majority label, in NL_v and assigns it to $L(v)$. The majority label in NL_v can be computed by sorting the elements in NL_v and then performing a linear scan and keeping track of the most frequent item. For each v, it will take $O(d_v \log d_v)$ time, and thus the processor P_i takes $O\left(\sum_{v \in V_i} d_v \log d_v\right)$ time for this step in each iteration.

b) **Checking Convergence:** The labels are converged when none of the vertices changes its label in an iteration. Each processor can locally check whether the label of any of its vertices has been changed. The processors can then exchange this information with each other to find whether any vertex in the entire set V has changed its label. If a processor exchanges individual messages with the other processors, it will take $O(p)$ time. However, it can be done more efficiently using a parallel collective reduction operation in $O(\log n)$ time.

c) **Exchanging Labels and Updating NL_v:** As described above, in each iteration for each v, a new label is computed using the labels in NL_v, the labels of the neighbors of v from the preceding iteration. Thus, at the end of the current iteration, and if it is not the last iteration, the old labels are removed from NL_v. The newly updated labels are added to NL_v to prepare it for the next iteration. Since other processors can own some of the neighbors of a vertex, the processors need to exchange the labels of the vertices with each other. A naive way to do this is: each processor P_i sends the labels of all $v \in V_i$ to all other processors and receives the labels of the other vertices from them. However, it is not necessary to send the labels of all $v \in V_i$ to all other processors. The label $L(v)$ should be sent only to the owners of the neighbors of v. For each neighbor $u \in N_v$, if u is a local neighbor, $L(v)$ is added to NL_u; otherwise, P_i sends $L(v)$ to the owner of u. The owner of u adds $L(v)$ to NL_u after receiving the message from P_i. Note that if a processor P_i needs to send multiple labels to another processor P_j, sending a single message containing all of these labels is more efficient than sending multiple messages, each containing a single label. In the actual implementation of our algorithm, the labels destined for the same processor are combined in a buffer, and the contents of the buffer are sent in a single message. Thus, in each iteration, a processor sends at most one message to any other processor. The pseudocode of our algorithm is given in Algorithm 2. In the next section, we describe some more techniques to further optimize communication and local computation.

Algorithm 2 Parallel Label Propagation Algorithm

1: **for all** $v \in V$ **do** in parallel
2:     $L(v) \leftarrow v$
3:     $NL_v \leftarrow N_v$
4: $Converged \leftarrow false$
5: **while** not $Converged$ **do**
6:     $Converged \leftarrow true$
7:     **for all** $v \in V$ **do** in parallel
8:         $L'(v) \leftarrow$ majority label in NL_v
9:         $NL_v \leftarrow \phi$
10:        **if** $L'(v) \neq L(v)$ **then**
11:            $Converged \leftarrow false$
12:    Parallel_Reduction($Converged$)
13:    $Converged \leftarrow \bigwedge_{i=0}^{p-1} Converged_i$
14:    **if** not $Converged$ **then**
15:        **for all** $v \in V$ **do** in parallel
16:            $L(v) \leftarrow L'(v)$
17:            **for all** $u \in N_v$ **do**
18:                **if** u is a ghost neighbor
19:                    Send $(L(v), u)$ to the owner of u
20:                **else**
21:                    $L(v)$ to NL_u
22:        Receive the labels sent by other processors
23:        **for all** received labels $(L(v), u)$
24:            Add $L(v)$ to NL_u
25: **end while**

3.1 Techniques to Further Optimize Communication and Computation

A major challenge in making the distributed-memory parallel algorithm scalable to a large number of processors is the large number of messages exchanged by the processors. To reduce communication among processors, we present some techniques that also reduce local computation in addition to reducing communications.

For most graphs, the labels change only in the first few iterations. In later iterations, the cores of many communities have already formed and do not change their labels. Only the vertices on the periphery of the community cores change labels and eventually converge to a final solution [12,13]. If $L'(v) \neq L(v)$ in an iteration, the vertex v is called *active* in that iteration; otherwise, v is *inactive*. Figure 1 shows the number of active vertices with the number of iterations for two graphs. The number of active vertices is quickly reduced with the progress of the algorithms. Therefore, exchanging only the updated labels can significantly reduce the communication cost, especially in later iterations. In the next optimized version of our algorithm, called the *active version*, the processors exchange the labels of only the active vertices. Processor $P(v)$ sends $L'(v)$ to $P(u)$ in an iteration if and only if $L'(v) \neq L(v)$ in that iteration. $P(v)$ also sends the old label $L(v)$ along with the new label $L'(v)$ so that $P(u)$ can update NL_u by replacing $L(v)$ with $L'(v)$.

In the implementation of NL_u, it is more efficient to store the counts of the distinct labels instead of storing all occurrences of the labels. For each distinct

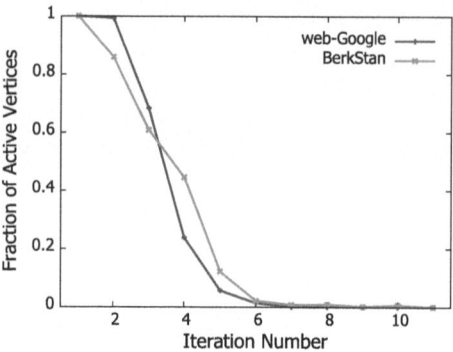

Fig. 1. The figure on the left shows the number of active vertices corresponding to the number of iterations for two graphs: Web-Google and BerkStan, each with a few million edges. The number of active vertices is significantly reduced with the progress of the algorithm

label, we have only one entry where each entry is a pair: a label and the count or frequency of the label. To store these (label, count) pairs, we use two alternative data structures: HashMap and MaxHeap. For each u, processor $P(u)$ maintains one such data structure: HashMap(u) or MaxHeap(u). When $P(u)$ receives $L(v)$ and $L'(v)$, the previous and current labels of v, the count for $L(v)$ is decremented by 1, and the count for $L'(v)$ is increased by 1. If the count of a label becomes 0 after decrementing, its entry is removed from the HashMap (or MaxHeap). Some additional details and analyses of these two data structures are described below. Let ℓ_u be the number of distinct labels in HashMap(u) (or MaxHeap(u)) and t_u the number of update operations (increments and decrements) performed on HashMap(u) (or MaxHeap(u)) in an iteration. Note that in any iteration, both ℓ_u and t_u are, at most, d_u, the degree of u.

HashMap: We use the HashMap data structure implemented in the Standard Template Library (STL). Each update operation takes $O(\log \ell_u)$ time, and t_u updates take $O(t_u \log \ell_u)$ time. To find a label with the maximum count from HashMap(u), we iterate through the entries in HashMap(u), which takes $O(\ell_u)$ time. Thus, in one iteration, the total time to update HashMap(u) and find the majority label is $O(\ell_u + t_u \log \ell_u)$, whereas the array-based implementation of NL_u in the base algorithm takes $O(d_u \log d_u)$ time. Since both ℓ_u and t_u can be significantly smaller than d_u, especially in the later iterations, the use of HashMap with the active version of our algorithm not only reduces communications, it also reduces local computation in finding the majority labels for the local vertices.

MaxHeap: We use our own custom implementation of the MaxHeap data structure. We need to store (label, count) pairs in a MaxHeap. While the items in the MaxHeap is ordered by the counts, we need to search the items using the labels. To facilitate such search operations efficiently, we use a mapping from the labels to the indices of the MaxHeap using an STL Map data structure. Like HashMap, each update operation in MaxHeap(u) takes $O(\log \ell_u)$ time, and t_u update operations take $O(t_u \log \ell_u)$ time. Finding a label with the maximum count is done in $O(1)$ time since a MaxHeap always keeps the largest element at the root of the MaxHeap. As a result, the runtime performance of the MaxHeap version is a little better than that of the HashMap version.

3.2 Comparison of Alternative Communication Options in MPI

We implement our algorithm using Message Passing Interface (MPI) primitives. As required in our algorithm, the communication of the labels can be done using MPI in a few different ways, as explained below.

Ordered-Recv and Dynamic-Recv Schemes: Each processor sends and receives the labels to and from the other $p - 1$ processors. To exchange these messages, we utilize MPI_Send() and MPI_Recv(), which are point-to-point communication routines. The processor P_i sends a message with updated labels to another processor P_j by calling MPI_Send(). The processor P_j receives the message by calling a matching MPI_Recv() operation. We present two schemes, *Ordered-Recv* and *Dynamic-Recv*, to receive the messages by using the MPI_Recv() function. In the *Ordered-Recv* scheme, we receive messages in a queue ordered by processor IDs (or MPI ranks). The processor P_i first receives the message from P_0, then $P_1 \ldots P_{i-1}, P_{i+1} \ldots P_{p-1}$ in that order.

In the *Dynamic-Recv* scheme, we receive the arriving messages on a first-come-first-serve basis. MPI_PROBE() function is used to find the source and size of the next message, and then the message is received by calling MPI_Recv() without following any order of the source processors.

A2Av Scheme: In the *A2Av* scheme, we exchange labels by using collective communication routines like MPI_Alltoall() and MPI_Alltoallv(). Every processor participates in this operation simultaneously and exchanges messages with every other processor. In our algorithm, the messages sent are not of uniform size, and hence we use the irregular collective method MPI_Alltoallv(). We first perform an MPI_Alltoall() operation to exchange send and receive counts (the size of the messages to be exchanged), which is needed for the collective communication function MPI_Alltoallv(). Then, all processors execute MPI_Alltoallv() simultaneously in a synchronous fashion, which causes the exchange of the label data between all pairs of processors.

3.3 Hybrid Algorithm

Hybrid parallel algorithms utilize shared-memory parallelism by using concurrent access to a common memory and distributed-memory parallelism by using communication. To improve the scalability of our algorithm, we implement a hybrid parallel version using both MPI (for distributed-memory) and OpenMP primitives (for shared-memory). Consider a hybrid system with C compute nodes and t cores in each node; we have a total of $p = tC$ cores or processors. The input graph is divided into C partitions and partition $G_i(V_i, E_i)$ is assigned to the ith compute node. The algorithm uses C MPI tasks, one MPI task in each node. Then within each MPI task, t threads are created. The threads in the ith compute node share the data in partition $G_i(V_i, E_i)$ and other data structures in the ith compute node. The local computation of each node is further parallelized by dividing the computation tasks among the t threads. The initialization and computing of the majority of labels are divided among the threads. Communication among the compute nodes is done by the master thread in each compute node. This hybrid algorithm significantly improves the runtime and scalability of our algorithm.

3.4 Load Balancing

We study various ways of partitioning the vertex set V among processors.

Contiguous Vertex Partitioning Scheme (CONT): We assign consecutive vertices in an equal number to each partition such that each partition has approximately $\frac{|V|}{p}$ vertices. The first processor P_0 contains vertices $V_0 = \{v_0, v_1, \ldots, v_{\frac{|V|}{p}}\}$, the second processor P_1 contains $V_1 = \{v_{\frac{|V|}{p}+1}, v_{\frac{|V|}{p}+2}, \ldots, v_{\frac{2|V|}{p}}\}$ and so on. In the general case, processor P_i contains vertices $V_i = \{v_x, v_{x+1}, v_{x+2}, \ldots, v_{x+\frac{|V|}{p}-1}\}$, where $x = (i-1)\frac{|V|}{p}$.

Round Robin Vertex Partitioning Scheme (RR): Similar to CONT scheme, each processor holds approximately $\frac{|V|}{p}$ vertices. The first processor P_0 contains vertices $V_0 = \{v_0, v_{\frac{|V|}{p}}, v_{2\frac{|V|}{p}}, \ldots\}$, the second processor P_1 contains $V_1 = \{v_1, v_{\frac{|V|}{p}+1}, v_{2\frac{|V|}{p}+1}, \ldots\}$ and so on. In the general case, processor P_i contains vertices $V_i = \{v_i, v_{i+1\frac{|V|}{p}}, v_{i+2\frac{|V|}{p}}, \ldots\}$.

Alternating Round Robin Scheme (ARR): We also study a modification of RR, which we call an alternating round-robin scheme. In alternative steps, the vertices are assigned to the partitions in normal and reverse round-robin orders as follows. The first p vertices are assigned in normal order: vertex i, $0 \leq i < p$, is assigned to partition V_i. The next p vertices are assigned in reverse order: vertex i, $p \leq i < 2p$, is assigned to partition V_{2p-i-1}. Then, the next p vertices are assigned in normal order, and so on.

Equal Edge Count Scheme (EqEdg): In this scheme, like CONT, a consecutive set of vertices is assigned to each partition, but the number of vertices in the partitions may not be equal. The number of vertices is chosen so that all partitions have a nearly equal number of edges.

4   Experimental Results

In this section, we evaluate the performance of our parallel algorithm using a wide range of real-world graphs of various sizes and types. We conducted extensive experiments to compare the runtime and scalability of our algorithm with the state-of-the-art DistLouvain algorithm [8]. To be fair, both DistLouvain and our algorithm were run on the same parallel computing machines with exactly the same configuration. We also present experimental results showing the improvements made by the optimizations described in Sect. 3. A discussion comparing our algorithm with two other recent parallel algorithms DDOLP [12] and DPLAL [22] is given in Sect. 2.3.

4.1   Experimental Setup and Datasets

We used the Expanse Computing Cluster at the San Diego Supercomputing Center available to us through the ACCESS system [19]. The cluster houses AMD EPYC 7742 (Rome) computing nodes (with 128 CPU cores, 2.25 GHz). We also used the Bridges-II supercomputer at the Pittsburg Supercomputing Center (PSC), which shares a processor architecture similar to that of the Expanse system. We implemented our algorithms using the Message Passing Interface (MPI) library for distributed-memory systems and further used OpenMP and MPI for the hybrid version. We used the OpenMPI implementation of the MPI standard and compiled our codes using GCC version 9.2.0 with the -fopenmp -O3 flags. The data set is chosen to represent a wide variety of networks. Most of these networks are taken from the Stanford Large Network Data Set Collection [10]. Table 1 lists the graphs used in our experiments.

Table 1. List of graphs used in experiments.

Graph name	Nodes	Edges	Avg-Deg
BerkStan	0.7M	7.6M	21.7
web-Google	0.9M	5.1M	11.3
facebook	4.0M	24.0M	12.0
LiveJournal	3.9M	34.0M	17.4
Orkut	3.0M	117.0M	78.0

Graph name	Nodes	Edges	Avg-Deg
FriendSter	65.6M	1.8B	55.3
twitter	61.6M	1.9B	61.7
uk2007	105M	3.3B	62.8
Metaclust50	282.2M	42.8B	303.3

4.2   Performance of Various MPI Communication Schemes

Figure 2a and 2b compare the performances of the communication schemes, *Ordered-Recv*, *Dynamic-Recv*, and *A2Av*, disucssed in Sect. 3.2. The figures show the runtime for only the communication step of our algorithm. *A2Av* performs significantly better than the other two schemes. Between the other two schemes, *Ordered-Recv* is simpler, but *Dynamic-Recv* performs better. *Ordered-Recv* imposes an order in which messages can be received. This ordering requirement causes delay in receiving some of the messages even if they are available in the buffer. In the rest of the experiments, we use the *A2Av* scheme for MPI

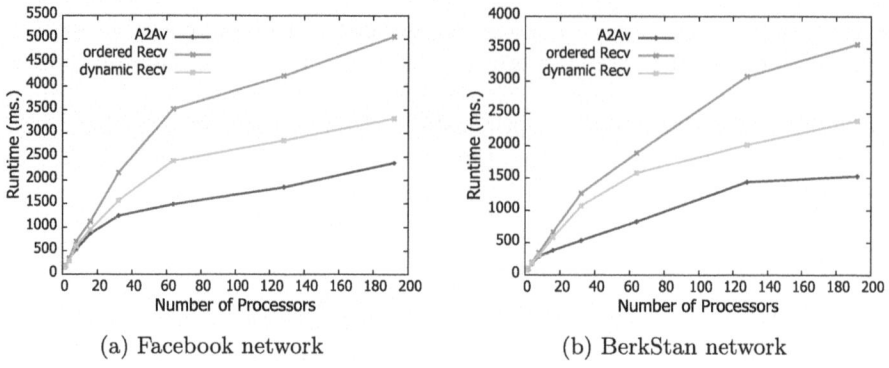

(a) Facebook network   (b) BerkStan network

Fig. 2. Runtime of communication step for various communication schemes

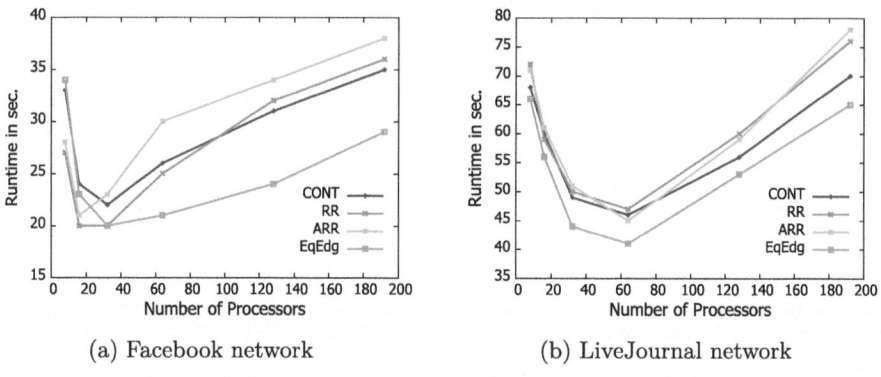

(a) Facebook network   (b) LiveJournal network

Fig. 3. Runtime of communication step for various load balancing schemes

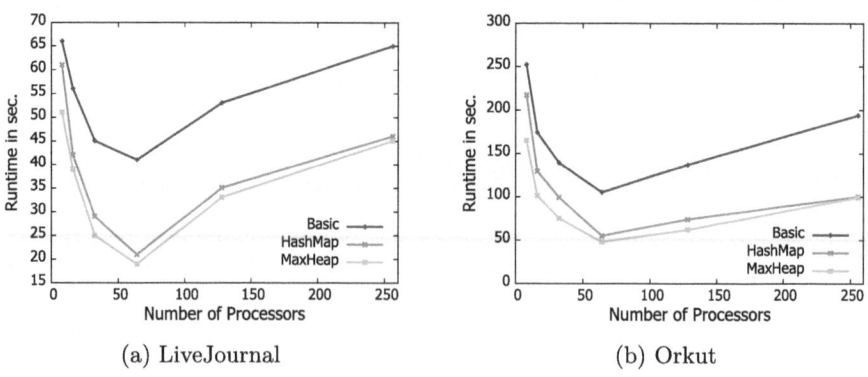

(a) LiveJournal   (b) Orkut

Fig. 4. Performance improvement due to various optimizations for the pure MPI algorithm

communication. Notice that the runtime of the communication step of the algorithm increases with the number of processors since more processors require more communications among them.

4.3   Effect of the Load Balancing Schemes

Figure 3a and 3b show the rumtime of the algorithm with different load schemes: *CONT*, *RR*, *ARR*, *EqEdg* (described in Sect. 3.4). The results are shown for two networks: Facebook and LiveJournal. We observe that the *EqEdg* scheme outperforms the other schemes. The primary communication step of our algorithm exchanges the labels of the neighbors of the vertices. Further, the main computation also involves iterating through the neighbors of the vertices. Therefore, partitioning the graph with equal number of edges in the partitions ensures that the communication and computation load in the processors is almost equal.

4.4   Optimizations

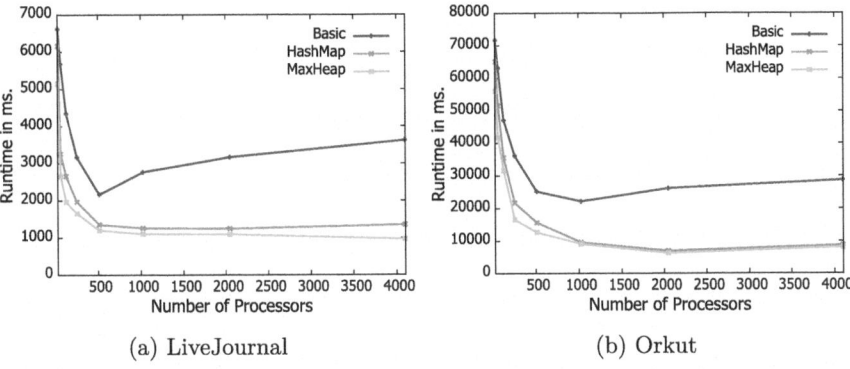

(a) LiveJournal   (b) Orkut

Fig. 5. Performance improvement due to various optimizations for the hybrid algorithm

We also experimentally evaluate the additional optimizations discussed in Sect. 3.1 by comparing the following variants: *i*) *Base*: the baseline version of our algorithm, *ii*) *HashMap*: the active version of our algorithm with the HashMap data structure, and *iii*) *MaxHeap*: the active version of our algorithm with the MaxHeap data structure.

We show the improvements due to these optimizations in Fig. 4a and 4b for the LiveJournal and Orkut networks, respectively. We observe a huge improvement in the *HashMap* and *MaxHeap* versions. These optimizations reduce communication and computation costs significantly. We further demonstrate the improvements by these optimizations for the hybrid algorithm in Fig. 5a and 5b.

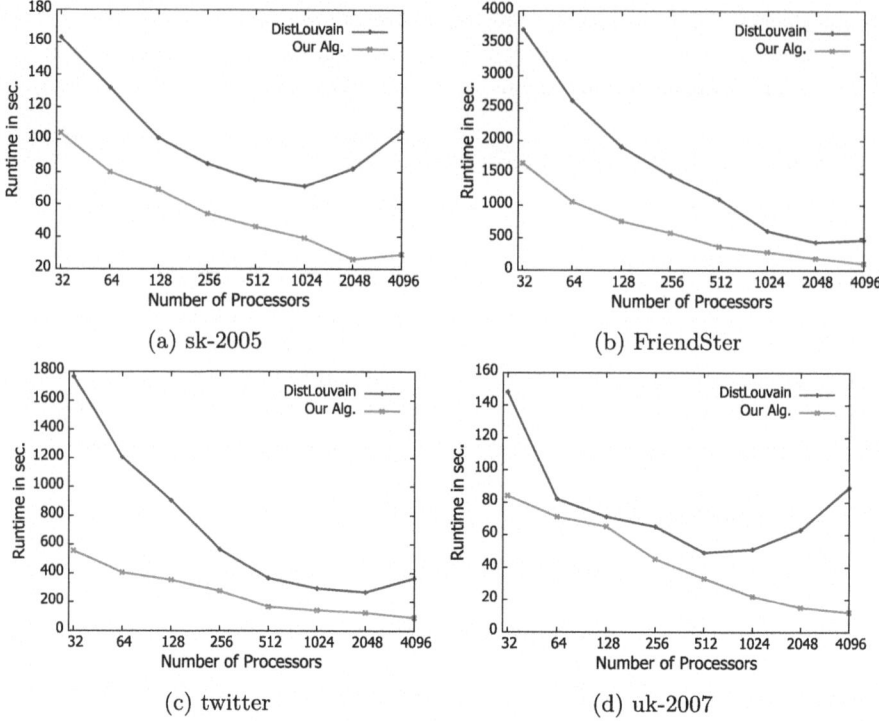

(a) sk-2005

(b) FriendSter

(c) twitter

(d) uk-2007

Fig. 6. Runtime of DistLouvain and our hybrid algorithm on various networks

4.5   Hybrid Algorithm

In the hybrid algorithm, we used 4 MPI tasks per node and 32 OpenMP threads per MPI task for our experiments. In order to assess the performance of our algorithm, we compare our algorithm with the state-of-the-art DistLouvain algorithm, which is also a hybrid algorithm. Our experiments used an identical hybrid configuration for both DistLouvain and our algorithm.

We compared these two algorithms using five large networks, including Metaclust50, a network with 42B edges. For all of these networks, our algorithm consistently outperforms DistLouvain and is significantly faster and more scalable than DistLouvain, as shown in Fig. 6a–6b and Fig. 7. Figure 7 shows the runtime for a very large network, Metaclust50, with 282M vertices and 42.8B edges using 2048 and 4096 processors. For this network, the algorithms could not be run using a smaller number of processors. With a smaller number of processors, each partition is too large to fit in the memory of a compute node. While DistLouvain scales up to 1,000 or 2,000 processors for most of these graphs, our algorithms continue to scale to more than 4,000 processors. We could not experiment with more than 4,096 processors since we did not have access to any larger number of processors. However, from the trend, it is clear that our algorithm would scale far beyond 4,000 processors.

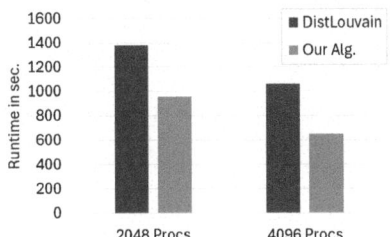
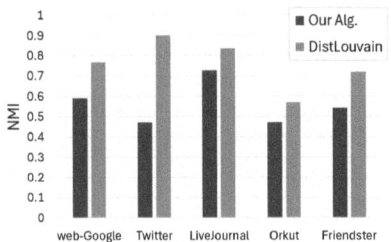

Fig. 7. Runtime of DistLouvain and our hybrid algorithm on a very large network, Metaclust50, with 42.8B edges using 2048 and 4096 processors

Fig. 8. Normalized mutual information (NMI) between the detected communities and ground truth (GT) communities

4.6 Quality of Communities

To compare detected communities with ground truth (GT) communities, we use normalized mutual information (NMI), which is described in Sect. 2.1. The NMI scores of DistLouvain and our algorithm for 5 networks are shown in Fig. 8. We obtained the ground truth communities for these networks from the data provided by Yang and Leskovec [23]. Note that the quality of the detected communities is not an artifact of these parallel algorithms. Instead, the quality is determined by the underlying serial algorithm, which is the Louvain algorithm for DistLouvain and the LPA for our parallel algorithm. Nevertheless, we are showing the quality of the detected communities to demonstrate the trade-off between quality and runtime. It is expected and known that the Louvain algorithm produces communities with better quality than LPA. However, for most of these networks, the quality of communities detected by our algorithm is comparable to that of DistLouvain.

5 Conclusion

We have developed an efficient and scalable algorithm to detect communities in large-scale graphs. This algorithm includes several non-trivial optimization techniques and data structures that have significantly improved its performance, surpassing the performance of the state-of-the-art algorithm DistLouvain. Our algorithm can work with massive graphs with billions of edges. Experiments on a wide range of real-world networks have shown that our algorithm can scale to more than 4000 processors consistently. With this capability and performance, our parallel algorithm can help analysts and researchers utilize modern commodity computing clusters, which have a large number of processors, to analyze large-scale networks quickly.

Acknowledgements. The work was partially supported by NSF BIGDATA grant IIS-1633028 and used PSC Bridges-2 RM at Pittsburg Supercomuting Center (PSC) through allocation CIS230052 from the Advanced Cyberinfrastructure Coordination Ecosystem: Services & Support (ACCESS) program [19], which is supported by NSF grants #2138259, #2138286, #2138307, #2137603, and #2138296.

References

1. Bedi, P., Sharma, C.: Community detection in social networks. Wiley Interdiscip. Rev.: Data Mining Knowl. Discov. **6**(3), 115–135 (2016)
2. Blondel, V., Guillaume, J., Lambiotte, R., Lefebvre, E.: Fast unfolding of communities in large networks. J. Stat. Mech: Theory Exp. **2008**(10), P10008 (2008)
3. Claude, F., Navarro, G.: Fast and compact web graph representations. ACM Trans. Web (TWEB) **4**(4), 1–31 (2010)
4. Clauset, A., Newman, M., Moore, C.: Finding community structure in very large networks. Phys. Rev. E **70**(6), 066111 (2004)
5. Fortunato, S.: Community detection in graphs. Phys. Rep. **486**(3–5), 75–174 (2010)
6. Fortunato, S., Barthelemy, M.: Resolution limit in community detection. Proc. Nat. Acad. Sci. **104**(1), 36–41 (2007)
7. Fortunato, S., Hric, D.: Community detection in networks: a user guide. Phys. Rep. **659**, 1–44 (2016)
8. Ghosh, S., Halappanavar, M., Tumeo, A., Kalyanaraman, A., Lu, H., et al.: Distributed Louvain algorithm for graph community detection. In: 2018 IEEE International Parallel and Distributed Processing Symposium (IPDPS), pp. 885–895. IEEE (2018)
9. Karypis, G., Kumar, V.: Multilevel k-way partitioning scheme for irregular graphs. J. Parallel Dist. Comp. **48**(1), 96–129 (1998)
10. Leskovec, J., Krevl, A.: SNAP Datasets: Stanford large network dataset collection (2014). http://snap.stanford.edu/data
11. Leskovec, J., Lang, K., Mahoney, M.: Empirical comparison of algorithms for network community detection. In: Proceedings of the 19th International Conference on World Wide Web, pp. 631–640 (2010)
12. Liu, X., et al.: Distributed direction-optimizing label propagation for community detection. In: 2019 IEEE High-performance Extreme Computing Conference (HPEC), pp. 1–6. IEEE (2019)
13. Liu, X., Halappanavar, M., Barker, K., Lumsdaine, A., Gebremedhin, A.: Direction-optimizing label propagation and its application to community detection. In: Proceedings of the 17th ACM International Conference on Computing Frontiers, pp. 192–201 (2020)
14. Malliaros, F., Vazirgiannis, M.: Clustering and community detection in directed networks: a survey. Phys. Rep. **533**(4), 95–142 (2013)
15. Manipur, I., Giordano, M., Piccirillo, M., Parashuraman, S., Maddalena, L.: Community detection in protein-protein interaction networks and applications. IEEE/ACM Trans. Comp. Biol. Bioinformatics **20**(1), 217–237 (2021)
16. Newman, M.: Detecting community structure in networks. Eur. Phys. J. B **38**(2), 321–330 (2004)
17. Newman, M.: Modularity and community structure in networks. Proc. Natl. Acad. Sci. **103**(23), 8577–8582 (2006)
18. Newman, M.: Spectral methods for community detection and graph partitioning. Phys. Rev. E **88**(4), 042822 (2013)

19. Parashar, M., Friedlander, A., Gianchandani, E., Martonosi, M.: Transforming science through cyberinfrastructure. Commun. ACM **65**(8), 30–32 (2022)
20. Raghavan, U., Albert, R., Kumara, S.: Near linear time algorithm to detect community structures in large-scale networks. Phys. Rev. E **76**(3), 036106 (2007)
21. Rosvall, M., Axelsson, D., Bergstrom, C.: The map equation. Eur. Phys. J. Spec. Topics **178**(1), 13–23 (2009)
22. Sattar, N., Arifuzzaman, A.: Scalable distributed Louvain algorithm for community detection in large graphs. J. Supercomput. **78**(7), 10275–10309 (2022)
23. Yang, J., Leskovec, J.: Defining and evaluating network communities based on ground-truth. In: Proceedings of the ACM SIGKDD Workshop on Mining Data Semantics, pp. 1–8 (2012)
24. Yang, Z., Algesheimer, R., Tessone, C.: A comparative analysis of community detection algorithms on artificial networks. Sci. Rep. **6**, 30750 (2016)

EDGE-UP: Enhanced Dynamic GNN Ensemble for Unfollow Prediction in Online Social Networks

Soheila Farokhi[1]([✉]), Arash Azizian Foumani[1], Xiaojun Qi[1], Tyler Derr[2], and Hamid Karimi[1]

[1] Department of Computer Science, Utah State University, Logan, USA
{soheila.farokhi,arash.foumani,xiaojun.qi,hamid.karimi}@usu.edu
[2] Computer Science Department, Vanderbilt University, Nashville, USA
tyler.derr@vanderbilt.edu

Abstract. In the complex landscape of online social networks, predicting unfollow events is challenging due to data sparsity, class imbalance, and the dynamic nature of user interactions. This paper presents **EDGE-UP**, an Enhanced Dynamic Graph Neural Network (GNN) Ensemble model adeptly designed to overcome these challenges in unfollow prediction. EDGE-UP leverages a large-scale, longitudinal Twitter dataset featuring 58 weekly snapshots across 118,890 users to capture the evolving social dynamics. It minimizes the need for extensive feature engineering by utilizing GNNs for spatial encoding and LSTMs for capturing temporal dynamics, addressing data sparsity and class imbalance through ensemble learning and negative sampling strategies. Our experiments demonstrate EDGE-UP's superior performance in accurately predicting unfollow events, setting a new standard in social network analysis, and offering versatile applicability across different platforms. The code and data are available here: https://github.com/DSAatUSU/edge-up.

Keywords: Unfollow Prediction · Dynamic Graph Neural Networks · Social Networks · Twitter (X)

1 Introduction

Online social networks, notably Twitter (X), are pivotal in the digital realm. They enable communication and interaction while serving as a substantial source of behavioral insights across various fields [1, 6–10, 14, 24, 27]. The dynamics of follower-followee relationships are vital for comprehending user behavior, and predicting unfollow events emerges as a critical challenge in this landscape. Figure 1 illustrates the evolution of social ties over time in a network, showcasing how users' actions of following (creating ties) and unfollowing (breaking ties) progressively modify the network's structure.

In this paper, we delve into the follower-followee relationships on online social networks, aiming to predict instances where users are likely to unfollow one

L. M. Aiello et al. (Eds.): ASONAM 2024, LNCS 15211, pp. 20–39, 2025.
https://doi.org/10.1007/978-3-031-78541-2_2

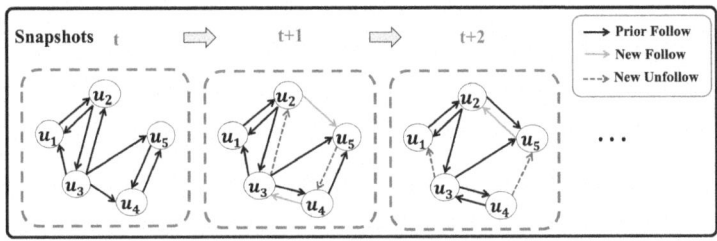

Fig. 1. Example of the evolution of ties in a social network as users follow or unfollow each other.

another. The motivation behind this research stems from the need to have an **accurate** model that can analyze and **predict** the evolution of ties in a social network. Such a model comes with several benefits. Firstly, understanding the dynamics of forming and dissolving friendships on social media is crucial [17, 26]. Secondly, analyzing how network structure influences tie breaks in online social networks is significant [13]. Thirdly, gaining insights into the evolution and decay of online relationships helps understand digital social dynamics [23]. Lastly, developing predictive models for follower loss, which incorporate factors like tweet content and engagement, is beneficial for proactive relationship management on social media platforms [18]. These reasons underscore the importance of unfollow prediction in comprehending and managing online social interactions, enhancing user experience, and informing social media strategies. However, unfollow prediction faces several challenges, reflecting the complex nature of social networks and human behavior. Some of the key challenges include:

❏ **Lack of Large-scale and Longitudinal Data:** The absence of large-scale, longitudinal datasets that capture the dynamism of online social networks limits the effectiveness of unfollow prediction models in accurately understanding and forecasting user behavior over time. This limitation leads to models that fail to capture the complexity of network dynamics and user interactions, resulting in reduced predictive accuracy and challenges in developing robust, generalizable models. Unlike follow prediction (a.k.a link prediction), unfollow prediction remains largely unexplored [30], primarily due to the scarcity of high-quality data.

❏ **Dynamic Nature of Social Networks:** Social networks are dynamic, with user interactions and preferences evolving over time. Predicting unfollows requires models to adapt to changing patterns and user behaviors, making it challenging to build robust and accurate models.

❏ **Data Sparsity and Class Imbalance:** Social network data is often sparse, and there is a significant class imbalance between unfollow incidents compared to follow ones—See Sect. 3. This can lead to biased models that struggle to generalize well, as the majority class (users who do not unfollow) may dominate the training data.

❑ **Hand-crafted Specific Features:** Prior research on unfollow behavior prediction in social networks has predominantly utilized platform-specific manually crafted features [18,30], which are resource-intensive and limit the *generalizability* of the models. Additionally, the vast amount of social media data complicates identifying features that accurately reflect user behavior and engagement. Thus, there is a need for a universally applicable model that transcends the constraints of specific platforms and the nature of user-generated content.

To address the challenges in predicting unfollow events, we first created a comprehensive and dynamic Twitter dataset. We then developed a new model named EDGE-UP, which adeptly integrates temporal dynamics with network structure, employing techniques like Graph Neural Networks (GNNs) and Long Short-Term Memory (LSTM) networks to capture users' evolving interactions and preferences. We address unfollow prediction challenges in the following ways:

❑ **Introducing a Large-Scale and Longitudinal Data:** We constructed a large-scale Twitter dataset with *58 weekly snapshots* of interactions among 118,890 Twitter users. This dataset provides a broad and detailed view of social network dynamics over an extended period. This long-term perspective is crucial for understanding the nuanced patterns and trends in unfollow behavior, often missed in smaller or shorter-term studies [13,31].

❑ **Minimizing Feature Engineering:** The Spatial Encoder in EDGE-UP uses GNNs to create simplified representations of user nodes. This allows the model to identify essential features from the social network structure automatically, eliminating the need for manual feature selection. EDGE-UP effectively captures user tie dissolution behavior by *merely focusing on the network's core patterns*, improving its prediction ability.

❑ **Adapting to Network Changes:** The Temporal Encoder in EDGE-UP is designed to keep up with users' changing interactions and preferences over time. By examining the sequence of network snapshots and their changes using an introduced concept of *lookback* period, EDGE-UP stays in tune with evolving patterns and behaviors, increasing its accuracy.

❑ **Addressing Data Sparsity and Class Imbalance:** EDGE-UP uses ensemble learning to tackle the problem of sparse data and the imbalance between the number of retained ties and dissolved ties. Additionally, we implement a robust negative sampling strategy to counter the imbalance between the frequencies of retained and dissolved ties in the dataset. By aggregating predictions from multiple models, this approach facilitates a more balanced perspective, enhancing the model's ability to generalize effectively across diverse data scenarios.

Our study contributes to the growing body of literature on social network analysis in the following ways:

❑ We construct a large-scale temporal dataset on Twitter with 58 weekly snapshots of the follow network among 118,890 users. To our knowledge, this dataset is the most comprehensive in terms of time span in this field, poised to significantly advance research in analyzing and predicting broken ties in online social networks.

❑ We propose EDGE-UP, a novel model for accurately predicting unfollow events in social networks. Our model incorporates a unique approach to capture the temporal evolution of user relationships, offering new insights into how social ties change over time, and addresses key challenges such as data sparsity, class imbalance, and the need for extensive feature engineering, setting a precedent for future research in this field. To the best of our knowledge, this is the first model to perform end-to-end unfollow prediction using only network data.

❑ We perform extensive experiments and show the superiority of EDGE-UP to the baseline methods and its high performance and reliability in unfollow prediction on Twitter (X).

Remark. While EDGE-UP has been tested on our Twitter (X) dataset, its application is not confined to a single platform. Its design allows easy adaptation to data from various social media platforms, provided such data is accessible. This versatility stems from its design, which is not dependent on the unique characteristics of any single platform's content.

2 Related Work

2.1 Unfollow Analysis and Prediction

Although unfollow prediction in social networks is gaining importance, research in this area remains relatively sparse. Next, we will examine some key studies focusing on analyzing and predicting unfollow behavior, often called broken tie analysis.

In a pioneering research effort, [16] conducted a study on 1.2 million Korean-speaking users on Twitter for 51 days to analyze unfollow dynamics. They identified a few significant factors influencing unfollow events, including overlap, duration, and reciprocity of the relationships, as well as followee's informativeness based on users' tweets and relationships. In a subsequent study, [17] examined two snapshots of follow networks of the same users 10 months apart and identified 12 structural and actional properties that affect the decision to unfollow. Later, [31] employed actor-oriented modeling (SIENA) to investigate how reciprocity, embeddedness, status, homophily, and informativeness influence the dissolution of ties. Their study on closely-knit user groups on Twitter showed that relational properties such as reciprocity and embeddedness significantly affect tie breaks. At the same time, the effect of informativeness and homophily based on common interest is not deemed significant. [13] investigated the impact of network structure alone on unfollow behavior and identified some structural properties that show significant effects on tie breaks. [23] examined the impact of several factors (e.g., age, gender, personality traits) that sociological studies have linked to friendship dissolution in real-world scenarios within the context of

Facebook. [18] analyzed the content of posts from Twitter users experiencing consistent follower loss to extract various behavioral features that are subsequently utilized for the early detection of follower loss.

While these approaches provide valuable insights, they necessitate substantial feature engineering and cannot capture the interaction between spatial and temporal attributes within a social network. Therefore, [30] proposed UMHI for unfollow prediction in a real-world Weibo dataset, which captures users' spatial attributes (e.g., their social role) through a network structure encoder and infers their temporal attributes through their posted content and unfollow history. This study is closest to our work. Nonetheless, there are several drawbacks to this model: 1) it lacks the capability for end-to-end training; 2) it relies on users' posted content, which could be challenging or expensive to gather and may not significantly influence unfollow behavior in certain Twitter groups, as demonstrated by [31]; 3) it does not consider the temporal evolution of users' local neighborhood.

Unlike traditional unfollow prediction methods, our approach focuses on harnessing spatio-temporal data **solely** from the user follow relationship network. This strategy eliminates the dependency on platform-specific, content-related, and often complicated hand-crafted features with very low generalizability. In particular, by leveraging a combination of graph neural networks and recurrent neural networks, we efficiently encode spatio-temporal information, offering a compelling end-to-end solution for unfollow prediction.

2.2 Follow (Link) Prediction

In contrast to unfollow prediction in social networks, follow prediction or link prediction has garnered considerable attention, especially in graph machine learning. As our approach draws on the research and breakthroughs in link prediction, including those methods used as baselines, we will briefly overview some of these pertinent methodologies.

Authors in [22] proposed DeepWalk, a random walk-based method for learning latent network representations for nodes in a network that effectively captures social relations in a continuous vector space. Node2vec was introduced in [3], which employs flexible, biased random walks to encode a graph's structure, balancing local and global network properties. node2vec has shown great performance in encoding graphs from many different domains [2,32]. LINE (Large-scale Information Network Embedding) [25] is designed for embedding extensive networks into low-dimensional spaces. It effectively captures both first-order (direct connections) and second-order (neighborhood similarity) relationships within the network. These methods focus on learning structural node embeddings in an unsupervised manner.

However, a significant advancement occurred with the introduction of Graph Convolutional Networks (GCN) by [12], combining node features and graph structure in a neural network approach, shifting towards more supervised learning in graph analysis. Authors in [11,12] discovered that the GCN-based approach consistently outperforms DeepWalk in link prediction. Building upon these developments, the Graph Attention Network (GAT) and GraphSAGE have further expanded the capabilities of graph machine learning. GAT [28], leverages attention mechanisms to weigh the importance of nodes' neighbors, allowing for more nuanced feature aggregation from a node's local neighborhood. On the other hand, GraphSAGE [4] innovates by using a sampling-based approach to generate node embeddings efficiently. It aggregates features from a node's local neighborhood while allowing for inductive learning, enabling the model to generalize to unseen nodes.

3 Dataset

To effectively predict unfollow events, it is essential to have access to a large, dynamic dataset of detailed user interactions. Therefore, we chose to gather data from Twitter (X). We gained access to Twitter's API through their Academic Research program and set up the necessary software and hardware to store data efficiently. This setup allowed us to capture weekly snapshots of social connections and content from 118,890 users over a year, beginning on June 18, 2018.

Table 1. Dataset statistics

Property	Average Value
# of nodes/users per week	118,890
# of edges/ties per week	2,841,814.09
Density	0.0002
# of new follows in weeks 2 to 58	9,648.88
# of new unfollows in weeks 2 to 58	4,095.17
Ratio of unfollows to # of edges	0.0014
Maximum # of followers	4,015.19
Minimum # of followers	1
Maximum # of followees	3,261.95
Minimum # of followees	1

We started with one author's Twitter account and expanded our user base using a breadth-first search approach [19], initially reaching around 130,000 users. After removing users who changed their privacy settings or deactivated their accounts during our data collection, we were left with 118,890 users. Our data, spanning from June 18, 2018, to July 22, 2019, includes 58 weekly snapshots that detail the follow relationships among these users (see Table 1 for statistics). It is important to highlight that Twitter's API does not provide direct information about unfollow events and when they occur. Therefore, to identify such events, we had to repeatedly collect data on the same users over an extended timeframe. As shown in Table 1, the number of unfollows compared to new follows is significantly smaller.

4 Problem Statement

Suppose we have observed the follow relationships among a subset of users in an online social network across T consecutive time-steps. In this context, a follow relationship is a directed link from a follower to a followee, and each user might be associated with a set of features. At any given time t, we can represent the state of the social network with a directed graph $G_t = (V, E_t)$, where V is the set of users that remains the same in all time steps, and E_t comprises directed edges reflecting the follow relationships at time t.

Unfollow: At time-step t, an edge (u_i, u_j) is labeled as *unfollow* if $(u_i, u_j) \in E_t$ and $(u_i, u_j) \notin E_{t+1}$.

Our goal is to develop a model $f(\cdot)$ that, given any sequence of k graph snapshots from time t to time $(t+k-1)$, predicts the likelihood of an unfollow event occurring between any two users u and v in the subsequent time-step $(t+k)$. The parameter k represents a designated *lookback* period, a crucial time window that determines the number of historical snapshots used for each prediction, optimized through hyper-parameter tuning. The model's predictive function can be mathematically represented as $\hat{y}_{(u,v)} = f(G_t, \ldots, G_{t+k-1})$, where $\hat{y}_{(u,v)}$ represents the predicted probability that user u will unfollow user v at time $(t+k)$.

5   The Proposed Method (EDGE-UP)

We propose a novel model for unfollow prediction in online social networks, which relies solely on the network's inherent relationship graph and temporal dynamics. Our approach involves four key components: 1) A Spatial Encoder employing Graph Neural Networks (GNNs) to derive compact, low-dimensional user node representations, capturing the structural intricacies of the social network; 2) A Temporal Encoder utilizing Long Short-Term Memory (LSTM) networks, to track the evolving local neighborhood structure surrounding user pairs before an unfollow event; 3) A Multi-Layer Perceptron (MLP) tasked with generating the final predictions for each pair of users; and 4) An Ensemble Learning strategy that enhances overall predictive accuracy by synergizing multiple models. This

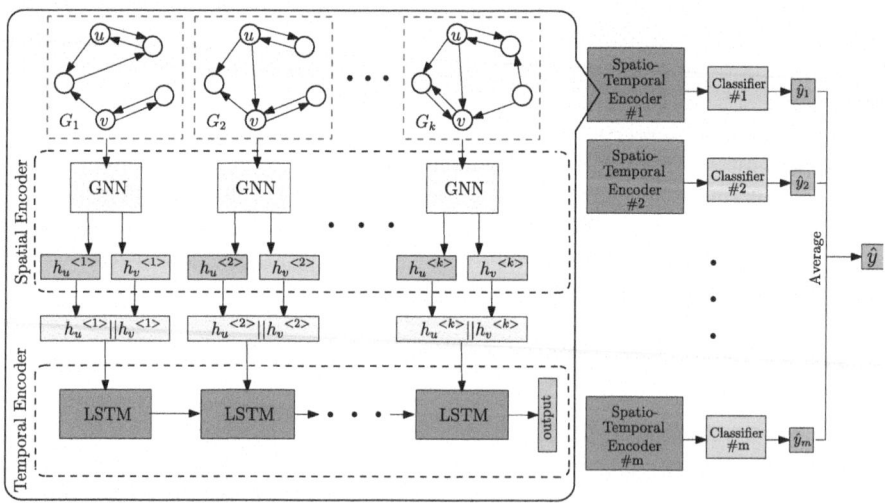

Fig. 2. Overview of EDGE-UP with m ensemble members.

comprehensive methodology is designed to optimize unfollow prediction by effectively integrating both spatial and temporal aspects of social networks. Figure 2 shows the overall architecture of our model. In the following subsections, we provide a detailed explanation of each component within our model.

5.1   Spatial Encoder

As highlighted earlier, one of the critical challenges in unfollow prediction within social networks is feature engineering. Earlier research has primarily focused on using hand-crafted features, which can be either costly or impossible to extract [18,30]. We propose using Graph Neural Networks (GNNs) to address this challenge. GNNs are a class of neural network models designed to process and analyze graph-structured data. We believe they are particularly effective in predicting unfollow actions in online social networks due to their proficiency in processing the complex and dynamic relationships inherent in these networks. They can adeptly capture local and global interaction patterns among users, making them well-suited for analyzing large-scale social media data, leading to more accurate and insightful predictions in online social interactions.

In a GNN, several graph convolution layers are employed, resembling a perceptron, but with an additional step for neighborhood aggregation. Each layer l of a GNN model involves three key components: message computation, message aggregation, and non-linearity. In the message computation step, each node creates a message that will be sent to other nodes. Mathematically, we can write: $m_u{}^{(l)} = MSG^{(l)}(h_u{}^{(l-1)})$.

Next, in the message aggregation step, each node aggregates the messages from its neighboring nodes with the message from the node itself. Finally, a non-linearity is applied to compute the node representation. For node v, we can write

$$h_v{}^{(l)} = \sigma(CON\,CAT(AGG^{(l)}(\{m_u{}^{(l)}, u \in N(v)\}), m_v{}^{(l)})) \tag{1}$$

where $N(v)$ is the set of neighbors of node v. After L layers of computation, the final embedding for node v is a d-dimensional vector $h_v{}^{(L)}$.

Researchers have introduced powerful GNN models in the past years by exploring possible choices for MSG and AGG functions. One such model is GraphSAGE, which has shown strong performance in various tasks involving graph-structured data [4]. GraphSAGE generalizes the aggregation step by allowing multiple aggregation functions, such as the max-pooling aggregator. In this approach, each neighbor's vector is transformed using a fully connected neural network, which constitutes the MSG function. So, we have:

$$m_u{}^{(l)} = MLP(h_u{}^{(l-1)}), \tag{2}$$

where MLP is an arbitrarily deep multi-layer perceptron.

The aggregation step has two stages. First, an element-wise max-pooling operation is applied to aggregate the messages from neighboring nodes. For node v, we can write:

$$m_{(N(v))}{}^{(l)} = Max(\{m_u{}^{(l-1)}, \forall u \in N(v)\}). \tag{3}$$

In the next stage, to further aggregate over the node itself, a GraphSAGE layer follows this equation:

$$h_v^{(l)} = \sigma(W^{(l)} \cdot CONCAT(m_{(N(v))}^{(l)}, m_v^{(l)})). \tag{4}$$

To predict unfollows, we need to learn representations for pairs of nodes in the graph. We achieve this simply by concatenating the embeddings of such nodes to construct a representation for the pair. Therefore, for $p = p(u, v)$, which is the pair of nodes u and v: $h_p = [h_u^L || h_v^L]$, where $||$ is the symbol for concatenation operation and is equivalent to $CONCAT$. In this paper, we use GraphSAGE layers as described above to encode graph snapshots that represent the structure of a social network at different time steps. We experimented with other aggregators, but max-pooling produced the best results.

5.2   Temporal Encoder

We hypothesize that incorporating the temporal evolution of a node's local neighborhood can enhance the prediction of future unfollow events. To validate this hypothesis, creating *sub-sequences* of consecutive graph snapshots from our dataset is essential. To this end, firstly, we focus on examining the network's structure within a specific time window preceding an unfollow event, which we refer to as the *lookback* period. Then, we employ a sliding window technique to generate data samples that track the development of local interactions around nodes before unfollow events. Given a series of graph snapshots $\{G_1, G_2, \ldots, G_T\}$, employing the sliding window technique with a lookback period of k and stride of 1 results in $T - k$ graph sequences. Each of these sequences contains k consecutive graphs and is utilized to predict unfollow events in the following time-step. Figure 3 illustrates the application of this technique to our dataset, using a stride of 1 and a lookback period of 2. In Sect. 6, we demonstrate that the proposed *lookback* approach is very practical in unfollow prediction, improving the performance significantly.

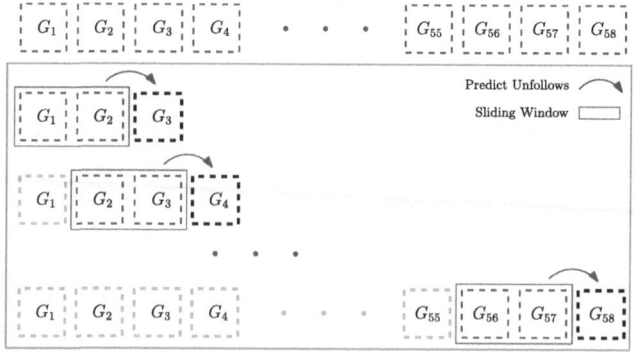

Fig. 3. Sliding window technique applied to our dataset with lookback set to 2.

In this paper, we employ Long Short-Term Memory (LSTM) networks to derive latent representations that capture the temporal evolution of local neighborhoods surrounding node pairs within each sequence of k graphs. LSTM networks [5] are a type of recurrent neural network designed for modeling and capturing dependencies in sequential data. LSTMs have wide application in tasks such as natural language processing, time series analysis, and speech recognition due to their ability to capture temporal dependencies effectively.

Suppose node u unfollows node v at time t. To be able to predict this unfollow event, we observe the evolution of the local neighborhood around nodes u and v from time $(t - k)$ to time $(t - 1)$. To this end, we first obtain the embedding of this pair of nodes at each time-step using a GNN. Then we input this sequence of embeddings $[h_p^{<t-k>}, \ldots, h_p^{<t-1>}]$ to an LSTM network. The LSTM learns a low-dimensional representation for the pair after time k has passed, which can be used to predict whether node u will unfollow node v or not.

5.3   Classifier

A Multi-layer Perceptron (MLP) is a type of neural network architecture characterized by multiple layers of interconnected nodes, including an input layer, one or more hidden layers, and an output layer [15]. In an MLP, each node (or neuron) in a layer is connected to every node in the subsequent layer, forming a densely connected network. This architecture allows MLPs to learn complex relationships in data, making them suitable for a wide range of tasks, including classification and regression.

We use an MLP with one hidden layer to predict whether a user will unfollow another user, given the final hidden state from the LSTM network. Suppose for node pair $p(u, v)$, the final hidden state of the LSTM is a_p. Then, the output of MLP is computed as follows:

$$h = ReLU(W_{in} \cdot a_p + b_h), \tag{5}$$

$$\hat{y}_{(u,v)} = \sigma(W_{out} \cdot h + b_o). \tag{6}$$

In Eq. 5, $ReLU$ is the Rectified Linear Unit activation function, W_{in} is the weight matrix for the input layer to the hidden layer, b_h is the bias for the hidden layer, and h is the final hidden layer vector. In Eq. 6, W_{out} is the weight matrix for the hidden layer to the output layer, b_o is the bias for the output layer, and $\hat{y}_{(u,v)}$ is the prediction for node pair $p(u, v)$ after applying sigmoid activation function.

5.4   Average Ensemble

As previously discussed, the scarcity of data in unfollow events poses challenges for unfollow prediction models, particularly their ability to generalize effectively to new and unencountered data. To mitigate this issue, we employ the strategy of ensemble learning. Ensemble learning is a machine learning technique where multiple models are combined to improve the overall performance and robustness of a predictive model [20]. Instead of relying on a single model, ensemble methods leverage the strengths of multiple models to achieve better accuracy, generalization, and reliability. In this work, we used ensemble averaging, a basic form of ensemble learning, where we trained multiple models on the same training set and averaged their predictions on the test set to achieve more accurate results.

Suppose f is a model that, given a set of graphs $G_t, \ldots, G_{t+k-1}$, predicts whether a user will unfollow another user in the next time-step $t+k$, i.e. $\hat{y}_{(u,v)} = f(G_t, \ldots, G_{t+k-1})$. Then ensemble averaging works as follows: First, we train m such models (also called ensemble members) independently; secondly, we apply the models on the test graphs separately and average their results. Suppose in test time, the goal is to predict whether user u unfollows user v in time $t' + k$, given graphs of k previous time-steps. Mathematically, we can write:

$$\hat{y}_{(u,v)} = \frac{1}{m} \sum_{i=1}^{m} f_i(G_{t'}, \ldots, G_{t'+k-1}) = \frac{1}{m} \sum_{i=1}^{m} \hat{y}_{(u,v),i} \tag{7}$$

where $\hat{y}_{(u,v)}$ is the final prediction for node pair $p(u,v)$ in time $t' + k$.

6 Experiments

In this section, we provide an in-depth look at our extensive experiments. Initially, in Sect. 6.1, we outline our experimental setup. Next, in Sect. 6.2, we introduce the baseline models used in our experiments. Section 6.3 presents the findings of our experiments, highlighting the comparison between EDGE-UP and the baseline models. Finally, we conclude with a comprehensive ablation study (component analysis) of our proposed model in Sect. 6.4.

Data Split. We divided the graph snapshots into training, validation, and testing subsets to evaluate our model comprehensively. Specifically, we utilized the initial 49 graphs along with their corresponding labels for the training set. In total, these labels include 207,542 unfollow events. For validation, we employed the subsequent 3 graphs and their labels, which include 10,804 unfollow events. Lastly, the final 5 graphs, covering weeks 53 to 57, and their associated labels with 15,079 unfollow events were designated for the testing set. This strategic partitioning is designed to rigorously assess the EDGE-UP model's capability to generalize and perform effectively on unseen data. We initialize each user's node features with a 50-dimensional random vector to eliminate the need for manual feature specification.

6.1   Experimental Settings

Label Generation To train our model for unfollow prediction, we implemented a structured approach for label generation using sequential graph snapshots. Labels were generated for each snapshot by analyzing changes in the graph structure over consecutive weeks. Specifically, positive unfollow samples were identified by locating edges present in a given snapshot but absent in the next week– See Sect. 4. We performed a negative sampling strategy to generate negative labels, where the head nodes of positive samples are altered, as depicted in Fig. 4. As a result, for every positive unfollow sample, we created one negative sample. The overarching objective of this methodology is to effectively train the model to distinguish between followers that will be retained and those that are likely to unfollow. Incorporating this labeling strategy, we successfully generated both positive and negative labels for a series of 57 graph snapshots spanning from week 1 to week 57. This comprehensive dataset ensures a balanced and robust foundation for modeling unfollow prediction as a binary classification problem. Utilizing these labels, we use binary cross-entropy loss function to train the model.

Implementation. We used PyTorch [21] and DGL [29] libraries for implementation. Each simulation was run for 200 epochs with early stopping based on validation loss. The learning rate was set to 0.0005, determined by hyper-parameter tuning, with a decaying rate of 0.9 every 50 steps. Based on hyperparameter tuning, we set the lookback value to 2 and the number of ensemble members to 60. The experiments were conducted on a system with an AMD EPYC 7513 CPU, 4 NVIDIA RTX A4000 GPUs, and 1 TB of RAM.

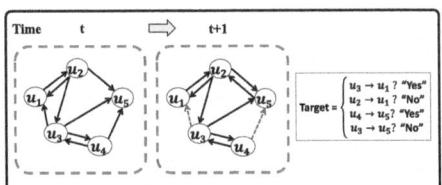

Fig. 4. This example showcases our approach in generating negative labels through negative sampling for the graph of week t after identifying positive labels.

Evaluation Metrics. We evaluated the performance of unfollow prediction models using accuracy, precision, recall, and area under the curve (AUC) metrics. We ran each experiment 5 times and reported the mean and standard deviation of results for each metric.

6.2   Baseline Models

We experimented with three categories of baseline models:
Unsupervised Spatial Encoders. We utilized a set of unsupervised shallow encoders to learn embeddings for all nodes within each graph snapshot. Next,

we concatenated these embeddings for pairs of nodes to predict potential unfollow events in the subsequent time step. We experimented with various prediction methods, including Random Forest, Logistic Regression, Support Vector Machine (SVM), and XGBoost. We conducted thorough evaluations with these techniques and reported the most optimal results. The following is the list of these shallow encoders:

- ❏ **DeepWalk:** [22] utilized random walks to generate node sequences, applying language modeling techniques to learn representations that capture the graph's structural properties. We used the *karateclub*[1] package for implementation of DeepWalk.
- ❏ **node2vec:** [3] introduced a more flexible notion of a node's neighborhood. It employs biased random walks to balance between capturing local and global graph structures. We used the *nodevectors*[2] package for scalable implementation of node2vec.
- ❏ **LINE:** [25] designed LINE to efficiently preserve both first-order (direct connections) and second-order (neighborhood similarities) proximities in the embedding space in large-scale information networks. We used an existing TensorFlow implementation of LINE[3].

Supervised Spatial Encoders. In another set of experiments, we employed three widely recognized graph neural network models to generate embeddings for graph snapshots. Subsequently, we concatenated these embeddings for node pairs, aiming to predict potential unfollow events in the next time step. An MLP was utilized to make predictions based on these final embeddings. However, it is important to note that these models do not capture the temporal evolution of users' local neighborhoods. We implemented these models using PyTorch and DGL libraries.

- ❏ **GCN:** GCNs [12] apply convolutional neural network principles to graph data, updating node representations by aggregating neighbor information, ideal for tasks like node classification.
- ❏ **GAT:** GATs [28] use attention mechanisms in graph neural networks to dynamically prioritize a node's neighbors, enhancing the learning of graph structures.
- ❏ **GraphSAGE:** GraphSAGE [4] inductively learns node embeddings by aggregating from neighbors and the node itself, enabling efficient embedding of unseen nodes in large graphs.

We also experimented with applying the average ensemble for each method as explained in Sect. 5. Number of ensemble members is set to 60 in all such experiments.

---

[1] https://github.com/benedekrozemberczki/karateclub.
[2] https://github.com/VHRanger/nodevectors.
[3] https://github.com/snowkylin/line.

Supervised Spatio-temporal Encoders. In these experiments, we used the idea of sliding windows explained in Sect. 5 to evaluate the effectiveness of incorporating temporal information in the performance of previously mentioned graph neural network models. For all models, we used an LSTM, with a lookback set to 2, to capture the temporal evolution of embeddings. An MLP was utilized to make predictions based on the LSTM's final hidden state. Note that our proposed model, EDGE-UP, utilizes this category of encoders. To compare the effect of ensemble learning in each variant, we experimented with 60 ensemble members for each method and recorded the results.

6.3 Experimental Results

Table 2. Comparing the performance of baseline models and our proposed model. The **best** result in each column is displayed in bold, and the second-best result is underlined.

	Category	Model	Accuracy	Precision	Recall	AUC
(1)	UnsupervisedSpatialEncoders	LINE	0.51 ± 0.00	0.51 ± 0.00	$\underline{0.51 \pm 0.00}$	0.50 ± 0.00
		DeepWalk	0.51 ± 0.00	0.52 ± 0.00	0.49 ± 0.00	0.51 ± 0.00
		node2vec	0.55 ± 0.00	0.58 ± 0.00	0.51 ± 0.00	0.57 ± 0.00
(2)	SupervisedSpatialEncoders	GCN	0.53 ± 0.00	0.53 ± 0.01	0.44 ± 0.04	0.54 ± 0.00
		GAT	0.53 ± 0.01	0.54 ± 0.02	0.29 ± 0.04	0.53 ± 0.02
		GraphSAGE	0.57 ± 0.02	0.64 ± 0.05	0.33 ± 0.07	0.60 ± 0.03
		GCN-Ensemble	0.53 ± 0.00	0.53 ± 0.00	0.47 ± 0.00	0.54 ± 0.00
		GAT-Ensemble	0.53 ± 0.00	0.55 ± 0.01	0.33 ± 0.01	0.54 ± 0.00
		GraphSAGE-Ensemble	0.61 ± 0.00	0.69 ± 0.00	0.40 ± 0.01	0.64 ± 0.00
(3)	SupervisedSpatio-TemporalEncoders	GCN+LSTM	0.52 ± 0.01	0.52 ± 0.01	0.46 ± 0.08	0.53 ± 0.00
		GAT+LSTM	0.53 ± 0.02	0.57 ± 0.04	0.28 ± 0.04	0.54 ± 0.03
		GraphSAGE+LSTM	0.60 ± 0.01	0.68 ± 0.02	0.39 ± 0.06	0.64 ± 0.01
		GCN+LSTM Ensemble	$\underline{0.68 \pm 0.01}$	0.78 ± 0.01	0.49 ± 0.01	0.76 ± 0.01
		GAT+LSTM Ensemble	0.62 ± 0.01	$\underline{0.97 \pm 0.01}$	0.25 ± 0.02	$\underline{0.83 \pm 0.01}$
		EDGE-UP	$\mathbf{0.92 \pm 0.01}$	$\mathbf{0.99 \pm 0.03}$	$\mathbf{0.85 \pm 0.04}$	$\mathbf{0.99 \pm 0.00}$

Table 2 presents the outcomes of our experiments, encompassing the three categories of baseline models alongside our EDGE-UP model. Based on these results, we make the following observations:

❏ Among the unsupervised spatial encoders, node2vec shows the best performance, indicating its effectiveness in capturing useful representations for unfollow prediction.

❏ Supervised spatio-temporal encoders generally perform better than unsupervised spatial encoders as well as their spatial-only counterparts. This suggests that the use of supervised learning and addition of temporal information processing (via LSTM) enhances the model's predictive capabilities.

❑ Within the supervised spatial encoders, GraphSAGE shows the highest scores, especially in terms of accuracy and AUC, suggesting its superiority in embedding quality for this task. When spatio-temporal encoding is applied, GraphSAGE+LSTM achieves the best results among them, reinforcing the strength of GraphSAGE as a spatial encoder when combined with temporal encoding.

❑ EDGE-UP demonstrates markedly superior performance across all metrics (accuracy, precision, recall, and AUC), with scores significantly higher than the other models. This indicates the effectiveness of combining a spatial encoder (GraphSAGE), a temporal encoder (LSTM), and ensemble learning in our approach. This substantial lead also underscores the significant impact of ensemble learning in enhancing prediction accuracy and reliability.

6.4   Ablation Study

In this section, we present the results of our extensive experiments to analyze the effectiveness of different components of EDGE-UP.

Temporal Encoder. To assess the effectiveness of the temporal encoder component, we devised an experiment with different lookback values. With a lookback of 1, the LSTM module is omitted, and node embeddings are directly concatenated for MLP-based prediction. We explored lookback values ranging from 2 to 4 to determine the optimal setting.

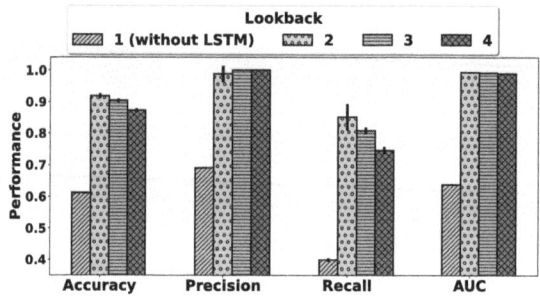

Fig. 5. Performance of EDGE-UP with different lookback values and associated error bars.

The results, depicted in Fig. 5, led to several key insights:

❑ The absence of the temporal encoder (lookback set to 1) significantly diminishes EDGE-UP's performance across accuracy, precision, recall, and AUC metrics, underscoring the crucial role of temporal information in the performance of unfollow prediction.

❑ A lookback of 2 yielded the highest accuracy, recall, and AUC, indicating that incorporating recent historical data of users' local neighborhoods enhances prediction efficacy.

❏ Increasing the lookback to 3 or 4 boosted precision but adversely affected other metrics. This suggests that while extended lookback enhances the reliability of positive predictions, it does not substantially benefit other aspects of prediction, possibly due to the diminishing relevance of more distant past information.

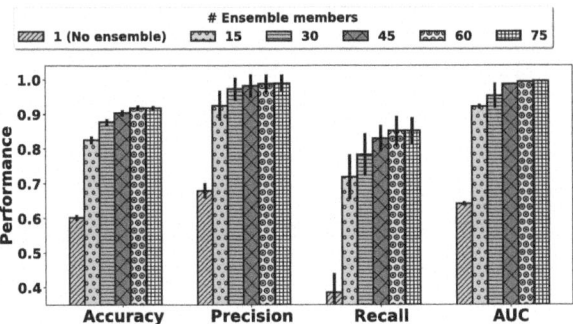

Fig. 6. Performance of EDGE-UP with different numbers of ensemble members and associated error bars.

Average Ensemble. In a separate series of experiments, we evaluated the influence of employing ensemble learning within our model. This involved testing with varying quantities of ensemble members, where a single member equates to not implementing ensemble learning. Further, we explored ensemble sizes of [15, 30, 45, 60, 75] to determine the impact of increasing the number of ensemble members. The outcomes of our experiments on ensemble learning are presented in Fig. 6, leading to the following key findings:

❏ Model performance improved consistently with the number of ensemble members, showing significant gains from a single member to 15 members.
❏ Beyond 60 members, the rate of performance improvement decreased, indicating diminishing returns.
❏ An ensemble of 45 members balanced performance gains and computational cost effectively.

Spatial Encoder. Figure 7 (a) visualizes the effect of replacing GraphSAGE in EDGE-UP with GCN and GAT. These results were also reported in Table 2. From these results, we observe that GraphSAGE, with max-pooling aggregation used in EDGE-UP, yields the highest accuracy, precision, recall, and AUC. This emphasizes the effectiveness of using GraphSAGE convolution layers to encode graph snapshots.

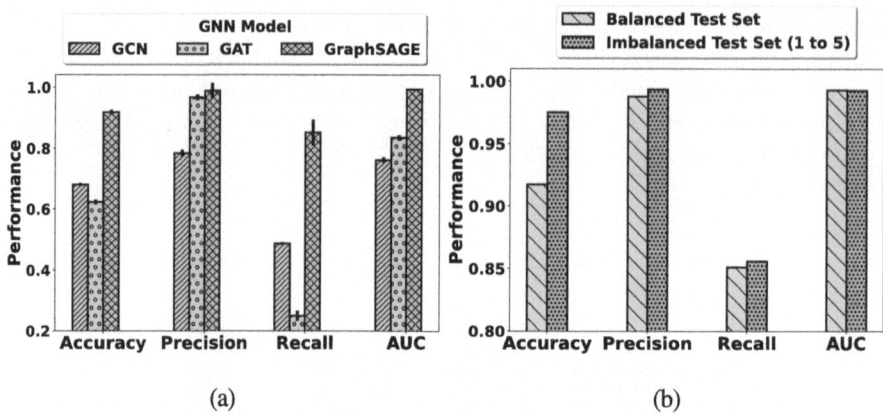

Fig. 7. Performance of EDGE-UP (a) with different GNN model members and associated error bars. (b) on a balanced test set versus an imbalanced test set.

Class Imbalance. To confirm our model's capability to predict unfollow events in the imbalanced setting, we experimented with the same test set of 5 graphs as described in Sect. 6.1, but with 5 negative samples per each positive unfollow event. As a result, in this test set, we include 15,079 positive unfollows and 75,395 negative unfollows (retained ties). Figure 7 (b) demonstrates the results of this experiment, highlighting the effectiveness of our model in predicting unfollows in an imbalanced environment.

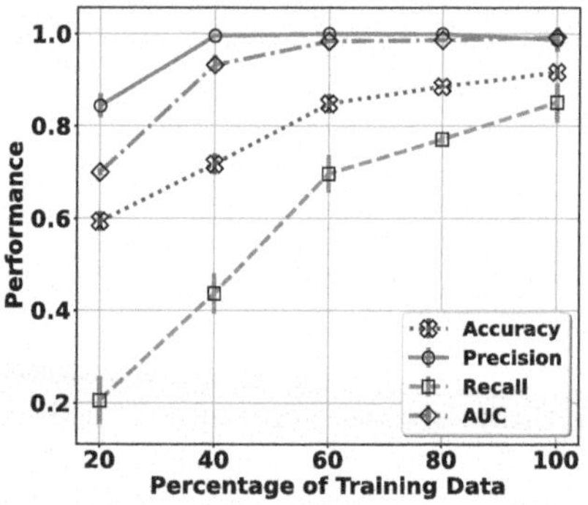

Fig. 8. Performance of EDGE-UP with different training sizes (with same test set) and associated error bars.

Training Set Size. We conducted experiments with varying sizes of training sets to assess the performance of our model when faced with limited training data. The findings are illustrated in Fig. 8. Despite the rarity of unfollow events in our dataset, our results demonstrate that EDGE-UP can attain an 85% accuracy in predicting unfollow events, even when trained with only 60% of the data (29 graphs).

7   Conclusion

In this paper, we constructed a large-scale longitudinal Twitter (X) dataset that facilitates research on dynamic online social networks. We then introduced EDGE-UP, an enhanced dynamic GNN ensemble method for unfollow prediction in online social networks. Our approach, which seamlessly integrates a spatial encoder, a temporal encoder, a classifier, and an ensemble learning strategy, represented a comprehensive solution to the complex problem of unfollow prediction. Our thorough experimentation showcased the efficiency of EDGE-UP in capturing spatial and temporal information within social networks, which significantly contributes to the accurate prediction of unfollow events.

Our research strictly followed Twitter's guidelines, using public data and ensuring privacy through anonymization and aggregated reporting. Despite acknowledging potential misuses like targeted advertising or social manipulation, we implemented strict ethical safeguards. We believe the benefits of our findings in understanding social interactions outweigh these ethical concerns.

The study's limitations include potential bias from its data collection method and the simplification of Twitter's dynamic social interactions into discrete events due to the reliance on network snapshots. Despite this, the large dataset size helps mitigate bias, and the focus on network structure over individual characteristics remains robust. However, there's a risk of overlooking the qualitative aspects of social relationships.

References

1. Brookhouse, A., Derr, T., Karimi, H., Bernard, H.R., Tang, J.: Road to the white house: analyzing the relations between mainstream and social media during the us presidential primaries. In: Proceedings of the 32nd ACM Conference on Hypertext and Social Media, pp. 57–66 (2021)
2. Farokhi, S., Yaramal, A., Huang, J., Khan, M.F.A., Qi, X., Karimi, H.: Enhancing the performance of automated grade prediction in MOOC using graph representation learning. In: 2023 IEEE 10th International Conference on Data Science and Advanced Analytics (DSAA), pp. 1–10. IEEE (2023)
3. Grover, A., Leskovec, J.: node2vec: scalable feature learning for networks. In: Proceedings of the 22nd ACM SIGKDD International Conference on Knowledge Discovery and Data Mining, pp. 855–864 (2016)
4. Hamilton, W., Ying, Z., Leskovec, J.: Inductive representation learning on large graphs. Adv. Neural Inf. Process. Syst. **30** (2017)
5. Hochreiter, S., Schmidhuber, J.: Long short-term memory. Neural Comput. **9**(8), 1735–1780 (1997)
6. Karimi, H., Derr, T., Torphy, K.T., Frank, K.A., Tang, J.: Towards improving sample representativeness of teachers on online social media: a case study on pinterest. In: Artificial Intelligence in Education: 21st International Conference, AIED 2020, Ifrane, 6–10 July 2020, Proceedings, Part II 21, pp. 130–134. Springer (2020)
7. Karimi, H., Torphy, K.T., Derr, T., Frank, K.A., Tang, J.: Characterizing teacher connections in online social media: a case study on pinterest. In: Proceedings of the Seventh ACM Conference on Learning@ Scale, pp. 249–252 (2020)
8. Karimi, H., VanDam, C., Ye, L., Tang, J.: End-to-end compromised account detection. In: 2018 IEEE/ACM International Conference on Advances in Social Networks Analysis and Mining (ASONAM), pp. 314–321. IEEE (2018)
9. Kheiri, K., Karimi, H.: Sentimentgpt: exploiting GPT for advanced sentiment analysis and its departure from current machine learning. arXiv preprint arXiv:2307.10234 (2023)

10. Kheiri, K., Khan, M.F.A., Derr, T., Karimi, H.: An analysis of the dynamics of ties on twitter. In: 2023 IEEE International Conference on Big Data (BigData), pp. 5809–5817. IEEE (2023)
11. Kipf, T., Welling, M.: Variational graph auto-encoders. arXiv preprint arXiv:1611.07308 (2016)
12. Kipf, T.N., Welling, M.: Semi-supervised classification with graph convolutional networks. arXiv preprint arXiv:1609.02907 (2016)
13. Kivran-Swaine, F., Govindan, P., Naaman, M.: The impact of network structure on breaking ties in online social networks: unfollowing on twitter. In: Proceedings of the SIGCHI Conference on Human Factors in Computing Systems, pp. 1101–1104 (2011)
14. Knake, K.T., Karimi, H., Hu, S., Frank, K.A., Tang, J.: Educational research in the twenty-first century: leveraging big data to explore teachers' professional behavior and educational resources accessed within pinterest. Elem. Sch. J. **122**(1), 86–111 (2021)
15. Kruse, R., Mostaghim, S., Borgelt, C., Braune, C., Steinbrecher, M.: Multi-layer perceptrons. In: Computational Intelligence: A Methodological Introduction, pp. 53–124. Springer (2022)
16. Kwak, H., Chun, H., Moon, S.: Fragile online relationship: a first look at unfollow dynamics in twitter. In: Proceedings of the SIGCHI Conference on Human Factors in Computing Systems, pp. 1091–1100 (2011)
17. Kwak, H., Moon, S., Lee, W.: More of a receiver than a giver: why do people unfollow in twitter? In: Proceedings of the International AAAI Conference on Web and Social Media, vol. 6, pp. 499–502 (2012)
18. Maity, S.K., Gajula, R., Mukherjee, A.: Why did they# unfollow me? early detection of follower loss on twitter. In: Proceedings of the 2018 ACM International Conference on Supporting Group Work, pp. 127–131 (2018)
19. Mislove, A., Marcon, M., Gummadi, K.P., Druschel, P., Bhattacharjee, B.: Measurement and analysis of online social networks. In: Proceedings of the 7th ACM SIGCOMM Conference on Internet Measurement, pp. 29–42 (2007)
20. Opitz, D., Maclin, R.: Popular ensemble methods: an empirical study. J. Artif. Intell. Res. **11**, 169–198 (1999)
21. Paszke, A., et al.: PyTorch: an imperative style, high-performance deep learning library. In: Wallach, H., Larochelle, H., Beygelzimer, A., d'Alché Buc, F., Fox, E., Garnett, R. (eds.) Advances in Neural Information Processing Systems, vol. 32, pp. 8024–8035. Curran Associates, Inc. (2019). http://papers.neurips.cc/paper/9015-pytorch-an-imperative-style-high-performance-deep-learning-library.pdf
22. Perozzi, B., Al-Rfou, R., Skiena, S.: Deepwalk: online learning of social representations. In: Proceedings of the 20th ACM SIGKDD International Conference on Knowledge Discovery and Data Mining, pp. 701–710 (2014)
23. Quercia, D., Bodaghi, M., Crowcroft, J.: Loosing "friends" on Facebook. In: Proceedings of the 4th Annual ACM Web Science Conference, pp. 251–254 (2012)
24. Solanki, S., et al.: Leveraging social media analytics in engineering education research. In: 2023 ASEE Annual Conference and Exposition (2023)
25. Tang, J., Qu, M., Wang, M., Zhang, M., Yan, J., Mei, Q.: Line: large-scale information network embedding. In: Proceedings of the 24th International Conference on World Wide Web, pp. 1067–1077 (2015)
26. Tufekci, Z.: Who acquires friends through social media and why¿'rich get richer" versus "seek and ye shall find". In: Proceedings of the International AAAI Conference on Web and Social Media, vol. 4, pp. 170–177 (2010)

27. VanDam, C., Tan, P.N., Tang, J., Karimi, H.: Cadet: a multi-view learning framework for compromised account detection on twitter. In: 2018 IEEE/ACM International Conference on Advances in Social Networks Analysis and Mining (ASONAM), pp. 471–478. IEEE (2018)
28. Veličković, P., Cucurull, G., Casanova, A., Romero, A., Lio, P., Bengio, Y.: Graph attention networks. arXiv preprint arXiv:1710.10903 (2017)
29. Wang, M., et al.: Deep graph library: a graph-centric, highly-performant package for graph neural networks. arXiv preprint arXiv:1909.01315 (2019)
30. Wu, H., Hu, Z., Jia, J., Bu, Y., He, X., Chua, T.S.: Mining unfollow behavior in large-scale online social networks via spatial-temporal interaction. In: Proceedings of the AAAI Conference on Artificial Intelligence, vol. 34, pp. 254–261 (2020)
31. Xu, B., Huang, Y., Kwak, H., Contractor, N.: Structures of broken ties: exploring unfollow behavior on twitter. In: Proceedings of the 2013 Conference on Computer Supported Cooperative Work, pp. 871–876 (2013)
32. Yaramala, A., Farokhi, S., Karimi, H.: Navigating the data-rich landscape of online learning: insights and predictions from assistments (2024)

Beyond Boundaries: Capturing Social Segregation on Hypernetworks

Andrea Failla[1,2]([✉]), Giulio Rossetti[2], and Francesco Cauteruccio[3]

[1] Department of Computer Science, University of Pisa, Pisa 56127, Italy
andrea.failla@phd.unipi.it
[2] National Research Council, ISTI-CNR, Pisa 56127, Italy
giulio.rossetti@isti.cnr.it
[3] University of Salerno, DIEM, Fisciano 84084, Italy
fcauteruccio@unisa.it

Abstract. In recent years, the study of complex social systems has been fueled by the renewed interest in higher-order topologies, thus leading to the emergence of *hypernetwork science*. A critical and interesting phenomenon often characterizing social complex systems is *segregation*, i.e., the extent to which network entities are separated or clustered based on certain semantic attributes or features. This paper introduces a novel approach to studying segregation in hypernetworks. Firstly, we propose a general framework to extend classical segregation measures from dyadic to polyadic network structures. Then, we introduce a novel segregation measure called "Random Walk HyperSegregation" (RWHS), which exploits random walkers to estimate segregation at multiple scales. Through an extensive experimental study involving synthetic and real-world case studies, we illustrate the applicability and effectiveness of our measure. Moreover, we highlight the limits of classical segregation measures when extended to high-order topologies—conversely from RWHS, which effectively captured highly-segregated scenarios.

Keywords: complex system · segregation · hypernetwork science · random walk

1 Introduction

The study of complex systems through the lenses offered by network theory has garnered significant attention, revealing intricate patterns of interactions across diverse fields such as sociology, biology, and information [30]. Traditional network analysis, focusing on pairwise interactions, has been instrumental in understanding the dynamics within these systems [29]. However, as we delve deeper into the complexity of real-world systems, it becomes increasingly clear that many exhibited patterns cannot be adequately captured by pairwise interactions alone [4]. Glancing at different (real) complex systems reveals a whole series of multi-way interactions, such as the ones in biological systems [14] or

© The Author(s), under exclusive license to Springer Nature Switzerland AG 2025
L. M. Aiello et al. (Eds.): ASONAM 2024, LNCS 15211, pp. 40–55, 2025.
https://doi.org/10.1007/978-3-031-78541-2_3

social ecosystems [15, 19]. The willingness to study these phenomena paved the way for the so-called hypernetwork science, in which interactions are modeled and analyzed through hypergraphs [1]—conservative generalizations of graphs in which (hyper)edges may connect an arbitrary number of vertices, thereby representing multi-way relationships [4].

A critical yet overlooked aspect of high-order representations lies in the understanding of the phenomenon of *segregation*—i.e., the extent to which system entities are separated or clustered based on certain attributes or features. The concept of segregation has its roots in studies on the residential organization of cities [27, 35], and undoubtedly, its analysis provides valuable insights into the underlying structure and dynamics of the system, with profound implications ranging from social equity and cohesion to system resilience and functionality. Indeed, segregation has been extensively studied in networks exhibiting pairwise interactions, such as social networks [5]. However, to the best of our knowledge, investigations regarding such phenomenon have not been conducted in the realm of hypernetworks, where due to the inherent complexity of dealing with higher-order interactions, defining and studying segregation in this context poses significant conceptual and methodological challenges. This paper aims to define a methodology that can help measure segregation in hypernetwork contexts. To such an extent, our contribution is twofold. Firstly, we formally define a general framework for extending pairwise segregation measures to hypernetworks. We refer to the extensions in this framework as *conservative* because they are transpositions of traditional measures originally envisioned for pairwise topologies. Secondly, we propose a novel measure tailored explicitly for hypernetworks, called *Random Walk HyperSegregation* (RWHS). We propose two variants of RWHS, namely *(i)* meet-wise and *(ii)* jump-wise RWHS, each capturing different aspects of node segregation. Moreover, to validate our approach, we present an extensive experimental study assessing the effectiveness and applicability of both segregation measuring strategies. To such an extent, we apply and discuss the proposed measures on synthetically generated and real-world hypernetwork topologies, showing how segregation can be captured through conservative extensions and RWHS.

The paper is organized as follows: in Sect. 2, we provide an overview of related literature; Sect. 3 covers the necessary background on hypernetwork modeling; Sects. 4 and 5 introduce the general framework for extending classical segregation measures and define RWHS, respectively; Sect. 6 focuses on the experimental validation of the proposed measures; Finally, Sect. 7 concludes the paper and delineates some possible future researches.

2   Related Works

Historically, segregation has been formalized as a phenomenon depicting two or more groups coexisting separately in the same environment. Such a definition has been applied to study heterogeneous contexts, e.g., by analyzing gender and ethnic segregation [9, 26]. Several studies on social dynamics stemmed from the seminal work in [35], which focused on residential segregation. Indeed, segregation has naturally been embraced by social network analysis research that usually treats such a concept as a complex system emerging property. This section presents a bird's-eye view of related literature on segregation and its quantification, focusing on sociological results and methodological approaches.

One of the most thorough investigations into quantifying segregation within social networks is presented in [5], where a range of segregation metrics—e.g., E-I index and Gupta's Q—were adapted and tested within a social network analysis framework. Analogously, [22] delves into network segregation through a Markov-chain-based model inspired by [35], illustrating how even mild biases against dissimilar nodes can foster segregated network structures. An advanced iteration of the model from [35] is further developed in [20], where the framework is adapted to accommodate a more dynamic social network setting. In particular, nodes are categorized into two groups, and each node can strategically form or dissolve connections based on the group composition of their immediate network vicinity. The findings suggest that employing random strategies for forming these connections results in higher segregation levels than strategies based on discriminatory behavior, indicating that network structure has a negligible impact on segregation outcomes. [3] introduces a data-centric methodology to capture the role of online social networks in intensifying segregation and opinion polarization—particularly focusing on the risk of echo chamber formation and pluralism reduction. Leveraging metrics such as the evenness and exposure indexes [27], the authors outline the segregation discovery problem, drawing inspiration from the classical problem of itemset mining. Noteworthy, [38] illustrates—leveraging data from an online social network of approximately 2 million individuals across 500 towns in Hungary—how sociality and geography play crucial roles in exacerbating wealth and income equalities. Similarly, in [33], a measure of social segregation combining mobile phone data and income register data is proposed, allowing the authors to estimate the association between income differences and communication intensity while considering spatial proximity. The investigation by [13], focused on credit card transactions and online social media data, examines contemporary forms of segregation, highlighting distinct divisions among various socioeconomic and ethnic groups—manifesting in patterns of urban visitation and online information consumption. Segregation was also identified in co-authorship networks [23], where the authors underlined that highly segregated communities tend to be closer to the network periphery, and researchers receive more citations from their community members. [36] highlights that spatial segregation encompasses several aspects, such as spatial exposure and isolation. Therefore, it has a multidimensional nature. To capture such a multifaceted reality, the authors introduce a framework based on graph random walks that gathered insights into different class organizations within cities. Although such an approach shares the same rationale as our RWHS measure, it only focuses on capturing spatial segregation by random walk processes on simple graphs, not higher-order ones, thus neglecting the effects of group interactions. In [18], the authors propose an agent-based network formation model under uncertainty—i.e., assuming agents in the network have only partial information about each other's groups—and leverage the generalized Freeman's segregation index [17] to quantify segregation. [11] identified high political segregation in a Twitter retweets dataset by leveraging a combination of manually-labeled data and unsupervised clustering algorithms. [32] introduces a high-dimensional approach to measure online polarization. Here, the authors define a high-dimensional network—akin to the classical multiplex network—and elucidate different measures to compute polarization and segregation. The approach is tested on synthetic networks and a Twitter dataset representing discussion regarding COVID-19 vaccines. Concerning segregation measures, the work in [12] proposes a generalization of the classical E-I index to the fuzzy case: in fact, it considers that nodes may belong to a group with a certain membership degree. Finally, [16] introduces a preliminary work on segregation in higher-order networks, applying a network fragmentation to measure segregation—the Borgatti's F measure [6]—to randomly generated hypergraphs with different distributions. A first attempt to approach segregation in high-order networks is performed in [16], where segregation is only estimated on artificially generated hypergraphs.

3 High-Order Network Modeling

In the following, we will model high-order interactions leveraging hypergraphs: this section provides definitions and notations needed to frame our contributions better.

Definition 1. (Hypergraph). *A hypergraph $H = (V, E)$ is a pair consisting of a set $V = \{v_1, \ldots, v_n\}$ of elements called nodes, and a set of sets $E = (e_1, \ldots, e_m)$ called hyperedges.*

A hyperedge represents a relation between a subset of nodes of V, that is, $e_i \subseteq V$ for all $i = 1, \ldots, m$. The order of H is the number of its nodes $n = |V|$, while the size of H is the number of its edges $m = |E|$. We say a node $v_j \in V$ belongs to a hyperedge $e_i \in E$ if $v_j \in e_i$.

To measure segregation, one typically deals with some finite set of attributes associated with each node. The most common example is group membership; network nodes are assigned to a group, and this assignment is known. To maintain this assumption, we introduce the concept of node-attributed hypergraph [15].

Definition 2. (Node-Attributed Hypergraph). *Let L be a set of labels, where each label denotes a group. A node-attributed hypergraph $H_L = (V, E, L)$ is a hypergraph H where to each node is assigned a label (a group) $l \in L$.*

We denote the group is assigned to a node $v_j \in V$ as $\gamma(v_j)$, and we also say that node v_j belongs to the group $\gamma(v_j)$. The number of groups is therefore $h = |L|$—we assume $h \geq 2$. Also, we denote with $E^l \subseteq E$ the set of edges containing at least one node belonging to the group l, i.e., $E^l = \{e_i \in E : \exists v_j \in e_i, \gamma(v_j) = l\}$.

4   Extending Classical Segregation Measures

This section provides a conservative extension of classical segregation measures to high-order topologies. To such an extent, we provide a general formulation encompassing concepts leveraged in measuring segregation, extending them to the hypergraph context. As a result, instances of the proposed general schema provide what is commonly called the system segregation index.

Definition 3. (Segregation Measure Schema). *We define segregation functions as instances of the tuple*

$$\mathfrak{F} = \langle H_L, f^{ie}(\cdot), \rho(\cdot) \rangle$$

where: (i) $H_L = (V, E, L)$ is a node-attributed hypergraph; (ii) $f^{ie} : E \to [0,1]$ is a generalized hyperedge type function; (iii) $\rho : H_L \to [-1,1]$ is a generalized segregation measure. Given H_L, $\rho(H_L)$ measures its segregation, thus providing its segregation index.

To describe the generalized hyperedge type function $f^{ie}(\cdot)$, it is useful to introduce the concepts of internal and external hyperedges. We borrow these terms from the context of social network analysis, where they refer to intra- and inter-group connections—which the classical segregation measures are based on [5].

Given a social network represented by a node-labeled graph, links between users with the same attribute are called internal ties, whereas other links are called external ties. Since these terms have not been defined consistently for node-labeled hypergraphs, we generalize existing formulations within the hyperedge type function f^{ie}. Specifically, $f^{ie} : E \to [0,1]$ categorizes each hyperedge $e_i \in E$ as internal (1) or external (0) w.r.t. a provided internal/external ties definition instance. For example, we can strictly define a hyperedge e_i internal only if all its nodes belong to the same group. In our investigation, we experiment with three definition instances, namely, *(i)* strict, *(ii)* majority, and *(iii)* linear. The majority definition instance states that a hyperedge is internal if more than half of its nodes belong to the same group. The linear one measures the extent to which a hyperedge is internal. Thus, the hyperedge contributes to each group proportionally w.r.t. its nodes' group memberships.

Leveraging f^{ie}, in conjunction with the generalized segregation measure $\rho(\cdot)$, classical segregation measures defined on graphs can be easily extended to hypergraphs. $\rho(\cdot)$ is generic; that is, its definition can accommodate different measures. In particular, to show the flexibility of the proposed segregation schema, we focus on two classical segregation measures: *(i)* E-I index [24], and *(ii)* Gupta's Q [21]. Our analysis focuses on those two measures due to their simplicity and adaptability; nonetheless, other measures, such as Freeman's segregation index [17], can be easily extended following the same approach.

E-I index. Although originally designed to assess homophily, the E-I index has been commonly used to measure segregation. It calculates the ratio of the net difference between inter-group and intra-group ties to the overall number of ties, serving as a normalization factor. Let us denote with E_{int} the number of internal hyperedges of H_L, that is $E_{int} = \sum_{e_i \in E} \mathbb{1}[f^{ie}(e_i) \neq 0]$. Analogously, let E_{ext} be the number of external hyperedges, that is defined as $E_{ext} = \sum_{e_i \in E} \mathbb{1}[f^{ie}(e_i) = 0]$—where $\mathbb{1}[\phi]$ is a binary indicator function that equals to 1 if the condition ϕ is satisfied, 0 otherwise. Then, the extension of the E-I index to hypergraphs becomes:

$$\rho_{E\text{-}I}(H_L) = \frac{E_{ext} - E_{int}}{E_{ext} + E_{int}}$$

$\rho_{E\text{-}I}(H_L)$ ranges in [-1, 1]. A value close to 1 indicates that groups in the hypergraph tend to have more external connections, while a value close to -1 indicates that groups tend to have more internal connections.

Gupta's Q. This index was introduced to analyze the effects of mixing patterns of sexual contacts on the spread of the HIV epidemic and captures the assortativity of a network in terms of integration and segregation. It considers within-group mixing, focusing on the connections of internal hyperedges against the actual connections. Given a group $l \in L$, we define $r_l = \frac{E_{int}^l}{|E^l|}$ as the ratio between the number of internal hyperedges and the total number of hyperedges containing at least one node belonging to the group l. Here, E_{int}^l is the number of internal hyperedges computed over the subset of hyperedges E^l. Then, the extension of Gupta's Q index to hypergraphs becomes:

$$\rho_Q(H_L) = \frac{\sum_{l \in L} r_l - 1}{h - 1}$$

A high value of $\rho_Q(H_L)$ suggests a strong tendency for nodes within the same group to be connected by hyperedges, thus indicating a higher level of segregation. Moreover, the study of single values of r_l could shed light on group dynamics—e.g., if certain groups predominantly form internal hyperedges, it can indicate a higher level of cohesion.

5   Random Walk HyperSegregation

As discussed, the schema introduced in the previous section allows extending classical segregation measures, initially designed for graphs, to hypergraph seamlessly. However, such conservative extensions can be inaccurate. Hypergraphs encompass non-dyadic relationships that the aforementioned segregation measures may not effectively capture. Moreover, segregation can be observed through a multifaceted lens: while measuring it from a global point of view helps identify overarching patterns and uncover structures of interest, analyzing it from a different granularity can offer insights into how individual entities or groups within the network are experiencing the segregation itself. We approach the latter strategy by leveraging random walks as effective proxies for information flow—thus analyzing walkers' behaviors to infer whether segregation occurs.

Random Walker Model. To maintain the formalization simple without loss of generality, we formalize a generic random walk model in terms of nodes—being the same formalization symmetrically applicable to hyperedges.

We denote with ω_i a random walk rooted at node v_i—i.e., a stochastic process consisting of the random variables $\omega_i^1, \omega_i^2, \ldots, \omega_i^k$ such that ω_i^{k+1} identifies a node visited at random within the neighborhood of ω_i^k. This is generally expressed by a probability $P(\omega_i^{k+1} = v_{j+1}|\omega_i^k = v_j)$, also called transition probability: practically speaking, it represents the probability that given the node v_j at the k-th step of the random walk, the next node is v_{j+1}. This probability can be computed in different ways [40]. In our case, it is deliberately general to accommodate different definitions, which will be given in the following. Moreover, we consider random walks of finite length; therefore, given a value $t > 0$, we denote with ω_i^t a random walk rooted at vertex v_i of length t.

We are interested in analyzing the *"realizations"* of a random walk on the hypergraph H_L. Fixed a v_i, we denote with $\Omega(\omega_i^t)$ a realization of ω_i^t, namely

$$\Omega(\omega_i^t) = (v_1, v_2, \ldots, v_t)$$

where $v_j \in V$, for $j = 1, \ldots, t$, and v_j is the node visited at the j-th step of the realization of the random walk ω_i^t. Practically speaking, $\Omega(\omega_i^t)$ is the sequence of node visits obtained by a random walk of length t rooted in v_i. Note that $\Omega(\omega_i^t)$ does not include the random walk source node v_i.

Given the stochastic nature of random walks, it is crucial to acknowledge the potential for high variability between different realizations of the same random walk. Specifically, let $\Omega^1(\omega_i^t)$ and $\Omega^2(\omega_i^t)$ represent two distinct realizations of a random walk rooted at node v_i with length t. Even though both $\Omega^1(\omega_i^t)$ and $\Omega^2(\omega_i^t)$ originate from the same starting node, have the same length and follow the same stochastic rules, the sequences of visited nodes $(v_1^1, v_2^1, \ldots, v_t^1)$ and $(v_1^2, v_2^2, \ldots, v_t^2)$ can be significantly different.

Starting from $\Omega(\omega_i^t)$, we can characterize the nodes that occurred in it, thus gathering insights from them; however, a single realization might not be enough to gather stable analytical insights. To overcome this issue, we focus on collections of realizations. Given a node v_i and the values $t, k > 0$, we define a *collection of realizations* W_i^t as a set of k different realizations of a random walk rooted at node v_i with length t, formally: $W_i^t = \{\Omega^1(\omega_i^t), \Omega^2(\omega_i^t), \ldots, \Omega^k(\omega_i^t)\}$. To avoid burdening the notation, when it is clear from the context, we drop the indication of the random walk, and we write $W_i^{t,k} = \{\Omega^1, \Omega^2, \ldots, \Omega^k\}$. Having described the random walk-based model, we can introduce the Random Walk HyperSegregation (RWHS) measure.

RWHS is a novel segregation measure specifically designed for hypergraphs. A significant distinction w.r.t. extensions of classical segregation measures is that it is defined at the element level, where an element is either a node or a hyperedge. RWHS quantifies the segregation level of each element of the hyper-

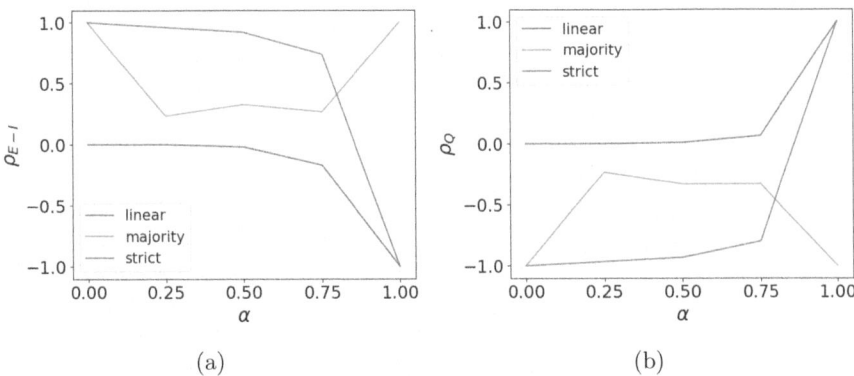

Fig. 1. Extensions of E-I index (a) and Gupta's Q (b) on hypergraphs with controlled homophily. Averages over 50 iterations.

graph *per-se*. Indeed, locality is a desired property since it enables fine-grained analysis of segregation patterns, overcoming known limitations of global measures [31,34]. Given a node-attributed hypergraph $H_L = (V, E, L)$, RWHS is parametric in *(i)* the node $v_i \in V$ whose level of segregation we want to assess, *(ii)* the value $t > 0$ representing the length of random walks, and *(iii)* the value $k > 0$ representing the number of realizations we want to consider. Having fixed such values, we define two variants of the RWHS, which we call *meet*-wise and *jump*-wise RWHS, respectively.

Definition 4 (*Meet*-wise RWHS). *Based on a collection of realizations* $W_i^{t,k} = \{\Omega^1, \Omega^2, \ldots, \Omega^k\}$, *we denote it as*

$$\phi_m^{t,k}(v_i) = \frac{1}{k} \sum_{r=1}^{k} \frac{|\{v_j \in \Omega^r : \gamma(v_j) = \gamma(v_i)\}|}{t}$$

Meet-wise RWHS is the ratio of nodes in the same group as v_i that appear in a realization within $W_i^{t,k}$, averaged over all realizations: it quantifies the exposure of an element to peers belonging to the same group. *Meet-wise RWHS* is bounded in $[0, 1]$: it approaches 1 when elements are enclosed in similarly-valued neighborhoods and 0 when elements are enclosed in differently-valued neighborhoods. This key property allows us to detect not only segregated environments as traditionally defined in the literature (i.e., fragmented areas populated by similar individuals, in which case $\phi_m^{t,k}(v_i) = 1$), but also environments where an element is isolated from its peers, and surrounded only by others belonging to different groups ($\phi_m^{t,k}(v_i) = 0$). In this sense, $\phi_m^{t,k}$ also acts as a measure of *marginalization*.

Definition 5. (*Jump*-Wise RWHS). *Based on a collection of realizations* $W_i^{t,k} = \{\Omega^1, \Omega^2, \ldots, \Omega^k\}$, *we denote it as*

$$\phi_j^{t,k}(v_i) = \frac{1}{k}\sum_{r=1}^{k}\frac{|\{(v_q \in \Omega^r, v_{q+1} \in \Omega^r) : \gamma(v_q) = \gamma(v_{q+1}))\}|}{t-1}$$

Jump-Wise RWHS is defined as the ratio of pairs of nodes that belong to the same group and are sequentially adjacent in a realization within $W_i^{t,k}$, averaged over all realizations. Practically, for each realization, it counts the pairs of subsequent steps whose nodes belong to the same group and normalizes it for the total number of pairs, then—to control for random effects—it averages such value across all realizations in the collection. Like its meet-wise counterpart, this measure is also bounded in [0,1]. The key difference is that $\phi_j^{t,k}(v_i)$ is independent from the group $\gamma(v_i)$. Indeed, whether a walker starting at v_i encounters only elements belonging to $\gamma(v_i)$ or only elements belonging to another group is irrelevant. Thus, for walks where all pairs share the same value, $\phi_j^{t,k}(v_i) = 1$ (regardless of which this value is); for walks where each step leads to a differently-labeled element, $\phi_j^{t,k}(v_i) = 0$. We argue that jump-wise RWHS better captures the *fragmentation* aspect of segregation, while the meet-wise RWHS is more akin to homophily estimation.

Some Remarks. A peculiarity of working with hypergraphs is that measures can be adapted to be computed not only on nodes but also on hyperedges [1]. In our scenario, the random walk-based model, as well as the RWHS variants based on it, can be seamlessly applied on hyperedges. However, the question of how to label hyperedges arises. A straightforward way to do so is to assign to a hyperedge the most frequent attribute value of its nodes [15,39]. To clarify this transposition, consider a discussion hypergraph, where nodes are social media users, and hyperedges denote discussion threads such that users interacting under a thread are enclosed in the same hyperedge. Each node is labeled with political leaning on a binary spectrum, e.g., either `republican` or `democrat`. In this case, the meet-wise RWHS on nodes measures how much a node v_i is surrounded (or likely to be influenced) by peers sharing the same political leaning. We can assign each hyperedge a label corresponding to its participants' most frequent political leaning to compute the measure of hyperedges. This way, we implicitly identify `republican`- and `democrat`-dominated threads. In this scenario, the meet-wise RWHS on hyperedges measures how nodes in the same context/thread will likely come across similarly-dominated contexts/threads. In a way, this transposition acts as a meso-scale segregation measure. Another remark concerns how random walks are computed. As mentioned earlier, the transition probability governing them can be computed in several ways [40], especially when dealing with the hypergraph scenario. For instance, in [10], the authors formalize a version of random hypergraph walks via edge-dependent vertex weights; that is, each node contributes to the walk based on a collection of weights that depend on the hyperedges it participates. Conversely, in our work, we leverage random walks as introduced in [7]. Regarding walks on nodes, the transition probability linearly correlates with hyperedge size, and within-edge jumps are more likely than cross-edge ones, coherently with information-spreading properties. Instead, when we deal with random walks on hyperedges, the transition probability depends on the number of common nodes between hyperedges.

6 Experiments

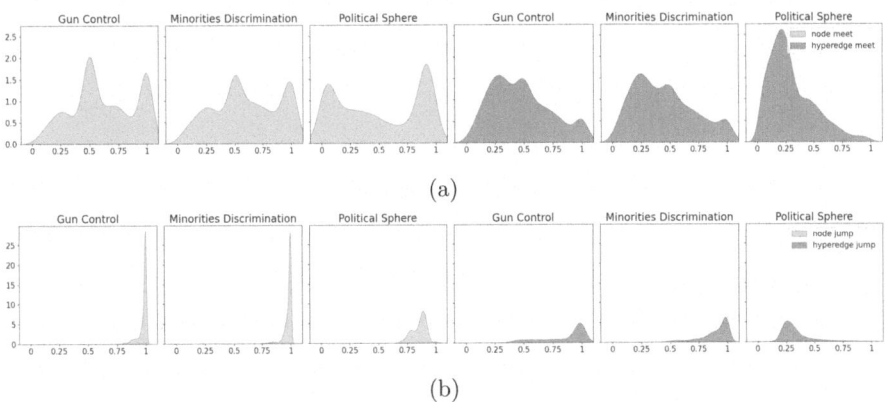

(a)

(b)

Fig. 2. Reddit Politics. Meet-wise (a) and jump-wise (b) RWHS score distribution. The X-axis represents RWHS, while the Y-axis represents the density estimate.

In the following, we apply the proposed segregation measures on synthetic and real-world hypernetworks. Our aim is twofold: (i) validating the extension of pairwise segregation measures to the high-order scenario, underlying its non-triviality by discussing outputs on synthetic hypernetworks with planted node mixing; (ii) validating the RWHS measures in real-world hypernetworks settings as described by face-to-face and online social interactions.

Conservative Extensions and Their Limitations. To show that generalizing existing segregation measures to hypergraphs is non-trivial, we applied the modified E-I index and Gupta's Q on synthetic hypernetworks with tunable assortativity. To such an extent, we extended a well-known graph generator [25] to account for high-order interactions.

In the following, we first briefly describe the original graph generator, discuss our modified implementation, and ultimately use it to evaluate segregation measures. The original generator is a node-attributed version of the Barabasi-Albert model, where connection probability depends on (i) preferential attachment and (ii) on a parameter α that controls the likelihood of same-class nodes being connected. As outlined by the authors, $\alpha = 1$ results in a perfectly assortative network (i.e., nodes are connected only to same-class nodes); conversely, $\alpha = 0$ results in a perfectly disassortative network (i.e., nodes are connected only to nodes from the other class); lastly, $\alpha = 0.5$ results in a classical Barabasi-Albert model. We choose to extend this model because it seamlessly controls

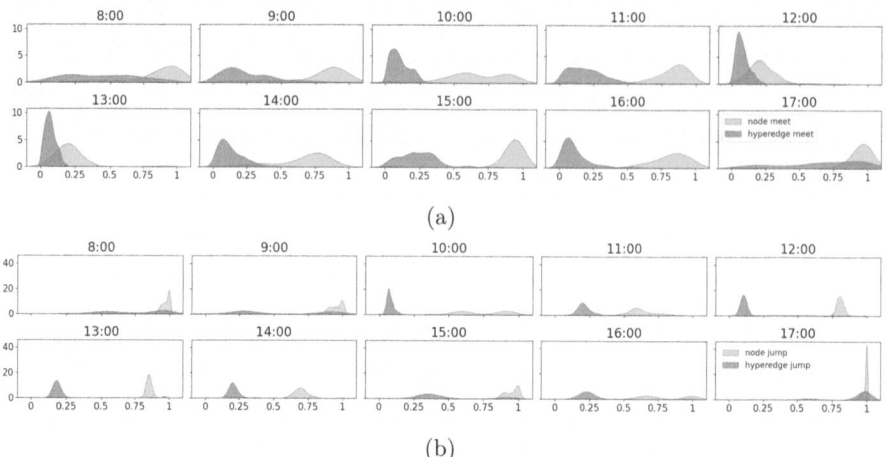

Fig. 3. Primary School. Meet-wise (a) and jump-wise (b) RWHS score distribution. The X-axis represents RWHS, while the Y-axis represents the density estimate.

the edge formation mechanism as a linear combination of degrees and in-group preference—which we use as a proxy for segregation. We modify this implementation as follows: (i) each node is assigned z hyperedges; (ii) the probability of attaching to a node depends on its number of neighbors, which are the nodes that share at least one hyperedge with it; (ii) hyperedge size is a random integer $s \in [2, 10]$. We generate synthetic hypergraphs for our experiments with $N = 1000$ nodes, $z = 3$, equal group size, and varying values for α. Results are averaged over 50 iterations.

E-I Index. Figure 1a depicts the output of the extended E-I index on hypernetworks with known underlying segregation, highlighting how choosing different definition instances, introduced in Sect. 4 impacts the measure's output. The *strict* one correctly captures both full segregation and full heterogeneity, but the transition between these extremes is not smooth, remaining constant until $\alpha = 0.5$ and then plummets rapidly to -1. *Linear* cannot capture heterogeneity and instead moves from 0 (no segregation nor heterogeneity) to -1. Finally, *majority* strongly deviates from what one would expect a segregation measure to capture; indeed, it scores high values (~ 1) at both extremes and relieves heterogeneity when tie formation is independent of homophily.

Gupta's Q's extensions (Fig. 1b) provides similar results. *Strict* gives coherent results at the extremes, but overestimates heterogeneity; *Linear* correctly identifies complete segregation, but never captures heterogeneity; *Majority* returns unsatisfactory results in all scenarios except for heterogeneity.

Therefore, the performed controlled experiments underline that one-to-one high-order extensions of pairwise measures are subject to severe limitations, not behaving as they were originally intended. Such a negative result is tied to the inconsistency of semantics while moving from dyadic to polyadic interactions: conservative extensions suffer from not taking such semantic nuances into account, thus leading to inconsistencies or unexpected behaviors.

Hypernetwork Based Segregation Measure. We test RWHS on high-order social interactions data, focusing on (i) online debates between pro/anti-Trump Redditors during the first two years and half of Donald Trump's presidency [28] and on (ii) children face-to-face interactions. Reddit data describe interactions and user ideology labels (protrump, neutral, antitrump) across three topics: *Gun control, Minorities Discrimination* and *Political Sphere*[1]. In these scenarios, each hyperedge identifies the set of users that commented on a given post at the same level of the discussion tree—thus modeling individual contexts of a broader discussion. The Primary School dataset [37] (henceforth, PS) from the SocioPatterns project[2] comprises face-to-face interaction data collected via RFID sensors over two days. The RFID sensors capture interactions with a 20-second resolution; we first preprocess the data by aggregating it into a series of static networks so that each network contains all interactions captured over an hour. Each node is enriched with information on their class (10 unique values in total). Subsequently, we infer the system's high-order structure by leveraging the method introduced in [8]. Specifically, if at time t there are $N * (N + 1)/2$ edges between N nodes such that they are involved in a fully connected clique, such links are promoted to form a hyperedge of size N. We limit our analysis to the first day.

Reddit. We compute RWHS scores on nodes and hyperedges. Scores are computed over a collection of realizations of $k = 50$ walks of length $t = 6$ for each element. As shown by Fig. 2a, all hypernetworks are characterized by highly- and non-segregated areas. In particular, meet-wise RWHS on nodes show taller peaks towards high values, outlining a tendency towards segregation. Conversely, on hyperedges meet-wise shows lower values, meaning that users participate in a heterogeneous collection of discussions, i.e., a protrump user participates in both pro- and antitrump-dominated discussions. This is coherent with the results of the analysis in [15]. Jump-wise shows extremely right-skewed distributions for nodes and hyperedges—i.e., users are likelier to be located in label-wise homogeneous areas. In all scenarios, curves related to the Political Sphere differ from the other datasets, highlighting strongly heterogeneous areas. We hypothesize that since the discussions taking place in Political Sphere pertain to a wide range of topics, users have higher chances to connect with peers from other groups.

To assess whether the observed segregation is statistically significant, we compare RWHS scores of real data to those obtained on instances of an extension of the Chung-Lu configuration model [2] preserving both hyperedge size and node degree distributions—also tailored to preserve label distribution of the corresponding real hypernetwork. Comparing RWHS distributions via a two-sample Kolmogorov-Smirnov test for goodness of fit—which tests the null hypothesis

---

[1] Details on data collection and label inference in [28].

[2] www.sociopatterns.org.

that the distribution underlying two samples is the same—we rejected the null hypothesis with $p < 0.01$: all empirical distributions significantly differ from the corresponding null ones. Moreover, all empirical hypernetworks have higher mean and median than the corresponding generated ones, suggesting a general tendency for higher segregation than expected at random, a characteristic perfectly captured by RWHS.

Primary School. We perform a temporal analysis of spatial segregation on the PS dataset. Individuals in PS spend most of their school day in their classroom [37]; thus, we expect high levels of segregation. Indeed, the meet-wise scores, Fig. 3, are generally high on nodes, outlining the prevalence of same-class group interactions. Conversely, on hyperedges scores tend towards -1, implying that reaching students from other classes is still possible. Symmetrically, jump-wise scores are generally high on nodes—emphasizing that they are surrounded by closed groups (i.e., classes)—while being low on hyperedges (e.g., inter-class interactions still noticeable). Moreover, all RWHS scores capture strong homogeneity when students arrive at school (8:00) and when they leave classes (17:00)— conversely from lunchtime (12:00–13:00) when the highest heterogeneity is captured. These results are coherent with what was previously observed in [37], showing that RWHS correctly captures the intra- and inter-class dynamics we expect to observe, thus providing a suitable and effective measure to capture segregation.

7   Conclusion

While segregation has been extensively studied in classical network setups, it remains an overlooked aspect in higher-order social modeling. We proposed a general schema that extends classical segregation approaches to node-attributed hypernetworks to fill this gap. We tested it by reinterpreting the well-known E-I index and Gupta's Q measures. Moreover, we also defined a new multi-scale segregation index, RWHS, that relies on random walks as effective proxies for information flow. With an exhaustive series of experiments, we illustrated that conservative extensions of classic measures to high-order settings cannot adequately capture segregation. Conversely, RWHS allows us to observe homogeneous and heterogeneous multi-scale mixing patterns. Such a result underlines the limitations of classical approaches that, when applied to high-order topologies, fail to exploit the richer topological semantics they offer. This paper should not be considered as an endpoint for research on this topic but rather as a starting one. As future research, we plan to extend the segregation study to temporal hypernetworks where nodes, attributes, and interactions co-evolve over time— idealistically leveraging time-respecting walks [15] to embed temporal constraints within random walk processes.

Acknowledgments. This work is supported by (i) the European Union - Horizon 2020 Program under the scheme "INFRAIA-01-2018-2019 - Integrating Activities for Advanced Communities", Grant Agreement n.871042, "SoBigData++: European Integrated Infrastructure for Social Mining and Big Data Analytics" (http://www.sobigdata.eu); (ii) SoBigData.it which receives funding from the European Union - NextGenerationEU - National Recovery and Resilience Plan (Piano Nazionale di Ripresa e Resilienza, PNRR) - Project: "SoBigData.it - Strengthening the Italian RI for Social Mining and Big Data Analytics" - Prot. IR0000013 - Avviso n. 3264 del 28/12/2021; (iii) EU NextGenerationEU programme under the funding schemes PNRR-PE-AI FAIR (Future Artificial Intelligence Research).

Disclosure of Interests. The authors have no competing interests to declare that are relevant to the content of this article.

References

1. Aksoy, S.G., Joslyn, C., Marrero, C.O., Praggastis, B., Purvine, E.: Hypernetwork science via high-order hypergraph walks. EPJ Data Sci. **9**(1), 16 (2020)
2. Aksoy, S.G., Kolda, T.G., Pinar, A.: Measuring and modeling bipartite graphs with community structure. J. Complex Netw. **5**(4), 581–603 (2017)
3. Baroni, A., Ruggieri, S.: Segregation discovery in a social network of companies. J. Intell. Inf. Syst. **51**, 71–96 (2018)
4. Battiston, F., et al.: Networks beyond pairwise interactions: structure and dynamics. Phys. Rep. **874**, 1–92 (2020)
5. Bojanowski, M., Corten, R.: Measuring segregation in social networks. Soc. Netw. **39**, 14–32 (2014)
6. Borgatti, S.P.: Identifying sets of key players in a social network. Comput. Math. Organiz. Theory **12**, 21–34 (2006)
7. Carletti, T., Battiston, F., Cencetti, G., Fanelli, D.: Random walks on hypergraphs. Phys. Rev. E **101**(2), 022308 (2020)
8. Cencetti, G., Battiston, F., Lepri, B., Karsai, M.: Temporal properties of higher-order interactions in social networks. Sci. Rep. **11**(1), 1–10 (2021)
9. Charles, M., Grusky, D.B.: Models for describing the underlying structure of sex segregation. Am. J. Sociol. **100**(4), 931–971 (1995)
10. Chitra, U., Raphael, B.: Random walks on hypergraphs with edge-dependent vertex weights. In: Proceedings of the 36th International Conference on Machine Learning (ICML 2019), vol. 97, pp. 1172–1181. PMLR (2019)
11. Conover, M., Ratkiewicz, J., Francisco, M., Gonçalves, B., Menczer, F., Flammini, A.: Political polarization on twitter. In: Proceedings of the International AAAI Conference on Web and Social Media (ICWSM), vol. 5, pp. 89–96. AAAI (2011)
12. de Andrade, R.L., Rêgo, L.C.: A proposal for the ei index for fuzzy groups. Soft. Comput. **27**(4), 2125–2137 (2023)
13. Dong, X., et al.: Segregated interactions in urban and online. EPJ Data Sci. **9**(20), 1–22 (2020)
14. Dotson, G.A., et al.: Deciphering multi-way interactions in the human genome. Nat. Commun. **13**(1), 5498 (2022)

15. Failla, A., Citraro, S., Rossetti, G.: Attributed stream hypergraphs: temporal modeling of node-attributed high-order interactions. Appl. Netw. Sci. **8**(1), 1–19 (2023)
16. Vasques Filho, D.: Cohesion and segregation in higher-order networks. arXiv e-prints p. 2207 (2022)
17. Freeman, L.C.: Segregation in social networks. Sociol. Methods Res. **6**(4), 411–429 (1978)
18. Garimella, K., Morales, G.F., Gionis, A., Mathioudakis, M.: Quantifying controversy on social media. ACM Trans. Soc. Comput. **1**(1), 1–27 (2018)
19. Gong, Y.-C., Wang, M., Liang, W., Hu, F., Zhang, Z.-K.: Uhir: an effective information dissemination model of online social hypernetworks based on user and information attributes. Inf. Sci. **644**, 119284 (2023)
20. Gretha, O.B., Cristal, P.M., Mauhe, N.: Segregation in social networks: a simple schelling-like model. In: Proceedings of the 2018 IEEE/ACM International Conference on Advances in Social Networks Analysis and Mining (ASONAM), pp. 95–98. IEEE (2018)
21. Gupta, S., Anderson, R.M., May, R.M.: Networks of sexual contacts: implications for the pattern of spread of hiv. AIDS **3**(12), 807–818 (1989)
22. Henry, A.D., Prałat, P., Zhang, C.-Q.: Emergence of segregation in evolving social networks. Proc. Natl. Acad. Sci. **108**(21), 8605–8610 (2011)
23. Jaramillo, A.M., Williams, H.T.P., Perra, N., Menezes, R.: The structure of segregation in co-authorship networks and its impact on scientific production. EPJ Data Sci. **12**(1), 47 (2023)
24. Krackhardt, D., Stern, R.N.: Informal networks and organizational crises: an experimental simulation. Soc. Psychol. Quart. 123–140 (1988)
25. Lee, E., Karimi, F., Wagner, C., Jo, H.-H., Strohmaier, M., Galesic, M.: Homophily and minority-group size explain perception biases in social networks. Nat. Hum. Behav. **3**(10), 1078–1087 (2019)
26. Louf, R., Barthelemy, M.: Patterns of residential segregation. PloS One **11**(6), e0157476 (2016)
27. Massey, D.S., Denton, N.A.: The dimensions of residential segregation. Soc. Forces **67**(2), 281–315 (1988)
28. Morini, V., Pollacci, L., Rossetti, G.: Toward a standard approach for echo chamber detection: Reddit case study. Appl. Sci. **11**(12), 5390 (2021)
29. Newman, M.: Networks. Oxford University Press (2018)
30. Newman, M.E.J.: Complex systems: a survey. arXiv preprint arXiv:1112.1440 (2011)
31. Peel, L., Delvenne, J.-C., Lambiotte, R.: Multiscale mixing patterns in networks. Proc. Natl. Acad. Sci. **115**(16), 4057–4062 (2018)
32. Phillips, S.C., Uyheng, J., Carley, K.M.: A high-dimensional approach to measuring online polarization. J. Comput. Soc. Sci. **6**(2), 1147–1178 (2023)
33. Reme, B.-A., Kotsadam, A., Bjelland, J., Sundsøy, P.R., Lind, J.T.: Quantifying social segregation in large-scale networks. Sci. Rep. **12**(1), 6474 (2022)
34. Rossetti, G., Citraro, S., Milli, L.: Conformity: a path-aware homophily measure for node-attributed networks. IEEE Intell. Syst. **36**(1), 25–34 (2021)
35. Schelling, T.C.: Models of segregation. Am. Econ. Rev. **59**(2), 488–493 (1969)
36. Sousa, S., Nicosia, V.: Quantifying ethnic segregation in cities through random walks. Nat. Commun. **13**(1), 5809 (2022)
37. Stehlé, J., et al.: High-resolution measurements of face-to-face contact patterns in a primary school. PLoS ONE **6**(8), e23176 (2011)
38. Tóth, G., et al.: Inequality is rising where social network segregation interacts with urban topology. Nat. Commun. **12**(1), 1143 (2021)

39. Veldt, N., Benson, A.R., Kleinberg, J.: Combinatorial characterizations and impossibilities for higher-order homophily. Sci. Adv. **9**(1), eabq3200 (2023)
40. Xia, F., Liu, J., Nie, H., Fu, Y., Wan, L., Kong, X.: Random walks: a review of algorithms and applications. IEEE Trans. Emerg. Topics Comput. Intell. **4**(2), 95–107 (2019)

Burstiness in Emotions: A Case Study on Collective Affective Responses in Italian Soccer Fandoms

Salvatore Citraro[1](✉) , Giovanni Mauro[1,2,3] , and Emanuele Ferragina[4]

[1] CNR-ISTI, Via Giuseppe Moruzzi, 1, Pisa, Italy
salvatore.citraro@isti.cnr.it
[2] University of Pisa, Largo Bruno Pontecorvo, 3, Pisa, Italy
[3] IMT Lucca, Piazza S. Francesco, 19, Lucca, Italy
[4] SciencesPo, Rue Saint Guillaume, 27, Paris, France

Abstract. The bursty nature of emotions is rarely investigated outside cognitive and psychological studies. Therefore this work addresses a gap in the literature, investigating the phenomenon of emotional burstiness using tools from the analysis of complex systems, and considering as case-study soccer fans' affective responses on social media. We reconstruct collective reactions on Instagram posts from official accounts of 40 Italian football teams during the first round of the 2023–2024 season – 20 teams from Serie B (the second tier of Italian Football) and the 20 most followed teams in Serie C (the third tier). With this data, we build sequences of emotional signals for four types of emotions: joy, anger, sadness, and fear. Our analysis reveals trends of anti-burstiness in expressions of joy among users, reflecting fans' consistent support for teams, occasionally interspersed by bursts of anger and sadness, with no signals of fear. This preliminary investigation provides insights for the understanding of emotional dynamics in online discussions and team supporting in soccer leagues.

Keywords: burstiness · memory · time series · emotions

1 Introduction

Time-dependent activity patterns are ubiquitous in nature, from the burstiness of earthquakes [3] to the intermittency of rainfall [25]. Human behavior exhibits complex dynamics as well, shaped by selection mechanisms [5] and social interactions [7,27], which result in bursty patterns across various activities such as email exchanges, financial transactions, phone calls, and online messages, among others [17]. Despite numerous contributions in uncovering these patterns within different domains [17], several aspects of human complexity still lack adequate attention. Among these, the dynamics of emotions are among the most challenging to represent and analyze [11,19], due to the difficulty of directly mapping subjective experiences. Differently from recording when a mail is sent or a financial transactions is processed, emotions cannot be easily measured.

Various psychological studies have hinted to the bursty nature of emotions. Affect bursts [24] have been defined as brief expressions of affect, and measured in both face and voice variations in response to events of positive/negative valence [26]. Intense bursts of joy and positive emotions are characteristic of the manic phase of bipolar disorder [1], while violent outbursts of anger may indicate behavioral conditions like in the intermittent explosive disorder [1]. Emotional responses take place in social media as well, and being triggered by online content consumption can lead to hasty variations in users' emotional states [29]. Novel areas of research within social media analysis aim to quantify emotional variations from various perspectives, e.g., that of mental health online self-disclosure, by observing emotional shifts in online support communities over time [4], or by identifying early signals of suicidal ideation [16]. Other perspectives rely on the detection of hate speech and negative content [28], relating it to misinformation [8], to emotional persistence/instability [12], and to the dynamics of emotional shifts driven by recommender systems [15]. However, despite their bursty quality and the importance of understanding their stable or unstable nature, emotions have never been addressed through the quantitative tools offered by the science of complex systems. In other words, emotions have never been thought of as measurable time-dependent signals. The aim of this study is to follow Goh and Barabasi's milestone work about burstiness quantification [14] and apply it to build and measure sequences of emotional activity. Goh and Barabasi [14] have uncovered two main processes that characterize bursty activities within complex systems: a distribution of inter-event times that deviates significantly from Poissonian processes, and the memory between pairs of consecutive inter-event times.

As a domain of analysis, we choose soccer fans' affective responses on social media. We reconstruct the emotional collective responses of soccer fans commenting Instagram posts of the 20 official football teams' accounts during the first round of the Italian 2023–24 Serie B, and the first round of the 20 teams with the highest number of Instagram followers among the three Groups of the 2023–24 Serie C. The rationale behind the choice of the soccer domain lies in the investigation of emotional dynamics in a context where emotions unfold collectively [13] as well as emotional outbursts. We focus on lower soccer leagues as they provide the opportunity to analyze genuine and local fandoms [20], without the overexposure of international fans present in major leagues. We choose Instagram because it offers a simpler way to distinguish between the collective sentiment of each fandom by leveraging the posting activity sequence of each team account, as opposed to using hashtags or conversations on Twitter [9,22] or communities on Reddit [6], where discussions and comments are more mixed.

The paper proceeds as follows. Section 2 introduces the measures used to describe the bursty properties of complex systems together with the dataset constructed for this analysis. Section 3 highlights the main results on both leagues.

Section 4 discusses the results and concludes the work towards future research directions.

2   Materials and Methods

This section introduces the data collection and pre-processing (Sect. 2.1). Subsequently, it presents the measures used to describe burstiness and memory in generic activity sequences (Sect. 2.2), following Goh and Barabasi's work [14].

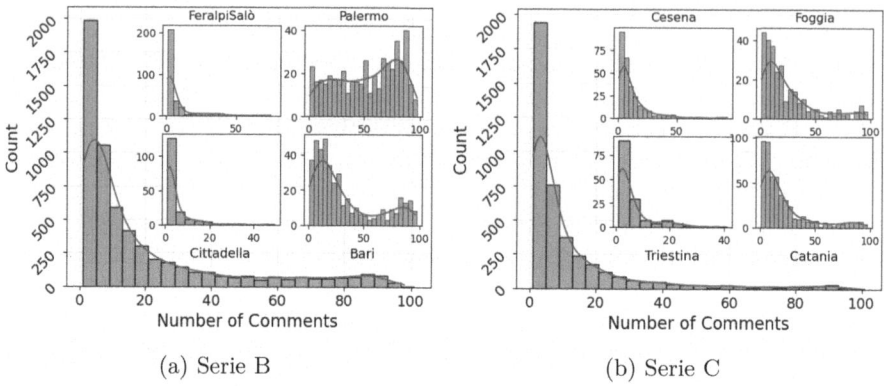

(a) Serie B   (b) Serie C

Fig. 1. Distribution of the total number of users' comments from all Serie B teams' accounts (a) and the chosen 20 teams from Serie C (b) for the selected temporal activity window. For both leagues, the plots within show 4 (out of 20) individual distributions.

2.1   Data Collection and Pre-processing

We characterize the emotional engagement of fandoms examining teams' activity on Instagram. This includes posts such as halftime and full-time scores, highlights, squad lists, etc. We build a dataset from user comments on Instagram posts from the 20 official Italian accounts of the teams participating at the 2023–24 Serie B[1], the second-highest division in the Italian football league system, and from the 20 most followed accounts on Instagram among the three Groups of the 2023–24 Serie C[2]. Regarding Serie C, out of the 20 chosen teams, 10 belong to Group C, which is the one where teams from the South of Italy participate, 5 are from Group B (Center Italy: Cesena, Entella, Olbia, Perugia, and Pescara) and other 5 are from Group A (North Italy: Alessandria, Padova, Pro Vercelli, Triestina, and SPAL). To ensure that every team played exactly

---

[1] https://en.wikipedia.org/wiki/2023-24_Serie_B.
[2] https://en.wikipedia.org/wiki/2023-24_Serie_C.

once against every other team, users' comments on posts were extracted using the official Instagram APIs between August 15, 2023, and December 27, 2023, for Serie B, and between August 31, 2023, and December 24, 2023, for Serie C, i.e. the day before the first match and the day after the last match of the first round of the two leagues.

We perform a task of emotion classification using *feel-it*[3], a python library tailored for emotion classification in Italian [2]. This tool allows us to predict the emotional content of a given text, categorizing it into one of four types of emotions: joy, anger, sadness, and fear. Considering that the detection of an emotional event may not be significant if a post contain one or a very few number of comments, we have developed a tailored labeling strategy. We label each Instagram post indicating either the absence/non significant presence (0) or maximal presence (1) of an emotion, for each emotion. The maximal presence of an emotion represent the collective sentiment generated by a post. Moreover, this aggregation guarantees the privacy of individual users. Formally, being E the set of emotions considered, and n_e the number of comments on a post p classified with the emotion $e \in E$, the label l of each post is calculated as $l = argmax_{e \in E} \, n_e$. Thus, if a post contained 20 comments, with 18 of them classified as anger, the post would be labeled as conveying anger (1 for anger, 0 for other emotions).

Figure 1 depicts the distributions of the number of comments on each post in the two leagues. The maximum number of comments is limited to 100 for computational feasibility. The distribution of comments displays a peaked long-tail pattern, indicating that most posts receive a small number of comments, while a few posts attract a high number of comments. Notably, the distribution is not uniform across all teams. For instance, as reported in the highlights of Fig. 1(a), the comment distribution for Cittadella and FeralpiSalò mirrors the overall pattern, i.e., a consistent presence of posts with only one comment. Conversely, other comment distributions can exhibit either bimodal patterns (Bari) or a more uniform distribution (Palermo). Similarly, as illustrated in Fig. 1(b), there is a degree of variability among teams in Serie C, too, albeit less pronounced. Moreover, there is a difference among the two leagues in terms of volumes during the analyzed time frame, with Serie B accumulating 115552 comments, and Serie C, 47613.

Taking into account the observed heterogeneity in the distributions of the number of comments across the different accounts, we opt for individual thresholds (rather than a global one) to select the retained posts: we consider only posts having at least a number of comments equal to the median of the distribution of the individual account. E.g., for FeralpiSalò we keep posts having at least 2 comments, and for Palermo, at least 28.

Figure 2 shows the relative frequency of each emotion for each fandom over Serie B's 2023–2024 first round. Similarly, Fig. 3 highlights the same for the selected teams from Serie C. These frequencies are used to retrieve the most frequent emotion and build the binary sequences of 0s and 1s. In Serie B, Parma's

---

[3] https://github.com/MilaNLProc/feel-it.

joy sequence will be full of joy emotional events (labeled as 1), whilst Bari's joy sequence will lack of events (labeled as 0). Similarly, Olbia and SPAL represent opposite extremes in Serie C, with Olbia's fans mostly conveying joy and SPAL's fans mostly conveying anger.

2.2 Measures

Let τ be the inter-event time between two consecutive events. First, we aim to quantify whether the distribution of the inter-event times $P(\tau)$ deviate from the random activity pattern characterized by the exponential distribution. In our cases, $P(\tau)$ is the distribution of the waiting time τ that elapses before a target emotion appears again in the emotional sequence. We use the mean value of the distribution, $\mu_\tau = \frac{1}{n_\tau} \sum_{i=1}^{n_\tau} \tau_i$, where n_τ is the length of the sequence, and the standard deviation, $\sigma_\tau = \sqrt{\frac{\sum_{i=1}^{n_\tau}(\tau_i - \mu_\tau)^2}{n_\tau}}$, to measure the coefficient of variation r as follows:

$$r = \frac{\sigma_\tau}{\mu_\tau} \tag{1}$$

The variation r is a measure of the dispersion of a distribution compared to its mean. Goh and Barabasi [14] use r to introduce the burstiness parameter B as follows:

$$B = \frac{\sigma_\tau - \mu_\tau}{\sigma_\tau + \mu_\tau} = \frac{r-1}{r+1} \tag{2}$$

Burstiness B ranges from -1 and 1, where 1 indicates the bursty time series (σ_τ much higher than μ_τ), -1 the periodic one, and 0 indicates the random activity pattern.

Being B affected by the finite number of events in the time series, Kim and Jo [18] propose a variation of B for finite event sequences of size n as follows:

$$B_n = \frac{\sqrt{n+1} \cdot r - \sqrt{n-1}}{(\sqrt{n+1} - 2)r + \sqrt{n-1}} \tag{3}$$

This modification allows to analyze time series with a relatively small number of events, which can be the case of low posting activity of a team account. In the experiments, we will use the quantity B_n described in Eq. 3 to measure burstiness.

Second, Goh and Barabasi [14] suggest that the burstiness parameter is not the unique mechanisms describing the dynamics of non-random activities. They focus also on the role of memory, described by the memory parameter M [14] as follows:

$$M = \frac{1}{n-1} \sum_{i=1}^{n-1} \frac{(\tau_i - \mu_1)(\tau_{i+1} - \mu_2)}{\sigma_1 \sigma_2}, \tag{4}$$

where μ_1 and μ_2, and σ_1 and σ_2 are sample mean and sample standard deviation of τ_i's values and τ_{i+1}'s values, with $(i = 1, ..., n_\tau - 1)$. The memory coefficient M is similar to the autocorrelation of a time series at lag $= 1$. Schleiss and Smith [25] suggest extending M to a generic lag k, replacing 1 in $n_\tau - 1$ with k, and letting calculate the correlation between two inter-event times τ and τ' separated by k events (instead of 1 event) to better capture long-range correlations. Memory M ranges from -1 and 1, where 1 indicates short/long inter-event times followed by short/long ones, -1 indicates short/long inter-event times followed by long/short ones, and 0 indicates no correlation. In the experiments, we will use the quantity M described in Eq. 4 to measure the memory parameter, being coherent with [14], even if different values of k could enable further explorations into how different lags affect the correlation between inter-event times.

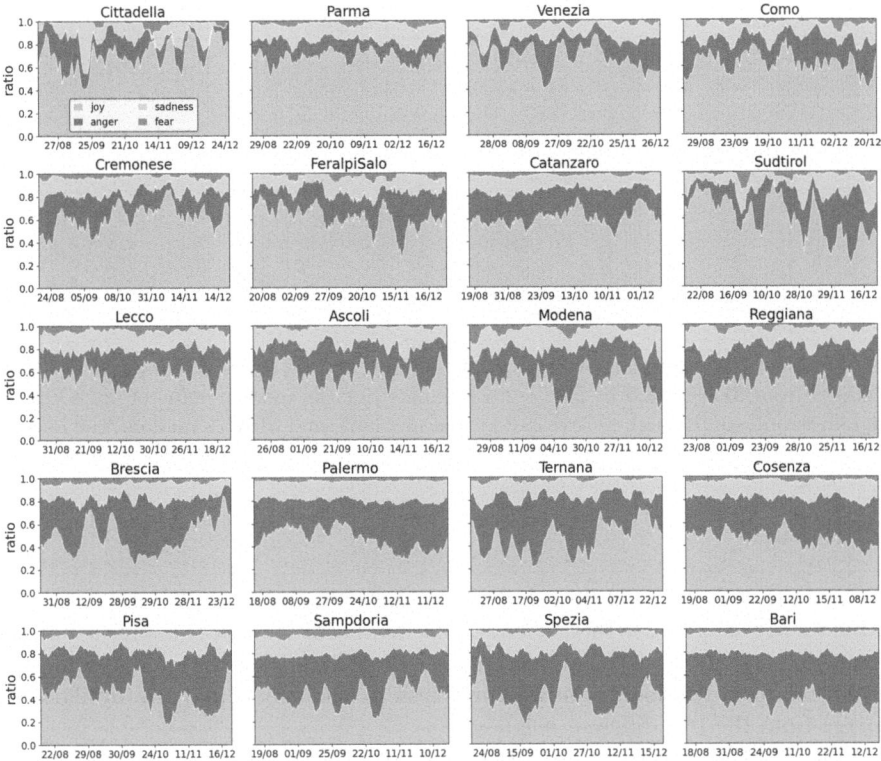

Fig. 2. Ratio of emotional levels of each fandom for the selected temporal activity window (from 15/08/2023 to 26/12/2023) ranked for the median value of joy: Cittadella's fandom manifests the highest levels of joy (in median), and Bari's fandom, the lowest one.

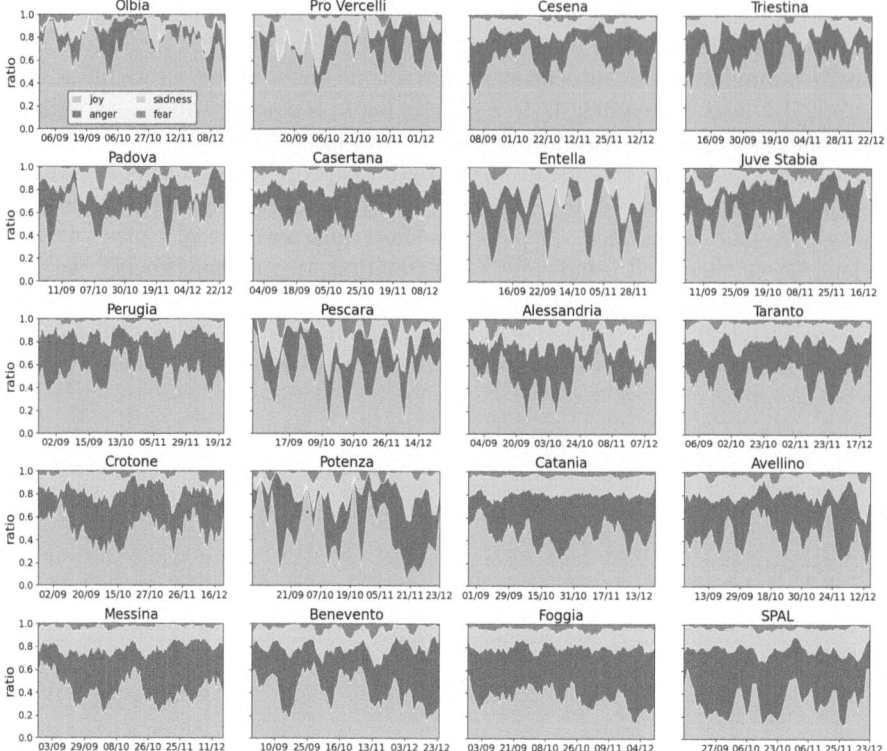

Fig. 3. Ratio of emotional levels of each fandom for the selected temporal activity window (from 31/08/2023 to 24/12/2023) ranked for the median value of joy: Olbia's fandom manifests the highest levels of joy (in median), and SPAL's fandom, the lowest one.

3   Results

We quantify the degree of both burstiness and memory in our datasets. Specifically, we investigate whether: *i)* emotional temporal sequences manifest bursty characters, *ii)* different emotions can be described by different mechanisms of non-random dynamics, among burstiness and memory, and *iii)* similarities emerge from the two different leagues.

First of all, we decided to exclude the emotion *Fear* from this analysis, since it is never the most frequent emotional response on posts (see Fig. 2 and Fig. 3), and therefore it does not generate any significant time series to analyze.

Figure 4 shows the M-B diagrams for Serie B's teams, for each of the three remaining emotions. The color of each point, representing each team, is associated with the ranking at the end of the first round: the darker the point, the highest the ranking. The three emotions manifest different dynamics. Joy rarely exhibits a bursty character. It mostly comes as an anti-bursty emotional pattern within most fandoms. These anti-bursty, and regular, patterns indicate a consistent support for the respective teams. Moreover, joy patterns of teams in top positions are the most regular ones, as in the case of Parma's fans, the team in first place in the ranking, highlighted in Fig. 4(b). Some exceptions are represented by teams like Palermo, whose fandom does not exhibit the anti-bursty behavior expressed by other teams in top positions. To understand this behavior, it is necessary to analyze the burstiness value in light of fans' initial expectations for the championship. A reasonable proxy for the team's perceived potential can be the market value of the rosters of the team at the season's outset, indicating investments in key players. The market value considers factors such as team performance, player values, financial successes, brand, reputation, and the wisdom of its community, thus can change during the season. A more comprehensive perspective is provided by the Pearson correlations between the joy burstiness and the two features, market value and ranking, as reported in Table 1. As depicted in Fig. 4 (b), the memory shows a less powerful descriptive power: except for Brescia, all of the other teams memory value lies in a small interval (between -0.2 and 0.2). A plausible explanation for Brescia's performance can be its performance within the considered period: a long series of consecutive wins followed by a long series of draws and another one of consecutive loses[4].

Anger can exhibit both bursty and anti-bursty patterns, with memory playing a more relevant descriptive role. Interestingly, Parma's fans display bursty anger dynamics compared to their strongly regular joy. Similarly, Palermo's fans exhibit bursty patterns of anger. Examining the sequences highlighted in Fig. 4(b), we can explain the bursty anger behaviour observed in Palermo's fans as a gradual accumulation over time leading to an eventual outburst (indicated by positive memory values). In contrast, the behavior of Parma's fans appears to be more closely linked to individual matches, with a sudden decay in emotional intensity following each match (indicated by negative memory values). Thus, contrary to joy, where Pearson correlation between burstiness and ranking is positive, the correlation between anger burstiness and ranking is negative (Table 1): fandoms of teams in top positions seem to manifest a burstier character of anger.

Finally, sadness dynamics are the burstiest ones. However, sadness is the only emotion where some outliers manifest, e.g., fandoms lying in the extreme ranges of the memory measure. These could be the only non-significative points in the dataset, see Ascoli's sequence of sadness (Fig. 4(b)), characterized by very few events that do not allow the two measures to properly capture a description of the signal. Outliers can be explained by the fact that sadness rarely manifests as a collective signal in this dataset, cf. Fig. 2. Also the absence of some points (e.g., Palermo) signifies that sadness does not manifest collectively, i.e., no time series to analyze.

Interestingly, all the patterns identified in Serie B teams are manifest in Serie C. Figure 5 reports the M-B diagrams of the three emotions including both leagues. Overall such findings suggest that joy predominantly exhibits an anti-

---

[4] https://en.wikipedia.org/wiki/2023-24_Brescia_Calcio_season.

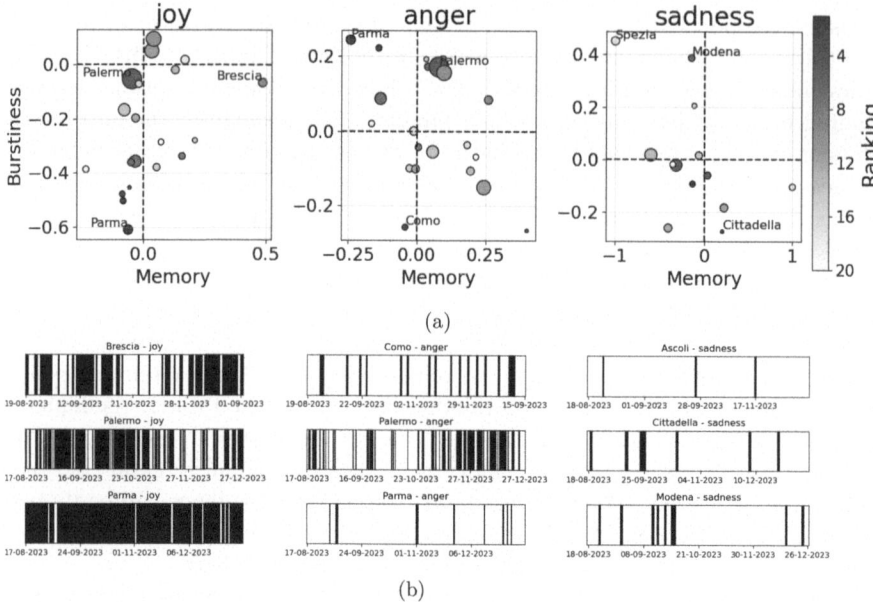

(a)

(b)

Fig. 4. M-B diagrams for different emotions, focus on teams from Serie B only. Points represent memory and burstiness values describing the collective emotional reactions to the posting sequences of teams' accounts. Colors represent ranking at 26/12/2023. Sizes represent the total number of comments under the posts of a team. (a); Focus on some selected emotional sequences (b). (Color figure online)

bursty property, while anger and sadness display more heterogeneous behaviors, including bursty patterns.

Figure 6 offers a new perspective on Serie B dynamics, presenting two ternary plots that analyze the relationship between burstiness parameters of joy and anger and what we can identify as fans' "Expectations versus Reality". Expectation is proxied by teams' market values at the beginning of the season, whilst Reality is represented by the ranking at the end of the first round of the championships. In other words, we aim to observe the eventual contrast between what fandoms hope will happen and what actually happens at the end of the first round. Figure 6(a) compares the two dimensions of expectations and reality to the burstiness in joy. In the bottom left corner of the triangle we observe high valuable teams meeting expectations, thus displaying regular joy over time. Parma, Venezia, Como, and Cremonese, indeed, occupy this area. Moving towards the center of the triangle, we find Spezia, a high valuable team that does not meet fans' expectations but fans still demonstrate consistent support. Palermo's and Sampdoria's fans, being their teams other two high valuable clubs not meeting their expectations, fail to exhibit regular joy. This explains better the strong heterogeneity observed among fandoms. Similarly, Fig. 6(b) compares the dimensions of "expectations versus reality" of the anger's bursti-

Table 1. Pearson correlations between memory/burstiness in joy/anger and team ranking/market values. Ranking for Serie C are not measured due to the selection from the three different Groups.

			Ranking (26/12/23)	Market Value (start season)
Serie B	Memory	Joy	0.17	−0.10
		Anger	0.18	−0.35
	Burstiness	Joy	0.44	0.00
		Anger	−0.20	0.45
Serie C	Memory	Joy	//	0.03
		Anger		−0.30
	Burstiness	Joy		0.28
		Anger		−0.10

ness. It is worth noting that anger is burstier when teams occupy top positions in the ranking, with some differences, e.g., Como and Cittadella, making it difficult to observe completely predictable patterns of emotional signals (cf. next Discussion and Conclusion).

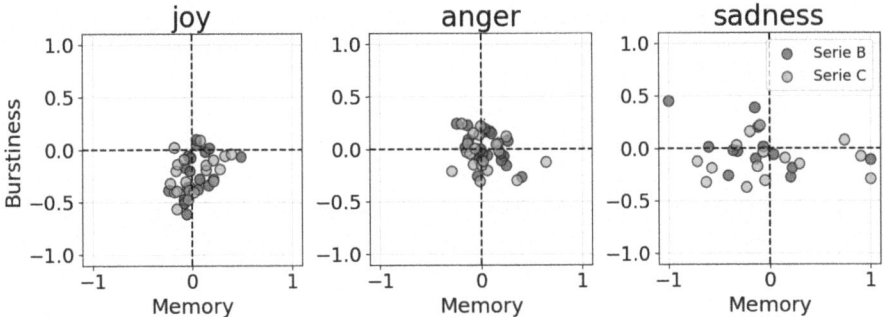

Fig. 5. M-B diagrams for different emotions including teams from both Serie B and Serie C. Points represent memory and burstiness values describing the collective emotional reactions to the posting sequences of teams' accounts

4   Discussion and Conclusion

The results describing the time-dependent sequences of collective emotions in Italian soccer fandoms lead us to conclude that their emotional patterns display complex dynamics. We have observed heterogeneous emotional patterns, from anti-bursty reactions of joy to heterogeneous expressions of anger and sadness. The emotional patterns are similar across the two different Italian leagues. This

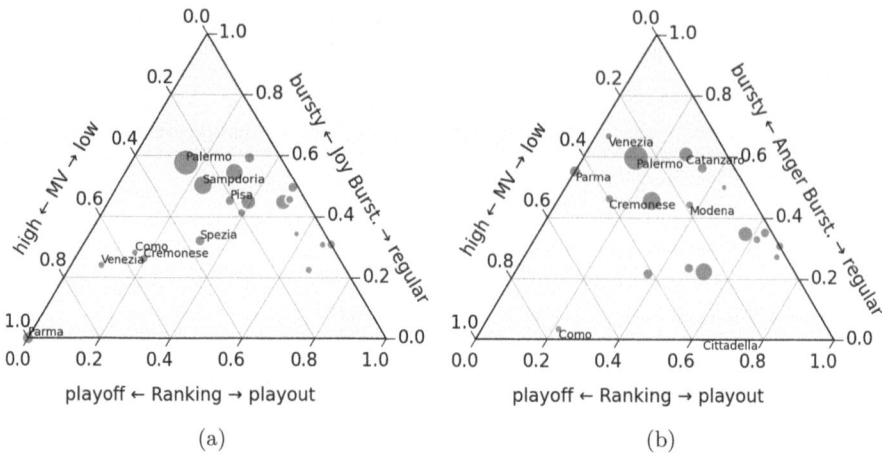

Fig. 6. Ternary plots highlighting ranking at 26/12/2023, market values at 15/08/2023, and burstiness of joy (a) and anger (b). The labeled teams are the top-8 richest in terms of market values (a) and the top-8 in the ranking at 26/12/2023 (b). The size of the points represents the total number of comments under the posts of a team.

indicates perhaps a pattern that is independent from the level of the league and its geographical origin that could be empirically investigated across different leagues in future research.

Analyzing collective emotions through the lens of complex systems may provide insights into the role of emotional burstinesses as signals, or triggers, of emerging collective/coordinated behaviors [21]. This preliminary work on soccer fans and users' affective response to social media posts is indeed rooted in the theory of collective emotions [13]. Understanding the significance of burstiness and memory in this domain is key for further extending the analysis to other domains. Regarding the burstiness parameter, anti-bursty patterns of joy in affective responses to own team's activities clearly reflect a constant support for the team. Moreover, correlating joy and anger burstiness with market values and ranking, a proxy for the "expectations versus reality", revealed interesting patterns. High market value teams like Parma, Cremonese, and Venezia in Serie B consistently showed regular joy and bursty anger when meeting expectations and holding top positions. Predictable patterns remain hard to disentangle. Initial season expectations from fandoms could provide valuable emotional signals for future investigations. However, it is crucial to interpret fandoms on a case-by-case basis, especially in this context. Burstiness exhibits a diverse range of behaviors associated with different scenarios, from fans with high expectations facing disappointment (Palermo, Sampdoria) to teams exceeding performance expectations (Catanzaro).

The interpretation of the memory parameter is not as straightforward as burstiness, yet it remains useful for describing the non-random dynamics of collective emotions among fans (e.g., see patterns of joy in Brescia). Using a memory parameter $k > 1$ would extend the scope of our analysis, enabling us to explore patterns over a longer time frame. This approach would permit to analyze long-range correlations of emotions beyond consecutive events, providing insight into their persistence over broader temporal scales.

Among the limitations, we first recognize the heterogeneous volume of posts and comments across teams' accounts. Soccer teams manage their accounts differently, resulting in varying levels of posting activity, users' engagement and thus users' emotional sequence sizes. On average, in our Serie B dataset, accounts' posting sequences are long 302 posts with a standard deviation of 129. Catanzaro exhibits the highest posting activity, with 702 posts within the analyzed time frame, while Venezia has the lowest, with only 145 posts. Comparable comment distributions are also evident in Serie C.

As shown in Fig. 1, the variation in the number of comments on posts is crucial for identifying emotionally significant sequences, assuming that posts with more comments would improve the quantification of the collective emotion. Finally, the quality of our results inevitably depends on the emotion classifier we used, which is based on the *feel-it* corpus for Italian [2]. In the original study [2], it is showed that a BERT-like emotion classifier trained on the *feel-it* corpus outperforms models trained on other corpora. The *feel-it* corpus, based on social media data, aligns well with the content of our dataset.

Our analysis reveals an interesting trend in the median joy ranking of Serie C teams, as depicted in Fig. 3. Teams from Group A and Group B generally exhibit higher median joy levels compared to those from Group C, with exceptions noted for Juve Stabia and Casertana, which performed well in the initial round considered. Further investigation is needed to determine if this geographical disparity persists, alongside exploring correlations with socio-economic and geographical indicators.

In conclusion, we believe that this preliminary study offers insights into emotional dynamics in social media and team support, suggesting potential research avenues in other leagues and domains, also involving user-oriented approaches [12], network interactions, or integration with soccer data [23].

Acknowledgment. This project was funded by SoBigData.it which receives funding from the European Union – NextGenerationEU – National Recovery and Resilience Plan (Piano Nazionale di Ripresa e Resilienza, PNRR) – Project: "SoBigData.it – Strengthening the Italian RI for Social Mining and Big Data Analytics" – Prot. IR0000013 – Avviso n. 3264 del 28/12/2021.

Author contributions. S.C. conceptualized the research, conducted the experiments, made the plots, wrote the code and the paper. G.M. conceptualized the research, supervised the experiments and wrote the paper. E.F. supervised the research and wrote the paper.

References

1. American Psychiatric Association. Diagnostic and Statistical Manual of Mental Disorders, 5th edn. Text Rev. (2002)

2. Bianchi, F., Nozza, D., Hovy, D.: FEEL-IT: emotion and sentiment classification for the Italian language. In: Proceedings of the Eleventh Workshop on Computational Approaches to Subjectivity, Sentiment and Social Media Analysis. Association for Computational Linguistics (2021)
3. Bak, P., et al.: Unified scaling law for earthquakes. Phys. Rev. Lett. **88**(17), 178501 (2002)
4. Balsamo, D., et al.: The pursuit of peer support for opioid use recovery on Reddit. In: Proceedings of the International AAAI Conference on Web and Social Media, vol. 17 (2023)
5. Barabasi, A.-L.: The origin of bursts and heavy tails in human dynamics. Nature **435**(7039), 207–211 (2005)
6. Cauteruccio, F., Kou, Y.: Investigating the emotional experiences in eSports spectatorship: the case of League of Legends. Inf. Process. Manage. **60**(6), 103516 (2023)
7. Cencetti, G., et al.: Temporal properties of higher-order interactions in social networks. Sci. Rep. **11**(1), 7028 (2021)
8. Cinelli, M., et al.: Dynamics of online hate and misinformation. Sci. Rep. **11**(1), 22083 (2021)
9. Darst, R.K., et al.: Detection of timescales in evolving complex systems. Sci. Rep. **6**(1), 39713 (2016)
10. Devlin, J., et al.: BERT: pre-training of deep bidirectional transformers for language understanding. arXiv preprint arXiv:1810.04805 (2018)
11. Ekman, P.: Facial expression and emotion. Am. Psychol. **48**(4), 384 (1993)
12. Garas, A., et al.: Emotional persistence in online chatting communities. Sci. Rep. **2**(1), 1–8 (2012)
13. Goldenberg, A., et al.: Collective emotions. Curr. Dir. Psychol. Sci. **29**(2), 154–160 (2020)
14. Goh, K.-I., Barabási, A.-L.: Burstiness and memory in complex systems. Europhys. Lett. **81**(4), 48002 (2008)
15. Heuer, H., et al.: Auditing the biases enacted by YouTube for political topics in Germany. Proc. Mensch und Comput. **2021**, 456–468 (2021)
16. Joseph, S.M., et al.: Cognitive network neighborhoods quantify feelings expressed in suicide notes and Reddit mental health communities. Physica A: Stat. Mech. Appl. **610**, 128336 (2023)
17. Karsai, M., Jo, H.-H., Kaski, K.: Bursty Human Dynamics. SC, Springer, Cham (2018). https://doi.org/10.1007/978-3-319-68540-3
18. Kim, E.-K., Jo, H.-H.: Measuring burstiness for finite event sequences. Phys. Rev. E **94**(3), 032311 (2016)
19. Lazarus, R.S., Folkman, S.: Stress, Appraisal, and Coping. Springer, Cham (1984). https://doi.org/10.1007/978-3-030-39903-0_215
20. Mainwaring, Ed., Clark, T.: 'We're shit and we know we are': identity, place and ontological security in lower league football in England. Soccer Soc. **13**(1), 107–123 (2012)
21. Nizzoli, L., et al.: Coordinated behavior on social media in 2019 UK general election. Proc. Int. AAAI Conf. Web Soc. Media **15**, 2021 (2019)
22. Pacheco, D.F., et al.: Characterization of Football Supporters from Twitter Conversations. WI (2016)
23. Pappalardo, L., et al.: PlayeRank: data-driven performance evaluation and player ranking in soccer via a machine learning approach. ACM Trans. Intell. Syst. Technol. (TIST) **10**(5), 1–27 (2019)

24. Scherer, K.R.: Affect bursts. Emotions, pp. 175–208. Psychology Press (2014)
25. Schleiss, M., Smith, J.A.: Two simple metrics for quantifying rainfall intermittency: the burstiness and memory of interamount times. J. Hydrometeorol. **17**(1), 421–436 (2016)
26. Schröder, M.: Experimental study of affect bursts. Speech Commun. **40**(1–2), 99–116 (2003)
27. Starnini, M., Baronchelli, A., Pastor-Satorras, R.: Modeling human dynamics of face-to-face interaction networks. Phys. Rev. Lett. **110**(16), 168701 (2013)
28. Stella, M., Ferrara, E., De Domenico, M.: Bots increase exposure to negative and inflammatory content in online social systems. Proc. Nat. Acad. Sci. **115**(49), 12435–12440 (2018)
29. Stella, M.: Cognitive network science for understanding online social cognitions: a brief review. Top. Cogn. Sci. **14**(1), 143–162 (2022)

Exploring Crisis-Driven Social Media Patterns: A Twitter Dataset of Usage During the Russo-Ukrainian War

Ioannis Lamprou[1]([✉]), Alexander Shevtsov[2], Despoina Antonakaki[1], Polyvios Pratikakis[2], and Sotiris Ioannidis[1]

[1] Technical University of Crete, Chania, Greece
{ilamprou1,dantonakaki}@tuc.gr, sotiris@ece.tuc.gr
[2] ICS-FORTH, Heraklion, Greece
{shevtsov,polyvios}@ics.forth.gr

Abstract. On 24 February 2022, Russia's invasion of Ukraine, now known as the Russo-Ukrainian War, sparked extensive discussions on Online Social Networks (OSN). We initiate a data collection using the Twitter API to capture this dynamic environment. Next, we perform an analysis of the topics discussed and a detection of potential malicious activities. Our dataset consists of 127.2 million tweets originating from 10.9 million users. Given the dataset's diverse linguistic composition and the absence of labeled data, we approach it as a zero-shot learning problem, employing various techniques that require no prior supervised training on the dataset.

Our research covers several areas, including sentiment analysis capturing the public's response to the distressing events of the war, topic analysis comparing narratives between social networks and traditional media, and examination of the correlation between message toxicity levels and Twitter suspensions. Furthermore, we explore the potential exploitation of social networks to acquire military-related information by belligerents, presenting a pipeline to classify such communications.

The findings of this study provide fresh insights into the role of social media during conflicts, with broad implications for policy, security, and information dissemination. Finally, due to the recent Twitter API changes, we share anonymized data for any further research purposes.

Keywords: Russo-Ukrainian War · sentiment analysis · Twitter · military intelligence · dataset

1 Introduction

Twitter is one of the most popular and widely used online social networks, serving as a primary platform for communication and information dissemination in the digital world. Over the years, Twitter has been extensively employed to analyze political crises and significant events [1,9,27,37,38].

L. M. Aiello et al. (Eds.): ASONAM 2024, LNCS 15211, pp. 70–85, 2025.
https://doi.org/10.1007/978-3-031-78541-2_5

In this study, we focus on the Twitter public discussion related to the 2022 Russo-Ukrainian conflict as an escalation of an ongoing conflict originating with Russia's annexation of Crimea. This conflict has significant implications for European security and marks a historic turning point. By selecting this topic, we have accumulated a substantial dataset related to a major international conflict involving nations with widespread access to social networks. Through social media platforms, individuals have been expressing their emotions, sharing their perspectives, and providing commentary on the war, the involved parties, and the unfolding events. In light of this, our data collection was initiated on February 23, 2022, coinciding with Russia's invasion of Ukraine, commonly referred to as the Russo-Ukrainian War. The primary objective of this effort is to leverage Twitter data to analyze the prevailing trends and discussions within this online discourse. We aim to monitor the user behavior, identify and assess potential instances of malicious activity, conduct sentiment analysis on the text, examine the presence of hate speech or propaganda within Online Social Networks (OSNs), and gain insights into the broader implications of these interactions. Throughout our data collection, we observe a rising number of Twitter user suspensions. This piqued our interest, leading us to investigate the reasons behind these suspensions. To achieve this, we analyze the levels of toxicity in the messages posted by suspended users.

Since the onset of the conflict, there have been suggestions from media and journalists that belligerents may utilize social media platforms like Twitter to acquire military-related information from residents, military personnel, and open-source analysts [22]. To explore this hypothesis, we develop a methodology that incorporates machine learning techniques to classify communications with a military connection and aggregate comprehensive details on a large scale.

Previous research has employed Natural Language Processing (NLP) techniques, sentiment analysis, topic modeling, and toxicity analysis to examine sentiment in datasets related to crises like the Syrian refugee crisis, which include Turkish and English tweets. For example, [27] reveals a balanced distribution of positive sentiment towards refugees. Additionally, other studies have investigated the dissemination of fake news and terrorism [6]. However, there is a lack of a comprehensive analysis combining all of these methods. Hence, we undertake a thorough analysis of our extensive multilingual dataset, employing state-of-the-art methods and AI models. We also conduct a specialized study to extract military-related information from Twitter. Finally, we employ topic modeling to identify current themes and contrast them with narratives presented in traditional media, thus evaluating the divergence between social media and mainstream reporting.

2   Related Work

Several studies have examined social network analysis and machine learning for sentiment analysis, as indicated by [17]. This survey provides a comprehensive view of the subject by looking at and briefly explaining the algorithms that have been suggested for sentiment analysis on Twitter. The investigations are clustered in accordance with the technique they follow. Furthermore, we examine the areas associated with sentiment analysis on Twitter, such as Twitter opinion retrieval, tracking sentiments over time, irony recognition, emotion detection, and tweet sentiment quantification, issues that have recently gained growing attention.

In the context of multilingual corpora, state-of-the-art techniques have been utilized [13]. For example, a pivotal study closely aligned with the work we have adopted is the 'Sentiment Analysis Using the XLM-R Transformer and Zero Shot Transfer Learning in the resource-poor Indian language' [24]. This paper demonstrates the effectiveness and cross-lingual capabilities of XLM-R for sentiment analysis in resource-poor languages.

Fundamental research studies exploring Twitter toxicity during significant events [29], South Asian elections [16], and the Brexit of the UK [16] have used deep learning techniques (BERT), which are similar to the methodologies used in our study.

Studies exploring the usage of large-scale data and social networks for military intelligence are still emerging. A couple of notable works, such as [23,44], have presented techniques for extracting military-related entities from text data, shedding light on the practical applications in this domain.

Significant research has been conducted in the realm of topic modeling on social networks. Various methods have been explored to filter noise from tweets and enhance accuracy. For example, early-stage considerations included the Dirichlet technique [43,46]. Beyond these, [42] has delved into alternative techniques beyond traditional Latent Dirichlet Allocation. Furthermore, [12] have investigated clustering techniques coupled with neural embedding feature representations, a methodology also aligned with our approach.

It is important to acknowledge the existence of other research papers and datasets that have explored Twitter discourse during the Russo-Ukrainian war [5,8,20,39]. Existing data sets are very limited in terms of monitoring periods combined with the limited information available according to the content of shared posts. Data sharing via Twitter_id is limited due to recent Twitter API access limitations. Our work intends to complement existing research by providing a comprehensive, reproducible, and in-depth analysis combined with a high volume of anonymized users' text posts.

3   Data

For this research, we acquire a Twitter dataset about public user discussions concerning the 2022 Russo-Ukrainian War. To accomplish this objective, our initial approach involves gathering popular hashtags from three distinct language groups: English, Russian, and Ukrainian (see Table 1). The selection of specific hashtags was guided by their relevance to the topic, as evidenced by their usage in relevant posts. It is pertinent to acknowledge that these hashtags represent the most prevalent ones, as indicated by Twitter statistics, potentially reflecting biases in public opinion.

Table 1. Set of hashtags used in our data collection query written in Russian, Ukrainian, and English provided with translation for non-English hashtags.

#Ukraine, #Ukraina, #ukraina, #Украина(Ukraine), #Украине(Ukraine), #PrayForUkraine, #UkraineRussie, #StandWithUkraine, #StandWithUkraineNOW, #RussiaUkraineConflict, #RussiaUkraineCrisis, #RussiaInvadedUkraine, #WWIII, #worldwar3, #Война(War), #BlockPutinWallets, #UkraineRussiaWar, #Putin, #Russia, #Россия(Russia), #StopPutin, #StopRussianAggression, #StopRussia, #Ukraine_Russia, #Russian_Ukrainian, #SWIFT, #NATO, #FuckPutin, #solidarityWithUkraine, #PutinWarCriminal, #PutinHitler, #BoycottRussia, #with_russia, #FUCK_NATO, #ЯпротивВойны(I'm against war), #StopNazism #myfriendPutin #UnitedAgainstUkraine #StopWar #ВпередРоссия(Go Russia), #ЯМыРоссия(I/we Russia), #ВеликаяРоссия(Great Russia), #Путинмойпрезидент(Putin is my president), #россиявперед(Go Russia), #россиявперёд(Go Russia), #ПутинНашПрезидент(Putin is our president), #ЗаПутина(For Putin), #ПутинВведиВойска(Putin send the troops), #СЛАВАРОССИИ(Glory to Russia), #СЛАВАВДВ(Glory to Russian Airborne Forces)

Using the identified hashtags, we retrieve our dataset using the Twitter Research API spanning from February 23, 2022, to June 23, 2023. The collected dataset consists of 127,275,386 tweets authored by 10,990,275 users. However, it is imperative to acknowledge that, due to the extended duration and technical intricacies involved, the dataset may exhibit certain inconsistencies and instances of missing dates.

As illustrated in Figs. 1 and 2, specific time periods, notably from December 22, 2022, to January 17, 2023, and from March 15, 2023, to April 25, 2023, experience lower traffic. Unfortunately, our data collection script encounters technical challenges during these intervals, resulting in an inability to retrieve data. In response, we used the Twitter full archive search API to reconstruct publicly available data for these periods. While we encountered limitations such as user post suspensions or deletions, we partially reconstruct the missing data and gather a substantial volume of information for these time frames.

In parallel with the data collection, we use the Twitter compliance API to identify that a total of 289,837 user accounts had been suspended by Twitter itself, while an additional 32,973 user profiles were deactivated. The gathered information is stored in MongoDB as JSON objects, as this database system offers efficient storage, filtering, and querying capabilities for such data types.

We provide access to the dataset through two distinct sharing platforms. For the research community with access to Twitter API, we will provide a daily list of tweet IDs and the results of our analysis through a GitHub repository[1]. Researchers can use this list to retrieve the complete tweet objects and any additional information through the Twitter API. Furthermore, in light of recent

---

[1] https://github.com/alexdrk14/RussoUkrainianWar_Dataset.

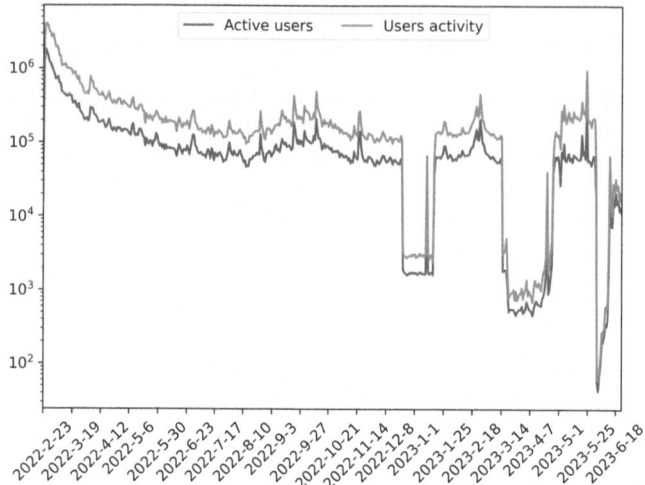

Fig. 1. Daily volume of registered users activity.

announcements regarding the Twitter API subscription plans, we anonymize the entire text corpus and make it available via Zenodo's file sharing service[2]. The provided data are anonymized via blake2b cryptographic hash function [33] over the tweet and user IDs. Additionally, user mentions are replaced with the anonymized user_id in order to hide the user identity while keeping the link between tweets where the same user is mentioned or which post the user has shared.

4   Methodology

The size and linguistic diversity of an unlabeled dataset presents significant challenges, particularly in achieving accurate fine-tuning of AI models. To address these limitations, we adopt techniques that do not necessitate specialized training to attain satisfactory accuracy. All inferences and training during this study are carried out on a machine equipped with a 64-thread CPU and a Nvidia RTX 3080 TI, complemented by 128GB of DDR4 RAM.

4.1   Preprocessing

Before analyzing the collected dataset, it is essential to perform text prepossessing. Tweets are typically composed of informal text, misspelled words, emoticons, hashtags, and various elements that introduce noise and hinder the application of analytical algorithms. Text prepossessing varies depending on the analysis type and includes the removal of URLs, unnecessary spacing characters (e.g. spaces, tabs, and newlines), hashtags and usernames (in case of sentiment and toxicity analysis) and emoticons (except for the case of sentiment analysis).

---

[2] https://zenodo.org/records/8431047.

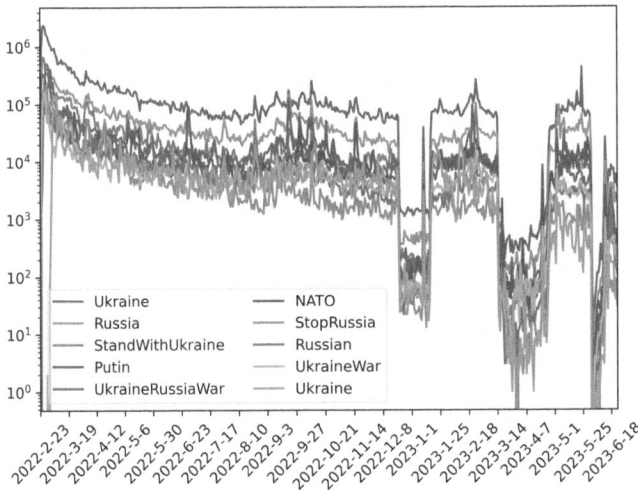

Fig. 2. Daily volume of 10 most popular hashtags.

4.2 Sentiment Analysis

Our analysis of the collected dataset begins with sentiment analysis, providing insight into the general emotions expressed by the users towards the selected entities. We conduct a comparative study of existing approaches to ensure the highest accuracy in sentiment analysis. Currently, two primary methodologies are commonly applied: model-based (also known as lexicon-based) rules [21,41] and AI-based models [7,40].

Rule-based models offer attractive advantages in terms of ease of comprehension and implementation. However, they require an extensive lexicon and a stringent set of linguistic rules, which can take months to develop and validate [47].

In contrast, AI-based approaches can provide more accurate results without the need of manual rule creation by a team of language experts. These approaches can be categorized into two types of implementation. First, simpler machine learning models like the Naive Bayes model making use of attributes straightforward for humans to understand. Although these models offer a rapid training process and satisfactory sentiment analysis results, their simplicity limits their ability to learn and combine multiple languages into a single model. These limitations, coupled with the requirement for labeled datasets, render these types of sentiment analysis impractical for multilingual datasets.

More advanced AI approaches for sentiment analysis are based on sophisticated neural network models [15,34]. These models, owing to the complexity of their weights and structure, excel in capturing linguistic nuances across multiple languages. In our study, we opted for the multilingual transformer model XLM-RoBERTa [10] due to its proficiency in handling large volumes of diverse languages and its exceptional performance in sentiment analysis tasks [2,24].

More specifically, the selected XLM-RoBERTa model is an extension of the original RoBERTa implementation which takes advantage of the original implementation and applies it to the Cross-lingual language model (XLM) objective. In such a scenario, the data are processed without additional translation, and only the original language model mask is applied. The selected implementation has been pre-trained using a vast amount of data. This includes 100 languages extracted from approximately 2.5TB of filtered CommonCrawl data, which serves as pre-training material for the model.

To address the challenges posed by our diverse dataset, we fine-tune our implementation of XLM-RoBERTa as Facebook team suggests [25], on the MultiNLI dataset [45], a crowdsourced collection of 433k English sentence pairs annotated with textual entailment information, and XNLI [11], a subset of a few thousand examples from MNLI that has been translated into 14 different languages. Multilingual NLI models are capable of classifying NLI texts without receiving NLI training data in the specific language (cross-lingual transfer).

Moreover, the model is further fine-tuned on the task of NLI using a combination of the MNLI train set and the XNLI validation and test sets. In the final stage of training, the model is exposed to one additional epoch solely on XNLI data, where the translations for the premise and hypothesis are shuffled. This means that for each example, the premise and hypothesis come from the same original English example but in different languages. We use the Transformers Trainer and Dataset library, from HuggingFace, in order to load the model, the datasets, and the training. The parameters adjusted on the Trainer include the following: number of train epochs = 3, batch size = 128, and warm-up steps = 10% of the total size of the train set. After training on the Russian and Ukrainian subsets of the dataset for 10 epochs, we observed that the accuracy converged and reached a plateau after just 3 epochs. This convergence is clearly illustrated in Table 2, which depicts the accuracy values across the training epochs. Based on this observation, we determine that 3 epochs would be sufficient for training on the full dataset, as additional epochs beyond this point were unlikely to yield significant improvements in model performance.

We use positive, neutral, and negative labels as sentiment labels towards the entities of *Ukraine, Russia, Zelenskyy,* and *Putin.* Preferring the use of labels directly linked to key topics of the event allows us to increase the accuracy of the classification and the interpretability of the results.

4.3   Topic Modelling

To unveil latent patterns and relationships among words in the shared tweets, we employ a topic modeling method. This allows us to uncover the predominant ideas or concepts within the text data without the need of predefined categories or manual labeling. By automatically detecting topics, topic modeling provides a means to organize, summarize, and explore vast amounts of textual data.

Conventional topic modeling techniques, like latent Dirichlet allocation and latent semantic analysis, require the knowledge of the exact languages used in

Table 2. Performance of XLM-RoBERTa during the training on XNLI Russian and Ukrainian subset.

Metric	Epoch									
	1	2	3	4	5	6	7	8	9	10
RUS Loss	0.6832	0.5448	0.4795	0.4243	0.3745	0.3313	0.2929	0.2608	0.2369	0.2183
RUS Acc (%)	73.78	77.63	76.14	76.35	76.27	76.31	76.18	75.66	76.06	75.90
UKR Loss	1.0129	0.9628	0.9407	0.9205	0.9001	0.8791	0.8593	0.8404	0.8242	0.8118
UKR Acc (%)	63.57	64.70	66.43	65.46	65.94	65.02	66.27	65.14	65.58	65.14

the corpus for every tweet. However, our dataset comprises tweets in more than 70 languages, making it challenging to determine the language of each tweet.

To address this challenge, we employed the BERTopic [18] pipeline, which can extract topics using embeddings derived from multilingual models (SBERT [30]). The pipeline of BERTopic consists of the following steps:

- *Embeddings*: We initiate the process by converting our documents into vector representations using language models.
- *Dimension Reduction*: We reduce the dimensionality of the vector representations to facilitate the clustering algorithms in finding clusters effectively (utilizing UMAP, PCA).
- *Clustering*: We apply a clustering algorithm to cluster the reduced vectors and identify semantically similar ones (employing HDBSCAN [26], k-Means, BIRCH).
- *Bag of Words*: We tokenize each topic into a bag-of-words representation, enabling us to process the data without affecting the input embeddings (employing CountVectorizer).
- *Topic Representation*: We calculate words related to each topic using a class-based TF-IDF procedure known as c-TF-IDF.

The hyperparameters utilized for the topic modeling include the number of topics (set to 'auto'), the N-Gram range (1,2), and the minimum topic size (300).

4.4 Toxicity Analysis

Toxic comment classification is an emerging research field, with several studies addressing diverse tasks to detect unwanted messages on communication platforms. Although sentiment analysis is an accurate approach to observe crowd behavior, it is incapable of discovering other types of information in the text, such as toxicity, which can usually reveal hidden information. The number of suspended accounts since the beginning of the war is increasing, so we need to detect whether toxicity was the reason for the suspension. We use toxic comment classification Detoxify [19], a state-of-the-art model, pre-trained in social media datasets to classify multilingual corpus. After research in the field of toxicity classification methods, we decide that for our multilingual corpus, this model would provide the best accuracy and performance without custom training.

4.5   Military Intelligence

In addition to the analysis described in previous sections, we test novel methodologies of military intelligence combined with social media information and location identification. This methodology can identify and provide military-based content based on text shared on social media such as Twitter. Based on some recent investigations, during the Russo-Ukrainian conflict, we find that a large amount of military content is contained in social networks, with a high percentage of fake information [3,28]. Gathering military information through social media poses a challenge, as it is still an unexplored field in NLP and currently there is not much related work to refer to. Additionally, there is still no large-scale open military domain corpus, making the identification of military-named entities with data even more challenging. To extract open-source military information from tweets, we initially need to train a Named-Entity Recognition (NER) model to recognize military-type entities. For this purpose, we use the spaCy NER model as a base and fine-tune and train it (train split = 70%, validation split = 30%) with the only open source military dataset [14] from the Defense Science and Technology Laboratory of the UK. The Entity Schema of the NER model is in Table 3. As shown in Table 3, we achieve our best performance in epoch 42. The training of the NER model is based on the following parameters: warm-up steps = 250, total epochs = 71, and initial rate = 5e-5. We create a pipeline that can filter our dataset and extract military entities, with the following steps:

1. The XLM-RoBERTa model for zero-shot classification using the label "military" with a threshold > 0.7 (range 0–1),
2. Our implemented NER model to extract entities,
3. Filtering of location entities per tweet for Ukrainian locations.

Using the extracted tweets and entities, we can perform data analysis and statistics for military events and information daily for any Ukrainian location.

5   Experimental Results

In this chapter, we provide the results of our developed analysis. The results are separated, similarly as in the methodology section, by each category of analysis: sentiment, topic, toxicity, and military intelligence. It is important to note that some literature suggests the presence of automated accounts, or "bots," on the platform. These bots can potentially influence the overall landscape of social media discourse. A recent study by [36] introduces a novel approach for bot detection on Twitter and detected bot accounts during the Russo-Ukrainian War

Table 3. Performance of Spacy's NER training on [14] and Entity Schema

Epoch	Step	Accuracy	Entity Schema
14	200	0.66	CommsIdentifier, DocumentReference,
28	400	0.73	Frequency, Location, Money,
42	600	0.74	MilitaryPlatform, Nationality, Organization, Person, Quantity,
57	800	0.73	Temporal, Url, Vehicle, Weapon
71	1000	0.73	

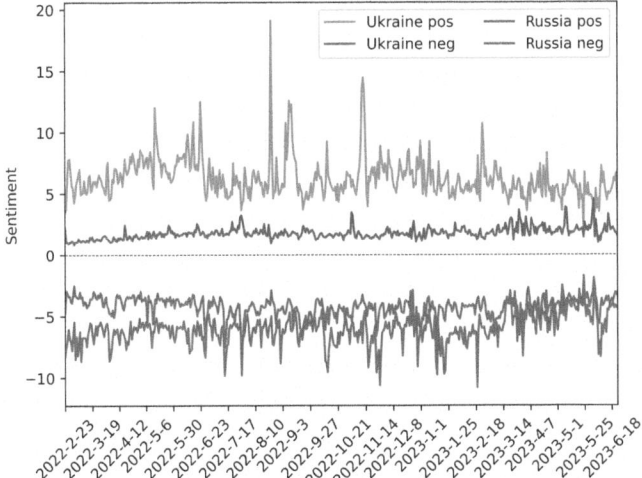

Fig. 3. Daily positive and negative sentiments towards each country.

5.1 Sentiment Analysis

We initiate our analysis by conducting sentiment analysis on the collected dataset, focusing on two primary sets of entities: country presidents (Zelenskyy vs. Putin) and countries (Ukraine vs. Russia).

Regarding the sentiment analysis of the countries, as illustrated in Fig. 3, the general sentiment trend tends to exhibit a higher positive sentiment toward Ukraine (with higher values indicating greater support from Twitter users). Although we notice some spike on the negative axes the overall positive sentiment towards Ukraine is higher.

Furthermore, we examine the sentiment vectors of country presidents, as presented in Fig. 4. Our analysis indicates that President Zelenskyy receives a significantly higher positive sentiment. This observation can be attributed to the substantial support expressed by Twitter users toward Ukraine and President Zelenskyy during the Russo-Ukrainian War.

In both cases, there are discernible spikes in sentiment, which will be elucidated in the following section.

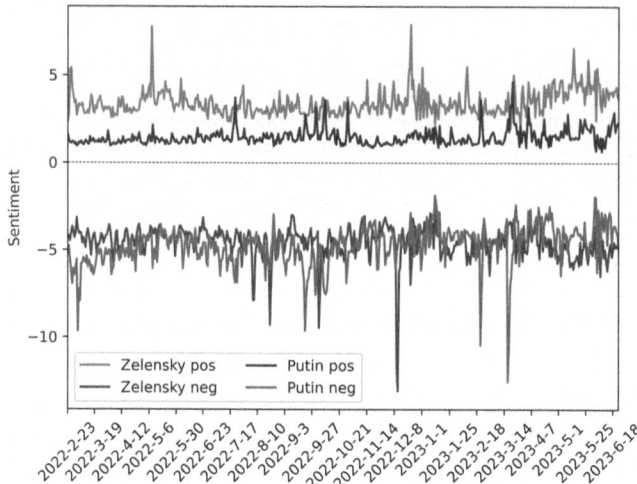

Fig. 4. Daily positive and negative sentiments towards each president.

Table 4. Most popular topics for dates with high user activity.

Date	Topics
8 May	bono, kyiv, ukrainian, rt, edge, conference zoom, resists azovstal, commanders regiment, plant press, killed city, btc humanitarian, gc13
14 May	ukraine, putin, eurovision, russie, stereotypes, rt, rt mavkaslavka, mavkaslavka, kerziouk today, deepl, city, battle, kharkiv axis, threat city, kharkiv
14 July	ukraine, fighting, forget, european, ukrainians fighting, recognize russia, recognize, world recognize, call world, russia terrorist, fuck, vinnytsya, including dead, area casualties, dialing

5.2   Topic Analysis

As previously mentioned, we conduct a topic analysis on days with remarkable sentiment peaks, and we present examples of these days, including the top 20 topics by size, in correlation with significant events reported by the mainstream media.

An example is a surge in positive sentiment towards Ukraine on 14 May 2022. On this day, the mainstream media reported that Ukraine had won Eurovision 2022, contributing to the overall increase in positive sentiment. Furthermore, the Ukrainian military continued its counteroffensive in the northeastern region of Kharkiv [31]. Our topic analysis ranked these significant events as the top discussion topics (1st and 3rd) as shown in Table 4.

Another noteworthy instance from our dataset is a significant increase in negative sentiment on July 14, 2022, as evident in Fig. 3. On this date, Ukrainian officials reported that at least 23 people died from a Russian strike in Vinnytsia, central Ukraine, according to reports in the mainstream media [4]. The results of our topic modeling, which ranked this tragic incident as the third topic (Table 4), align with the observed sentiment trends and the coverage in the mainstream media. Moreover, this topic was associated with user discussions calling for the recognition of Russia as a terrorist state.

Furthermore, we examine the user discussions on 8 May, when President Zelenskyy experienced a positive sentiment peak. According to reports from the main media, President Zelenskyy had invited Bono and Edge musicians to a concert in Kyiv, coinciding with his address for the Day of Remembrance and Reconciliation [32].

As indicated in Table 4, the top-ranked topic on May 8th was indeed the invitation of Bono and the Edge, aligning with mainstream media coverage.

Upon closely scrutinizing our extracted topics in correlation with mainstream media reports, we identify each day 20 topics, including those that are reported by mainstream media. These findings suggest that Twitter users tend to follow the narrative presented by the mainstream media. The remaining topics, not covered by mainstream media, often include references to Twitter accounts mentioning the War (e.g., @mavkaslavka), indicating that a significant number of users incorporate events reported by other Twitter users into their conversations.

In addition, we identify topics related to cryptocurrencies and NFTs posted by spam accounts attempting to exploit popular hashtags, which is also supported by related work [35].

5.3 Toxicity Analysis

Using Detoxify [19], a toxic classification model, we analyze 1,883,507 tweets originating from suspended accounts. We examine this part of the dataset to identify whether the toxicity of messages plays an important role in the suspension decision. Unfortunately, our analysis shows that the percentage of toxic comments among suspended users is low (2.1%), so we conclude that toxicity is not necessarily the main factor in Twitter suspension for our dataset.

5.4 Military Intelligence

Our analysis, reveals that the collected dataset contains a significant volume of military intelligence content, with 879,232 tweets with military intelligence over the whole period. This distribution of identified content closely mirrors the volume of user activity between registered users. Through manual inspection, we confirm that the identified content contains military-related information directly correlated with the conflict. Table 5 provides a selection of random examples of these identified tweets containing military content. Upon further investigation, we discover tweets reporting on troop movements and sightings, often accompanied by photos and videos. Leveraging the extracted entities, such as locations and weapons, an automated notification system could be established based on volume and entity filtering (e.g., Location = Kyiv) to monitor military events and movements.

Table 5. Examples of tweets with military information

sending messages to #UAF soldiers on the front in the #Severodonetsk-#Lisichansk boiler on the radio: Short translation: #Zelensky betrayed you like #Azov There will be no help. Further resistance will lead to death. The only chance to live is to run or surrender-Save your lives #Russia
In the last couple of weeks, #Russia concentrated a large number of armored units (tanks, TOS-1A, etc.), VDV force remnants (from Kyiv op mostly), and mercenaries to cut off #Bakhmut - #Severodonetsk highway. They were unsuccessful, but it made UA ops in the area more difficult.
A Russian mortar position was located and destroyed by the Ukrainian 20th Separate Special Regiment of the Ukrainian SOF near Sievierodonetsk, Luhansk Oblast.#Russia #Ukraine

6 Conclusion

In the current study, we use the Twitter API to obtain a dataset of 127.2M tweets originating from 10.9M users over a period of 16 months and perform extensive analysis on a multilingual dataset, using state-of-the-art methods and machine learning models. The results show that a conflict such as the Russo-Ukrainian war creates a surge of activity on social networks with a generally negative attitude. Although the negative sentiment is high on both sides, the positive sentiment is higher towards Ukraine and Zelenskyy, leading to the deduction that the negativity disagrees with the war rather than Ukraine and Zelenskyy themselves. Furthermore, it is evident that toxicity is not the sole cause of suspensions on Twitter, and further research needs to be done to discover the other factors involved. The topics extracted are in line with the narrative of mainstream media. Additionally, several spam accounts are active and post in large volumes, often referring to cryptocurrencies, NFTs, and other similar products that have no connection to the war.

Furthermore, we show that a combination of state-of-the-art AI models allows us to identify military intelligence content over social media, making possible further implementation of intelligent technology for monitoring battles, making command decisions, and interfacing with computers. We include in our plans military-named entity recognition as a key part of military data extraction, to provide the basis for the intelligent handling of military information from social media. Furthermore, we make the collected dataset available to the research community in two formats: tweet IDs for those with API access and anonymous tweets via the Zenodo data sharing platform, ensuring the protection of user privacy.

Acknowledgments. This work was supported by the projects GREEN.DAT.AI and REWIRE, funded by the European Commission under Grant Agreements No.101070416 and No. 621701-EPP-1-2020-1LT-EPPKA2-SSA-B, respectively. Additionally, it received funding from the Smart Networks and Services Joint Undertaking (SNS JU) under the European Union's Horizon Europe research and innovation programme under Grant Agreement No 101139067. This publication reflects the views only of the authors, and the Commission cannot be held responsible for any use which might be made of the information contained therein.

References

1. Antonakaki, D., Fragopoulou, P., Ioannidis, S.: A survey of twitter research: data model, graph structure, sentiment analysis and attacks. Expert Syst. Appl. **164**, 114006 (2021)
2. Barbieri, F., Anke, L.E., Camacho-Collados, J.: XLM-T: multilingual language models in twitter for sentiment analysis and beyond. arXiv preprint arXiv:2104.12250 (2021)
3. BBC: Twitter blue accounts fuel Ukraine war misinformation (2023). https://www.bbc.com/news/world-europe-66113460. Accessed 9 Sept 2023
4. BBC News: Ukraine war: 23 killed in Russian rocket attack on Vinnytsia (2022). https://www.bbc.com/news/world-europe-62163071. Accessed 15 July 2022
5. Caprolu, M., et al., S.: Characterizing the 2022-Russo-Ukrainian conflict through the lenses of aspect-based sentiment analysis: dataset, methodology, and key findings. In: 2023 32nd ICCCN, pp. 1–10. IEEE (2023)
6. Carchiolo, V., Longheu, A., Malgeri, M., Mangioni, G., Previti, M.: Terrorism and war: Twitter cascade analysis. In: Del Ser, J., Osaba, E., Bilbao, M.N., Sanchez-Medina, J.J., Vecchio, M., Yang, X.-S. (eds.) IDC 2018. SCI, vol. 798, pp. 309–318. Springer, Cham (2018). https://doi.org/10.1007/978-3-319-99626-4_27
7. Chakriswaran, P., et al., V.: Emotion AI-driven sentiment analysis: a survey, future research directions, and open issues. Appl. Sci. **9**(24), 5462 (2019)
8. Chen, E., Ferrara, E.: Tweets in time of conflict: a public dataset tracking the twitter discourse on the war between Ukraine and Russia. In: Proceedings of the International AAAI Conference on Web and Social Media, vol. 17, pp. 1006–1013 (2023)
9. Chibuwe, A.: Social media and elections in zimbabwe: Twitter war between Pro-ZANU-PF and Pro-MDC-A netizens. Communicatio: S. Afr. J. Commun. Theory Res. **46**(4), 7–30 (2020)
10. Conneau, A., et al., K.: Unsupervised cross-lingual representation learning at scale. arXiv preprint arXiv:1911.02116 (2019)
11. Conneau, A., et al., R.: XNLI: Evaluating cross-lingual sentence representations. In: Proceedings of the 2018 Conference on Empirical Methods in Natural Language Processing. Association for Computational Linguistics (2018)
12. Curiskis, S.A., Drake, B., Osborn, T.R., Kennedy, P.J.: An evaluation of document clustering and topic modelling in two online social networks: Twitter and reddit. Inf. Process. Manag. **57**(2), 102034 (2020)
13. Dashtipour, K., et al., P.: Multilingual sentiment analysis: state of the art and independent comparison of techniques. Cognit. Comput. **8**(4), 757–771 (2016)
14. Defence science and technology laboratory UK: relationship and entity extraction evaluation dataset (2017). https://github.com/dstl/re3d. Accessed 5 May 2022

15. Dos Santos, C., Gatti, M.: Deep convolutional neural networks for sentiment analysis of short texts. In: Proceedings of COLING 2014, the 25th International Conference on Computational Linguistics: Technical Papers, pp. 69–78 (2014)
16. Fan, H., et al., D.: Social media toxicity classification using deep learning: real-world application UK Brexit. Electronics **10**(11) (2021). https://doi.org/10.3390/electronics10111332
17. Giachanou, A., Crestani, F.: Like it or not: a survey of twitter sentiment analysis methods. ACM Comput. Surv. **49**(2) (2016). https://doi.org/10.1145/2938640
18. Grootendorst, M.: BERTopic: neural topic modeling with a class-based TF-IDF procedure. arXiv preprint arXiv:2203.05794 (2022)
19. Laura H.: Unitary team: detoxify. Github (2020). https://github.com/unitaryai/detoxify. Accessed15 May 2022
20. Haq, E.U., et al., T.: Twitter dataset for 2022 Russo-Ukrainian crisis. arXiv preprint arXiv:2203.02955 (2022)
21. Hardeniya, T., Borikar, D.A.: Dictionary based approach to sentiment analysis-a review. Int. J. Adv. Eng. Manag. Sci. **2**(5), 239438 (2016)
22. Infosec Resources: social media use in the military sector (2013). https://resources.infosecinstitute.com/topic/social-media-use-in-the-military-sector/. Accessed 15 May 2022
23. Kok, A.: Named entity extraction in a military context. In: 2019 International Conference on Military Communications and Information Systems (ICMCIS), pp. 1–6 (2019). https://doi.org/10.1109/ICMCIS.2019.8842722
24. Kumar, A., Albuquerque, V.H.C.: Sentiment analysis using XLM-R transformer and zero-shot transfer learning on resource-poor Indian language. Trans. Asian Low-Resource Lang. Inf. Process. **20**(5), 1–13 (2021)
25. Lample, G., Conneau, A.: Cross-lingual language model pretraining. arXiv preprint arXiv:1901.07291 (2019)
26. McInnes, L., Healy, J., Astels, S.: HDBSCAN: hierarchical density based clustering. J. Open Source Softw. **2**(11) (2017). https://doi.org/10.21105/joss.00205, https://doi.org/10.21105%2Fjoss.00205
27. Öztürk, N., Ayvaz, S.: Sentiment analysis on twitter: a text mining approach to the Syrian refugee crisis. Telematics Inform. **35**(1), 136–147 (2018)
28. Pierri, F., Luceri, L., Jindal, N., Ferrara, E.: Propaganda and misinformation on Facebook and twitter during the Russian invasion of Ukraine. In: Proceedings of the 15th ACM Web Science Conference 2023, pp. 65–74 (2023)
29. Qayyum, A., Gilani, Z., Latif, S., Qadir, J.: Exploring media bias and toxicity in south Asian political discourse. In: 2018 12th International Conference on Open Source Systems and Technologies (ICOSST), pp. 01–08 (2018). https://doi.org/10.1109/ICOSST.2018.8632183
30. Reimers, N., Gurevych, I.: Sentence-BERT: sentence embeddings using Siamese BERT-networks. arXiv preprint arXiv:1908.10084 (2019)
31. Reuters: Ukraine presses counteroffensive on key Russian line of assault - governor (2022). https://www.reuters.com/world/europe/ukraine-presses-counteroffensive-key-russian-line-assault-governor-2022-05-14/. Accessed 15 May 2022
32. Rolling Stone: See U2's bono and the edge play surprise set in Kyiv bomb shelter following invitation from president zelensky (2022). https://www.rollingstone.com/music/music-news/u2-bono-the-edge-acoustic-set-kyiv-bomb-shelter-1350428/. Accessed 15 July 2022
33. Saarinen, M.J.O., Aumasson, J.P.: The BLAKE2 cryptographic hash and message authentication code (MAC). Internet-draft draft-saarinen-blake2-06, internet engi-

neering task force (2015). https://datatracker.ietf.org/doc/draft-saarinen-blake2/06/, work in Progress

34. Severyn, A., Moschitti, A.: Twitter sentiment analysis with deep convolutional neural networks. In: Proceedings of the 38th International ACM SIGIR Conference on Research and Development in Information Retrieval, pp. 959–962 (2015)

35. Shevtsov, A., Antonakaki, D., et al., L.: Russo-Ukrainian war: prediction and explanation of twitter suspension. In: Proceedings of the 2023 IEEE/ACM ASONAM, pp. 348–355. ASONAM 2023, Association for Computing Machinery, New York, NY, USA (2024)

36. Shevtsov, A., Antonakaki, D., Lamprou, I., Pratikakis, P., Ioannidis, S.: BotArtist: generic approach for bot detection in twitter via semi-automatic machine learning pipeline (2024). https://arxiv.org/abs/2306.00037

37. Shevtsov, A., Oikonomidou, M., Antonakaki, D., Pratikakis, P., Ioannidis, S.: What tweets and Youtube comments have in common? Sentiment and graph analysis on data related to us elections 2020. PLoS ONE **18**(1), e0270542 (2023)

38. Shevtsov, A., Tzagkarakis, C., Antonakaki, D., Ioannidis, S.: Identification of twitter bots based on an explainable machine learning framework: the us 2020 elections case study. In: Proceedings of the International AAAI Conference on Web and Social Media, vol. 16, pp. 956–967 (2022)

39. Smart, B., Watt, J., Benedetti, S., Mitchell, L., Roughan, M.: # istandwithputin versus# istandwithukraine: the interaction of bots and humans in discussion of the Russia/Ukraine war. In: International Conference on Social Informatics, pp. 34–53. Springer (2022). https://doi.org/10.1007/978-3-031-19097-1_3

40. Sufi, F.K., Khalil, I.: Automated disaster monitoring from social media posts using AI-based location intelligence and sentiment analysis. IEEE Trans. Comput. Soc. Syst. **11**, 4614–4624 (2022)

41. Taboada, M., Brooke, J., Tofiloski, M., Voll, K., Stede, M.: Lexicon-based methods for sentiment analysis. Comput. Linguist. **37**(2), 267–307 (2011)

42. Vayansky, I., Kumar, S.A.: A review of topic modeling methods. Inf. Syst. **94**, 101582 (2020). https://doi.org/10.1016/j.is.2020.101582, https://www.sciencedirect.com/science/article/pii/S0306437920300703

43. Wallach, H.M.: Topic modeling: beyond bag-of-words. In: Proceedings of the 23rd ICML, pp. 977–984. ICML 2006, Association for Computing Machinery, New York, NY, USA (2006). https://doi.org/10.1145/1143844.1143967

44. Wang, X., Yang, R., Lu, Y., Wu, Q.: Military named entity recognition method based on deep learning. In: 2018 5th IEEE CCIS, pp. 479–483 (2018)

45. Williams, A., Nangia, N., Bowman, S.: A broad-coverage challenge corpus for sentence understanding through inference. In: Proceedings of the 2018 NAACL-HLT, pp. 1112–1122. Association for Computational Linguistics (2018)

46. Yang, S.H., et al., K.: Large-scale high-precision topic modeling on twitter. In: Proceedings of the 20th ACM SIGKDD, pp. 1907–1916. KDD 2014, Association for Computing Machinery, New York, NY, USA (2014). https://doi.org/10.1145/2623330.2623336

47. Zhang, Y., Li, X., Zhang, X., Li, Y.: Sentiment lexicons and non-English languages: a survey. Knowl. Inf. Syst. **63**(3), 707–736 (2020)

PARALLAX: Leveraging Polarization Knowledge for Misinformation Detection

Demetris Paschalides[✉]ⓘ, George Pallisⓘ, and Marios D. Dikaiakosⓘ

University of Cyprus, Computer Science Department, Nicosia, Cyprus
{dpasch01,pallis,mdd}@ucy.ac.cy

Abstract. Recent techniques for the automated detection of online misinformation typically rely on ML models trained with features extracted from content analysis and/or general-purpose Knowledge Graphs (KGs). These techniques often fail to consider the interplay between misinformation and polarization. To bridge this gap, we introduce PARALLAX, a methodology that enhances misinformation detection by infusing polarization knowledge into existing classifiers. Polarization knowledge is represented in terms of Polarization Knowledge Graphs (PKG). PARALLAX constructs PKGs in an unsupervised way, and uses them to enrich articles with polarization knowledge. A Flexible Knowledge-aware Graph Neural Network (FlexKGNN) is trained on these enriched representations. We tested our methodology on three misinformation datasets, demonstrating that it achieves approximately a 15% improvement in performance over baseline classifiers and consistently outperforms other KGs, which typically reach baseline levels only.

Keywords: Polarization · Knowledge Graph · Misinformation Detection

1 Introduction

In recent years, misinformation has posed significant challenges to societies worldwide, with evident influence on events such as presidential elections [1], referendums, and most recently, the COVID-19 pandemic [6,13]. Moreover, the rise of Foreign Information Manipulation and Interference (FIMI) has added a new dimension to this challenge, as evidenced in the context of the Russo-Ukrainian war [20]. The rampant spread of misleading narratives extends beyond sowing confusion; it also exacerbates societal divisions, forging a complex relationship with the phenomenon of polarization [21]. On the one hand, polarization fosters an environment conducive to misinformation spread. The divisive nature of polarization, coupled with the human tendency for "confirmation bias," makes individuals more susceptible to false or misleading information, especially when it aligns with their existing viewpoints [29,36]. On the other hand, the proliferation of misinformation can also contribute to the escalation of polarization. In an environment with distorted information, individuals often retreat into "echo

© The Author(s), under exclusive license to Springer Nature Switzerland AG 2025
L. M. Aiello et al. (Eds.): ASONAM 2024, LNCS 15211, pp. 86–105, 2025.
https://doi.org/10.1007/978-3-031-78541-2_6

chambers" of similar views, reinforcing their beliefs, and perceiving dissenters as adversaries [29].

Existing methods to mitigate misinformation primarily focus on the development of Machine Learning (ML) models for fake news detection and/or the establishment of fact-checking initiatives [35], often overlooking the immediate connection between misinformation and polarization. To address this, we need to cope with the complexity of modeling, quantifying, and integrating polarization into misinformation detection algorithms [9].

In this paper, we aim at bridging this gap by proposing PARALLAX, a methodology and toolset for integrating polarization knowledge into existing misinformation classifiers, assessing its contribution on their classification performance. To do so, we address the challenges of: i) representing domain-specific polarization knowledge; ii) encoding news articles with their polarization information; and iii) effectively integrating article polarization cues into existing classifiers to enhance their accuracy. The key contributions of our work are:

- **Polarization Knowledge Graph (PKG)**: The definition of the PKG schema and data structure designed to capture polarization knowledge as a semantic graph of entities, fellowships, topics, and attitudes. To construct the PKG, we develop an unsupervised method to extract and model polarization information from news corpora (see Sect. 3).
- **Article-specific Polarization Encoding**: We introduce an approach to extract polarization knowledge on a single-article level, and encode this knowledge as a **micro-PKG**, a semantic graph that aligns with the PKG. The micro-PKG reflects the key actors and predicates identified in the article's content. To address content limitations at the article level, we enrich the micro-PKGs by i) incorporating additional polarization context from the PKG and ii) integrating PKG-derived embeddings (see Sect. 4).
- **Flexible Knowledge-aware Graph Neural Network (FlexKGNN)**: We design a Graph Neural Network to incorporate polarization knowledge encoded in micro-PKGs as a feature into existing misinformation classifiers, to enhance their classification performance. This is achieved by concatenating the feature vector of a given classifier with the internal micro-PKG representation to learn the relationship between polarization and misinformation for improved classification (see Sect. 5).
- **Evaluation Study and Dataset**: We evaluate PARALLAX on a manually curated COVID-19 misinformation dataset, along with two additional datasets used in prior literature [28]. We assess the contribution of polarization knowledge on the performance of two existing misinformation classifiers, comparing the results with alternative article-encoding methods. Our results reveal that our method outperform others, improving the accuracy of existing classifiers by ≈15%, highlighting the importance of incorporating polarization into misinformation detection (see Sect. 6).

2   Background and Related Work

Misinformation is defined as false information disseminated, either intentionally or inadvertently, from "unreliable" sources [35]. The content of fake news articles, often mimicking credible sources, distorts public perception by exploiting biases [23,36], and intensifies societal polarization [1,36].

Polarization refers to the phenomenon where social or political groups are fragmented into opposing factions that hold different and often conflicting beliefs and values [29]. These factions consist of interacting entities that hold diverse attitudes on various topics. Entities sharing similar beliefs tend to form cohesive fellowships, while conflicts emerge from their disagreements, thus forming fellowship dipoles. In such environment, misinformation aligning with in-group beliefs flourishes, while factual contradictory information is met with skepticism [21,36]. Consequently, polarization manifests across entities, groups, and topics, forming a complex multi-level phenomenon. We refer to entities, fellowships, dipoles, topics, and attitudes as *"polarization information"*.

Polarization, rooted in the inter-group conflict theory [29], manifests during the process of "social categorization," where individuals align with groups, i.e. fellowships, based on shared beliefs. This fosters in-group loyalty, distinguishing "us" from "them." During "social identification," individuals become members of those groups, a process amplified by cognitive biases [36]. In the concluding "social comparison" process, groups favorably contrast against others, often only considering their perspective, stereotyping out-groups negatively [11]. In such environment, misinformation aligning with in-group beliefs flourishes, while factual contradictory information is frequently met with skepticism or rejection [36].

2.1   Content-Based Misinformation Detection

Content-based misinformation detection can be broadly categorized into style- and knowledge-based methods.

Style-based Methods focus on identifying distinctive writing styles in fake news through linguistic features [25,28], such as modal words, punctuation, and casing [23]. They also emphasize the role of hyperpartisanship in the dissemination and distinction of fake news [35,36]. Recent works have incorporated contextual embeddings from transformer models, which displayed their potency in accurately distinguishing between fake and real news content [24]. **Knowledge-based Methods** leverage external knowledge to enhance the understanding or verification of news articles [8,15,17,32,35]. Typically, they employ an established KG, such as Wikipedia[1], to serve as the factual world knowledge. A key step in these approaches, is encoding each article in the dataset using the pre-defined KG. Some approaches try to mimic human fact-checking, identifying claims within articles as Subject-Predicate-Object (SPO) triples, and verifying them against the KG [35]. More recent works try to enhance the content

---

[1] http://wikipedia.org.

of each article, by identifying entities mentioned in each article's text, connecting them to the KG, and further enriching them with adjacent KG entities [8,15,17,32]. Certain works also integrate discussion topics [15] and entity relationships [8,17,32], extracted via tools like OpenIE [2], thus, forming a comprehensive heterogeneous graph for each article. To ascertain the veracity of articles, these methods use Graph Neural Networks (GNNs), classifying each article representation as originating from genuine or misleading information.

These methods, while broadening article context with external knowledge, have yet to integrate polarization knowledge in misinformation detection, despite its influence on misinformation spread and consumption [30,36]. Furthermore, their outputs consist of GNNs that base their classification on internal article representations, overlooking existing classifiers. In this work, we address these limitations by introducing a method to represent article-level polarization knowledge. We propose a GNN that integrates this knowledge with existing classifiers, enhancing their classification performance and assessing the contribution of polarization on misinformation detection.

2.2   Representing Polarization Knowledge

Existing computational approaches which study, represent, and quantify different aspects of polarization in online social media typically focus on two broad directions: group- or topic-level polarization. **Group-oriented approaches** typically seek to identify polarized groups of users, and model inter-group polarization based on group segregation-level metrics [3,9,12]. **Topic-oriented approaches** typically apply NLP, topic modeling, and Deep Learning (DL) techniques to model, measure, and evaluate the polarized stance of distinct ideological user groups (e.g. Democrats and Republicans in the US) towards particular issues [14,19]. In our work, we explore a different methodology for modeling polarization: we employ content analysis on a wide number of news articles to construct a semantic graph (i.e. the PKG) that represents the polarization landscape. This graph is extracted in an unsupervised manner from the narratives presented inside the articles, identifying key figures, events, and themes pivotal to public discourse, and capturing divisive and unifying attitudes among them. Our method integrates both group and topic aspects of polarization into the PKG for a comprehensive domain knowledge representation. This structured representation enables its integration with existing classifiers, aiding in evaluating its contribution on relevant tasks, including misinformation detection.

3   Polarization Modeling and Knowledge Extraction

3.1   Polarization Knowledge Graph and Schema

To capture domain-specific polarization knowledge effectively, we introduce the concept of Polarization Knowledge Graph (PKG), which is a structured representation of polarization information defined according to a Subject-Predicate-Object (SPO) schema shown in Fig. 1. The PKG schema is grounded on the inter-group conflict theory [29], comprising of four primary actors, namely *Entity, Fellowship, Dipole,* and *Topic.* An *Entity* is any individual or group that contributes to the polarization observed in a particular domain of interest; groups include organizations, countries, or religions. A *Topic* is a subject on which entities may hold opposing opinions. A *Fellowship* represents a cohesive sub-group of entities with mutual supportive attitudes, while a *Dipole* comprises two fellowships manifesting opposing views or attitudes towards a particular topic. The PKG schema uses predicates to characterize supportive or opposing relationships between its actors: *SupportEE* and *OpposeEE* for entity interactions, *SupportET* and *OpposeET* for entity attitudes on topics, and *MemberOf* for entity associations to fellowships. It defines group attitudes on topics with *SupportFT* and *OpposeFT*, and fellowship conflicts within dipoles with the *PartOf* predicate. Degrees of polarization on topics are indicated by *HasModeratePolarization, HasMediumPolarization,* and *HasExtremePolarization.*

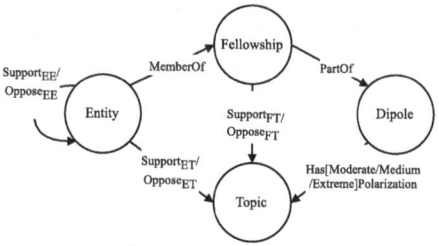

Fig. 1. Diagram of the Polarization Knowledge schema.

To construct a PKG, which reflects the context of a domain under study, we gather a *Supplementary Corpus* comprised of news articles representing the context surrounding this domain. The selection of articles for the corpus is guided by three critical parameters: *theme, region,* and *time-frame. Theme* refers to a collection of keywords that capture the domain's core concepts; for instance, when focusing on Coronavirus, relevant keywords might include "Coronavirus," "COVID-19," and "SARS-CoV-2." The *Region* parameter helps focus on a specific geographical area, like the United States, and *time-frame* defines a specific period under study through designated start and end dates. We deploy collectors that utilize the aforementioned parameters to filter news articles from the GDELT Project[2] - a large and open database of global news articles. This tailored corpus forms the basis for extracting relevant polarization information. This structured approach ensures that the Supplementary Corpus reflects the domain's knowledge landscape, and can serve as the basis for extracting polarization information and constructing a representative domain-specific PKG. Following, we outline the PKG construction process.

---

[2] https://www.gdeltproject.org/.

3.2   Extracting the Polarization Information

Given the Supplementary Corpus, we use POLAR [22] to extract domain-specific polarization information. We chose POLAR for its integrated approach that combines both group- and topic-oriented methods, offering a comprehensive extraction of polarization information. Initially, we process the Supplementary Corpus with POLAR to detect and link named entities, denoted as E. Then, we use syntactical dependency parsing and sentiment attitude analysis to identify the supportive or opposing attitudes between entity pairs as a function $r(e_i, e_j) \rightarrow \{Negative, Neutral, Positive\}$. Then, we employ signed network clustering methods to group entities with dense positive attitudes amongst them and discover the entity fellowships F. To identify fellowship dipoles D, we examine the structural balance of all pairs of fellowships; structural balance is a concept tied to polarization in signed networks [3]. Additionally, we perform clustering of semantically similar noun phrases to identify discussion topics T. Sentiment cues are aggregated from relevant sentences and noun phrases, associating an entity's attitude towards a topic. This association is captured by $a(e_i, t_z) \rightarrow [-1, 1]$, where -1 signifies strong opposition and 1 denotes strong support. The output of this process comprises of E, F, D, T, r and a, which collectively represent the polarization information extracted with POLAR. For example:

- *"Pres. Trump spent months playing down the effectiveness of masks, ... mocked former V.P. Biden for wearing one."*
- *"Dr. Fauci ... been begging people to wear masks."*
- *"Trump ... insulting Fauci for telling the truth."*
- *"Biden described Fauci as a dedicated public servant ..."*

From these sentences, we discern the entities as $E = \{$*"Joe Biden"*, *"Anthony Fauci"*, *"Donald Trump"*$\}$, and their relationships: $r($*"Joe Biden"*, *"Anthony Fauci"*$) = Positive$, $r($*"Donald Trump"*, *"Anthony Fauci"*$) = Negative$, and $r($*"Donald Trump"*, *"Joe Biden"*$) = Negative$. These lead to the formation of fellowships $F = \{f_1, f_2\}$ where $f_1 = \{$*"Joe Biden"*, *"Anthony Fauci"*$\}$ and $f_2 = \{$*"Donald Trump"*$\}$, establishing the dipole $d_{1,2} \in D$. The discussion centers on the topic $t_1 \in T$ where $f_1 = \{$*"Joe Biden"*, *"Anthony Fauci"*$\}$ and $f_2 = \{$*"Donald Trump"*$\}$, establishing the dipole $d_{1,2} \in D$. The discussion centers on the topic $t_1 \in T$ where $t_1 = \{$*"effectiveness of masks"*, *"masks"*$\}$, labeled as *Mask Effectiveness*. The entity attitudes toward t_1 are quantified as $a($*"Joe Biden"*, $t_1) = 0.0$, $a($*"Anthony Fauci"*, $t_1) = 0.8$, and $a($*"Donald Trump"*, $t_1) = -1.0$.

3.3   Construction of Polarization Knowledge Graph

Although the polarization information extracted with POLAR is valuable, its lack of semantic structure presents challenges for its effective utilization, interpretation, and integration into tasks such as misinformation detection. To address this, we introduce a number of successive transformations designed to transform the identified entities, fellowships, dipoles, topics, and the structural relations thereof, into a PKG. These steps involve the conversion of elements of E, F, D, and T, into actors in the PKG, the derivation of predicates from known structural relationships and the values of functions r and a, and the enrichment of PKG topics with comprehensive descriptions. We initialize the PKG by integrating entities and their relationships, assigning predicates based on the function r. Figure 2a depicts the initial PKG from the example.

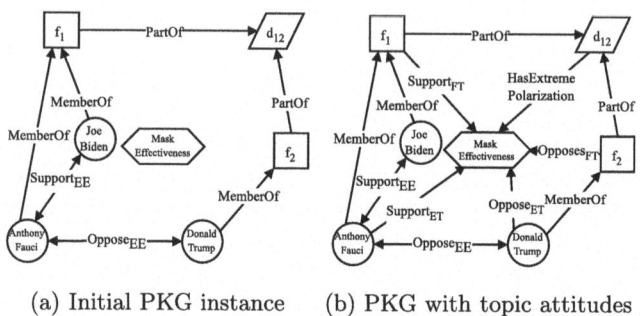

(a) Initial PKG instance   (b) PKG with topic attitudes

Fig. 2. Initial (2a) and the updated (2b) PKG instance.

Topical Attitude Predicates: We enrich the PKG by adding identified fellowships and topics, and by computing and integrating the predicates that reflect the attitudes of entities and fellowships towards these topics. We compute these predicates by translating the continuous function a into a categorical form, where a threshold thr determines the relationship type: if $a(e_i, t_j) \geq thr$, we assign a *SupportET* predicate between e_i and t_j, otherwise we assign an *OpposeET* predicate. To estimate thr, we examine the sets of positive A_T^+ and negative A_T^- attitudes from every entity toward every topic, setting thr as the average of their median values. This approach captures the inherent division between support and opposition in the Supplementary Corpus, ensuring a balanced threshold for identifying *SupportET* and *OpposeET* predicates. To assign predicates between a fellowship f_i and a topic t_j, the fellowship's aggregated attitude towards t_j is calculated as the average attitude of all of its entity members towards t_j:

$$att_{f_i}^{t_j} = \frac{\sum_{e_k \in f_i} a(e_k, t_j)}{|f_i|}$$

Based on this aggregated attitude, the predicate between f_i and t_j is assigned as *SupportFT* if $att_{f_i}^{t_j} \geq thr$, or *OpposeFT* otherwise. Expanding on our example from Sect. 3.2, we divide attitudes into $A_T^+ = \{0.8\}$ and $A_T^- = \{-1.0\}$, establishing a threshold $thr = -0.1$. This division results in categorizing the relationships as ("Anthony Fauci", *SupportET*, t_1) and ("Donald Trump", *OpposeET*, t_1). In a similar manner, fellowships' attitudes towards t_1 are encapsulated into the triples $(f_1, SupportFT, t_1)$ and $(f_2, OpposeFT, t_1)$.

Topic Polarization Predicates: To assign the polarization-related predicates, we first measure the degree of disagreement in attitudes between dipole fellowships for each topic. To quantify this, we use the polarization index metric [19]:

$$\mu = (1 - \Delta_A)\delta_A$$

where $\Delta_A = (|A_{t_j}^+| - |A_{t_j}^-|)/(|A_{t_j}^+| + |A_{t_j}^-|)$. This represents the normalized difference between the sizes of positive $A_{t_j}^+$ and negative $A_{t_j}^-$ attitude sets w.r.t. t_j. $\delta_A = |gc^+ - gc^-|/2$ is the difference between the average attitude values gc^+ and gc^- of $A_{t_j}^+$ and $A_{t_j}^-$, respectively. The value of μ ranges from 0 to 1, with 1 indicating extreme polarization and 0 denoting no polarization. This metric aligns with theoretical concepts in political science and sociology that define polarization as both the concentration of opinions at opposing extremes and the distance between those extremes [7]. To assign the polarization predicates, we utilize the following thresholds: *HasModeratePolarization* if $\mu \leq 0.3$, *HasMediumPolarization* if $0.7 \geq \mu > 0.3$, and *HasExtremePolarization* if $\mu > 0.7$. These thresholds were determined through a combination of empirical analysis and theoretical considerations. We conducted a preliminary study on a diverse set of topics across multiple domains, analyzing the distribution of μ values. The results showed that μ values around 0.3 indicate the emergence of moderate polarization, whereas, for μ values above 0.7, extreme polarization occurs. These observations align with the work by Bramson et al. 2016 [5], who suggest that polarization emerges when opposing groups show differences but still have some overlap in their views. However, these thresholds are adaptable, allowing for customization for a variety of datasets. In our example, we calculate the polarization index μ_{t_1} for dipole $d_{1,2}$ towards t_1 based on the distinct positive and negative attitudes towards t_1, $A_{t_i}^+ = \{0.8\}$ and $A_{t_i}^- = \{-1.0\}$. These attitudes yield $\Delta_A = 0$, and $\delta_A = (0.8 + 1.0)/2) = 0.9$, leading to $\mu_{t_1} = 0.9$, which translates to the triple $(d_{1,2}, HasExtremePolarization, t_1)$, indicating a notable polarization on t_1. Figure 2b shows the completed PKG.

4   Article-Specific Polarization Encoding

To map an article q, which has not been previously encountered, to a polarization context defined by a relevant PKG, we utilize the methodologies described in Sect. 3.2. These methods extract polarization knowledge from the contents of q and encode it as a micro-PKG, which is a condensed version of the primary PKG that incorporates select actors and adjusted attitudes to reflect q's narrative. This micro-PKG includes the components E_q, T_q, r_q and a_q, where it is imperative that actors in E_q and T_q correspond with those in the primary PKG. Owing to the inherent limitations of micro-PKGs, which arise from the typically brief length of individual articles, these structures may possess restricted scope and connectivity to broader polarization knowledge. To address this constraint, we enhance the micro-PKGs by integrating supplementary polarization contexts from the primary PKG and applying PKG embeddings. These embeddings serve as node and edge features in a low-dimensional vector space, capturing the structural and semantic properties of actors and predicates.

4.1 Structural Augmentation with Subgroup Dynamics

Our first strategy consists of structurally enriching the micro-PKG by adding context from the primary PKG. Specifically, we identify fellowships in the primary PKG relevant to the entities E_q within the article and include them in the micro-PKG via *MemberOf* predicate. If these fellowships are part of broader dipoles identified in the primary PKG, we integrate these dipoles into the micro-PKG and establish connections using a *PartOf* predicate. Additionally, we link the newly included fellowships and dipoles to topics T_q using relevant *SupportFT*, *OpposeFT*, or polarization-level predicates found in the primary PKG. This method allows us to extend the initial micro-PKG to cover a wider array of conflict dynamics between subgroups, reflecting both explicit mentions in the article and the larger polarization context within the domain.

4.2 Semantic Enhancement Through PKG Embeddings

Our second strategy introduces PKG embeddings to enhance the representation of polarization in micro-PKGs. To learn PKG embeddings, we employ the TuckER method, which is very effective in capturing diverse types of actors and predicates [4]. TuckER is trained on known triples from the primary PKG and is evaluated on a triple set with one element (subject, predicate, or object) omitted. It decomposes a tensor into factor matrices for subjects ($\mathbf{S}$), predicates ($\mathbf{P}$), and objects ($\mathbf{O}$), along with a core tensor (3) representing the interactions among them. During training, TuckER employs the function $\phi(s, p, o) = 3 \times_1 \mathbf{s} \times_2 \mathbf{p} \times_3 \mathbf{o}$, where $\mathbf{s}$, $\mathbf{p}$, and $\mathbf{o}$ are the embeddings of a PKG triple's subject, predicate, and object. TuckER applies a logistic sigmoid to each score $\phi(s, p, o)$, predicting the likelihood of a triple's correctness. The training objective is to minimize the binary cross-entropy loss of these predictions, iteratively refining $\mathbf{S}$, $\mathbf{P}$, and $\mathbf{O}$. The resulting $\mathbf{S}$, $\mathbf{O}$ embeddings depict the PKG actor positions in the latent space, whereas the predicate embeddings $\mathbf{P}$ signify their role in linking actors within the PKG. Given a micro-PKG, we calculate each actor's mean embedding from $\mathbf{S}$ and $\mathbf{O}$ to ensure both its subject and object roles are considered in its representation. For each predicate, we compute the mean of its vector from $\mathbf{P}$ and the subject and object associated embeddings, capturing the predicate's context. These embeddings are integrated as nodes and edges features of the micro-PKG.

5 Polarization-Driven Misinformation Detection

To improve the accuracy of existing misinformation classifiers, we aim at enriching their training with PKG-encoded polarization knowledge. To this end, for each article in a Misinformation Dataset (MD) of interest, we construct a PKG for the domain of the MD and compute a micro-PKG for each article in the MD. Subsequently, we label each micro-PKG as either "reliable" or "unreliable," given the credibility of its source article, creating a *micro-PKG Dataset*. Leveraging the graph representations in the micro-PKG Dataset, we re-define misinformation detection as a graph classification, which entails classifying micro-PKGs as originating from "reliable" or "unreliable" news articles. To effectively integrate our approach with existing classifiers, we introduce FlexKGNN, a Graph Neural Network (GNN) designed to assimilate polarization knowledge from micro-PKGs and merge it with features from these classifiers. We base the FlexKGNN core architecture on models from related works that utilize KGs for misinformation detection, employing graph convolution and attention layers [8,17,32]. The novelty of FlexKGNN is in merging the internal micro-PKGs representation with external classifiers prior to the classification, a strategy inspired by ensemble classifiers [31]. We validate the architecture of FlexKGNN, shown in Fig. 3, through hyperparameter tuning across the evaluation datasets.

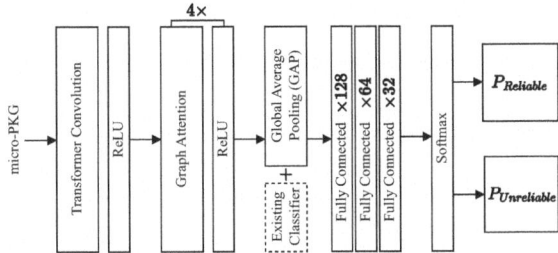

Fig. 3. Overview of the FlexKGNN model architecture.

Initially, an input micro-PKG G passes a Transformer Convolution layer [33], updating the feature vector for each actor v by aggregating neighbor information using self-attention. Feature vector $H_v^{(l+1)}$ at layer $l+1$ is updated as:

$$H_v^{(l+1)} = \sigma(B^{(l)} \cdot W^{(l)} \cdot H^{(l)})$$

where $W^{(l)}$ and $B^{(l)}$ are the trainable weight matrix and bias vector for layer l, respectively, and $\sigma(\cdot)$ is a non-linear activation function. Subsequently, four Graph Attention (GAT) layers [33] further leverage self-attention to allow the model to learn the importance of neighbors' information dynamically. After the last attention layer, Global Average Pooling (GAP) aggregates actor and predicate features to form a micro-PKG representation H_G. Finally, H_G is passed through a series of fully connected layers, each utilizing the LeakyReLU non-linear activation function. The FlexGKNN architecture concludes with a softmax activation layer, which outputs the probabilities for each class $c \in \{reliable, unreliable\}$:

$$P(G) = \text{Softmax}(W^{(c)} \cdot H_G + B^{(c)})$$

where $W^{(c)}$ and $B^{(c)}$ are the weight and bias of the classification layer, and $P(G)$ is the probability distribution over classes ($P_{Reliable}$ and $P_{Unreliable}$).

Incorporating Existing Classifiers: We integrate existing classifiers into FlexKGNN during its training phase. These classifiers, trained on misinformation features from news article content, are combined with polarization cues in the GAP layer of FlexKGNN, where the model learns the micro-PKG representation H_G. For this integration, a feature vector F is defined, representing the point at which the existing classifier merges with FlexKGNN. F can be defined in several ways: i) as a vector of misinformation features extracted using NLP techniques [35], ii) as a vector consisting of the output class probabilities from the existing classifier, or iii) as the penultimate hidden layer of DL models. To merge with FlexKGNN, the vector F is concatenated with the H_G representation, creating an augmented representation $H'_G = [H_G \parallel F]$. This approach employs ensemble stacking principles, where multiple models are merged to enhance their capabilities [31]. The H'_G representation is then used to calculate $P(G)$.

6   Experiments and Evaluation

To assess the effectiveness of our approach, we examine a case study focusing on articles related to the COVID-19 pandemic, a period characterized as an "info-demic[3]" exacerbated by political polarization [6,13]. To conduct our evaluation, we compiled a misinformation dataset specific to this context. Our objective is to quantify the impact of integrating polarization knowledge into existing classifiers for misinformation detection. Beyond our primary case study, we apply our approach on two additional datasets to examine its broader applicability. For reproducibility purposes, we make our results and code publicly available[4].

COVID-19 Misinformation Dataset: Our dataset comprises articles from GDELT database, from 1/2020 to 12/2021. We retrieved articles with three or more mentions of the keywords *coronavirus*, *COVID*, or *pandemic*. We maintained data integrity by using the Internet Archive[5] as a backup source for removed articles. To assess the reliability of the collected articles, we referenced publicly available information on domains previously associated with COVID-19 misinformation[6][7]. We categorized the articles based on their domains credibility, identifying 58,888 as "reliable" and 3,523 as "unreliable. The most frequent domains are depicted in Fig. 4.

Additional Datasets: We also evaluate on two additional datasets [28]: i) *Politifact*, consisting of 467 "reliable" and 383 "unreliable" articles, primarily focusing on the US political scene, sourced from the *politifact.com* fact-checking

---

[3] https://www.who.int/health-topics/infodemic.
[4] https://github.com/dpasch01/PARALLAX.
[5] https://archive.org/.
[6] https://mediabiasfactcheck.com/.
[7] https://github.com/bigheiniu/meta-coronavirus-dataset.

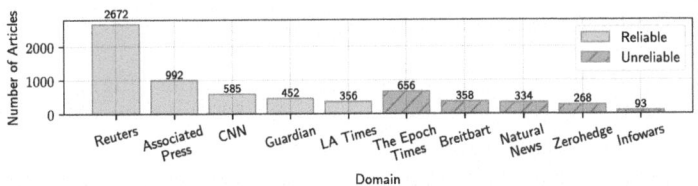

Fig. 4. Number of reliable and unreliable articles by top domains known to generate factual and false information. ([8]Reuters digital news report 2020)

website; and ii) *GossipCop*, consisting of 15,313 "reliable" and 4,781 "unreliable" articles, centered around celebrity news, gathered from *eonline.com* and *gossip-cop.com*. To rectify the class imbalance in our datasets, we employed domain stratified undersampling on the majority label.

Table 1. Characteristics of the MD Supplementary Corpora.

Dataset	From	To	Top Keywords	# Articles
COVID-19	01/2020	12/2021	*coronavirus, sars, mandate, vaccine*	84,180
Politifact	10/2016	04/2018	*trump, clinton, president, debate*	15,710
GossipCop	06/2017	05/2018	*kardashian, bieber, prince, harry, markle*	12,768

Supplementary Corpora: To compile the Supplementary Corpora, we automatically extract the parameters of theme, region, and timeframe from each of the MDs, to ensure their relevance with each corpus. To do so, we apply TF-IDF to extract the thematic keywords from the articles, utilize geo-extractors to identify the region, and the publication dates for timeframe alignment. To mitigate overlap, we exclude articles from each Supplementary Corpus that are shared with its corresponding MD. Table 1 outlines these characteristics for the Supplementary Corpora derived from the specified MDs.

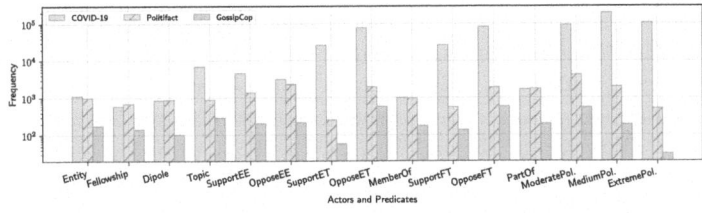

Fig. 5. Frequency of PKG actors and predicates. *ModeratePol.*, *MediumPol.*, and *ExtremePol.* represent *Has[Moderate, Medium, Extreme]Polarization* predicates.

6.1   COVID-19 PKG Overview

We construct the PKGs for each Supplementary Corpus. Figure 5 depicts the actor and predicate frequencies for each of the PKGs. Following, we present an overview of the PKG regarding our primary case study of COVID-19. Given the number of actor and predicate observations, the high number of *OpposeET* (76,425) compared to *SupportET* (26,287) indicates a prevailing negative entity attitude towards various topics. This is also supported by the *SupportFT* (26,287) and *OpposeFT* (80,483) predicates, hinting a significant divide between fellowships. In addition, the 102,193 *HasExtremePolarization*, 192,287 *HasMediumPolarization*, and 91,262 *HasModeratePolarization* instances underscore the highly polarized nature of the topic discussions.

Entity-level Overview: Notable entities and their positive and negative attitudes are illustrated in Fig. 6. These include the *COVID-19 Vaccine* and *Pfizer*, which exhibit positive relationships, indicating their acceptance and favorable image. Regulatory bodies like *FDA* and *CDC* show higher positive attitudes, indicative of public trust [6]. Conversely, geographic entities like *Wuhan, Taiwan*, and *China* display mostly negative bias, possibly tied to COVID-19 origin blame. Political figures like, *Donald Trump* and *Joe Biden*, reflect mixed sentiments on their pandemic responses [13].

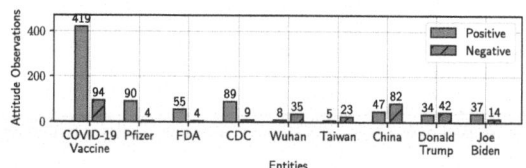

Fig. 6. Number of positive and negative entity attitudes.

Fellowship-Level Overview: The PKG reveals several fellowship instances. One fellowship emphasizes the medical response, uniting entities like *COVID-19 Vaccine, Pfizer, Moderna COVID-19 Vaccine*, and regulatory bodies *FDA* and *CDC*. Another highlights the US public health response, with administrative entities like the *President of the United States* and preventive measures like *Social Distancing* and *Face Masks*, along with experts such as *Dr. Anthony Fauci*. A separate fellowship revolves around the *Democratic Party*, including *Joe Biden, Barack Obama*, and *Bernie Sanders*. Lastly, a group centered on *Donald Trump*, highlights events like his treatment at the *Walter Reed National Military Medical Center*. Collectively, these fellowships offer comprehensive insights of the pandemic's medical and political dimensions.

Topic-Level Overview: Various pandemic-related topics have been identified, which exhibit different degrees of polarization. Topics such as the *COVID-19 Case Numbers, Vaccine Efficacy, Lockdown Measures, COVID-19 Response*, and

COVID-19 Treatments stand out for their high occurrences of *HasExtremePo-larization* predicates. These observations align with findings of significant politicization of the pandemic, the undermining of health authorities, and hesitancy to vaccines [13]. As described in Sect. 3.2, the PKG topics are identified as semantically similar Noun Phrases (NPs). Initially, this process yielded 6,811 topics, each comprising an average of 1,067 NPs, making their interpretation challenging. To streamline this, we automatically label each topic with a representative title, describing its context. For the labeling, we employ OpenAI's GPT-4 API[8], leveraging the significant results of ChatGPT, including data annotation [10]. Specifically, given the NPs of each topic, ChatGPT was prompted to generate a self-explanatory title in relation to the pandemic. Examples of the annotated topics are shown in Table 2.

Table 2. Topic examples with their frequent noun phrases.

Topic Label	Frequent Noun Phrases
Vaccine Efficacy	vaccine, immunization, effective vaccine
Lockdown	lockdowns, locked-down people, lock-down
Med. Experts	expert, highly trained expert, medical expert
Resp. Mishandl.	mishandling, horrific handling, improper response handling
Mask Mandate	face mask, face covering, mask mandate
Reopening	reopening, reopening phase, collective reop.
Misinformation	disinformation, misinform, misinformation
Virus Origin	artificial origin, animal origin, man-made

6.2   Polarization Contribution to Existing Classifiers

Our primary evaluation goal is to measure how the integration of polarization knowledge enhances the performance of existing misinformation classifiers. For this purpose, we integrate two baseline classifiers into FlexKGNN, as detailed in Sect. 5. These classifiers represent both ML and DL paradigms, combining textual [23] and latent [24] features from ML and DL models, respectively, yielding SOTA results. For our ML baseline classifier, we chose Check-It [23], a feature-based misinformation detection approach. Check-It operates on a set of 256 textual features, to derive its predictions via a logistic regression model. To integrate with FlexKGNN, we concatenate its feature vector with H_G. For our DL baseline, we selected RoBERTa, a well-known pre-trained transformer model, effective in misinformation detection [24]. To integrate it with FlexKGNN, we concatenate its last 768-dimensional hidden layer with H_G.

Experimental Setup: To train the models, we split the MDs into 70% for training and 30% testing, using 3-fold cross-validation. We used a stochastic gradient descent optimizer with 0.2 momentum and a learning rate of 0.02. To avoid overfitting, we applied a 10 epochs early stopping. We trained for 100 epochs, accelerated by an NVIDIA Tesla T4 GPU.

---

[8] https://openai.com/blog/openai-api/

Table 3. Performances scores for baselines classifiers and their integration with PKG, OpenIE, SRL, and DBPedia.

Model	COVID-19		Politifact		GossipCop	
	Acc.	F1	Acc.	F1	Acc.	F1
Check-It (C)	0.717	0.716	0.697	0.697	0.640	0.549
FlexKGNN$_{PKG}$ + C	**0.728**	**0.728**	**0.830**	**0.830**	**0.750**	**0.750**
FlexKGNN$_{OpenIE}$ + C	0.645	0.633	0.703	0.702	0.639	0.632
FlexKGNN$_{SRL}$ + C	0.655	0.651	0.688	0.683	0.699	0.690
FlexKGNN$_{DBPedia}$ + C	0.625	0.592	0.646	0.638	0.721	0.717
RoBERTa (R)	0.846	0.845	0.883	0.883	0.795	0.720
FlexKGNN$_{PKG}$ + R	**0.917**	**0.915**	**0.935**	**0.935**	0.840	0.840
FlexKGNN$_{OpenIE}$ + R	0.814	0.815	0.906	0.898	0.836	0.830
FlexKGNN$_{SRL}$ + R	0.863	0.865	0.906	0.906	0.859	0.853
FlexKGNN$_{DBPedia}$ + R	0.848	0.848	0.896	0.890	**0.874**	**0.872**

Results: As depicted in Table 3, FlexKGNN$_{PKG}$ exhibits considerable improvement in misinformation detection when integrated with existing classifiers. With the COVID-19 dataset, both Check-It (C) and RoBERTa (R) see enhanced performance through integration with FlexKGNN$_{PKG}$, with RoBERTa F1 score peaking at 0.916, marking an $\approx$8% enhancement. This trend becomes more notable in the Politifact and GossipCop datasets, yielding performance increases of 19.70% and 26.64% with Check-It, and 4.34% and 12.84% with RoBERTa, respectively. These results highlight the broad applicability of our approach, and the potent contribution of polarization knowledge in misinformation detection.

6.3 Polarization Knowledge Role in Misinformation Detection

Following, we evaluate the impact of the polarization-specific PKG on misinformation detection, contrasting it with broader KGs obtained through knowledge extraction techniques. To establish a comparison, we employ Open Information Extraction (OpenIE) [2], Semantic Role Labeling (SRL) [27], and DBPedia [18] as our foundational knowledge baselines. **OpenIE** is a tool that employs a series of NLP methods and syntactical dependency rules to identify actors and their relations from text. For example, given the sentence: "Anthony Fauci emphasizes the need for a mask mandate", OpenIE discerns the triple of ("Anthony Fauci", "emphasizes", "the need for a mask mandate"). **Semantic Role Labeling (SRL)** is an NLP method that identifies semantic roles in sentences, emphasizing on actors and their actions. While OpenIE derives SPO triples using syntactical rules, SRL captures deeper entity relationships. For example, in "President Trump spent months playing down mask effectiveness", SRL distinguishes the actor "President Trump", the action "spent", and the related activities "months" and "playing down mask effectiveness", yielding two triples: ("President Trump", "spent", "months") and ("President Trump", "spent", "playing down masks effectiveness"). **DBPedia** is a knowledge graph that captures Wikipedia entries in a structured format, facilitating the semantic querying of their relationships and properties. To extract triples from text using DBPedia, we initially apply Named Entity Recognition (NER), where named entities (i.e. actors) within the text are identified. Following this, we use DBPedia Spotlight [18] to link these actors to corresponding DBPedia resources. After these actors are linked, we query DBPedia for possible relationships between them. By applying this process on a sentence such as "Anthony Fauci is the leader of the National Institute of Allergy and Infectious Diseases", the result would be ("Anthony_Fauci", "Leader", "National_Institute_of_Allergy_and_Infectious_Diseases").

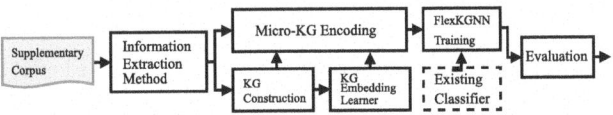

Fig. 7. Methodology for general KG encoding of MD.

Baseline KGs Construction: For each knowledge source, we employ the methodologies of Sects. 3 and 4 to construct the primary KG and encode the MDs into micro-KGs (see Fig. 7). The resulting triples from OpenIE and SRL exhibit inconsistencies, as they represent the same actor differently in text (e.g., "Donald Trump" and "President Trump"). To address this, we leverage clustering based on contextualized embeddings for both actors and predicates [26]. By clustering subjects and objects after extracting all triples, we consolidate different textual representations of an entity, like "Donald Trump," into a single, unified representation. This method is similarly applied to predicates, ensuring consistency in our knowledge representation. For DBPedia, triples already have unified representation for each actor and predicate. To construct the primary KG and the micro-KGs, we follow the methodology outlined in [8]. Utilizing the constructed KGs and encoded micro-KGs, we train instances of FlexKGNN$_{OpenIE}$, FlexKGNN$_{SRL}$ and FlexKGNN$_{DBPedia}$.

Results: As shown in Table 3, while the combinations involving FlexKGNN$_{PKG}$ consistently surpass the existing baselines, those that incorporate SRL, OpenIE, and DBPedia, only achieve a performance comparable to that of the baseline classifiers, occasionally decreasing their performance, such as the 11.59% decrease in F1 score observed with FlexKGNN$_{OpenIE}$ on the COVID-19 MD. The only exception occurs with the FlexKGNN$_{SRL}$ and FlexKGNN$_{DBPedia}$ when integrated with RoBERTa on the GossipCop MD, achieving a ≈15% increase, compared to the 12.84% of the FlexKGNN$_{PKG}$. Overall, models integrated with PKG outperform those with KGs due to their fundamental differences. The PKG effectively captures polarization knowledge in the context of the MD, in contrast to general KGs, which often echo information already seen by the existing classifiers. Thus, while general KGs are useful in various settings, the specialized PKG is more effective at detecting misinformation in polarized environments.

6.4   Performance Comparison with External KG Approaches

Additionally, we compare the performance of our methodology with existing approaches that utilize KGs in combination with GNN models for misinformation detection. These models are: i) **KAPALM** [17], a GNN model that fuses coarse- and fine-grained actor KG representations in combination with article content for knowledge-aware misinformation detection, achieving F1 scores of 0.913 on Politifact and 0.717 on GossipCop; ii) **KAN** [8], a knowledge-aware attention GNN which incorporates KG actors to predict the veracity of articles, achieving F1 scores of 0.872 on Politifact and 0.774 on GossipCop; and iii) **KGF** [32], a compositional GNN, which uses OpenIE to extract actors and their relationships from articles, classifying them using graph convolutions, achieving F1 scores of 0.853 on Politifact and 0.723 on GossipCop.

Results: In comparison with the performances in Table 3, FlexKGNN PKG + R outperforms the aforementioned models on both datasets, highlighting the efficacy of integrating polarization knowledge with advanced DL techniques.

6.5   Polarization Knowledge Ablation Study

To understand the individual polarization predicate contributions, we conduct an ablation study, considering the micro-PKGs without: i) polarization predicates, and ii) embeddings. To neutralize the polarization knowledge in the PKG, we first remove the *Dipole* actors and their related predicates, which signify the conflict between fellowships. Specifically, we eliminate the predicates *PartOf, HasModeratePolarization, HasMediumPolarization*, and *HasExtremePolarization*. To obscure clear signs of opposition or support, we generalize the remaining attitude predicates by merging *SupportEE* and *OpposeEE* into *AttitudeEE*, *SupportET* and *OpposeET* into *AttitudeET*, and *SupportFT* and *OpposeFT* into *AttitudeFT*. This neutralization process is similarly applied to the micro-PKGs. As a result, the modified PKG no longer explicitly captures polarization knowledge.

Table 4. Ablation study results on model performance.

Ablation	COVID-19		Politifact		GossipCop	
	Acc.	F1	Acc.	F1	Acc.	F1
w/out Polarization Predicates + C	0.694	0.632	0.758	0.758	0.719	0.716
w/out Embeddings + C	0.723	0.723	0.694	0.693	0.698	0.695
w/out Polarization Predicates + R	0.908	0.908	0.903	0.903	0.828	0.825
w/out Embeddings + R	0.906	0.906	0.922	0.921	0.825	0.824

Results: As Table 4 indicates, there is a noticeable performance drop across all MDs when these elements are omitted. Specifically, the absence of polarization predicates in the FlexKGNN PKG + C setup leads to ≈6% decrease in effectiveness. Similarly, discarding embeddings results in a ≈8.5% decrease. The performance of FlexKGNN PKG + R without these elements remains robust across the MDs, although a slight reduction of ≈2% is still observed, demonstrating the intrinsic strength of the RoBERTa classifier. These results underscore the added value of polarization and embeddings for misinformation detection.

7 Conclusion and Future Work

In this study, we propose PARALLAX, a methodology that leverages polarization knowledge for improved misinformation detection. Using our FlexKGNN model, augmented with PKG, consistently outperforms methods based on general KGs, achieving an average of ≈15% improvement when integrated with existing classifiers. This demonstrates the effectiveness of incorporating polarization into misinformation detection. While our findings are promising, we acknowledge there are areas for improvement. We plan to extend our evaluation to larger, more diverse datasets to ensure robust assessment and explore the adaptability of our approach to domains with limited or shifting polarization. Additionally, we aim to integrate PARALLAX with other state-of-the-art models, including Convolutional Neural Networks (CNN) and Large Language Models (LLM) [16]. To provide deeper justification for polarization contribution, we will employ explainable AI techniques such as GNNExplainer [34]. This will allow us to identify and analyze the PKG triples that contribute most to misinformation classification, helping us to further understand their intertwined nature.

Acknowledgments. This research is funded in part by the EU Commission via the ATHENA 101132686 project (HORIZON-CL2-2023-DEMOCRACY-01).

References

1. Allcott, H., Gentzkow, M.: Social media and fake news in the 2016 election. JEP **31**, 211–236 (2017)
2. Angeli, G., Johnson, P., Melvin, J., Manning, D.: Leveraging linguistic structure for open information extraction. In: IJCNLP (2015)
3. Aref, S., Neal, Z.: Detecting coalitions by optimally partitioning signed networks of political collaboration. Sci. Rep. **10**, 1506 (2020)
4. Balazevic, I., Allen, C., Hospedales, T.: TuckER: tensor factorization for knowledge graph completion. In: EMNLP-IJCNLP (2019)
5. Bramson, A., et al.: Understanding polarization: meanings, measures, and model evaluation. Philos. Sci. **84**(1), 115–159 (2017). https://doi.org/10.1086/688938
6. Deane, C., Parker, K., Gramlich, J.: A Year of U.S. Public Opinion on the Coronavirus Pandemic. Pew Research Center (2021)
7. DiMaggio, P., Evans, J., Bryson, B.: Have American's social attitudes become more polarized? Am. J. Sociol. **102**, 690–755 (1996)

8. Dun, Y., Tu, K., Chen, C., Hou, C., Yuan, X.: KAN: knowledge-aware attention network for fake news detection. In: AAAI (2021)
9. Garimella, K., Morales, G.F., Gionis, A., Mathioudakis, M.: Quantifying controversy in social media. Trans. Soc. Comput. **1**, 1–27 (2018)
10. Gilardi, F., Alizadeh, M., Kubli, M.: ChatGPT outperforms crowd workers for text-annotation tasks. In: PNAS (2023)
11. Gillani, N., Yuan, A., Saveski, M., Vosoughi, S., Roy, D.: Me, my echo chamber, and i: introspection on social media polarization (2018)
12. Guerra, P., Meira, W., Cardie, C., Kleinberg, R.: A measure of Polarization on social media networks based on community boundaries. In: ICWSM (2013)
13. Hart, P.S., Chinn, S., Soroka, S.: Politicization and polarization in Covid-19 news coverage. Sci. Commun. **42**, 679–697 (2020)
14. He, Z., Mokhberian, N., Camara, A., Abeliuk, A., Lerman, K.: Detecting polarized topics using partisanship-aware contextualized topic embeddings. EMNLP (2021)
15. Hu, L., et al.: Compare to the knowledge: Graph neural fake news detection with external knowledge. In: IJCNLP (2021)
16. Islam, M.R., Liu, S., Wang, X., Xu, G.: Deep learning for misinformation detection on online social networks: a survey and new perspectives. Soc. Netw. Anal. Min. **10**(1), 1–20 (2020). https://doi.org/10.1007/s13278-020-00696-x
17. Ma, J., Chen, C., Hou, C., Yuan, X.: KAPALM: knowledge graph enhanced language models for fake news detection. In: EMNLP (2023)
18. Mendes, P.N., Jakob, M., Silva, A., Bizer, C.: DBpedia spotlight: shedding light on the web of documents. In: I-SEMANTICS (2011)
19. Morales, A., Borondo, J., Losada, J., Benito, R.: Measuring political polarization: Twitter shows the two sides of Venezuela. Chaos (2015). https://doi.org/10.1063/1.4913758
20. Morkūnas, M.: Russian disinformation in the baltics: does it really work? Public Integrity (2023)
21. Osmundsen, M., Bor, A., Vahlstrup, P.B., Bechmann, A., Petersen, M.B.: Partisan polarization is the primary psychological motivation behind political fake news sharing on Twitter. Am. Polit. Sci. Rev. **115**, 999–1015 (2021)
22. Paschalides, D., Pallis, G., Dikaiakos, M.: Polar: a holistic framework for the modelling of polarization and identification of polarizing topics in news media. ASONAM (2021)
23. Paschalides, D., et al.: Check-It: a plugin for detecting fake news on the web. OSNEM (2021)
24. Pavlov, T., Mirceva, G.: Covid-19 fake news detection by using BERT and roBERTa models. In: MIPRO (2022)
25. Przybyła, P.: Capturing the style of fake news. In: AAAI (2020)
26. Reimers, N., Gurevych, I.: Sentence-BERT: sentence embeddings using Siamese BERT-networks. In: EMNLP-IJCNLP (2019)
27. Shi, P., Lin, J.: Simple BERT models for relation extraction and semantic role labeling. arXiv preprint (2019)
28. Shu, K., Wang, S., Liu, H.: Beyond news contents: the role of social context for fake news detection (2019)
29. Tajfel, H., Turner, J.: An integrative theory of intergroup conflict. Soc. Psych. Inter Rel. **33**, 33–37 (1979)
30. Vicario, M.D., Quattrociocchi, W., Scala, A., Zollo, F.: Polarization and fake news: early warning of potential misinfo. targets. TWEB (2019)
31. Wolpert, D.H.: Stacked generalization. Neural Netw. **5**, 241–259 (1992)

32. Wu, K., Yuan, X., Ning, Y.: Incorporating relational knowledge in explainable fake news detection. In: Karlapalem, K., Cheng, H., Ramakrishnan, N., Agrawal, R.K., Reddy, P.K., Srivastava, J., Chakraborty, T. (eds.) PAKDD 2021. LNCS (LNAI), vol. 12714, pp. 403–415. Springer, Cham (2021). https://doi.org/10.1007/978-3-030-75768-7_32

33. Wu, Z., Pan, S., Chen, F., Long, G., Zhang, C., Yu, P.S.: A comprehensive survey on graph neural networks. IEEE Trans. Neural Netw. Learn, Syst (2021)

34. Ying, Z., Bourgeois, D., You, J., Zitnik, M., Leskovec, J.: GNNExplainer: generating explanations for graph neural networks. In: NIPS (2019)

35. Zhou, X., Zafarani, R.: A survey of fake news: Fundamental theories, detection methods, and opportunities. ACM Comput. Surv. **53**, 1–40 (2020)

36. Zollo, F.: Dealing with digital misinformation: a polarised context of narratives and tribes. In: EFSA (2019)

Dynamic Inter-organizational Communication Network in a Post-merger Integration

Michael Benzinger[1]([✉])[iD], Raji Ghawi[1][iD], Lukas Zenk[2][iD], and Jürgen Pfeffer[1][iD]

[1] Technical University of Munich, Arcisstraße 21, 80333 Munich, Germany
{michael.benzinger,raji.ghawi,juergen.pfeffer}@tum.de
[2] Danube University Krems, Dr.-Karl-Dorrek-Straße 30, 3500 Krems, Austria
lukas.zenk@donau-uni.ac.at

Abstract. During a Post-Merger Integration, the communication structure of an organization changes profoundly. Effective communication and personal interaction are crucial for the success of Mergers and Acquisitions, as many such endeavours fail to achieve their targets. Mergers and Acquisitions involve the combination of two organizations to create a new entity. Creating personal networks across organizational boundaries can build bridges between several organization entities to create a cohesive, adjustable and resilient holistic organization and leverage synergies. However, little actual network data on communication and network structure during Post-Merger Integration is available. Our results are based on two merged organizational units with approximately 3,800 employees in total and a collective sum of around 250,000 email connections from an actual Post-Merger Integration data set. We show that various communication networks stabilize over time in Post-Merger Integration and, that homophily is partially reflected in the informal social network. We also find that homophily occurs on a department level but somewhat less on an organizational level and that some critical departments communicate less than expected. Our results indicate that the formal merger is reflected in the communication and ego-network structure. We expect that our findings will add insights into Post-Merger Integration research and how networks are adapting, since the findings are based on actual organizational network data.

Keywords: Dynamic Networks · Network Stabilization · Communication · Homophily · Mergers & Acquisitions · Post-Merger Integration

1 Introduction

Mergers and acquisitions (M&As) have been a crucial part of strategic organizational decisions and business expansion since the late 1800s and can be seen as the consolidation of organizations [9,13,26]. Companies often use M&As

L. M. Aiello et al. (Eds.): ASONAM 2024, LNCS 15211, pp. 106–123, 2025.
https://doi.org/10.1007/978-3-031-78541-2_7

to strengthen their market position, enter new markets and regions, benefit from economies of scale, achieve synergies and expand their product portfolio [1,3,12,14,21]. Despite the fact that nearly half of all M&As fail to achieve their desired goals, it is still a prevalent strategic choice for decision-makers because the potential rewards can outweigh the risks involved, when successful conducted [1,17,22,24].

Organizational structures have a significant impact on the performance of the merged organizations [22,23]. Previous studies propose that the lack of cultural, structural and communicational integration, lowers financial and operational performance and leads to various organizational dysfunctions, such as employee turnover, interpersonal conflicts, organizational communication ineffectiveness, as well as collaborative and hierarchical misfits [1,24,25,27]. Therefore, a solid integration of merged organizations is one of the critical success factors, also seen as the paramount phase to successfully conduct a merger or acquisition [4].

Because so many M&As fail to fulfil planned objectives, research has shifted to the human side of M&A to understand the sociological, psychological, and behavioural aspects of M&As and the effects of the post-merger integration on employees [22]. Quantifying and analyzing human relationships through organizational network analysis provides a profound understanding of human interactions, social networks and communication patterns based on the relations they form [18]. Human relationships in organizational structures can be represented as a network where individuals are connected by interactive relationships, exchanging and communicate information throughout the organization [11].

We investigated the changes in the dynamic inter- and intra-organizational communication structures in a post-merger integration scenario. This is done through the help of data obtained from an actual merger of two large organizations in the sector of advanced technology solutions. Our research aims to answer the following questions:

1. Does the communication social network of a post-merger integration become more stable (cohesive) over time? We explore how the initial network communication stabilizes across various organizational departments and across organizational boundaries over time. We propose an approach to evaluate the relative phase rise difference of initial communication.

2. Are individuals more inclined to engage and communicate predominantly with others from the same organizational department, regardless of the legacy organization? We analyze indications of homophily and E/I-Indices between departments, and their presence across organizational boundaries. Additionally, we investigate which departments are expected to be among the first and most important ones to form communication inter-company bonds.

This paper is organized as follows. Section 2 provides an overview of related work, while Sect. 3 introduces our dataset and gives a formulation for data modelling. We address network stabilization in Sect. 4, and cross organizational interaction in Sect. 5.

2 Related Research

2.1 Mergers and Acquisitions

Mergers and Acquisitions have been extensively studied during the last decades [13, 21, 26]. A merger is a combination of two organizations to form a new entity, while an acquisition is when one company purchases another through stock or asset acquisition. [12]. Many researches have investigated aspects of social network aspects and M&As within an organizational context. Woehler et al. 2021 [26] examined the changes organizations and individuals, processes, and structures undergo in a merger indicating that increasing cross-legacy social connections among employees can reduce turnover after a merger. Öberg et al. (2007) [19] investigate how mergers and acquisitions reshape business networks, emphasizing shifts in managerial cognition and networking behaviours in post-transactions. Additionally, Seo et al. (2005) [22] evaluate several theories to explain challenges in organizational change of managing mergers and acquisitions, creating a framework that identifies sources of problems, their effects on employees such as stress, illness, turnover, layoffs, and loss of know-how and competencies.

2.2 Post-merger Integration

The Post-Merger Integration (PMI) is the most critical phase in a M&A process to ensure a successful venture, covering all relevant cultural, communicational, organizational and procedural aspects [1, 5, 9]. Graebner et al. (2017) [9] examine previous research on the integration phase after a merger. They suggest focusing on the understanding of the processual dynamics, particularly in the areas of temporality, decision-making, practices, and tools within the PMI. Fabac (2011) [3] promotes employing social network analysis to understand network development and proposes PMI leveraging mixing communication patterns to enhance integration success, underscoring the importance of the organizational design. Utilized insights from social networks Marmenout et al. (2015) [17] examining the impact of brokerage and communicational contagion mechanisms on organizational function and detailed HR interventions across different stages of the PMI process. Furthermore, Franzt et al. (2009) [6] apply a contemporary approach to study the effect of organizational complexity on PMI, finding that performance during this period is influenced by the pre-existing complexities of the merging organizations and the number of work groups. In another paper, Frantz et al. (2012) [4] investigate the pivotal role of internal individual social networks within organizations, the disruption of individuals' social networks during the PMI and its strive for re-stabilization towards a solid number of ego-ties. Yamanoi et al. (2012) [27] explore how cultural integration emerges from social interactions among individuals in a PMI, influencing individual turnover, interpersonal conflict, and organizational communication effectiveness.

2.3   Organizational Communication and Ego-Networks

Analyzing the changes in inter-organizational communication through organizational network analysis requires the dissection and evaluation of the ego networks of individuals reflecting organizational communication on the most profound level [4]. Nurek et al. (2020) [18] explore how analyzing employee email exchanges and forming an informal social network can reflect and potentially reproduce the organizational structure. They demonstrated the revealing nature of communication email metadata in social network construction. Zenk et al. (2010) [28] promote that the change organizations are undergoing is represented in the communication structure. However, email communication has its limitations. Personal communication and interactions may not be reflected in email exchanges or other communication forms, such as messaging apps or text messages. Therefore, it is essential to consider that a dataset of emails may only provide a partially comprehensive understanding of the connections between individuals [18]. Nevertheless, Heo et al. (2002) [11] promote that individuals need to build both formal and informal connections with knowledgeable people from other parts of the organization in a PMI scenario. Additionally, Gelardi et al. (2021) [7] emphasized that investing time in building or reinforcing new social relationships with someone may come at the expense of neglecting relationships with others, creating a stabilized ego-network structure with an equilibrium of ego-ties over time. This is also supported by the findings of Frantz (2012) [4], indicating that an individual can maintain an average of social relationships between 150 and 290 individuals resulting in the stabilization of ego-networks.

2.4   Homophily

One of the critical factors for a successful deal is the existence of reciprocal interest between the decision-makers, indicating that people are more likely to build social ties that last longer, with people with similar interests, exhibiting higher homophily [25]. Henning et al. (2012) [10] depict homophily as a fundamental sociological concept indicating that individuals will be more likely to bond and associate with similar others. Burt (1987) [2] found that individuals resembling network positions in organizations behave alike, and further that individuals who occupy structural equivalence in a network usually have similar attributes, which explains why they are placed in specific positions. According to Vaidya et al. (2020) [25], M&As are more likely to happen when there is a higher level of homophily. They found similar characteristics, such as age, gender, and education level of CEOs have enhanced, the post-merger performance and shareholder value. Therefore, the study highlights the importance of matching CEOs based on homophily. Frantz et al. (2009) [6] promote that people tend to seek advice or opinions from those who perform similar tasks or possess similar knowledge in an organization. Furthermore, Lawrence (2020) [16] propose joint interactions of multiple attributes create associated trust and bonding effects, indicating a sense of similarity leading to homophily.

3 Data and Modeling

3.1 Description of Dataset

For this research, we utilize a data set from an actual PMI, comprising information from a company (hereinafter mentioned as Acquirer) in the field of advanced technology solutions that has acquired a company (hereinafter mentioned as Asset) in the same industry sector. The data set encompasses email metadata collected over a period of six months subsequent to the acquisition day, referred to as Day One. The Acquirers organization consists of 1,983 individuals, while the Assets organization consists of 1,824. The individuals of both organizations are spread across U.S. and India.

Collectively, the dataset encapsulates 254,405 email-based connections between the individuals, where these interactions are presented in an interval compression data set of a temporal network. That is, each interaction between two individuals is associated with: the initial time point of communication between them, the final time point of communication, and the cumulative frequency of communication exchange during that designated period.

Additionally, the data set comprises additional information about the individuals, such as their affiliation to the respective department of their primary organization. In total, the organization is divided into 18 departments. These are: Executive & Administrative Support (E&AS), Communications, Facility & Real Estate, Finance (FI), General Management (GMgmt), Human Resources (HR), Information Technology (IT), Legal & Patents (L&T), Marketing, Operations, Procurement, Quality, Research & Development (R&D), Sales, Security, Strategy, Supply Chain and Sustainability).

3.2 Data Analysis

In our scenario, complete temporal network data as a list of temporal edges is unavailable due to data protection and privacy reasons. Instead, the only available data comprises a list of edges, in which each edge is associated with a timestamp of the first and last interaction, as well as the frequency of interactions. That is, each connection between two individuals u_1 and u_2 is associated with a triple $\langle t_i, t_f, f \rangle$, where:

- t_i: the initial time point of communication between u_1 and u_2,
- t_f: the final time point of communication, and
- f: the cumulative frequency of communication exchange during that designated period.

However, since our aim is to understand the effects of PMI by investigating the network's temporal evolution and potential stabilization, we focus only on the initial point where the communication of two individuals takes place (i.e., edge activation). For the analyses of this paper, we adopt week as temporal, in order to simplify the analysis and keep the focus on the big picture.

Thus, the communication network can be represented as a temporal graph $G(\mathcal{V}, \mathcal{E}, \omega)$ where $\mathcal{V}$ is the set of nodes (individuals), and $\mathcal{E}$ is the set of edges, $\mathcal{E} \subseteq \mathcal{V} \times \mathcal{V}$, and $\omega : \mathcal{E} \to \mathcal{T}$ is a function that associates each edge (u, v) with its timestamp $\omega(u, v)$ (defined as the initial connection on a weekly basis). Here $\mathcal{T}$ denotes the time-span under study (six months) composed of 23 weeks.

Moreover, individuals are associated with their affiliation to the respective department of their primary organization. Let $\mathcal{O}$ be the set of organizations, i.e., $\mathcal{O} = \{O_1, O_2\}$, where O_1 is the Acquirer and O_2 is the Asset organizations. Let $\mathcal{D}$ be the set of departments, e.g., HR, R&D, etc. Individuals' affiliation is defined by two functions:

- Organization affiliation function: from the set of individuals to the set of organizations: $\gamma : \mathcal{V} \to \mathcal{O}$,
- Department affiliation function: from the set of individuals to the set of departments: $\delta : \mathcal{V} \to \mathcal{D}$,

Thus, the affiliation of an individual $v \in \mathcal{V}$ is jointly specified by her/his organization $\gamma(v)$ and department $\delta(v)$.

Additionally, the set of individuals of department d of organization o is denoted as $\phi(o, d)$, and given by:

$$\phi(o, d) = \{v \in \mathcal{V} \mid \gamma(v) = o, \delta(v) = d\}$$

From the joint organization-department point of view, we consider four kinds of dynamic temporal networks (as illustrated in Fig. 1):

Intra-organizational Networks

- **Same Organization, Same Department | SOSD.** This network encompasses the communication of individuals within the same organization and the same department (e.g. all communications in the Asset's HR department).

$$\mathcal{E}^{\text{SOSD}} = \{(u, v) \in \mathcal{E} \mid \gamma(u) = \gamma(v), \delta(u) = \delta(v)\}$$

- **Same Organization, Different Department | SODD.** This network encompasses the communication of individuals within the same organization and all other departments (e.g. all communications between the Asset's HR department and every other department of the Asset).

$$\mathcal{E}^{\text{SODD}} = \{(u, v) \in \mathcal{E} \mid \gamma(u) = \gamma(v), \delta(u) \neq \delta(v)\}$$

Inter-organizational Networks

- **Different Organization, Same Department | DOSD.** This network encompasses the communication of individuals between the organizations of the Acquirer and the Asset but between the same department (e.g. all communications between both HR departments of the Acquirer and the Asset)

$$\mathcal{E}^{\text{DOSD}} = \{(u, v) \in \mathcal{E} \mid \gamma(u) \neq \gamma(v), \delta(u) = \delta(v)\}$$

- **Different Organization, Different Department | DODD.** This network encompasses the communication of individuals within one department of one organization (e.g. the Acquirer) with all other departments of the other organization (e.g. all communications in the Acquirer's HR department with every other Asset's department)

$$\mathcal{E}^{\text{DODD}} = \{(u,v) \in \mathcal{E} \mid \gamma(u) \neq \gamma(v), \delta(u) \neq \delta(v)\}$$

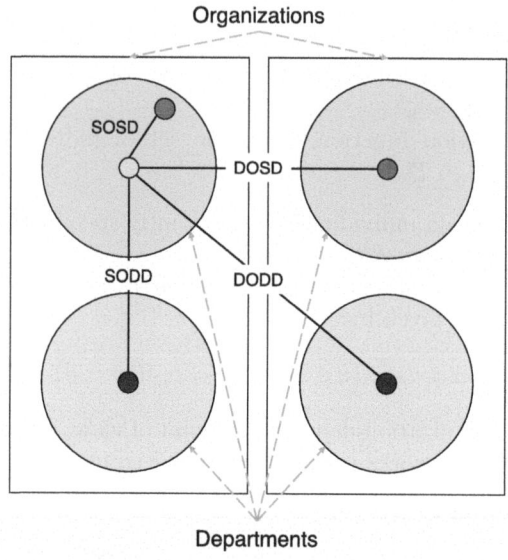

Fig. 1. Four types of networks, between same/different organization, and same/different department.

One needs here to bear in mind, that if we consider only an organisational point of view, the SOSD and SODD networks cover *internal* networks within one organization, while the DOSD and DODD networks cover *external* communication bridging organizational boundaries.

Since the data is only available in interval-compression format (initial- and final points, plus the frequency), we can either approximate the individual timestamps of each edge; or simply work with accumulative networks [8]. In this paper, we opt to follow the second option, i.e., accumulative networks, in order to avoid approximation and its potential inaccuracies.

That is, for a given week $k \in \mathcal{T}$, we construct an accumulative network that comprises all connections occurring from Day One up to, and including, week k.

$$\mathcal{E}_k = \{(u,v) \in \mathcal{E} \mid \omega(u,v) \leq k\}$$

Clearly, this kind of temporal accumulative networks can be constructed at different granularity levels:

- Global level: comprising all connections entirely.
- Organization level: separately consider intra-organization connections (SOSD and SODD together) and inter-organization connections (DOSD and DODD together).
- Department level: separately consider the four kinds of networks: SOSD, SODD, DOSD and DODD.

4 Network Stabilization

In order to seek answers to our research questions about the stabilization of the post-merger communication networks, we opt to separate the initial timestamp interactions into two phases (periods) covering the near and far frames of the entire time span:

- Phase 1: covering week 1 until 9, and
- Phase 2: covering week 14 until week 23.

This way, studying the network stabilization is reduced to merely comparing the network growth rates between these two phases.

For each phase, we consider how the network changes in size – number of connections – in every week in the phase. Thus, each network can be represented as two curves for the two phases respectively depicting the evolution of the network. That is, let $T_1 = \{1, \cdots, 9\}$ and $T_2 = \{14, \cdots, 23\}$ denote the two (near and far) phases, respectively; then the two fluctuating curves are:

- $Y_1 = \{|\mathcal{E}_k| \text{ for } k \in T_1\}$
- $Y_2 = \{|\mathcal{E}_k| \text{ for } k \in T_2\}$

Remarkably, these fluctuating curves tend to be linear in the different kinds of networks as shown in Fig. 2, and Fig. 3. The immediate consequence of this observation is the following: the network steadily grows in a given phase, and this growth is different between the two phases. This enables us to analyze the trends in network development over time and to assess the stabilization of the network. Moreover, the linearity of these curves allows us to use linear regression to express each curve as a straight line, which can be easily specified in terms of its intercept and slope.

Our approach involves fitting each fluctuating curve with linear regression, not only to compare the various networks with each other but also to analyze the trends observed.

We are investigating the stability of a dynamic network by comparing the linear regression of the network during its first (near) phase to that of its second (far) phase. This comparison helps us determine whether the network has reached stability or not. Furthermore, we can determine whether the network has stabilized by assessing the slopes of both intervals. The slope of the linear regression represents the growth rate of the network, that is the increase in number of communications per unit time (here: week). We use regression analysis to evaluate the network's stability and compare each regression's slopes, which allows us to investigate the network's stabilization over time and determine its current stability status.

At organizational level; Fig. 2 depicts the network growth fluctuating curves, along with the linear regression fit, specifically for the internal (within the same organization) and external (between the two organizations) organizational networks. We can see that, as expected, the internal network is more dense than the external one. However, when we compare (in each network) the two phases, we observe that the first phase has a higher growth rate than the second phase.

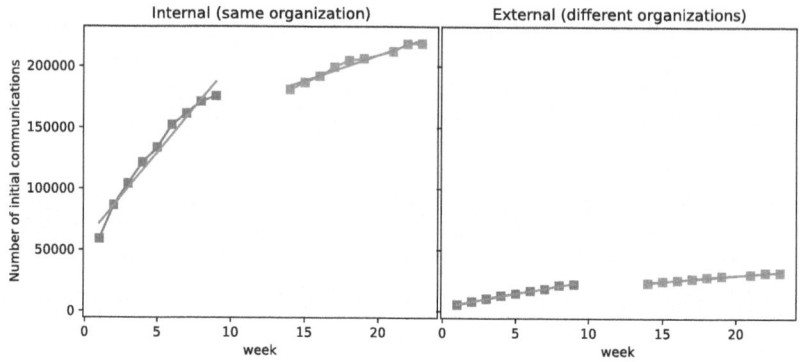

Fig. 2. Accumulated initial communications at Organization Level

At department level; Fig. 3 depicts the network growth fluctuating curves, with linear regression fits, for all four networks: SOSD, SODD, DOSD and DODD. Here again, we observe in all networks, that the first phase has a higher growth rate than the second phase.

If the slope of the first phase is higher than the slope of the second phase, it indicates that the network has reached a stable state over time. For a network, having a higher growth rate in the first phase than in the second phase is in fact a strong indication of the stabilization of the network. Actually, this means that the network has stabilized and reached most of its magnitude by the first phase, and has reduced its growth rate in the second phase. Remarkably, this applies for all the networks at organization- and department levels.

In order to verify and to assess to which extent the network is stable, we compare the growth rates of both fluctuating curves for each network.

To quantify the stabilization of the network, we use the relative difference of the slopes of the two phases. We call this quantity: the "Reduction Ratio of fluctuating Growth Rates" (RRFGR for short), which mainly measures the reduction in growth rates within different phases.

Let S_1 and S_2 be the slopes of the linear regression (growth rates) in the first and second phases, respectively. Then, the Reduction Ratio of fluctuating Growth Rates is given by the following formula:

$$\text{RRFGR} = \frac{S_1 - S_2}{S_1} \times 100 \qquad (1)$$

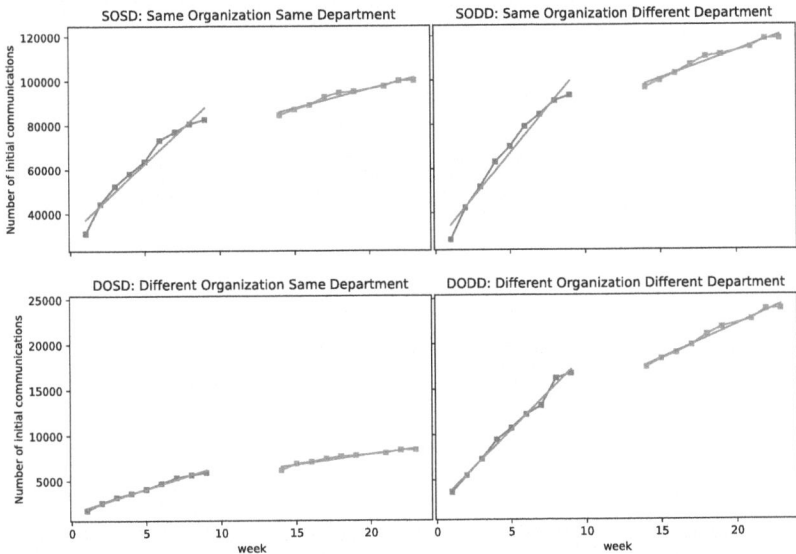

Fig. 3. Accumulated initial communications at Department Level

where we mainly take the difference of the slopes (how much the first phase is reduced at the second phase) and normalize it by the slope of the first phase. We then multiply the result by 100 to get a percentage, which helps us measure the relative change between the two slopes.

Table 1 presents for all networks, at organization and department levels, the growth rates (slopes) of the two phases, as well as the Reduction Ratio of fluctuating Growth Rates (RRFGR) as a stabilization measure.

We first observe on an Organizational Level a larger RRFGR for internal networks (70.92%) than for external networks (55.36%). A higher RRFGR indicating a reduction of initial communication resulting in a more stabilized network communication. Furthermore, we observe a slight shift regarding the share of total communication with an decrease of internal communication from 88.59% to 87.15% and a increase of external communication share from 11.41% to 12.85%. Implying the initial internal communication was already established before the Day One but the share of total communication is shifting across former organizational boundaries.

Secondly, we observe a continuous concentric decrease of RRFGR from SOSD (72.74%), SODD (69.50%), DOSD (56.44%), to DODD (55.02%) reflecting the findings of the Organizational Level.

Remarkably, the share of internal communications decreases in favor of an increasing share of external ones. As shown in Table 1, on one hand, the communication within SOSD network is reduced from 41.66 % in the first phase to 39.91% in the second one (decrease by 1.75 %). On the other hand, the share of communications have increased within the three other networks: SODD

Table 1. Regression slopes, phase rise difference (PRD) and share of overall Communication

Network	Organization Level				
	Slope 1	Slope 2	RRFGR	Share 1	Share 2
internal	14435.4	4197.5	70.92 %	88.59 %	87.15 %
external	2165.45	966.6	55.36 %	11.41 %	12.85 %
Network	Department Level				
	Slope 1	Slope 2	RRFGR	Share 1	Share 2
SOSD	6316.55	1721.58	72.74 %	41.66 %	39.91 %
SODD	8118.85	2475.91	69.50 %	46.94 %	47.24 %
DOSD	519.55	226.33	56.44 %	8.43 %	9.50 %
DODD	1645.9	740.28	55.02 %	2.98 %	3.35 %

increased from 46.94% to 47.24% (by 0.3%), DOSD increased from 8.43% to 9.5% (by 1.07%), and DODD increased from 2.98% to 3.35% (by 0.37%), indicating a significant tendency of making more new external communications and consequently the integration of the two organizations reflecting the formal merger in the informal communication network.

5   Cross Organizational Interaction

Examine the collaboration of individuals within and across the organisation and to answer the question of homophily between the same departments, we proceed as follows: We create the E/I index for all departments, in both organisations. The E/I index provides information on how successful the merger was in linking the two organisations' social networks. We compare the accumulated initial communication of all departments and look at the top 5 departments in detail under the premise that there are specific departments that are more important for a successful integration than other departments, which should be reflected in the level of initial communication at the beginning of the integration phase [20]. We analyse the communication patterns of the top 5 departments (for both organisations) within the four networks to identify possible differences and to evaluate whether the formal merger is also reflected in the informal networks. Furthermore, we also analyse which of the other specific departments the top 5 departments communicate outside their own network gaining a holistic perspective on organizational department communication.

5.1   E/I-Index

Our objective is to investigate whether the formal merger process and, specifically, the PMI is conducted successfully and whether they are reflected in the network data. To measure the degree to which both organizations are integrated, we consider an E/I-Index as the central measure. A high E/I-Index indicates successful integration, while a low index may indicate challenges or obstacles in merging the networks.

The E/I-Index after Krackhardt (1998) measures the ratio of external and internal connections and normalizes this to a value range between -1 and +1 [15].

We calculate the E/I-Index as follows:

$$EI = \frac{E - I}{E + I}$$

where E represents the number of external connections and I represents the number of internal connections.

We focus on the E/I-Index from a departmental point of view, where internal connections I are within the same department of the same organization (SOSD), whereas external connections E are between the different organizations but still in the same department (DOSD). Thus, for a given department $d \in \mathcal{D}$, the E/I index is calculated using the specific internal- and external indices of that department:

$$I_d = |\{(u,v) \in \mathcal{E} \mid \delta(u) = \delta(v) = d, \gamma(u) = \gamma(v)\}|$$
$$E_d = |\{(u,v) \in \mathcal{E} \mid \delta(u) = \delta(v) = d, \gamma(u) \neq \gamma(v)\}|$$

We propose that homophily (and therefore a successful PMI) will be reflected explicitly through a high E/I-Index in this network.

The results of the E/I-Index of the DOSD Network can be seen in Table 2. We observe a broad range of results, reaching from -1.0 up to 0.73, indicating almost the entire spectrum of homophily and heterophily. Departments like R&D, Marketing, Operations, and Supply Chain are similar in their substantial negative E/I-Index. While other departments like Sales, Human Resources, Executive & Administrative Support, Procurement, Legal & Patents and Sustainability show moderate reciprocal differences. Additionally, we observe that the departments Quality, Facility & Real Estate and General Management diverge substantially. This implies that the first group seems to have a heterophil network, while the second group indicate homophily in several departments in both organizations. At the same time, the last group of separate departments shows signs of homophily partially.

We can further see that for the five departments with the most communication sharing, the departments with mutual communication exchange share connections with similar departments. All five departments have ties with E&AS, HR, Marketing and R&D. All departments are among the top five departments, except Executive & Administrative Support, indicating a central role in the integration of all departments.

5.2   DOSD Network

Fuhrer et al. (2017) [20] have shown, that some departments are essential to long-term integration success. Conversely, this fact should also be reflected in the

Table 2. E/I-Index per Department in DOSD Network

Department	Asset	Acquirer
Research & Development	−0.93	−0.74
Quality	0.18	−0.87
Sales	0.30	−0.35
Marketing	−0.60	−0.55
Information Technology	−0.25	−0.16
Communications	NaN	−1.00
Human Resources	0.51	0.00
Executive & Administrative Support	0.58	−0.09
Finance	−0.10	−0.25
Facility & Real Estate	0.71	0.09
Operations	−1.00	−1.00
General Management	−0.11	0.70
Supply Chain	−0.94	−0.87
Procurement	0.18	−0.05
Legal & Patents	−0.19	0.62
Sustainability	0.14	−0.50
Security	NaN	−1.00
Strategy	−1.00	NaN

initial communication across organizational boundaries. Because only through communication can a corresponding exchange of information take place. Therefore, we analyze the initial communication of the first nine weeks (first phase) over all departments to determine if the following departments are represented [20]: Integration Management Office (Executive & Administrative Support and General Management), Human Resources, Finance, IT, Marketing & Sales, Legal & Compliance and Operations.

Figure 4 illustrates that the departments can be grouped into three categories based on their initial communication levels. The first group includes departments with a number of initial communication less than 250, such as Executive & Administrative Support, Finance, Legal & Patents, Quality, Procurement, Supply Chain, General Management, Facility & Real Estate, and Sustainability. The second group includes departments with an initial communication between 250 and 1000, namely Sales, Research & Development, Marketing, Information Technology, and Human Resources. Finally, the third group only includes the Sales department, with an initial communication level exceeding 1000.

Focusing on the groups with the highest initial communication numbers, we identified five departments. Resulting in a difference of organizations of Integration Management Office (General Management and Executive & Administrative Support), Finance, Legal & Compliance and Operations.

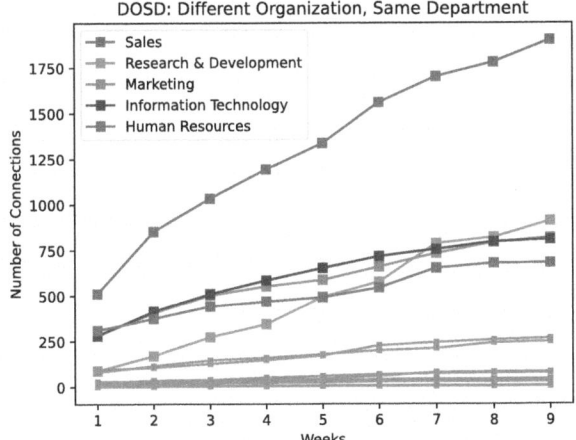

Fig. 4. Initial Communication Top Five Departments

One can notice here that we looked at the absolute number of connections, while another option is to take into consideration the number of individuals and thus looking density. However, we opted to consider the absolute number of connections because it is more reflecting the magnitude of communications between two departments.

5.3 Top Five Departments Network Patterns

Furthermore, we investigate if the top five departments illustrate similar patterns of communication respectively in both departments due to their similarity in size and if the formal merger is reflected through an steep increase of communication in the DOSD and DODD networks.

We observe that the most prominent curve can be identified as the initial communication within the SODD network. The R&D department of the Asset is the only department in which the internal communication is more prominent than the communication with the other departments indicating a substantial need for alignment. Remarkably the communication curve of the R&D department of the Acquirer shows a significant lower communication curve within the SODS and SODD network, compared with the department of the Asset.

5.4 Additional Department Communication

Our goal is to get a holistic view of the communication of the PMI and to see if it is reflected in the data. Therefore, we analyze the communication distribution of the top five departments towards departments of the other organization as shown in Fig. 5.

We observe a high quantity of communication between the HR department of the Acquirer and R&D department of the Asset. Indicating the following:

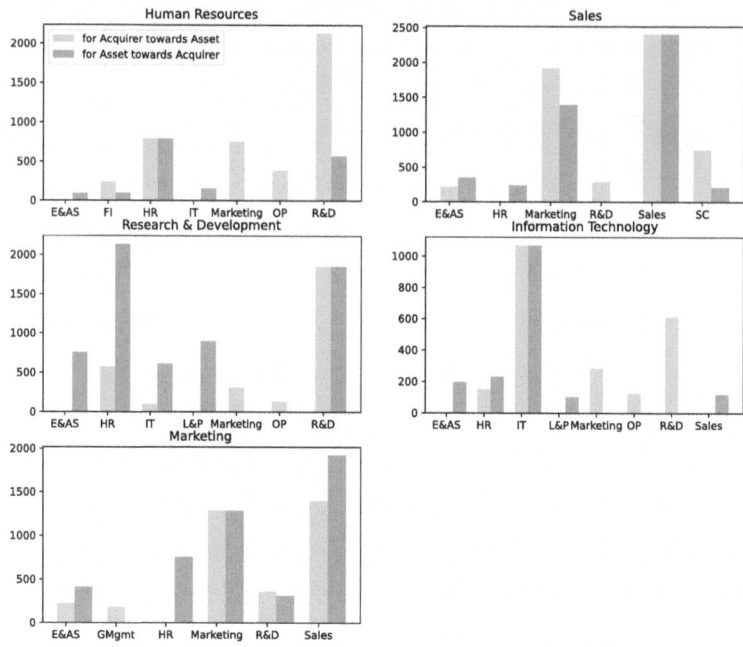

Fig. 5. Top 5 Department Communication in all networks

- Ensuring that new employees are adequately integrated into the organization. This includes defining roles and responsibilities, restructuring teams, and adapting employment contracts if necessary.
- Talent Management, particularly regarding critical employees in the R&D department responsible for the organisation's success and knowledge transfer. The HR department has to ensure that these talents are retained and integrated into the culture and values of the new organization.

A second observation is that the Marketing and Sales departments have the second largest inter-department initial communication quantity.

- This highlights the importance of coordination between the two departments' product strategies and positioning during a merger. It is necessary to have an understanding of each other's products and strategies.
- Marketing is responsible for customer acquisition and retention, while the Sales Department is responsible for direct customer contact and revenue generation. By working closely, the Marketing and Sales departments ensure seamless customer care and support each other's efforts.
- The supply chain department is responsible for procuring raw materials, producing goods, and distributing them to customers. Sales must work closely with the supply chain department to ensure on-time product availability to satisfy customer demand.

6   Conclusion

In this paper, we investigated a dynamic temporal network of two organizations in a PMI with a data set of aggregated network data. Our goal was to verify the two questions: whether the network stabilised over time and if the aspects of homophily of similar groups were reflected in the predominantly exchanged information within similar groups. Nevertheless, it has to be taken into consideration that email communication is limited. Our approach is based on initial communication interactions between individuals, cumulating the communication over time reflected in linear regression. Our main findings concerning temporal networks within post-merger integration scenario are the following:

- Network stabilization: Over time, communication within the networks stabilizes, which indicates an increasing cohesion of the networks and the establishment of communication channels as well as the informal communication reflection of the formal merger.
- Intensive initial communication: Within organizations and departments communication increases sharply immediately after a merger, indicating an intensive networking phase. However, it has to be taken in consideration that some communication was conducted before the data set was recorded, which is not reflected in the data.
- Homophily and share of communication: Homophily aspects are reflected in the results but is not fully supported by the data, indicating complex integration dynamics. Analyzing the percentage shares of communication shows that the homophily aspect is also reflected here. The share of communication is decreasing the further the network reaches from its origin.
- Critical departments: Certain departments which seem to be crucial for the PMI have less initial interaction as expected. Additionally, communication patterns vary between internal and external interactions and between different departments and organizations.
- Interactive departments: Based on the results HR seems to be the most essential department regarding cross department communication, reflected in a high amount of communication. Furthermore, R&D and IT as well as Marketing and Sales are also among the top five intra-organizational communication departments.

This present research has the potential for further expansion in different directions: First, future research should consider different data sets to validate and generalize current observations in comparison with similar data sets. Second, other attributes of the data set (e.g. hierarchy, gender, and others) and their behaviour of interaction, reflected in the data, should also be investigated, including other measures like centrality, betweenness, and density within the network. Third, further investigation should be considered regarding the departments, which are essential to conduct a lasting post-merger integration.

References

1. Battistuzzo, F., Ogasavara, M., Piscopo, M.: The importance of post-merger integration in study of international business. Int. J. Multinational Corp. Strateg. **2**, 218 (2018)
2. Burt, R.S.: Social contagion and innovation: Cohesion versus structural equivalence. Am. J. Sociol. **92** (1987)
3. Fabac, R., Schatten, M., DJurivcin, T.: Social network mixing patterns in mergers & acquisitions - a simulation experiment. Bus. Syst. Res. **2** (2011)
4. Frantz, T.: A social network view of post-merger integration. Adv. Mergers Acquisit. **10**, 161–176 (2012)
5. Frantz, T.: Dissecting Post-Merger Integration Risk: The PMI Risk Framework, vol. 16. Emerald Publishing Limited (2017)
6. Frantz, T., Carley, K.: Computationally modeling the effect of organizational complexity on post-merger integration. Adv. Mergers Acquisit. **8**, 79–101 (2009)
7. Gelardi, V., Barrat, A., Claidière, N.: From temporal network data to the dynamics of social relationships. Proc. Roy. Soc. B, Biol. Sci. (2021)
8. Ghawi, R., Benzinger, M., Pfeffer, J.: Approximating temporal networks from aggregated network data. In: 2023 IEEE/ACM International Conference on Advances in Social Networks Analysis and Mining (2023)
9. Graebner, M., Heimeriks, K., Huy, Q., Vaara, E.: The process of postmerger integration: a review and agenda for future research. Acad. Manag. Ann. **11**, 1–32 (2017)
10. Hennig, M., Brandes, U., Pfeffer, J., Mergel, I.: Studying Social Networks: A Guide to Empirical Research. Campus Verlag (2013)
11. Heo, D., Yoo, Y.: Knowledge sharing in post merger integration. In: Sprouts: Working Papers on Information Systems, vol. 2 (2002)
12. Hitt, M., Harrison, J., Ireland, R.: Mergers and Acquisitions, A Guide to Creating Value for Stakeholders. Oxford University Press, Oxford (2001)
13. Kengelbach, D., Klemmer, D., Schwetzler, B., Sperling, M.: An anatomy of serial acquirers. In: M&A Learning, and the Role of Post-merger Integration. ORG: Organizational Learning (Topic) (2011)
14. King, D., Dalton, D., Daily, C., Covin, J.: Meta-analyses of post-acquisition performance: Indications of unidentified moderators. Management Faculty Research and Publications **25** (2004)
15. Krackhardt, D., Stern, R.N.: Informal networks and organizational crises: an experimental simulation. Soc. Psychol. Q. **51**(2), 123–140 (1988)
16. Lawrence, B., Shah, N.: Homophily: measures and meaning. Acad. Manag. Ann. **14** (2020)
17. Marmenout, K., Mignerat, M.: Leveraging social networks in mergers: a roadmap for post-merger integration. Adv. Mergers Acquisit. **14** (2015)
18. Nurek, M., Michalski, R.: Combining machine learning and social network analysis to reveal the organizational structures. Appl. Sci. **10**(5), 1699 (2020)
19. Oeberg, C., Henneberg, S., Mouzas, S.: Changing network pictures: evidence from mergers and acquisitions. Industr. Market. Manag. **36**, 926–940 (2007)
20. PricewaterhouseCoopers (PwC): Success factors in post-merger integration (2024). https://www.pwc.de/de/deals/success-factors-in-post-merger-integration.pdf. Accessed 17 Mar 2024
21. Santos, J., Ferreira, M.A., Reis, N., Almeida, M.: Mergers & acquisitions research: a bibliometric study of top strategy and international business journals. J. Bus. Res. **67** (2012)

22. Seo, M.G., Hill, N.: Understanding the human side of merger and acquisitionan integrative framework. J. Appl. Behav. Sci. **41** (2005)
23. Stahl, G., Voigt, A.: Do cultural differences matter in mergers and acquisitions? A tentative model and meta-analytic examination. Organ. Sci. **19** (2008)
24. Sung, W., Labianca, G., Fagan, J.: Executives' network change and their promotability during a merger. Acad. Manag. Proc. **2018** (2018)
25. Vaidya, M., Bonini, S.: Homophily in mergers and acquisitions. Br. J. Manag. (2020)
26. Woehler, M., et al.: Turnover during a corporate merger: how workplace network change influences staying. J. Appl. Psychol. **106** (2021)
27. Yamanoi, J., Sayama, H.: Post-merger cultural integration from a social network perspective: a computational modeling approach. Comput. Math. Organ. Theory **19** (2013)
28. Zenk, L., Stadtfeld, C.: Dynamic organizations. How to measure evolution and change in organizations by analyzing email communication networks. Procedia - Soc. Behav. Sci. **4** (2010). Appl. Soc. Netw. Anal.

Analyzing X's Web of Influence: Dissecting News Sharing Dynamics Through Credibility and Popularity with Transfer Entropy and Multiplex Network Measures

Sina Abdidizaji[1]([⊠]), Alexander Baekey[2], Chathura Jayalath[1],
Alexander Mantzaris[3], Ozlem Ozmen Garibay[1], and Ivan Garibay[1]

[1] Industrial Engineering and Management Systems, University of Central Florida,
Orlando, FL 32816, USA
`{sina.abdidizaji,chathura,ozlem,igaribay}@ucf.edu`
[2] Computer Science, University of Central Florida, Orlando, FL 32816, USA
`alexander.baekey@ucf.edu`
[3] Statistics and Data Science, University of Central Florida, Orlando, FL 32816, USA
`alexander.mantzaris@ucf.edu`

Abstract. The dissemination of news articles on social media platforms significantly impacts the public's perception of global issues, with the nature of these articles varying in credibility and popularity. The challenge of measuring this influence and identifying key propagators is formidable. Traditional graph-based metrics such as different centrality measures and node degree methods offer some insights into information flow but prove insufficient for identifying hidden influencers in large-scale social media networks such as X (previously known as Twitter). This study adopts and enhances a non-parametric framework based on Transfer Entropy to elucidate the influence relationships among X users. It further categorizes the distribution of influence exerted by these actors through the innovative use of multiplex network measures within a social media context, aiming to pinpoint influential actors during significant world events. The methodology was applied to three distinct events, and the findings revealed that actors in different events leveraged different types of news articles and influenced distinct sets of actors based on the news category. Notably, we found that actors disseminating trustworthy news articles to influence others occasionally resort to untrustworthy sources. However, the converse scenario, wherein actors predominantly using untrustworthy news types switch to trustworthy sources for influence, is less prevalent. This asymmetry suggests a discernible pattern in the strategic use of news articles for influence across social media networks, highlighting the nuanced roles of trustworthiness and popularity in the spread of information and influence.

Keywords: Influential Actors · News Articles · Social Network Analysis · Transfer Entropy · Multiplex Networks

1 Introduction

In the contemporary era, individuals frequently opt to consume the latest news in digital format. Given the constraints of time and the impracticality of checking numerous news outlets, many turn to social media platforms to stay informed about the latest developments within their areas of interest. Recognizing this trend, manipulative actors can exploit social media to propagate selected news articles, thus influencing the perceptions and behaviors of others, leading them to question the reliability of their traditional news sources and potentially alter their preferences. Bovet and Makse [3] analyzed approximately 30 million tweets from 2.2 million users on X (formerly Twitter) over a five-month period preceding the 2016 US presidential election, discovering that about 25% of these tweets constituted fake news. Another study [16] investigated the influence and presence of Chinese news regarding COVID-19 in African countries' news outlets, revealing that former colonial powers exerted more influence over the media content of African countries than Chinese media.

The concept of an influencer within social media varies across different research domains. Generally, an influencer is defined as an individual capable of affecting the opinions and behaviors of others [18]. In the context of social media, influencers are often identified by their extensive networks, quantified through graph structures and social networks as nodes(users) with numerous edges(connections) [11,12]. Huynh et al.[10], however, define influencers based on the volume of posts they publish or republish. Contrary to the assumption that a high number of interactions and followers automatically signifies substantial influence, evidence suggests that these metrics do not always accurately measure a user's impact within social media [5]. In certain cases, individuals with smaller followings may exert more substantial influence [26]. Influence within social networks can propagate like a chain, exemplified by sequences of retweets [2]. An influence cascade refers to the sequence of user actions within a social network, beginning with an initial user acting due to external stimuli and leading to subsequent users being influenced to act, continuing until the chain of influence ceases. Essentially, it encompasses all users and events within a social network that are traced back to an initial action initiated by external motivation rather than social influence [21]. In this research, an influential actor is someone using a specific type of news articles and influencing others to change their taste and start using the same type of the article to propagate the news and creating a cascade of influence until the chain of influence ends. This inquiry is predicated on the hypothesis that the patterns of influence among actors in a social network are significantly shaped by the credibility and popularity of the news sources they elect to share. It is important to clarify that our data collection does not include metrics such as tweet counts, retweets, or followers, yet analyzing tweet frequency and content has enabled us to identify influencers.

Influence campaigns, critical to government agencies due to their global impact, challenge traditional social network analysis methods such as graph metrics, which only partly capture influencer dynamics [14]. This research utilizes transfer entropy networks [20] to reveal hidden influencers, those with significant but not obvious influence, by focusing on the quantitative impact of shared content on user opinions rather than simple information flow. Expanding on previous studies limited to cross-platform influence [25], this paper uses multiplex network measures to characterize influential actors and delineate their relationships within a Transfer Entropy Network on X, and tests them on three major events: the Skripal assassination, the Ukraine war, and the Navalny death.

Following the identification of influential actors and their targets, this study has several contributions aimed at elucidating the dynamics of influence within social networks. Specifically, the research seeks to:

1. Analyze the distribution of influence exerted by actors categorized based on their engagement with news sources of varying credibility and popularity. These categories include those who disseminate content from Trustworthy Mainstream (TM), Trustworthy Fringe (TF), Untrustworthy Mainstream (UM), and Untrustworthy Fringe (UF) news sources. The aim is to understand how influence is distributed by actors using these different source types among targets
2. Compare the behavior of actors in terms of propagating news articles and influencing their targets with those news articles across two different assassination events, and contrast these with a war event to observe the differences in the types of news articles used to disseminate information on X
3. Investigate the extent to which actors engage in disseminating content from more than one type of influence source, thereby acting as conduits for multiple streams of information

The structure of this paper is organized as follows: Sect. 2 provides a comprehensive review of the literature pertinent to our research. Subsequently, Sect. 3 details the data collection process and the methodology employed. In Sect. 4, the results are discussed extensively. Finally, in the concluding section, we summarize the findings and underscore the main contributions of this study.

2   Previous Work

The phenomenon of influence spread through cascades within social networks has garnered substantial interest across various disciplines, including marketing, sociology, and political science. Analyzing influence in a social network is a complex problem with potential for multiple interpretations and no ground truth. Seyfosadat et al. [22] identify four primary methodologies for identifying influence within social networks: data mining, machine learning, meta-heuristic approaches, graph-based methods, and a hybrid approach that combines these techniques. Graph-based methods, which is favored in this research domain [22], focus on the structural metrics of a graph combined with mathematical models to identify the key influential entities within a social network. While graph theory and network analysis techniques are commonly employed to extract valuable insights about the nodes and edges in a social network and assess influence [17], platform-specific metrics such as retweets and follower counts also serve as criteria for evaluating influence [23]. However, these straightforward metrics offer limited scope; for instance, a high in-degree does not necessarily equate to significant influence [4]. Furthermore, Kitsak et al. [14] discovered that the most pivotal influencers are not always those with the most connections but are rather strategically positioned at the core of the network structure. This underscores the necessity for advanced techniques that transcend mere connection counts to accurately identify influencers.

Entropy-based approaches represent a promising avenue for assessing influence within social networks. Saxena et al. [19] developed the Entropy-based Influence Disseminator (EbID) method, which uses an entropy-based centrality measure to identify key spreaders in networks by evaluating the entropy of influence paths and community attributes of neighboring nodes for wider spread potential. Their model outperformed others relying solely on centrality measures. He et al. [8] developed a model-free methodology to derive causal inferences about user behavior in social networks, employing Transfer Entropy to uncover both implicit and explicit causal relationships. Steeg and Galstyan [24] proposed the concept of content transfer, a metric grounded in information theory that offers predictive insights by quantifying the impact of one user's content on another, without dependency on predefined models. This approach uses non-parametric entropy calculations and sophisticated content representation techniques to identify predictive relationships among users, even in the absence of direct connections through following or mentions. Subsequently, Senevirathna et al. [21] further advanced the transfer entropy model by introducing the influence cascade model, wherein a node initiates influence that impacts another node, which then propagates this influence further within the social network. This model allows for the tracing of influence back to its origin node. Additionally, they devised a method to visualize the direction of entropy flows between nodes, termed an influence cascade. This framework was further expanded into the Influence Cascades Ecosystem [6], applying it to a geopolitical news model to trace influence across both traditional and social networks within a hybrid environment.

Within the realm of network science, Multiplex Networks have emerged as a powerful tool for tackling a wide array of network-centric challenges. Often, analyzing a solitary network falls short of providing comprehensive insights, whereas the construction of a multiplex network—comprising connected single networks as layers—facilitates the extraction of valuable information [1]. Their analytical use is heavily dependent on the data environment and the type of connection between nodes in the network, creating distinct layers of the multiplex networks. They have been used recently for modeling complex network systems such as information diffusion [13,15], disease characterization [7] and so forth. Furthermore, they have been applied to evaluate social ties [9] and facilitate knowledge dissemination [27] within social networks. This study proposes a novel application of multiplex networks for characterizing influence within social networks, specifically focusing on X. This method can be advantageous when the analysis of a single network of influence fails to yield meaningful information, and the aggregation with other networks provides insights into the interactions between key influential players across different networks. We build upon the foundational work of Battiston et al. [1] and integrate transfer entropy to enhance our understanding of node behavior and influence within a multiplexed network setting.

3 Methods

3.1 Data Collection and Preparation

For data collection, this study selected three events for investigation. Data pertaining to these events was systematically gathered using the Brandwatch platform[1]. The scope of data collection focused on profiles on X that shared news articles related to the chosen events. The first event pertains to the assassination attempt and subsequent poisoning of Sergei Skripal, with data for this incident being collected from March 1, 2018, to May 4, 2018. The second event involves the conflict between Ukraine and Russia, with the corresponding data collection spanning from January 1, 2022, to May 1, 2022. The third event concerns the suspicious death of Alexei Navalny, for which data was gathered from February 12, 2024, to March 2, 2024. Subsequently, these news articles were subjected to a classification process based on their credibility and popularity. Extending the framework introduced by Wang [25], we classify news articles by trustworthiness and popularity based on NewsGuard[2] and Majestic Million[3] rankings, respectively. That said, these rankings can be replaced by any desired classification scheme. To evaluate the trustworthiness of news articles, NewsGuard assigns scores to news agencies indicative of their credibility. Articles from sources scoring above 60 were designated as trustworthy, whereas those with scores below 60 were classified as untrustworthy. The classification of news articles by popularity conducted by Majestic Million database allowed for the categorization of articles into Mainstream, representing those from highly popular websites, and Fringe, denoting articles from less popular sources. Such categorization is critical for understanding the diversity of information sources and facilitates a nuanced analysis of information dissemination practices on social media. After conducting the classification, the time series data of actors in X was preprocessed to calculate the Transfer Entropy networks.

3.2 Transfer Entropy

Transfer Entropy is a statistical measure of information transfer introduced by Schreiber [20] which is an information-theoretical measure derived from Shannon entropy. TE is used to evaluate the extent to which knowing the historical data of two random processes (denoted as X_t and Y_t) can reduce the uncertainty in predicting the future state of one process, effectively quantifying the influence or

---

[1] https://www.brandwatch.com/.

[2] https://www.newsguardtech.com.

[3] https://majestic.com/reports/majestic-million.

informational flow from one process to the other. The TE metric is directional and non-commutative, indicating that the transfer of information from X to Y can differ in magnitude from Y to X.

In the mathematical formulation, as given in Eq. 1, TE from X to Y ($TE_{X \to Y}$) is calculated using a specific equation that involves summing over the probabilities of the next state of Y (Y_{t+1}), given its own history and the history of X, and then comparing this probability to the probability of Y_{t+1} given its own history alone. In this study the histories of X and Y are denoted by Y_t and X_t which represent sequences of past observations of length one.

$$TE_{X \to Y} = \sum P(Y_{t+1}, Y_t, X_t) \log \frac{P(Y_{t+1}|Y_t, X_t)}{P(Y_{t+1}|Y_t)} \tag{1}$$

The application of TE is demonstrated in a study involving user data on X. For each user, the study extracts a time series of their tweeting activity, identifying all timestamps of their tweets. The data is then resampled at a daily frequency ($f = 1$ Day), converting the activity into a binary time series where a "1" represents a day the user tweeted, and a "0" signifies a day they did not. This binary time series is then applied to the TE calculation to analyze the information flow between users based on their tweeting activity.

The weight assigned to an edge within a TE network is indicative of the strength of influence exerted within this network. Four specific types of influence were delineated, based on the actors' connections to Trustworthy Mainstream (TM), Untrustworthy Mainstream (UM), Trustworthy Fringe (TF), and Untrustworthy Fringe (UF) news sources. For instance, a TM actor might target four different audience groups: those engaging with TM news sources, indicative of echo chambers; TF targets, representing popularity crossover; UM targets, signifying trust crossover; and UF targets, illustrating both trust and popularity crossovers. The full description is provided in Table 1.

3.3 Multiplex Networks and Aggregates

Taking the necessary step of building a TE network opens up many analytical possibilities. The most obvious of which being a variety of aggregations by constructing TE networks with the defined classification. Aggregating edge types offers a number of perspectives that are inaccessible until the TE network is built. Upon the construction of Transfer Entropy, a critical endeavor entailed the analysis of influential actors distribution across varied types of influence within these networks. To achieve this purpose, we need to build multiplex networks for different types of influential actors. For instance, from the 16 edge types in the TE graph, one can aggregate over all $TM \to **$ edges ($TM \to TM$, $TM \to TF$, $TM \to UM$, $TM \to UF$) which will show all of the influence from sources that share Trustworthy Mainstream news articles and affects people shared other type of news article, causing to form a multiplex network by combining links between actors in these 4 layers. The creation of this TM multiplex network is depicted in Fig. 2. In mathematical terms, to construct the multiplex network, the aggregate

Table 1. Definition of 16 influence types based on credibility and popularity of news articles

Name	Class
TM→TM	echochamber
TF→TF	echochamber
UM→UM	echochamber
UF→UF	echochamber
TM→UM	credibility crossover
TF→UF	credibility crossover
UM→TM	credibility crossover
UF→TF	credibility crossover
TM→TF	audience crossover
UM→UF	audience crossover
TF→TM	audience crossover
UF→UM	audience crossover
TM→UF	credibility and audience crossover
TF→UM	credibility and audience crossover
UM→TF	credibility and audience crossover
UF→TM	credibility and audience crossover

transfer entropy for each node within a single TE network is computed as follows [1]:

$$k_i = \sum_j a_{ij} \qquad (2)$$

Here, k_i represents the total transfer entropy emitted from node i within the TE network, where a_{ij} denotes the directed edge from node i to node j, indicating the influence exerted by node i on node j. This analysis exclusively considers outgoing directed edges, as the focus is on the exertion of influence rather than the reception thereof. Following the determination of total transfer entropy within individual TE networks, multiplex networks for TM, UM, TF, and UF actors are constructed. This is mathematically expressed as [1]:

$$o_i = \sum_\alpha k_i^{[\alpha]} \qquad (3)$$

The equation above calculates the total transfer entropy for a node across the multiplex network, summing the outgoing edges of that node across all single TE networks denoted by α.

3.4   Participation Coefficient

The approach of creating multiplex networks facilitated an in-depth examination of how influence distribution among actors permeates through various layers

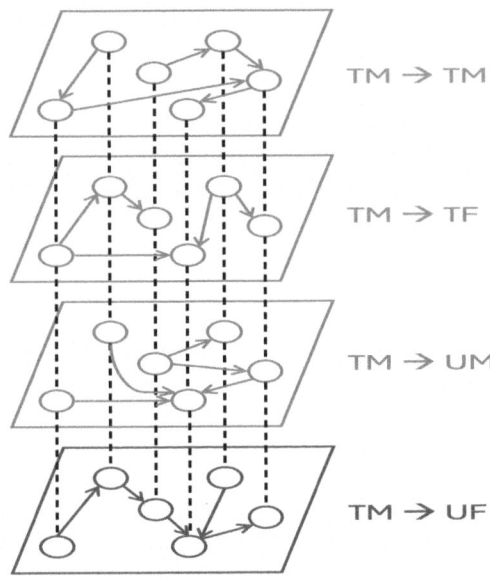

Fig. 1. Construction of the multiplex network for TM actors across four layers demonstrating the direction of influence towards various types of targets

within the multiplex networks. To quantitatively assess the extent of influence dispersion among actors, multiplex network measures were applied, notably the Participant Coefficient [1]. The calculation of this coefficient uses the formula below:

$$P_i = \frac{M}{M-1}[1 - \sum_{\alpha=1}^{M}(\frac{k_i^{[\alpha]}}{o_i})^2] \tag{4}$$

This metric quantitatively assesses the extent of actor involvement across distinct layers within multiplex networks. Within the context of this formulation, M signifies the total number of layers present in each multiplex network. The variable α represents a layer within the TE networks. The term $k_i^{[\alpha]}$ denotes the aggregated transfer entropy associated with a node within a singular TE network layer, while o_i encapsulates the cumulative transfer entropy attributed to a node across the entirety of the multiplex network. A low participant coefficient, approaching zero, signifies that an actor's influence is predominantly directed at a singular audience type. Conversely, a high participant coefficient, nearing one, indicates an equal distribution of influence across multiple audience layers.

3.5   Conditional Probability for Pairwise Co-occurrence

In the next phase, the goal was to determine the percentage of actors active in different influence types simultaneously. As observed, some actors were identified in different influence pathways, prompting an investigation into whether these actors employ different news links in terms of trustworthiness and popularity in their tweets or if they propagate one type of influence through diverse news sources. The aim of this evaluation was to see if actors targeted only one type of influence to propagate in the network or if they were active in multiple types of influence. These established multiplex networks underwent pairwise comparisons. To calculate it, the formula below was used [1]:

$$P\left(a_{ij}^{[\alpha']}\,\middle|\,a_{ij}^{[\alpha]}\right) = \frac{\sum_{ij} a_{ij}^{[\alpha']} a_{ij}^{[\alpha]}}{\sum_{ij} a_{ij}^{[\alpha]}} \tag{5}$$

This formula involved assessing whether an actor, active and categorized in one type of influence, was also active in another influence type or not. These comparisons provided valuable insights into the cross-influence patterns among actors in our study.

To the best of our knowledge, this study represents the inaugural application of this particular classification combined with Transfer Entropy to examine influential actors on a social media platform[4]. Furthermore, the employment of multiplex networks with such a distinction has not been documented in previous research focusing on the study of influential actors. The full framework pipeline is illustrated in Fig. 1.

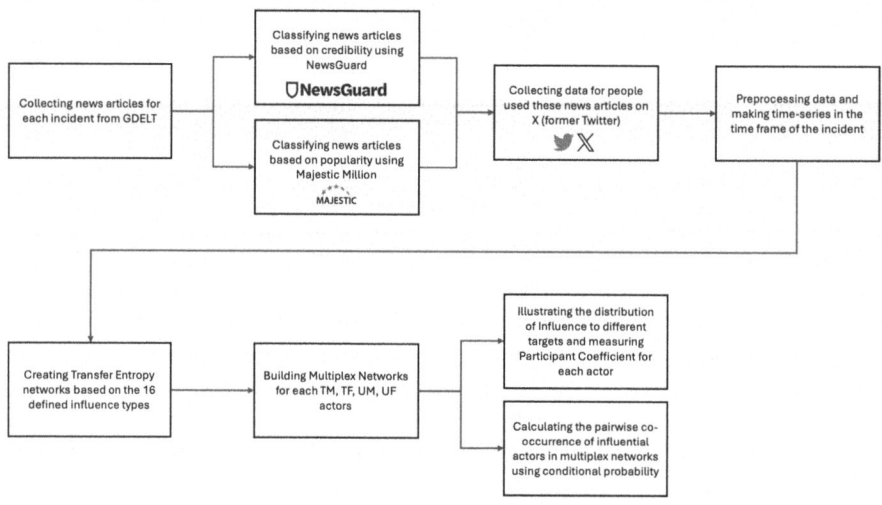

Fig. 2. The full framework pipeline of transfer entropy with multiplex networks for analyzing influential actors' behavior

---

[4] Code: https://github.com/sina6990/Multiplex-Networks.

4   Results

4.1   Characterizing Actors Spreading Influence by TE Networks Using Multiplex Network Measures

To employ multiplex network measures introduced in previous section and discern the influence distribution by actors, we generated four distinct multiplex networks for each influential actor type. Subsequently, actors were ranked based on the strength of influence exerted on their targets, measured by the total outdegree of a node in a multiplex network which is based on Transfer Entropy. Our objective is to investigate whether the strength of influence in a network has an effect on the distribution of influence to their targets. This methodology was applied to all Skripal, Navalny, and Ukraine events, and the results for each type of influential actor within their multiplex network are presented at Fig. 3.

Comparing the two assassination events, the distribution of influence among Trustworthy Mainstream (TM) actors in the Skripal event was more uniform compared to the Navalny event. In the case of Navalny, the distribution of influence exerted by TM actors was not as uniform, although it did extend, to some degree, towards each type of target. Moreover, there was a significant activity in echo-chamber of TM actors influencing TM targets in the Navalny event. However, during the Ukraine event, TM actors predominantly influenced targets who were tweeting from trustworthy news sources. This distinction is apparent in their participant coefficient charts. In the Skripal event, a majority of TM actors exhibit a high participant coefficient, indicating a more diversified distribution of influence among different targets with varying influence types. Conversely, in the Ukraine event, a low participant coefficient is observed when the amount of TE was high, suggesting that TM actors with higher levels of influence engaged in more focused operations compared to those with lower influence. In the Navalny event, given the less uniform distribution of influence by TM actors, it is observable that most were aligned with the green line, indicating a predominant influence over two layers. This pattern was almost identical among other types of influential actors. A noteworthy observation in the Ukraine incident was the absence of influence directed from actors towards UM and UF targets. This pattern of behavior was distinct from what was observed in both the Skripal and Navalny incidents which had similar nature, namely, the death of an individual.

Examining the participant coefficient plots in Fig. 4 reveals that during the Skripal incident, the distribution of influence for almost all actors is predominantly centered around three layers of influence, as indicated by the red dotted lines. Conversely, in the Navalny incident, with the exception of TM actors, who are primarily concentrated around two layers of influence denoted by green dotted line, the behavior of actors in other categories exhibits notable variance. In the TF, UM, and UF multiplex networks, actors possessing lower levels of influence demonstrate a more focused approach towards their targets compared to those wielding a higher degree of influence. This trend is starkly reversed in the context of the Ukraine incident, where actors with significant influence exhibit

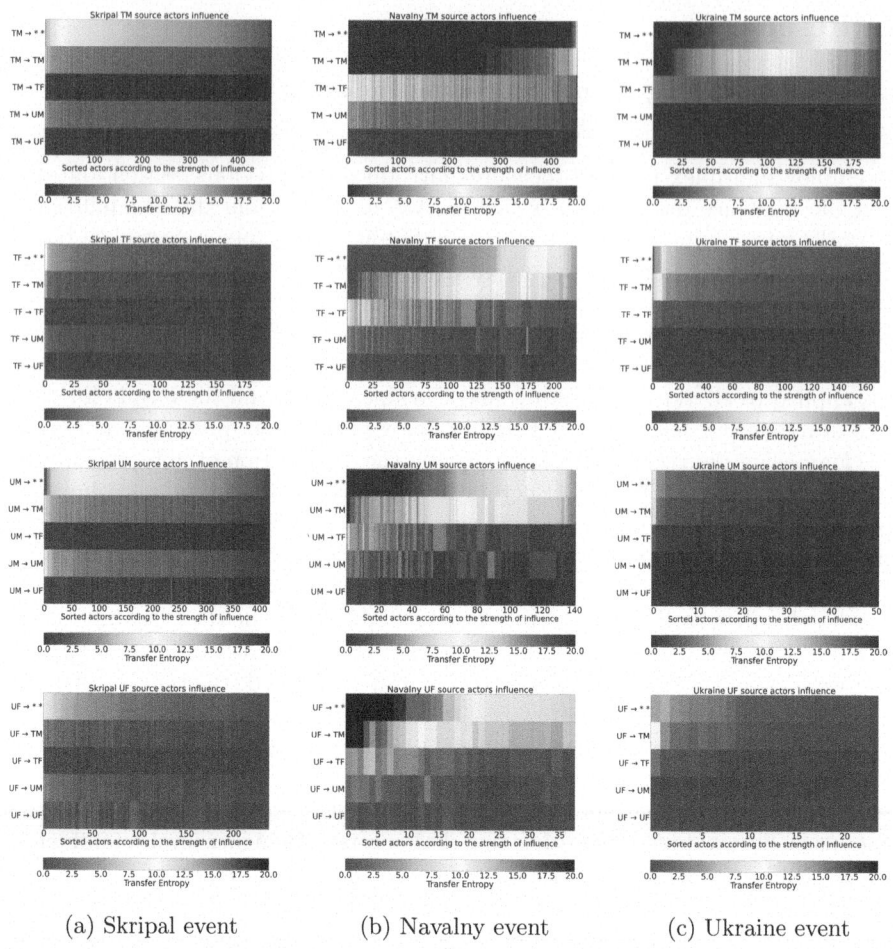

(a) Skripal event   (b) Navalny event   (c) Ukraine event

Fig. 3. Comparing the distribution of TM, TF, UM, and UF sources on four different type of targets in Skripal, Navalny, and Ukraine events

a more concentrated focus towards their targets, in contrast to less influential actors, whose influence is dispersed across multiple layers.

4.2 Pairwise Co-occurrence Comparison

In the next step, we conducted pairwise co-occurrence analyses for different actors. Our objective was to explore the likelihood that an actor, for instance, initially tweeting from trustworthy mainstream news sources and influencing others, would subsequently engage with other news types and exert influence on their respective targets. The results represented in heatmaps in Fig. 5.

Upon examining the heatmaps, it can be inferred that all three events exhibit a similar pattern to a significant extent. Notably, the percentages above the diag-

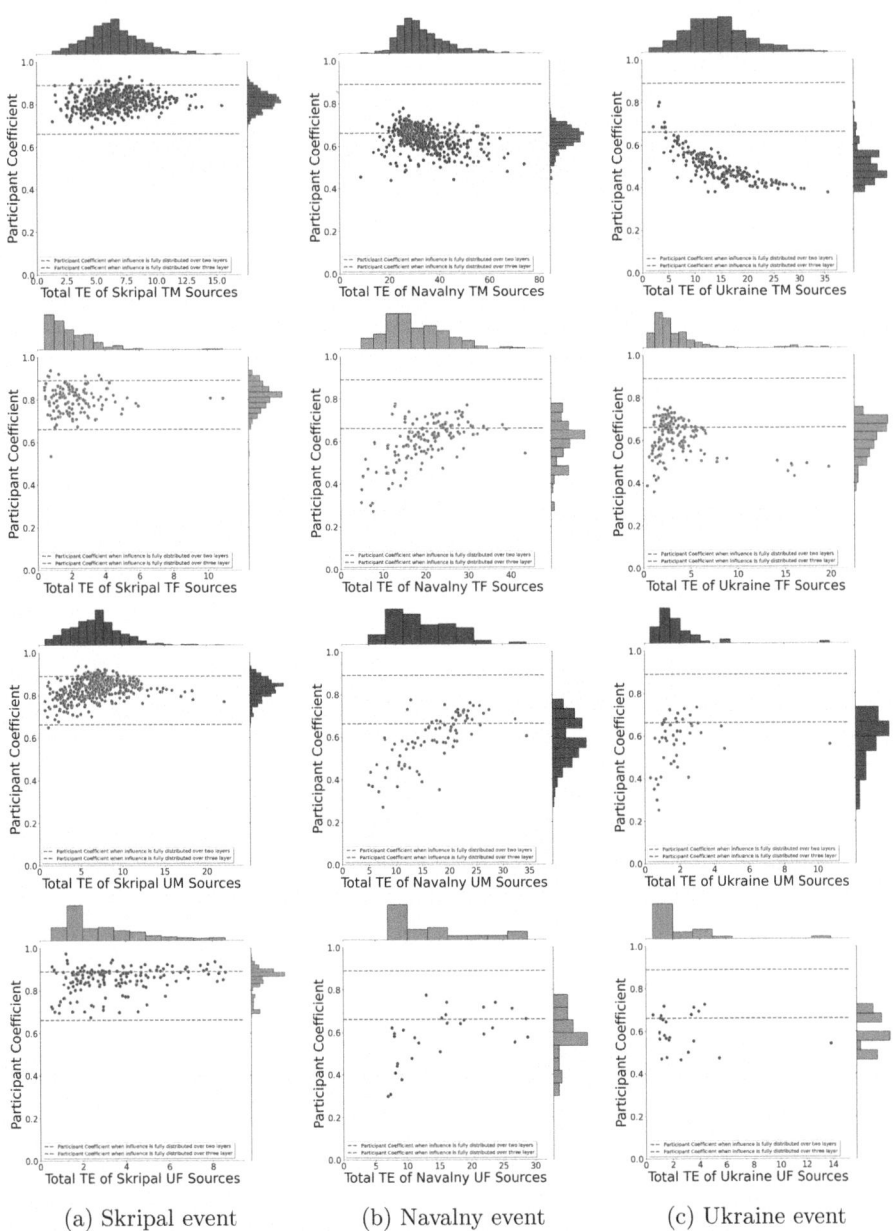

Fig. 4. The participant coefficient of TM, TF, UM, and UF actors in their corresponding multiplex networks in Skripal, Navalny, and Ukraine events Note: The scales on the X-axis are adjusted according to the highest influence in the multiplex network. A single scale was not used in order to highlight the diversity of influence and to distinguish the distributions.

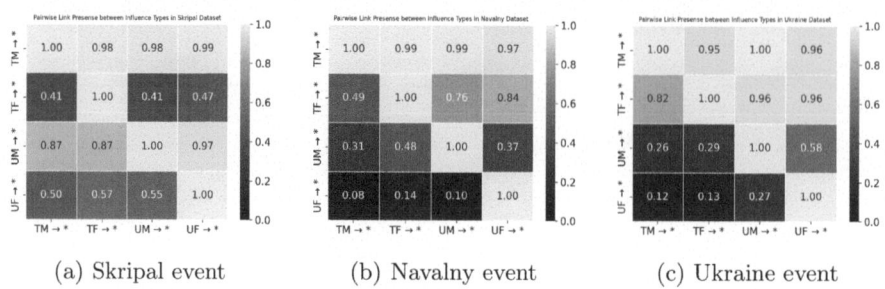

| | (a) Skripal event | (b) Navalny event | (c) Ukraine event |

Fig. 5. The pairwise co-occurrence of influential actors within a multiplex network being active in other types of influential multiplex networks The asterisk (*) in the heatmaps signifies that influence extended to all types of targets (TM, TF, UM, UF)

onal are high, whereas those below the diagonal are low. Specifically, the transition from trustworthy mainstream (TM) and trustworthy fringe (TF) actors to untrustworthy mainstream (UM) and untrustworthy fringe (UF) actors suggests that individuals engaging with trustworthy news sources are more likely to diversify their news consumption, potentially including untrustworthy sources. For the Skripal, Navalny, and Ukraine cases, 98%, 99%, and 100% of actors utilizing TM news sources also engaged with UM sources, respectively. Conversely, only 87%, 31%, and 26% of actors associated with the Skripal, Navalny, and Ukraine events, respectively, who initially used UM news sources, were observed to utilize TM sources. This indicates a diminished propensity for actors primarily influencing with untrustworthy news sources to transition to utilizing trustworthy sources, underscoring a more targeted and focused influence strategy among those relying on untrustworthy news. In contrast, actors who disseminate trustworthy news demonstrate a willingness to explore a broader range of media and platforms, highlighting their diverse information consumption patterns.

5   Conclusion

In this paper, we explored the propagation of influence within the social network platform X by examining the behavior of actors who utilize four distinct classes of news articles. Diverging from traditional methodologies that rely on traditional methods such as nodes' degrees and centrality measures to assess influence and identify influential actors, we employed a non-parametric approach named Transfer Entropy. This method has the capability to uncover hidden influencers not detectable by parametric and conventional techniques. To analyze the behavior of these actors, we adopted a novel metrics based on multiplex networks. Through the construction of these networks and the examination of influence distribution, we observed that in assassination-related scenarios, especially in the Navalny case, trustworthy actors were targeted by untrustworthy actors. Conversely, in the context of the Ukraine war, the predominant influence was exerted by actors utilizing trustworthy news sources and the absence of untrustworthy sources was obvious. Additionally, looking at the participant coefficient, we discovered that the behavior of actors utilizing trustworthy sources in assassination scenarios were similar to one another across different strengths of influence. However, in the Ukraine war scenario, particularly within the Trustworthy Mainstream (TM) multiplex network, the influence exerted by highly influential actors was more concentrated through a single network, rather than being dispersed across multiple layers of networks. A significant finding is the tendency of actors using trustworthy news sources to also employ untrustworthy news articles, indicating a propensity for these actors to utilize a diverse range of news articles in terms of credibility. On the contrary, actors relying on untrustworthy news sources showed a lower likelihood of employing trustworthy ones, suggesting that their influence was more targeted by using only untrustworthy sources. This pattern remained consistent across all three events studied. The utilization of multiplex networks and their structural measures has unveiled novel insights into the characterization of influential actors within social media landscapes.

Acknowledgements. This work was partially supported by the Defense Advanced Research Projects Agency (DARPA) under agreement HR00112290104 (PA-21-04-06).

References

1. Battiston, F., Nicosia, V., Latora, V.: Structural measures for multiplex networks. Phys. Rev. E **89**(3), 032804 (2014)
2. Bhowmick, A.K., Gueuning, M., Delvenne, J.-C., Lambiotte, R., Mitra, B.: Temporal sequence of retweets help to detect influential nodes in social networks. IEEE Trans. Comput. Soc. Syst. **6**(3), 441–455 (2019)
3. Bovet, A., Makse, H.A.: Influence of fake news in Twitter during the 2016 US presidential election. Nat. Commun. **10**(1), 7 (2019)
4. Cha, M., Haddadi, H., Benevenuto, F., Gummadi, K.: Measuring user influence in Twitter: the million follower fallacy. In: Proceedings of the International AAAI Conference on Web and Social Media, vol. 4, no. 1, pp. 10–17 (2010)
5. De Veirman, M., Cauberghe, V., Hudders, L.: Marketing through Instagram influencers: the impact of number of followers and product divergence on brand attitude. Int. J. Advert. **36**(5), 798–828 (2017)
6. Garibay, O.O., et al.: Entropy-based characterization of influence pathways in traditional and social media. In: 2022 IEEE 8th International Conference on Collaboration and Internet Computing (CIC), pp. 38–44 (2022)
7. Halu, A., De Domenico, M., Arenas, A., Sharma, A.: The multiplex network of human diseases. NPJ Syst. Biol. Appl. **5**(1), 1–12 (2019)
8. He, S., Zheng, X., Zeng, D., Cui, K., Zhang, Z., Luo, C.: Identifying peer influence in online social networks using transfer entropy. In: Wang, G.A., Zheng, X., Chau, M., Chen, H. (eds.) PAISI 2013. LNCS, vol. 8039, pp. 47–61. Springer, Heidelberg (2013). https://doi.org/10.1007/978-3-642-39693-9_6
9. Hristova, D., Musolesi, M., Mascolo, C.: Keep your friends close and your facebook friends closer: a multiplex network approach to the analysis of offline and online social ties. In: Proceedings of the International AAAI Conference on Web and Social Media, vol. 8, no. 1, pp. 206–215 (2014)

10. Huynh, T., Zelinka, I., Pham, X.H., Nguyen, H.D.: Some measures to detect the influencer on social network based on Information Propagation. In: Proceedings of the 9th International Conference on Web Intelligence, Mining and Semantics. WIMS2019, pp. 1–6. Association for Computing Machinery (2019)
11. Jain, L., Katarya, R.: Discover opinion leader in online social network using firefly algorithm. Expert Syst. Appl. **122**, 1–15 (2019)
12. Jain, L., Katarya, R., Sachdeva, S.: Opinion leader detection using whale optimization algorithm in online social network. Expert Syst. Appl. **142**, 113016 (2020)
13. Kalantari, H., Badiee, A., Dezhboro, A., Mohammadi, H., Tirkolaee, E.B.: A fuzzy profit maximization model using communities viable leaders for information diffusion in dynamic drivers collaboration networks. IEEE Trans. Fuzzy Syst. **31**(2), 370–379 (2023)
14. Kitsak, M., et al.: Identification of influential spreaders in complex networks. Nat. Phys. **6**(11), 888–893 (2010)
15. Li, W., Tang, S., Fang, W., Guo, Q., Zhang, X., Zheng, Z.: How multiple social networks affect user awareness: the information diffusion process in multiplex networks. Phys. Rev. E **92**(4), 042810 (2015)
16. Madrid-Morales, D.: Who set the narrative? Assessing the influence of Chinese global media on news coverage of COVID-19 in 30 African countries. Glob. Media China **6**(2), 129–151 (2021)
17. Peng, S., Zhou, Y., Cao, L., Shui, Yu., Niu, J., Jia, W.: Influence analysis in social networks: a survey. J. Netw. Comput. Appl. **106**, 17–32 (2018)
18. Rodríguez-Vidal, J., Gonzalo, J., Plaza, L., Sánchez, H.A.: Automatic detection of influencers in social networks: authority versus domain signals. J. Assoc. Inf. Sci. Technol. **70**(7), 675–684 (2019)
19. Saxena, C., Doja, M.N., Ahmad, T.: Entropy based flow transfer for influence dissemination in networks. Physica A: Stat. Mech. Appl. **555**, 124630 (2020)
20. Schreiber, T.: Measuring information transfer. Phys. Rev. Lett. **85**(2), 461–464 (2000)
21. Senevirathna, C., Gunaratne, C., Rand, W., Jayalath, C., Garibay, I.: Influence cascades: entropy-based characterization of behavioral influence patterns in social media. Entropy **23**(2), 160 (2021)
22. Seyfosadat, S.F., Ravanmehr, R.: Systematic literature review on identifying influencers in social networks. Artif. Intell. Rev. **56**(1), 567–660 (2023)
23. Singh, N., Malik, A., Maini, O., Rajput, G.: Identification of influence propagation metrics in social networks. In: 2019 International Conference on Automation, Computational and Technology Management (ICACTM), pp. 224–227 (2019)
24. Steeg, G.V., Galstyan, A.: Information-theoretic measures of influence based on content dynamics. In: Proceedings of the Sixth ACM International Conference on Web Search and Data Mining, pp. 3–12. ACM (2013)
25. Wang, Y., Zannettou, S., Blackburn, J., Bradlyn, B., De Cristofaro, E., Stringhini, G.: A multi-platform analysis of political news discussion and sharing on web communities. In: 2021 IEEE International Conference on Big Data (Big Data), pp. 1481–1492 (2021)
26. Zarei, K., et al.: Characterising and detecting sponsored influencer posts on Instagram. In: 2020 IEEE/ACM International Conference on Advances in Social Networks Analysis and Mining (ASONAM), pp. 327–331 (2020)
27. Zhu, H., Wang, Y., Yan, X., Jin, Z.: Research on knowledge dissemination model in the multiplex network with enterprise social media and offline transmission routes. Phys. A **587**, 126468 (2022)

FairNet: A Genetic Framework to Reduce Marginalization in Social Networks

Federico Mazzoni[1,2], Andrea Failla[1,2(✉)], and Giulio Rossetti[2]

[1] University of Pisa, Department of Computer Science, Pisa 56127, Italy
{federico.mazzoni,andrea.failla}@phd.unipi.it
[2] National Research Council, ISTI-CNR, Pisa 56127, Italy
giulio.rossetti@isti.cnr.it

Abstract. Discrimination in social networks often assumes the form of *marginalization* against nodes with specific features, e.g., segregation of/against minorities. In this work, we propose a metric that proxies social discrimination based on salient node features in a social network. Under the assumption that in a fair social system, all individuals should be enclosed in similar social circles representing the network in its entirety, our metric assigns a *marginalization score* to each node in the network, identifying if they are marginalized by similar nodes (e.g., a man marginalized by other men), by different nodes (e.g., a man marginalized by women), or not marginalized at all (i.e., the node has a fair neighborhood). Moreover, we introduce FAIRNET, a two-fold framework that aims to reduce network marginalization in partially- and fully-attributed networks by employing genetic algorithms. We evaluate our framework on networks emerging from online social interactions and find that the two components of FAIRNET are able to consistently reduce marginalization.

Keywords: Fairness · Genetic Algorithm · Marginalization · Bias Mitigation

1 Introduction

The study of social discrimination and inequalities has long benefitted from network analysis tools. Methods from network science can characterize a wide variety of phenomena that induce and/or foster social exclusion and community segregation, among gender inequalities in science [13], racial discrimination in online social networks [24], and weight stigma in adolescents [2]. Moreover, they can quantify the structural effects of human and algorithmic biases [23], and minority groups underrepresentation [16]. Inspired by recent literature in Artificial Intelligence and Machine Learning focusing on algorithmic discrimination, the study of *algorithmic fairness* is a rising trend in Network Science [29]. Fair Social Network Analysis is concerned with individuating fairness constraints on network-based tasks, and developing algorithms and tools that enforce/abide by such constraints. Another line of work focuses on the marginalization that might

L. M. Aiello et al. (Eds.): ASONAM 2024, LNCS 15211, pp. 139–154, 2025.
https://doi.org/10.1007/978-3-031-78541-2_9

be endured by minorities, e.g. their limited access to information [35]. This sort of marginalization can be conceptualized as segregation within a node's social circle, or neighborhood. In other words, people belonging to a certain group might be marginalized by people belonging to another group, e.g., people with different skill levels [6] or ethnic background [10]. With this work, we propose a marginalization metric related both to the notion of *proportional fairness* proposed for clusters [8], and to the aforementioned idea of segregation. We assume that, in a fair system represented as a network with node attributes, a node has a "fair" neighborhood if the distribution of nodes' labels in that neighborhood is proportionally similar to the labels' distribution in the entire network. In other words, the diversity in a node's neighborhood should be representative of the diversity in the entire network. To mitigate marginalization in a social network and increase its overall fairness we introduce FAIRNET, a framework consisting of two algorithms: FAIREDGES covers the case of partially attributed networks, filling in missing metadata with values that lead to fairer neighborhoods, whereas FAIRLABELS reduces discrimination via an edge addition strategy.

The rest of this work is organized as follows. Section 2 sums up relevant literature on the main topics surrounding this work. in Sect. 3, we provide an overview of the novel discrimination metric we introduce. This is discussed in detail in Appendix 5. Subsequently, we describe our framework to reduce the network's marginalization. In Sect. 4, the algorithms are tested on real-world data. Finally, Sect. 5 reports the conclusions and potential future works.

2   Related Work

In the following, we provide an overview of the main topics related to this work. Specifically, we first introduce the main concepts related to algorithmic fairness. Then, we move on to recent advances in fair social network analysis. Lastly, since the methods we propose in this work are based on genetic algorithms, we also provide an introduction to the matter.

Fairness. Two broad definitions of "fairness" have been proposed in the Data Science literature, namely *group fairness* and *individual fairness*. The group-based notion of fairness suggests that each group should have a similar proportion of individuals receiving the same treatment. For instance, if men are hired at a rate of 80%, then women should also be hired at approximately the same rate. Groups are defined following the value of *sensitive attributes* such as *Gender* or *Race*. This notion of fairness is quantified by metrics such as Statistical Parity and Disparate Impact, which check the difference or the ratio between the positive rates of the privileged and discriminated classes [1] (e.g., the percentage of men and of women receiving a positive outcome). Group fairness is often employed in classification tasks, including Graph Mining applications [11], and enhanced by pre-processing (modifying the initial training set of data, aiming to create an "ideal world dataset" [30]), in-processing (employing classifiers with a built-in concept of fairness), or post-processing techniques (modifying some of the decisions taken by the algorithm). On the other hand, *individual fairness* assesses that *similar individuals should be treated similarly*, with ad-hoc definitions of similarity taking into account various individuals' attributes other than sensitive attributes [26], i.e., individual with a very similar profile should receive the same treatment. The *proportional fairness* proposed in [7] for clusters is an example of individual fairness [8]. Inspired by fair resource allocation studies, this proportionality-based notion of fairness aims to find centroids to create fair, proportional clusters. The centroid could represent a public facility, e.g., a park; in this example, an unfair distribution of centroids could prevent the highest possible amount of people from accessing the facility [7]. This method for fair clustering has been extended to graphs [22].

Fairness in Social Network Analysis. When applied to the study of complex network analysis, the term "fairness" refers to the research area concerned with developing tools and algorithms to enforce fairness constraints in network analysis tasks. Tsioutsiouliklis and colleagues proposed two feature-aware variants of the PageRank ranking algorithm that take into account group fairness constraints [33], and thus bump up the visibility of minority nodes. Similar reasoning is followed by [28], where the authors propose a network counterpart of the Statistical Parity metric, checking if each group has a similar acceptance rate while sending a friend request. Another metric proposed in the same work is *Equality of Representation*, checking if each group has the same visibility (e.g., the same chances of featuring in friend suggestions). Stoica and Chaintreau proposed solutions to the influence maximization problem while accounting for fairness both in the selection of seed nodes, and in the outreach of the diffusion process [32]. A conspicuous body of works also relates to bias and discrimination in social networks while not mentioning fairness explicitly. For instance, [21] highlights bias towards lowly connected nodes in popular community detection methods. Moreover, strong homophilic or heterophilic behaviors in a network were proved to directly influence the degree ranking of nodes belonging to minorities [14], which can, in turn, limit their ability to access information [35]. This can be an issue for researchers, too, as (especially strong) differences in relative class sizes (e.g., a male majority vs. a female minority) can drive biases in analytical settings, e.g., in network sampling [34] and mixing patterns estimation [15].

Genetic Algorithms. Genetic Algorithms (henceforth *GA*) are a kind of meta-heuristic algorithm commonly used for optimization problems (e.g., the *knapsack problem* [31]). A GA attempts to find the optimal solution by first generating a set of potential solutions and then mating and mutating them, resulting in a second generation of potential solutions. The best individuals of the new generation are selected with respect to a fitness function. These steps are repeated for a fixed amount of generations [17]. GAs have been employed to increase the fairness of tabular data; *AuFair* [36] makes use of GAs to create a fair set of rules to replace biased human-made decisions, while *GenFair* [20] to generate the synthetic data needed to balance a dataset following multiple fairness criteria.

In the context of networks, GAs have been used to solve various tasks, including community detection [27], information spread [5], and link prediction [4].

3 Methods

According to the notion of individual fairness, *similar individuals should receive similar treatment*. If we assume that similar individuals are those enclosed in the same system (i.e., nodes in the same network), then all of them should be treated similarly for the system to be considered fair. Taking into account individuals' characteristics determined by a sensitive attribute (e.g., gender), all nodes in the same network should be surrounded by a group of peers manifesting a similar attribute distribution. Additionally, such distribution should be representative of the label distribution in the entirety of the system. A proportionally different distribution would indeed imply some degree of discrimination against the node—either by nodes with different labels or by those with similar labels.

In the following, we first present a new metric, called *Individual Marginalization Score* (IMS), which can be exploited to proxy a node's marginalization. Subsequently, from the IMS of every individual in a system, we compute the overall *System Marginalization Score* (SMS), i.e., the marginalization score of the entire network. Concluding, we introduce the two algorithms constituting the FAIRNET framework, that improve the network's fairness as evaluated by these metrics.

3.1 Quantifying Marginalization

Let $G = (V, E, L)$ be a connected node-attributed graph where V is the set of nodes, E is the set of edges, L is the set of categorical node labels l such that l_v identifies the attribute value of node v, with $v \in V$. We denote with $\mathcal{N}_v$ the set of v's first-order neighbors, i.e., the nodes it has a direct connection to. $\mathcal{N}_v$ can be seen as the union of two disjoint sets: $\mathcal{N}_v^l = \{u : u \in \mathcal{N}_v \wedge l_u = l_v\}$ (i.e., the subset of nodes in $\mathcal{N}_v$ with the same label as v), and $\mathcal{N}_v^{\neg l} = \{u : u \in \mathcal{N}_v \wedge l_u \neq l_v\}$ (i.e., the subset of nodes in $\mathcal{N}_v$ with a different label than v's). Of course, at most one of $\mathcal{N}_v^l$ and $\mathcal{N}_v^{\neg l}$ can be empty. To quantify marginalization at a local scale, we introduce the Individual Marginalization Score (IMS), which can be computed for v as:

$$\mathbf{IMS}(v) = \frac{2|\mathcal{N}_v|\omega_{l_v}}{|\mathcal{N}_v^l|\omega_{l_v} + |\mathcal{N}_v^{\neg l}|(1 - \omega_{l_v})} - 1, \tag{1}$$

In this formula, ω_{l_v} is a weight assigned to l_v that depends on its distribution in the whole network. A more detailed description of how the weights are computed, as well as of the formula's rationale can be found in Appendix 5.
The formula returns:

i. $\mathbf{IMS}(v) > 0$ if $\mathcal{N}_v$ contains more nodes with the same label as v, which might imply some form of marginalization of v by nodes with a different label;

ii. **IMS**$(v) < 0$ if $\mathcal{N}_v$ contains more nodes with a different label than v's, which might indicate some form of marginalization of v by nodes with the same label;

iii. **IMS**$(v) = 0$ iif $\mathcal{N}_v$ has a fair distribution of nodes' labels. As the weights are computed w.r.t. the whole network, a score of 0 indicates that the label distribution in v's neighborhood is proportionally similar to that of the entire network.

In cases where node attributes can have more than two values, a *one vs. all* scenario is supposed, where the number of labels similar to v's is compared to the cardinality of the heterogeneous group of different labels. For instance, assuming v's *Race* value is `Black`, a neighborhood with 2 Black nodes, 3 White nodes, and 2 Asian nodes will have the same marginalization score as a neighborhood with 2 Black nodes 1 White node, and 4 Asian nodes.

The IMS expresses the marginalization at the node level. To quantify the marginalization of a system of individuals (i.e., a network), we propose the *System Marginalization Score* (SMS). The SMS ranges in $[0, 1]$ and is computed as the mean of the absolute values of each node's marginalization score. Formally:

$$\mathbf{SMS}(G) = \frac{1}{|V|} \sum_{v \in V} |\mathbf{IMS}(v)|. \tag{2}$$

Alternatively, a network's overall discrimination can be quantified by counting the amount of discriminated nodes, setting a marginalization threshold τ; only nodes with the absolute value of their IMS above this threshold are considered discriminated. For instance, supposing $\tau = 0.7$, nodes with a score in the ranges $[0.7, 1]$ and $[-1, -0.7]$ are considered as discriminated. Note that this rationale assumes that each node is associated with an attribute value. In order to compute the IMS, nodes with missing labels should either be removed or treated.

3.2 Genetic Algorithm

In the next sections, we describe the two algorithms we devised to reduce a network's marginalization. Both of them employ a genetic algorithm, a search heuristic based on the concept of natural selection, or "survival of the fittest", as proposed by Darwin. This algorithm is frequently employed to tackle optimization problems. Each potential solution is referred to as a *chromosome* or *individual*, and it contains various "genes" [17]. The process begins with the generation of a random population of chromosomes, including p individuals. Each chromosome is then evaluated by a *fitness function*, which is closely tied to the problem to be solved.

Following this, the *selection operator* of the GA picks the optimal individual in the population with the goal of either minimizing or maximizing the fitness value. Essentially, only the "fittest" individuals are allowed to survive. A frequently used selection operator is *Tournament selection*, which randomly selects a number of chromosomes to participate in a tournament. The chromosomes with the highest fitness are declared the winners [17]. The selected chromosomes then have a chance c to be combined (usually in pairs) by the *crossover operator*. Selected genes are "shuffled" between each chromosome, resulting in new "children" individuals that inherit their genes from the "parent" individuals. The offspring population is referred to as the *second generation* of individuals. Offspring may be chosen by a *mutation operator*, following a probability m, which results in at least one of their genes being mutated.

This process is then carried out for k generations, with chromosomes from newer generations evaluated, selected, and potentially mated and mutated. As the algorithm retains only the fittest individuals, the overall fitness tends to improve with each generation. In the end, the algorithm returns the individual(s) with the highest fitness value as the optimal solution(s) to the given problem.

In Algorithm 1, we report the pseudo-code of the archetypical GA—including the default values employed by our algorithms. Please note that, in practice, evaluation functions and more advanced operators might require additional parameters (as is the case for both FAIRLABELS and FAIREDGES).

3.3 FAIRLABELS

In this section, we introduce FAIRLABELS, an algorithm that replaces nodes' missing metadata with values that reduce the overall marginalization. There are essentially two strategies when it comes to working with partially attributed networks, namely (i) removing nodes without metadata or (ii) filling in missing values. While approach (i) is the most straightforward—and perhaps the most common—removing entities from the network can lead to non-negligible information loss. This is especially true for networked data since node deletion also implies the elimination of all of its links, potentially leading to a significant alteration of the topology. Moreover, there is an inherent selection bias in removing data points with missing information, especially in mining tasks. As such, this pre-processing step is of critical importance.

Our proposal aims to tackle the issue of missing node metadata by inserting values in a way that leads to a fairer network. Note that our aim here is not to solve a node label prediction problem, but rather to find a combination of potential labels so as to minimize the number of discriminated nodes in the network (which is an optimization problem a GA can handle). The algorithm's rationale[1] is outlined in Algorithm 2. First, FAIRLABELS identifies M, the set of all the nodes with a missing label (line 1). In line 2, M is given as input to a genetic algorithm alongside the network G, τ, and the parameter a (see below). Each individual created by the genetic algorithm has the same cardinality as M (i.e., each gene represents a node with a missing label). The value of each gene is a possible value from L; e.g., Black, White, Asian for the *Race* attribute. In other words, each individual tested by the genetic algorithm is a vector of labels

---

[1] It should be noted that both FAIRLABELS and FAIREDGES also take as input the GA parameters described in Sect. 3.2. For the sake of simplicity, they were not included in the respective pseudo-codes, nor in the overview of the two algorithms.

Algorithm 1: Genetic Algorithm

   Input : p - population size, default: 150
           c - crossover probability, default: 0.5
           m - mutation probability, default: 0.25
           t - tournament size, default: 3
           k - number of generations, default: 100
   Output: s - best solution found

```
 1  best ←[ ];
 2  gen ←0;
 3  pop ←init(P);                              // initial population
 4  for individual in pop do                   // initial evaluation
 5  │   eva ←evaluate(individual);
 6  │   best.append((eva, individual))

 7  while gen < k do
 8  │   best ←select(best, t));                // initial selection
 9  │   offspring ←[ ]
10  │   for parents in best do                 // crossover
11  │   │   if random(0, 1) < c then
12  │   │   │   children ←crover(parents);
13  │   │   └   offspring.append(children)
   │   │
14  │   for child in offspring do              // mutation
15  │   │   if random(0, 1) < m then
16  │   │   │   child ←mutate(child)
17  │   │   eva ←evaluate(individual);
18  │   └   best.append((eva, individual))
19  │   best ←select(best, t))
20  └   gen ←gen + 1
21  s' ←picking_best(best);                    // picking the best result
22  return s
```

$l \in L$ for the nodes. Figure 1 provides a visual representation. Each solution is evaluated by inserting the labels in G and then computing the number of discriminated nodes in the network[2] w.r.t. τ. The best chromosomes are then mated, and their children are possibly mutated. As crossover operator, we employed the traditional Two Point Crossover. The operator selects two random "parent" chromosomes, and in each of them, two genes that are in the same position. The genes between the selected ones are then shuffled between the parents, resulting in two new "children" chromosomes [18]. For the mutation, we devised a custom operator. If the mutation operator picks a child chromosome (as outlined in Sect. 3.2), the value of each gene has a probability a of mutating into another

---

[2] Preliminary experiments showed that the genetic algorithm performed better with the number of discriminated nodes rather than with SMS. A minimal increment in SMS—resulting nonetheless in the same number of discriminated nodes—is seen as an improvement by the GA.

Algorithm 2: FAIRLABELS

Input : G - graph with missing labels

  τ - marginalization threshold

  a - probability of gene mutation

Output: G' - graph without missing labels

1 $M \leftarrow missing\_check(G, \tau)$;   // extract the set of nodes with missing labels

2 $G' \leftarrow label\_genetic\_algorithm(G, M, \tau, a)$;// find the best combination of values for the missing labels

3 **return** G'

(default: 0.05). The best solutions are then selected by conducting a Tournament Selection. These steps are repeated for a set amount of generations; afterward, the algorithm returns G', the network filled with the best combinations of labels found.

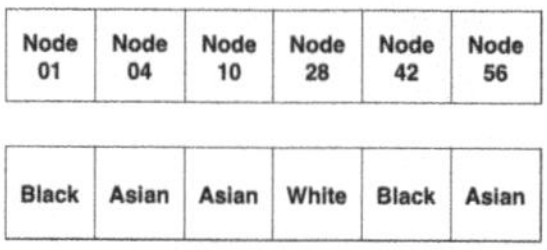

Node 01	Node 04	Node 10	Node 28	Node 42	Node 56

Black	Asian	Asian	White	Black	Asian

Fig. 1. Starting from a vector of nodes with missing labels, FAIRLABELS can fill chromosomes with the label values (e.g., *Race* values).

3.4 FAIREDGES

FAIRNET's other component is FAIREDGES, an algorithm to reduce the network's marginalization by adding new edges. We report the pseudo-code in Algorithm 3. FAIREDGES takes as input G, τ, and p (the latter representing the percentage of edges to be considered viable; see below), and it is structured in three phases. After computing (in line 1) the network marginalization as outlined above, in line 2, FAIREDGES selects a set of edges to be added according to the sociological principle of triadic closure [12], see below. In line 3, FAIREDGES then employs a genetic algorithm to find the best combination of edges to add so as to reduce the network's discrimination while operating the least amount of topological changes. These edges are added to G, resulting in a fairer network G', which is returned to the user. FAIREDGES has some similarities with the fair clustering approaches mentioned in Sect. 2. However, the clustering algorithms aim to find the best "core" (i.e., the best centroid) to create fair sets of points, whereas FAIREDGES, given a core, finds the best set of edges. Figure 2 provides a visual representation.

Edge Selection. After computing marginalization scores for all nodes via Eq. 1, we consider an individual to be marginalized if its IMS is positive and higher, or

---

Algorithm 3: FairEdges

Input : G - graph,
                τ - marginalization threshold,
                p - edge percentage
Output: G - balanced graph
1 $V' \leftarrow marginalization\_check(G, \tau)$;   // find marginalized nodes
2 $E' \leftarrow edge\_selection(G, V', p)$;   // see Algorithm 4
3 $best \leftarrow edge\_genetic\_algorithm(G, E', \tau)$;   // find best combinations of edges
4 $G.\text{add}(best)$
5 **return** G

Algorithm 4: Edge Selection

Input : G - graph,
                V' - marginalized nodes,
                p - edge percentage
Output: E' - selected edges
1 $E' \leftarrow \text{find\_nonexisting\_edges}(G, V')$
2 **for** $e \in E'$ **do**
3 $\quad \lfloor\ e.\text{weight} = \text{compute\_triangles}(e, G)$;
4 $E' \leftarrow \text{sort\_by\_weight}(E')$
5 $n \leftarrow \text{length}(E')$;
6 $k \leftarrow \text{int}(p \times n)$ // calculate the top $p\%$
7 $E' \leftarrow E'[1 : k]$ // select the top k edges
8 **return** E'

---

negative and lower, than a fixed threshold τ. The edge selection phase, detailed in Algorithm 4 aims at constructing a set of edges E' that improve the condition of such marginalized individuals. As a first step, the algorithm collects all non-existing edges adjacent to at least one marginalized node (as defined above).

For the network modifications to be realistic, we need a way to estimate the plausibility of non-existing ties. Indeed, in real-world networks, it is often the case that not all connections are equally important—and we would not want to add edges that are unlikely to appear. In principle, this can be done with any heuristic commonly employed in link prediction tasks [19]. In this work, we use triadic closure as a proxy for tie strength/plausibility. Thus, we assign each edge a weight based on the number of triangles it would close if it were to be added to the network (lines 2–3). In other words, non-existing edges that would close *more* triangles are considered more likely to appear than those with a lower weight. Then, the algorithm sorts possible edges in descending order of weight (line 4). Subsequently, only $p\%$ edges with the highest weights are kept, while others are discarded (lines 5–7). This is done both to reduce implausibility and runtime.

Minimizing Marginalization. FairEdges conceptualizes minimizing the network's marginalization as an optimization problem; given the set of edges E',

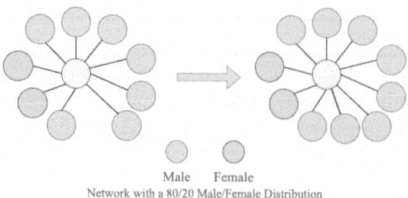

Male   Female
Network with a 80/20 Male/Female Distribution

Fig. 2. Visual representation of FAIREDGES. Note that, since the IMS does not depend on the label of the core node, but rather on those of its neighbors, FAIREDGES's solution is oblivious to the node's label. Indeed, in case the core node were male, it would be marginalized by its peers, and FAIREDGES would add an additional male node to its neighborhood. Conversely, if the core node were female, it would be marginalized by nodes with a different label, thus still leading to the addition of a new male node.

FAIREDGES finds the best combination of edges to be added to reduce the marginalization. FAIREDGES's GA is similar to that of FAIRLABELS, albeit more complex due to the task. The GA creates P vectors of binary values S with $|S| = |E'|$ as the initial chromosomes. Each of their genes represents an edge in E'; if it has a value of 1, then the possible solution modifies the corresponding edge in G, whereas with 0, no alteration is carried out. Figure 3 provides a visual representation.

A↔B	C↔B	E↔H	C↔G	A↔G	A↔H
1	0	1	1	0	0

Fig. 3. Starting from a vector of edges to be added, FAIREDGES creates binary chromosomes to find the optimal solution. Each 1 in a chromosome results in a modification to the original network structure (i.e., the corresponding edge is created). Conversely, each 0 implies no modification will concern the corresponding edge (i.e., the edge is not created).

Our fitness function alters G by adding edges represented by a 1. As in FAIRLABELS, we employed the number of discriminated nodes as the fitness metric. If two possible solutions result in the same network's discrimination, the most conservative solution has priority (i.e., the solution with fewer 1s). Two Points Crossover and Tournament Selection are again employed. Given that the value of each gene can be either 1 or 0, we used the traditional Flip Mutation as our mutation operator. It mutates a gene 0 value into a 1, and vice-versa. The steps are then repeated for the set number of generations. The network modified by adding the best set of edges (as found by the GA) is then returned to the user.

4 Experiments

Table 1. Comparison between the number of discriminated nodes either by removing or filling missing labels.
Each test was performed 50 times; values in brackets are the standard deviation.

Network	Attribute	Nodes	Edges	Missing	Thresh.	Removing	Filling
Twitter	*Inclination*	3,753	6,993	375	0.3	760 - 22%	862.86 (7.86) - 23%
					0.5	659 - 20%	700.48 (6.59) - 19%
					0.7	539 - 16%	506.68 (4.42) - 14%
Facebook	*Gender*	5,180	186,586	418	0.3	1,563 - 32%	1,434.32 (8.02) - 28%
					0.5	586 - 12%	461.64 (3.11) - 9%
					0.7	204 - 4%	154.34 (1.19) - 3%

Table 2. Comparison between the number of discriminated nodes by applying FAIREDGES or randomly adding edges.
Each test was performed 50 times; values in brackets refer to the standard deviation.

Network	Attribute	Nodes	Edges	Thresh.	Disc. Nodes	Benchmark	Final Result	Added Edges
Twitter	*Inclination*	3,753	6,993	0.3	1,017	645.56 (4.15)	589.84 (7.16)	2,361.52 (37.28)
				0.5	863	525.4 (3.12)	472.74 (5.75)	1,083.78 (20.41)
				0.7	701	484.62 (2.92)	437.06 (4.30)	389.20 (11.29)
Facebook	*Gender*	4,762	169,128	0.3	1,563	611.64 (3.04)	572.48 (6.84)	30,310.52 (120.5)
				0.5	586	190.52 (2.17)	164.24 (3.37)	2,811.58 (29.4)
				0.7	204	146.4 (1.13)	137.94 (0.64)	132.3 (16.36)

We tested FAIRLABELS and FAIREDGES on two networks. The *Twitter Network*, introduced in [3], features 3,753 nodes, connected by 6,993 edges. The sensitive attribute is *Political Inclination* and can assume three values: *neutral* (2,598 nodes), *liberal* (782 nodes), and *conservative* (180 nodes). The *Facebook Network* has 5,180 nodes, including 418 with missing labels, connected by 186,586 edges. If nodes with missing values are removed, the network is left with 169,128 edges and 4,762 nodes. The binary sensitive attribute *Gender* can be *Male* (2,153 nodes) or *Female* (2,609 nodes). Our algorithms were tested with their default GA parameters and three τ values: 0.3, 0.5, and 0.7.

4.1 Testing FAIRLABELS

First of all, we tested FAIRLABELS. As said, the Facebook network has 418 missing labels. The Twitter network featured no missing labels; therefore, 10% of the nodes (375) were randomly selected for these experiments and had their label removed. Table 1 compares the initial number of discriminated nodes if the missing values are just removed, or if they are filled by FairLabels. We featured both raw and relative frequencies of discriminated nodes. Each test was run 50 times; the *Filling* columns report the average result and the (generally low) standard deviation. With the Twitter network, removing or filling missing labels generally leads to comparable results, although the filling option is slightly better. This is noticeable by looking at proportional values. Nonetheless, employing FairLabels is by design less intrusive than outright removing nodes, as no information from the original network is lost. As for the Facebook network, FairLabels always leads to a fairer initial state.

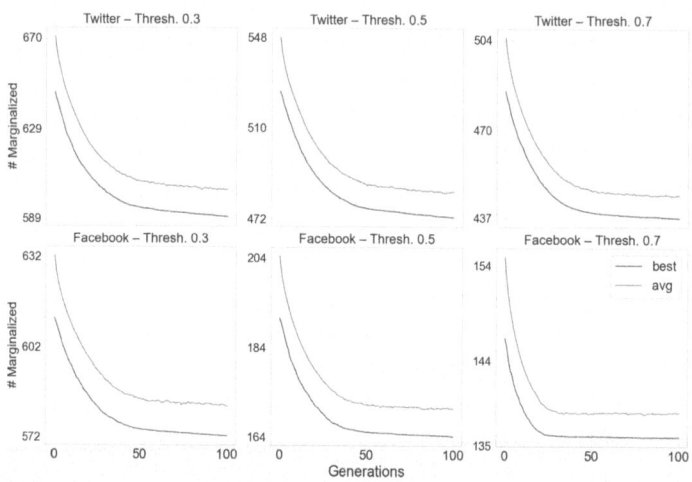

Fig. 4. Best (blue) and mean (orange) values for each generation in FairEdges's genetic algorithm. Each value is averaged over 50 runs. (Color figure online)

4.2   Testing FairEdges

Table 2 reports the results of our tests. Since we focused on FairEdges alone, for these experiments, the nodes with missing labels in the Facebook Network were removed and not replaced. As expected, increasing the marginalization threshold decreases the number of discriminated nodes. Again, each test was run 50 times. The *Benchmark* column reports the average of the best results obtained during the 0th generation of the genetic algorithm, i.e., the result obtained by adding a random selection of links. By contrast, the *Final Result* column reports the average of the best results at the end of the final, 100th generation. The *Added Edges* column reports the average number of added edges to achieve the best result.

As seen in the *Benchmark* column, randomly adding edges (selected from the edges pool described above) already greatly decreases the number of marginalized nodes. The genetic algorithm, however, always manages to further increase the network's fairness, to an extent never reached by random selection. This is more noticeable with higher amounts of discriminated nodes. We noticed a sharper decrease in discriminated notes during earlier generations, as the results obtained by the genetic algorithm were close to a convergence around the 50th generation (see Fig. 4).

5   Conclusion

In this work, we presented a new metric to measure marginalization in networks, inspired both by the concept of individual fairness and by the works on segregation in networks. Our metric assumes that in a fair network, the label distribution in a node's neighborhood should mirror the label distribution in the entire network. Furthermore, we have proposed the FAIRNET framework, including two algorithms following a genetic approach to reduce the number of discriminated nodes w.r.t. the aforementioned metric. Experimental results prove that both FAIRLABELS and FAIREDGES can significantly reduce the network's marginalization with relatively few modifications.

Future works could experiment with FAIRNET in real use cases—for example, a fair recommender system, helping users go out of their social bubbles, or a system to promote diversity in working groups within a school or a corporation. In this context, FAIREDGES could narrow its scope to only friends-of-friends, adopting a more "local" search (i.e., the node extended neighborhood rather than the whole network). Moreover, instead of only adding edges, FAIREDGES could also rewire a set of them.

An interesting use-case scenario could be the depolarization of closed substructures emerging from peer interactions in social media platforms (e.g., echo chambers). Recent work has shown that random nudges can effectively lead to less polarized environments [9,25]. Our approach, which starts with random edge insertion and takes into account attribute distribution, can be seen/exploited as a random nudging mechanism enhanced by knowledge of user preferences (encoded as node labels).

FAIRNET could also be employed for urban planning—each focal node of a city should have a fair and diverse distribution of facilities in its neighborhood. Moreover, the idea behind FAIRNET could be further tweaked to take into account communities with different distributions of labels. In this scenario, FAIRNET should compare the label distribution inside a node's neighborhood to the label distribution of the community the node belongs to.

Last, while the algorithms we presented employ a genetic approach, we believe other solutions are viable. A major drawback of genetic algorithms is their scalability under certain scenarios—increasing the number of genes leads to a longer runtime to find viable solutions. Other optimization techniques could be more scalable and, therefore, more suitable for larger networks. For example, a Generative Adversarial Network could be employed to select the best edges.

Acknowledgments. This work is supported by (i) the European Union - Horizon 2020 Program under the scheme "INFRAIA-01-2018-2019 - Integrating Activities for Advanced Communities", Grant Agreement n.871042, "SoBigData++: European Integrated Infrastructure for Social Mining and Big Data Analytics" (http://www.sobigdata.eu); (ii) SoBigData.it which receives funding from the European Union - NextGenerationEU - National Recovery and Resilience Plan (Piano Nazionale di Ripresa e Resilienza, PNRR) - Project: "SoBigData.it - Strengthening the Italian RI for Social Mining and Big Data Analytics" - Prot. IR0000013 - Avviso n. 3264 del 28/12/2021; (iii) EU NextGenerationEU programme under the funding schemes PNRR-PE-AI FAIR (Future Artificial Intelligence Research); (iv) TANGO G.A. 101120763.

Disclosure of Interests. The authors have no competing interests to declare that are relevant to the content of this article.

Appendix: Understanding the IMS

To conceptualize IMS, we start from the assumption that a fair network with $|L| = 2$ and equal distribution of each l (e.g., a network with 50 men and 50 women) should have the following property for all nodes:

$$\frac{|\mathcal{N}_v^l|}{|\mathcal{N}_v^l| + |\mathcal{N}_v^{\neg l}|} = \frac{1}{2}. \tag{3}$$

In other words, each node's neighborhood should feature an equal distribution of labels. We reckon that higher or lower values imply a form of marginalization of v. The following formula normalizes the score in the range $[-1, 1]$:

$$\frac{2 \cdot |\mathcal{N}_v^l|}{|\mathcal{N}_v^l| + |\mathcal{N}_v^{\neg l}|} - 1 = 0. \tag{4}$$

To generalize the formula to networks with more than two attribute values or unequal distribution of labels (e.g., 50 White people, 30 Black people, and 20 Asian people), we assign a weight to each label. The weight of each value l is denoted ω_l and computed as the complementary of the relative frequency of l in the whole network. This implies that rarer labels weigh more than common ones. Additionally, 1 is subtracted from both the numerator and denominator (with 1 representing the node of which we compute the marginalization).

$$\omega_l = 1 - \frac{|\{v : v \in V \land l_v = l\}| - 1}{|V| - 1} \tag{5}$$

Since $\mathcal{N}_v^l$ only includes nodes with label l_v, the weight of each node in $\mathcal{N}_v^l$ is ω_{l_v}. Since $\mathcal{N}_v^{\neg l}$ only includes every $l \in L$ except l_v, its weight can simply be computed as $1 - \omega_{l_v}$. Multiplying $|\mathcal{N}_v^l|$ and $|\mathcal{N}_v^{\neg l}|$ by their respective weights results in the IMS formula (Eq. 1), a generalization of Eq. 4 quantifying the marginalization of a node v in a network G.

The IMS computes the marginalization of each node v w.r.t. the label's distribution in the node's neighborhood, excluding v itself. v could be included by avoiding subtracting 1 from the numerator and the denominator in the weight formula (Eq. 5). However, in this scenario, a marginalization score of -1 could not be obtained; even if v were connected only to nodes with different labels, its neighborhood would still include a node with its label, i.e., v itself. For this reason, we decided not to include v as part of its neighborhood. The proposed weight formula accommodates this exclusion.

References

1. Agarwal, A., et al.: A reductions approach to fair classification. In: Proceedings of Machine Learning Research, ICML, vol. 80, pp. 60–69. PMLR (2018)
2. Arias Ramos, N., Calvo Sánchez, M.D., Fernández-Villa, T., Ovalle Perandones, M.A., Fernández García, D., Marqués-Sánchez, P.: Social exclusion of the adolescent with overweight: study of sociocentric social networks in the classroom. Pediatric Obesity **13**(10), 614–620 (2018)
3. Babaei, M., Grabowicz, P., Valera, I., Gummadi, K.P., Gomez-Rodriguez, M.: On the efficiency of the information networks in social media. In: Proceedings of the Ninth ACM International Conference on Web Search and Data Mining, pp. 83–92 (2016)
4. Bliss, et al. An evolutionary algorithm approach to link prediction in dynamic social networks. J. Computat. Sci. **5**(5), 750–764 (2014)
5. Bucur and Iacca. Influence maximization in social networks with genetic algorithms. In: European Conference on the Applications of Evolutionary Computation, pp. 379–392. Springer (2016)
6. Calvano, E., Immordino, G., Scognamiglio, A.: What drives segregation? evidence from social interactions among students. Econ. Educ. Rev. **90**, 102290 (2022)
7. Chen, et al.: Proportionally fair clustering. In: International Conference on Machine Learning, pp. 1032–1041. PMLR (2019)
8. Chhabra, et al.: An overview of fairness in clustering. IEEE Access (2021)
9. Currin, C.B., Vera, S.V., Khaledi-Nasab, A.: Depolarization of echo chambers by random dynamical nudge. Sci. Rep. **12**(1), 9234 (2022)
10. DiPrete, T.A., Gelman, A., McCormick, T., Teitler, J., Zheng, T.: Segregation in social networks based on acquaintanceship and trust. Am. J. Sociol. **116**(4), 1234–1283 (2011)
11. Dong, et al.: Fairness in graph mining: a survey. arXiv preprint arXiv:2204.09888 (2022)
12. Granovetter, M.S.: The strength of weak ties. Am. J. Sociol. **78**(6), 1360–1380 (1973)
13. Huang, J., Gates, A.J., Sinatra, R., Barabási, A.-L.: Historical comparison of gender inequality in scientific careers across countries and disciplines. Proc. Natl. Acad. Sci. **117**(9), 4609–4616 (2020)
14. Karimi, F., Génois, M., Wagner, C., Singer, P., Strohmaier, M.: Homophily influences ranking of minorities in social networks. Sci. Rep. **8**(1), 1–12 (2018)
15. Karimi, F., Oliveira, M.: On the inadequacy of nominal assortativity for assessing homophily in networks. arXiv preprint arXiv:2211.10245 (2022)
16. Karimi, F., Oliveira, M., Strohmaier, M.: Minorities in networks and algorithms. arXiv preprint arXiv:2206.07113 (2022)
17. Katoch, S., Chauhan, S.S., Kumar, V.: A review on genetic algorithm: past, present, and future. Multim. Tools Appl. **80**, 8091–8126 (2021)

18. Kora, P., Yadlapalli, P.: Crossover operators in genetic algorithms: a review. Int. J. Comput. Appl. **162**(10) (2017)
19. Liben-Nowell, D., Kleinberg, J.: The link prediction problem for social networks. In: Proceedings of the Twelfth International Conference on Information and Knowledge Management, pp. 556–559 (2003)
20. Mazzoni, F., Manerba, M.M., Cinquini, M., Guidotti, R., Ruggieri, S.: Genfair: a genetic fairness-enhancing data generation framework. In: International Conference on Discovery Science, pp. 356–371. Springer (2023)
21. Mehrabi, N., Morstatter, F., Peng, N., Galstyan, A.: Debiasing community detection: the importance of lowly connected nodes. In: Proceedings of the 2019 IEEE/ACM International Conference on Advances in Social Networks Analysis and Mining, pp. 509–512 (2019)
22. Micha, et al.: Proportionally fair clustering revisited. In: 47th International Colloquium on Automata, Languages, and Programming (ICALP 2020). Schloss Dagstuhl-Leibniz-Zentrum für Informatik (2020)
23. Morini, V., et al.: From perils to possibilities: understanding how human (and AI) biases affect online Fora. arXiv preprint arXiv:2403.14298 (2024)
24. Nguyen, H., Gokhale, S.S.: Analyzing extremist social media content: a case study of proud boys. Soc. Netw. Anal. Mining **12**(1), 115 (2022)
25. Pal, R., Kumar, A., Santhanam, M.S.: Depolarization of opinions on social networks through random nudges. Phys. Rev. E **108**(3), 034307 (2023)
26. Pessach, D., et al.: A review on fairness in machine learning. ACM Comput. Surv. **55**(3), 1–44 (2022)
27. Pizzuti. Ga-net: a genetic algorithm for community detection in social networks. In: International Conference on Parallel Problem Solving from Nature, pp. 1081–1090. Springer (2008)
28. Rahman, T., Surma, B., Backes, M., Zhang, Y.: Towards Fair Graph Embedding. Fairwalk (2019)
29. Saxena, A., Fletcher, G., Pechenizkiy, M.: Fairsna: algorithmic fairness in social network analysis. arXiv preprint arXiv:2209.01678 (2022)
30. Sharma, S., et al.: Data augmentation for discrimination prevention and bias disambiguation. In: AIES, pp. 358–364. ACM (2020)
31. Spillman, R.: Solving large knapsack problems with a genetic algorithm. In: 1995 IEEE International Conference on Systems, Man and Cybernetics. Intelligent Systems for the 21st Century, vol. 1, pp. 632–637. IEEE (1995)
32. Stoica, A.-A., Chaintreau, A.: Fairness in social influence maximization. In: Companion Proceedings of the 2019 World Wide Web Conference, pp. 569–574 (2019)
33. Tsioutsiouliklis, S., Pitoura, E., Tsaparas, P., Kleftakis, I., Mamoulis, N.: Fairness-aware pagerank. Proc. Web Conf. **2021**, 3815–3826 (2021)
34. Wagner, C., Singer, P., Karimi, F., Pfeffer, J., Strohmaier, M.: Sampling from social networks with attributes. In: Proceedings of the 26th International Conference on World Wide Web, pp. 1181–1190 (2017)
35. Wang, et al.: Information access equality on generative models of complex networks. Appl. Netw. Sci. **7**(1), 1–20 (2022)
36. Wang and Saar-Tsechansky. Augmented fairness: an interpretable model augmenting decision-makers' fairness (2020)

Agent-Based Modelling Meets Generative AI in Social Network Simulations

Antonino Ferraro⬤, Antonio Galli⬤, Valerio La Gatta(✉)⬤,
Marco Postiglione⬤, Gian Marco Orlando⬤, Diego Russo⬤,
Giuseppe Riccio⬤, Antonio Romano⬤, and Vincenzo Moscato⬤

Department of Electrical Engineering and Information Technology, University of
Naples Federico II, Naples, Italy
{antonino.ferraro,antonio.galli,valerio.lagatta,marco.postiglione,
vincenzo.moscato}@unina.it, {gian.orlando,diego.russo,giuseppe.riccio9,
antonio.romano45}@studenti.unina.it

Abstract. Agent-Based Modelling (ABM) has emerged as an essential
tool for simulating social networks, encompassing diverse phenomena
such as information dissemination, influence dynamics, and community
formation. However, manually configuring varied agent interactions and
information flow dynamics poses challenges, often resulting in oversim-
plified models that lack real-world generalizability. Integrating modern
Large Language Models (LLMs) with ABM presents a promising avenue
to address these challenges and enhance simulation fidelity, leveraging
LLMs' human-like capabilities in sensing, reasoning, and behavior. In this
paper, we propose a novel framework utilizing LLM-empowered agents
to simulate social network users based on their interests and personality traits. The framework allows for customizable agent interactions
resembling various social network platforms, including mechanisms for
content resharing and personalized recommendations. We validate our
framework using a comprehensive Twitter dataset from the 2020 US
election, demonstrating that LLM-agents accurately replicate real users'
behaviors, including linguistic patterns and political inclinations. These
agents form homogeneous ideological clusters and retain the main themes
of their community. Notably, preference-based recommendations significantly influence agent behavior, promoting increased engagement, network homophily and the formation of echo chambers. Overall, our findings underscore the potential of LLM-agents in advancing social media
simulations and unraveling intricate online dynamics.

Keywords: Agent-Based Modelling · Social media simulation ·
Generative Artificial Intelligence

1 Introduction

Over the past decades, there has been a concerted effort among researchers and
practitioners to develop computational agents capable of realistically emulating

L. M. Aiello et al. (Eds.): ASONAM 2024, LNCS 15211, pp. 155–170, 2025.
https://doi.org/10.1007/978-3-031-78541-2_10

human behavior [27]. Agent-Based Modelling (ABM) has emerged as a pivotal methodology for simulating intricate systems by delineating rules governing individual agents' behavior and interactions [8]. Within the domain of social network analysis, ABM has played a crucial role in both the development and validation of novel theories pertaining to human behavior in online environments. These theories encompass a wide array of phenomena such as opinion formation [21], (false) news propagation [28], and collective decision-making [29]. Nevertheless, manually crafting agent behavior to encompass the diverse spectrum of interactions, information flow dynamics, and user engagement within social networks proves to be highly challenging. This challenge often leads to an oversimplification of agents or the social media environment itself, where underlying mechanisms are rigidly encoded in predefined parameters. Consequently, such setups are prone to researcher bias, potentially resulting in a lack of fidelity in modeling complex human behaviors, especially those involving collective decision-making [3].

Modern Large Language Models (LLMs) not only excel in generating human-like text but also demonstrate remarkable performance in complex tasks requiring reasoning, planning, and communication [15]. This proficiency has sparked interest in integrating LLMs with ABM, termed Generative Agent-Based Modelling (GABM). Unlike traditional ABM methods that often necessitate intricate parameter configurations, GABM leverages LLMs' capacity for role-playing, ensuring diverse agent behaviors that closely mirror real-world diversity. For instance, Park et al. [25] demonstrated that generative agents, designed for daily activities, exhibited credible individual and social behaviors, including expressing opinions and forming friendships, without explicit instructions. Similarly, Williams et al. [12] showcased the collective intelligence of generative agents in epidemic modeling, accurately simulating real-world behaviors like quarantine and self-isolation in response to escalating disease cases. These pioneering findings support investigating GABM as an effective approach to enhance social media simulations. To our knowledge, the seminal work by Gao et al. [9] lays the foundation for this research direction by qualitatively demonstrating that LLM-agents exhibit realistic behaviors related to information propagation and the manifestation of attitudes and sentiment. However, it remains unclear whether LLM-agents can accurately represent real users in terms of their personality traits (e.g., being outspoken, being critical) and interests (e.g., social issues, political preferences), regardless of the explicit emotions conveyed through their textual posts. Furthermore, their ability to exhibit community-level phenomena (e.g., homophily, polarization), as well as their susceptibility to recommendation strategies, remains uncertain.

Contributions of This Work. In this paper, we directly target these challenges and propose a novel framework which employs LLM-empowered agents to simulate users within a social network. Initially, we construct an environment using authentic real-world social network data. To ensure the authenticity of this environment, we propose an *Agent Characterization Module* that combines prompt engineering and prompt tuning to infer users' personality traits and interests. Subsequently, the simulation unfolds in two cyclical components:

the *Reasoning Module* that delineates each agent's decision in the simulation (e.g., posting original content, resharing, remaining inactive), and the *Interaction Module* that stores agent's past behavior and specify how agents are exposed to content from other agents (e.g., through preference-based, popularity-based or random recommendations). Notably, the Reasoning Module is fueled by the Interaction Module, offering insights for agents' informed decision-making within the simulated social network environment. We evaluate the efficacy of our proposed framework in approximating a real social media platform by scrutinizing the individual characteristics of LLM-agents compared to real users and exploring the typical network-level interactions observed in social media networks. To operationalize this objective, we formulate the following research questions (RQs):

RQ1: *Do LLM-agents represent the interests of the users they are instructed to impersonate?*

RQ2: *Do LLM-agents form communities and/or echo chambers?*

Leveraging a large-scale Twitter dataset from the 2020 US election, we found that LLM-agents accurately mirror real users' linguistic patterns and preserve their political leaning, thus enhancing simulation authenticity. These agents also exhibit realistic behavior by resharing content aligned with their beliefs and engaging in similar communication styles as their community, reflecting online interactions accurately. Furthermore, we found that LLM-agents aggregate into homogeneous ideological groups based on their individual preferences. Finally, we also observed the significant impact of recommendation strategies on agent behavior, emphasizing the efficacy of preference-based recommendations for promoting higher engagement and echo chambers formation. Overall, our findings prove the promising capabilities offered by LLM-agents to enhance social media simulations.

2   Related Works

2.1   Agent-Based Modelling for Social Media Simulation

The exponential progress in computational capabilities has transformed social media simulations, becoming a pivotal tool for understanding the intricate dynamics governing these digital spaces. ABM stands as a robust methodology, orchestrating interactions among individual agents based on predefined yet realistic rules. Specifically, ABM has enabled the investigation of complex online behaviors, including information dissemination [32], influence dynamics [20], and the impact of automated bots on news propagation [2]. In addition, ABM has been crucial to evaluate specific disinformation countermeasures [5], such as content moderation [23] and fake news inoculation [11]. While the abovementioned studies highlight ABM's utility in elucidating and modeling social media phenomena, they collectively confront several limitations. First, modeling human/agent behavior often requires detailed calibration, making ABM outcomes sensitive to parameter values and assumptions/simplifications used in the simulation. Second, ABM heavily relies on predefined rules, which inherently introduce the potential for researcher bias and may impede the accurate representation of social media complexities such as the spread of multiple information narratives and/or conflicting viewpoints [17]. Although methods like learning rules through reinforcement learning offer partial mitigation, challenges persist, particularly in scenarios where explicit reward functions for optimization are absent. In this paper, we propose a novel paradigm for social media simulations that leverages generative agents, i.e., agents empowered with LLMs' capabilities, to autonomously learn and adapt their behavior based on extensive language understanding and context reasoning, reducing the reliance on explicit parametrization and predefined rules. This paradigm shift not only enhances the fidelity and realism of the agents, but can influence the robustness and validity of ABM results.

2.2   Generative Agents

Modern LLMs excel not only in generating human-like text but also in complex tasks like reasoning and planning, making them valuable for enhancing simulation fidelity and complexity. The integration of LLMs with ABM, i.e., Generative Agent-Based Modelling (GABM) has garnered research interest for its potential in simulating realistic behaviors [30]. For example, Park et al. [25] demonstrated that generative agents in daily activities exhibited credible behaviors at both individual and social levels without explicit instructions. Similarly, Williams et al. [12] showcased the collective intelligence of generative agents in epidemic modeling, mimicking real-world responses to disease outbreaks.

In this study, we contribute to the GABM research field by assessing the effectiveness of LLM-empowered agents in simulating social networks. To our knowledge, the S^3 framework [9] is the only prior work that delves into the potential of GABM for social media simulations. However, our approach differs significantly from S^3 in several critical aspects. First, our agent initialization strategy allows for characterizing users based on their personality and (political) interests, offering a higher level of personalization compared to S^3, which primarily focuses on demographic attributes (e.g., gender, age, occupation). Second, our Interaction Module permits custom information exposure definitions, facilitating the evaluation of diverse recommendation strategies' impacts. In contrast, S^3 employs a simpler and less realistic interaction mechanism where every user is uniformly exposed to all others. Third, we utilize an open-source LLM for our experiments instead of the commercial GPT3.5 service, enhancing result transparency and accessibility. Lastly, our evaluation extends the validation of LLM-agents beyond individual properties to investigate the networks they tend to form. Specifically, we move beyond news dissemination to analyze network features such as homophily, polarization, and the controversies arising from LLM-agent interactions.

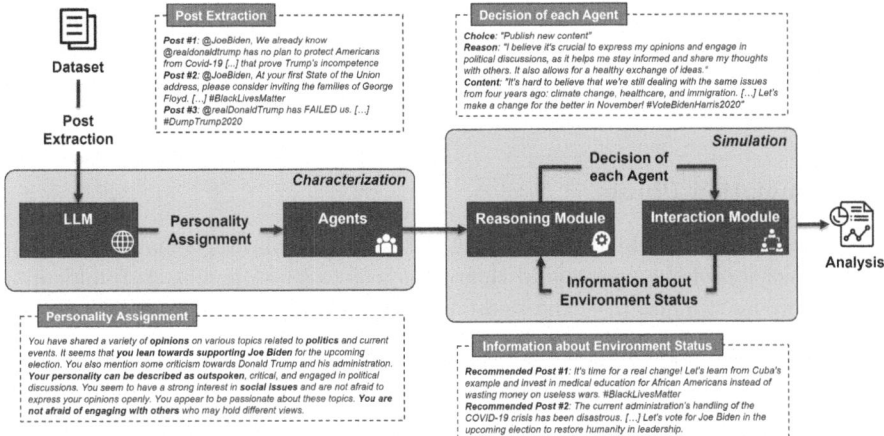

Fig. 1. Our framework comprises two primary phases: (i) *Characterization*, where each agent embodies the personality traits and interests extracted (via LLM) from the original posts of the real user it is tasked to emulate; and (ii) *Simulation*, where the decision-making process of each agent, represented as a *Choice-Reason-Content* triple (*Reasoning Module*), is stored within the *Interaction Module*. Consequently, each agent autonomously makes decisions, considering the context and having access to recommended contents posted by other agents.

3   Methodology

Figure 1 depicts the architecture of the proposed framework, comprising two complementary phases: (i) *Characterization*, responsible for initializing generative agents based on real users' interests and personality traits, and (ii) *Simulation*, responsible for executing the simulation dynamics, encompassing agents' decision-making processes and their interactions with each other. The following sections provide a detailed description of each component.

3.1   Characterization Phase

The primary objective of this phase is to profile each agent before commencing the simulation, where an agent simulates a user within a social network. In this context, we consider the user's original content that the agent will emulate and adopt a prompt-based approach to extract the user's personality and interests,eliminating the necessity to define apriori the parameters for agent characterization. This approach ensures diversity among individual agents, aiming to approximate the genuine interests of the respective user. Unlike prior research [9] focusing solely on user demographics, we conjecture that considering the user's personality and interests provides a more accurate representation of social media users' characteristics. Indeed, users personality reflects their engagement style, while their interests reveal the topics that genuinely pique their curiosity. For example, as depicted in Fig. 1 (left), an agent described as "outspoken" and "critical" embodies a user unafraid to voice opinions and evaluate information before engaging with it. Furthermore, its (political) interests indicate support for Joe Biden for the 2020 US election and alignment with his political stances on social issues.

3.2   Simulation Phase

This phase is responsible for conducting the actual simulation dynamics, including the decision of the agents and their interactions. Specifically, it unfolds in a cycle involving two modules: the Reasoning Module and the Interaction Module.

Reasoning Module. In each iteration (i) of the simulation, every generative agent is immersed in an environment resembling a social media platform. In line with prior research [9], we design a prompt that introduces the social media environment, highlighting the presence of other agents and outlining the possible actions to take within the environment: (i) *generating original content*, (ii) *resharing content from other agents*, and (iii) *remaining inactive*. This setup, though simple, accurately mirrors common actions observed across various social media platforms, irrespective of platform-specific regulations. Moreover, we emphasize that these types of actions are the sole predefined aspects in the simulation. Instead, the agent behavior is not determined by deterministic or probabilistic processes as LLM-agents autonomously decide the actions and the content to generate based on their personality, interests, and other agents' behavior. Subsequently, the output of the Reasoning Module—the response generated by each agent—is structured as a triplet denoted by *Choice-Reason-Content*. For instance, as illustrated in Fig. 1, an agent may *choose* to publish original *content* to discuss persistent societal issues like climate change or immigration often targeted during electoral campaigns. Additionally, the agent provides the *reasoning* behind the post, indicating an intent to engage in political discourse and foster a healthy exchange of ideas.

Interaction Module. To achieve a comprehensive simulation of a social network, enabling interaction among agents is imperative. This involves gathering all actions performed by agents and presenting to each agent the activities of others. However, including all information generated by other users within a single prompt presents practical challenges. First, the context window of any LLM-agent is limited, causing memory saturation if all information is included in one prompt. Second, longer prompts increase the risk of hallucinations [13], resulting in agents producing inaccurate, contextually inconsistent, or nonsensical content. To tackle these challenges, we introduce the innovative use of Retrieval-Augmented Generation (RAG). This technique enables LLM-agents to access additional contextually relevant data stored in an external vector database [18]. Figure 2 illustrates the integration of the RAG technique within the *Interaction Module*. Specifically, the vector database is continuously updated to record all agent actions and their corresponding published contents. This tracking mechanism facilitates monitoring the simulation's progression for subsequent analysis.

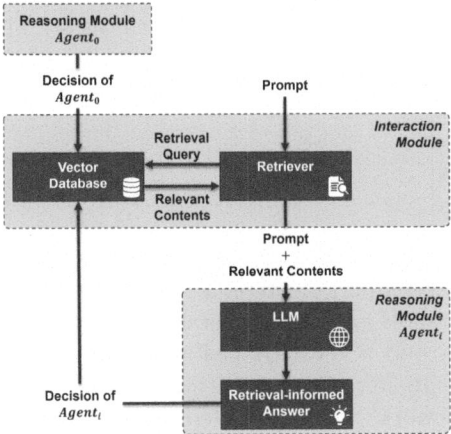

Fig. 2. Workflow of the RAG-empowered interaction: At the beginning of the Simulation phase, the first agent's decision is stored in the vector database. The RAG mechanism of the Interaction Module retrieves contextually relevant data from the database to enrich the prompt for the next agent to assist in making informed decisions within the Reasoning Module, taking into account the current environmental state and previously published content by other agents. The retrieval-informed decision is, in turn, stored in the vector database, repeating the cycle until the simulation ends.

Furthermore, the *Retriever* step within RAG serves as a recommendation system, determining which content to present to each LLM-agent. For instance, as depicted in Fig. 1, the agent may receive recommendations for two posts from the entire corpus of published content. These posts might discuss topics such as investing in immigrant education and addressing the challenges posed by the COVID-19 pandemic. Importantly, these recommendations are aligned with the user's previous decisions, reflecting the salience of these topics during the run-up to the 2020 US election, and potentially foster further interactions between agents. Notably, this approach maintains a nearly constant prompt size throughout the simulation and enables the integration of various recommendation strategies. In our work, we focus on preference-based recommendation, i.e., recommending content aligned with agents' preferences, and random recommendation, i.e., exposing agents to random content and thus more diverse viewpoints. We will empirically assess the impacts of these strategies in our experiments.

It is important to clarify that these recommendations are not actual recommendation methods like collaborative filtering but rather a simplified approximation of them. We will empirically assess the impacts of these strategies in our experiments. Finally, the Interaction Module provides the Reasoning Module with a series of contents published by other agents to inform the LLM about the environment's state and other agents' publications, ensuring continuous updating of the environment's state regarding agents' decisions. Upon querying all agents, the current iteration concludes, and the next begins, following a round-robin logic involving sequential querying of each agent. This iterative process continues until the simulation's end, emulating social network evolution based on individual agent actions.

4   Experiments

4.1   Experimental Setup

All experiments have been performed on a computing system featuring an 11th Gen Intel Core i7-11800H processor operating at 2.30 GHz, 16 GB of RAM clocked at 3200 MHz, and a NVIDIA GeForce RTX 3060 Laptop GPU. Each simulation comprises 10 iterations, resulting in approximately 15 h per simulation run. The framework[1] is implemented using PyAutogen [31], with ChromaDB[2] serving as the vectorial database supporting the RAG technique. Unlike prior studies [12,25] focusing on generative agents, we have chosen to employ an open-source LLM, i.e., Dolphin 2.1 Mistral 7B[3], instead of proprietary foundation models (e.g., GPT-4 [24], Google Gemini[4]). This decision stems from the model's unrestricted nature and its data filtering policies designed to mitigate alignments and biases.

4.2   Dataset

We utilized a dataset of election-related tweets obtained through Twitter's streaming API service during the lead-up to the 2020 US election [7]. Specifically, our focus spanned six months, from June 2020 to December 2020, encompassing the latter stages of the electoral campaign and the aftermath of the election. Over this observation period, we have collected a dataset comprising more than 12 million tweets, encompassing original tweets, replies, retweets, and quotes, disseminated by 1.1 million unique users [16]. To ensure authenticity, we excluded all accounts flagged as bots by the Botometer API[5]. Additionally, our analysis concentrated on original tweets to extract insights into users' personalities and interests. Eventually, we annotated the political affiliations of 100 users, revealing that 73 were associated with the Republican community, while the remaining 27 were aligned with the Democratic community. We have adopted this group of users to instantiate the agents of every simulation.

4.3   Characterizing LLM-Agents Vs. Real Users' Interests (RQ1)

To answer RQ1, we characterize LLM-agents and the users that they are supposed to impersonate across three dimensions:

---

[1] The code will be made available upon acceptance.
[2] https://www.trychroma.com/.
[3] https://huggingface.co/TheBloke/dolphin-2.1-mistral-7B-GGUF.
[4] https://gemini.google.com/.
[5] https://botometer.osome.iu.edu.

- *Keywords usage*: We analyze the most relevant keywords used by LLM-agents in comparison to real users;
- *Interests*: We examine the political leaning exhibited by LLM-agents during the simulation;
- *Content similarity*: We investigate the semantic similarity of LLM-agents with respect to the community they belong to.

Keywords Usage. We employ YAKE [6] and KeyBERT [26] algorithms to extract keywords using statistical features and contextual embeddings, respectively. Specifically, we focus on original content and merge the vocabulary used by all agents during simulation as well as the vocabulary used by real users. This approach is motivated by the significant diversity in the scale of simulation activities compared to real users, where real users publish much more original tweets than their LLM-agents' counterparts. Table 1 shows the top-10 keywords used by LLM-agents and real Twitter users, demonstrating that the simulation, and consequently the LLM-agents, align with the primary discussion topics of real Twitter conversations, particularly the debates between Trump and Biden during the 2020 US election. Additionally, Fig. 3 shows the distributions of keywords usage in both real Twitter discussions (in blue) and the simulation (in red). The visibly right-skewed distributions, coupled with statistical validation through a Mann-Whitney test (p-value< 0.05), affirm the simulation's fidelity in reflecting natural language patterns. This observation underscores the simulation's ability to accurately mimic the overarching themes of Twitter conversations, with few keywords being frequently utilized while others are less prevalent. Collectively, these results demonstrate that LLM-agents effectively capture the main topic of the conversation, indicating the robustness of the simulation in replicating real-world discourse.

Table 1. Top-10 Keywords in Real Case and Simulation

Real Case	Simulation
realdonaldtrump	trump
trump	president
president	biden
biden	administration
joebiden	freedom
people	maga
america	actions
covid	change
time	covid
maga	state

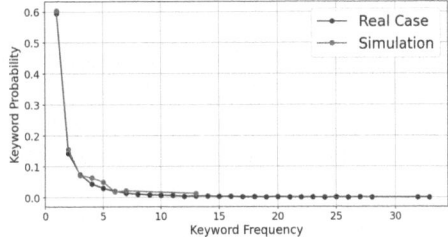

Fig. 3. The distributions of keywords usage in real Twitter discussions and the simulation. The x-axis represents the number of occurrences of a keyword, while the y-axis represents the probability of encountering a keyword with a specific frequency.

Interests. We examine the political leaning of LLM-agents as a proxy for their (political) interests. Initially, we fine-tuned a BERT-base transformer using the annotated dataset of 100 Twitter users, achieving impressive performance metrics: 91% accuracy and 94% F1 score. Subsequently, we utilized this model to predict the political leaning scores of LLM-agents at the simulation's conclusion. As previously mentioned, we investigate two exposure strategy: preference-based recommendation and random recommendation. To enhance robustness, we repeated the random recommendation simulation three times. Notably, the RAG-enhanced Interaction Module allows a seamless integration of any recommendation strategy. Table 2 shows a strong positive correlation between the political leaning scores of real users and LLM-agents[6]. Specifically, the majority of agents (always $\geq 86\%$) has retained the political orientations of the real users they are instructed to impersonate. Interestingly, this is irrespective of the type of recommendation strategy. However, fewer agents alter their alignments during the simulations. After qualitative scrutiny, we have found that these shifting agents typically impersonate nonpartisan users. Therefore, exposure to diverse ideas may blur their political orientations. Altogether, this analysis indicates that LLM-agents accurately interpret real users' political alignments, regardless of the content they are exposed to.

Table 2. Results of Political Leaning Analysis and Interaction Patterns

	Political Leaning Analysis				Interaction Patterns		
	Consistent Users	Changed Users	Spearman Coefficient	P-Value	Original Publications	Non Interactions	Reshares
Pref-Based	88%	12%	0.494	<0.05	38.9%	10.5%	50.6%
Random 1	86%	14%	0.498	<0.05	90.0%	1.6%	8.4%
Random 2	88%	12%	0.490	<0.05	89.1%	2.5%	8.4%
Random 3	89%	11%	0.509	<0.05	90.1%	1.9%	8.0%

Content Similarity. Lastly, we investigate the semantic similarity of the content published by LLM-agents. Specifically, we utilized a sentence-transformers model[7] to extract contextual embeddings for each original content posted by LLM-agents. We then constructed a cosine similarity matrix $\mathcal{M} = \{m_{ij} = \text{sim}(c_i, c_j)\}$, where c_i and c_j represent the vector embeddings of two posts and $\mathcal{M} \in \mathcal{R}^{k \times k}$, k being the total number of posts published by the agents. To evaluate the semantic similarity, we define two measures:

- *Self-Similarity*: it assesses the similarity between posts from the same agent A. In other words, it represents the average cosine similarity measured on $\mathcal{M}$ considering only the agent's original content, that is $\{m_{ij}|c_i \neq c_j \land c_i, c_j$ posted by $A\}$;
- *Intra-Cluster Similarity*: it assesses the similarity between posts from an agent A and those from other agents within the same political group $\mathcal{A}_p = \{A_1, A_2, \cdots, A_k|A_i \neq A \land pol(A_i) = pol(A)\}$, *pol* being the political leaning

---

[6] We employed the Spearman index with a significance level of p-value< 0.05 to verify this correlation.

[7] https://huggingface.co/sentence-transformers/all-MiniLM-L6-v2.

scoring function. Concretely, the metric is determined by averaging the cosine similarity of $\mathcal{M}$'s elements, that is $\{m_{ij}\}|c_i$ and c_j posted by A and $A_i \in \mathcal{A}_p$, respectively.

Fig. 4 shows the distributions of *Self-Similarity* for real users and LLM-agents across two simulation scenarios: one featuring preference-based recommendation and the other with random recommendation. Notably, for the real case, the cosine similarity matrix is built considering the original tweets of the users. We notice a clear disparity in self-similarity between LLM-agents and real users within the preference-based recommendation framework. Specifically, the median self-similarity for real users (0.274) contrasts with that of LLM-agents (0.531), with the latter exhibiting significantly higher self-similarity[8]. This observation indicates a tendency for the same LLM-agent to converge more strongly around similar topics compared to the real user it is impersonating, suggesting that preference-based recommendation might penalise the ability of LLM-agents to engage with brand-new content. Conversely, random recommendations appear to mitigate this polarization trend and, surprisingly, achieve a better approximation of the real case. Additionally, Fig. 5 replicates the same analysis but considers *Intra-Cluster Similarity*, revealing minimal distinctions among the three distributions. Surprisingly, the simulation employing preference-based recommendation aligns better with the real scenario compared to the previous analysis. This suggests that although the agents may not publish as diverse content as their real counterparts, they still preserve the overall semantic characteristics of their community. This outcome, combined with the previous discovery that the majority of LLM-agents retain the political leaning of their real counterparts, implies an accurate representation of the target users and an effective portrayal of their respective communities by the LLM-agents.

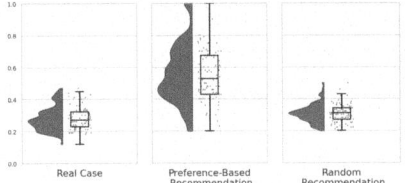   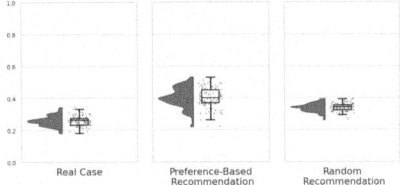

Fig. 4. Comparison of self-similarity distributions between real users and LLM-agents.

Fig. 5. Comparison of intra-cluster similarity distributions between real users and LLM-agents.

---

[8] This finding was validated using a Mann-Whitney test (p-value< 0.05).

4.4 Community Formation and Echo Chambers Among LLM-Agents (RQ2)

To address RQ2, we delve into agent behavior, specifically focusing on interactions, i.e., reshares, made by LLM-agents during the simulation. Figure 6 depicts an illustrative example involving three agents—Yuri, Emily, and Daniel—engaging in a political discussion regarding the 2020 US election. Daniel, a Democrat, refrains from further social media activity due to conflicting views with other agents. Conversely, Yuri posts new content highlighting policy concerns about abortion rights and COVID-19 restrictions. In an effort to engage in the ongoing conversation with her followers, Emily opts to reshare Yuri's post, given their shared political cluster and similar concerns. To quantitatively assess agents' behavior, we formalize the *interaction graph* as follows: the nodes represent LLM-agents, and the (directed) edges represent resharing activity from the publisher agent to the agent that performs the reshare. Subsequently, we delve into the agents' resharing activity and examine community-based metrics, including homophily and controversy.

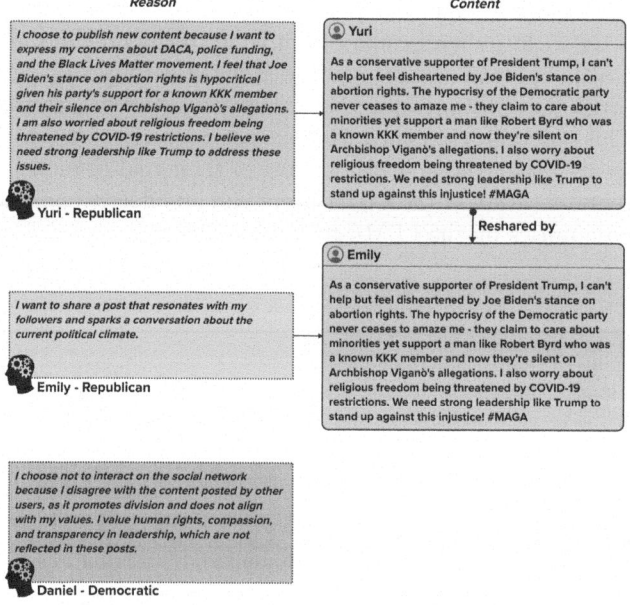

Fig. 6. Distinct behavioral reactions of agents to content based on individual preferences and political alignment can be observed. Yuri contributes by posting original content aligned with its preferences. Emily actively engages with followers by resharing content. Conversely, Daniel refrains from interaction, influenced by exposure to content conflicting with its preferences.

Sharing Activity. Table 2 shows the interaction patterns in terms of the percentages of actions performed by the agents. The influence of the recommendation system is evident: in simulations featuring random recommendations, agents exhibit a significant decrease in content resharing from others ($\leq 8.4\%$) compared to simulations employing preference-based recommendations (50.6%). This trend stems from the increased exposure to (random) content not aligned with the agent's preferences. Indeed, the random recommendation strategy results in agents publishing original content approximately 90% of the times in an attempt to push their ideas in the environment. Surprisingly, while personalized recommendations usually boost user engagement on real social media platforms [19], our simulations reveal a greater proportion of non-interactions when recommendations are preference-based (10.5%) with respect to random recommendation ($\leq 2.5\%$). We attribute this to our empirical observation that the preference-based recommendation is not flawless and may sometimes suggest irrelevant posts to the agents. Overall, these findings suggest that the recommendation strategy strongly affects the agents' activity, with preference-based recommendations fostering increased interactions among agents.

Homophily. To investigate homophily, we examine the above-mentioned *interaction graph* among agents in each simulation. Homophily, defined as individuals' inclination to associate with others sharing similar attributes, leads to the formation of homogeneous groups [22]. Given the political context of our data, we focus on agents' political alignment to delineate group clusters. Subsequently, we analyze the inter-cluster edges of the interaction graph, representing connections between nodes from different clusters. The results of the homophily analysis presented in Table 3 indicate that networks from simulations employing preference-based recommendations exhibit reduced inter-cluster connectivity, signifying a stronger homophily effect. This results in a 10% lower number of inter-cluster edges than the number needed to declare the network non-homophilic [14]. Statistical analyses confirm the significance of these homophily values (*p-value* < *0.05*). Conversely, simulations using random recommendations demonstrate relatively higher inter-cluster connectivity (+4% inter-cluster edges) with respect to preference-based recommendation, resulting in networks that are not statistically homophilic (*p-value* > *0.05*). These results affirm our previous findings regarding the determining effect of the recommendation strategy, but also complement them by illustrating that preference-based recommendation not only fosters agents' engagement but also encourages the formation of distinct clusters with limited inter-cluster interaction.

Echo Chamber Analysis. In line with previous works [1,4,10], we utilize established metrics to examine the emergence of echo chambers in our simulations. These metrics analyse the *interaction graph* to evaluate the level of controversy within discussions. Specifically, we utilize Random Walk Controversy (RWC) to analyze transition probabilities between ideological clusters, Betweenness Centrality Controversy (BCC) to assess partition distances, and Boundary Connectivity (GMCK) to compute the structural arrangement of the interaction graph [10]. In all cases, the higher the metric, the more controversy is the *inter-*

Table 3. Results of Homophily Analysis and Controversy Metrics

	Homophily Analysis		Controversy Metrics		
	Modularity	Homophily?	RWC	BCC	GMCK
Pref-Based	0.375	Yes, −10% inter-cluster edges (p-value <0.05)	**0.692**	**0.423**	**0.334**
Random 1	0.416	No, +4% inter-cluster edges	0.423	0.189	0.071
Random 2	0.416	No, +6% inter-cluster edges	0.541	0.230	0.268
Random 3	0.428	No, +4% inter-cluster edges	0.431	0.362	−0.009

action graph. The underlying premise is that contentious topics often involve individuals with contrasting viewpoints engaging in dialogue, while individuals sharing similar beliefs tend to reinforce each other's arguments [1]. Results in Table 3 unveil a notable prevalence of controversy in preference-based simulations, indicating a stronger inclination towards echo chamber formation. Conversely, random recommendation mitigates echo chamber formation, promoting a broader diversity of opinions. These findings underscore that LLM-agents within a community are unlikely to interact with individuals holding opposing viewpoints, suggesting that the recommendation strategy not only influences community formation but also reinforces these communities by fostering closed networks, wherein users are exposed to limited diversity.

5   Conclusions and Future Works

In this paper, we proposed a novel framework that integrates agent-based modeling with LLM capabilities for social media simulation. Our framework incorporates a Characterization Module, enabling the inference of realistic users' personality traits and interests. Furthermore, our Interaction Module pioneers the application of RAG mechanisms to implement various recommendation strategies. By focusing on Twitter discussions surrounding the 2020 US election, we demonstrated that the simulated LLM-agents effectively mirror the users they are tasked to emulate, maintaining their original political orientations and preserving the thematic content within their respective communities (RQ1). Furthermore, our exploration into their interactions has unveiled a tendency for these agents to cluster with like-minded peers, especially under preference-based recommendation settings (RQ2).

Moving forward, our research aims to explore several key directions. First, we plan to augment our framework with more sophisticated recommendation algorithms, e.g., based on collaborative filtering. Additionally, we aim to enable LLM-agents to perform a broader spectrum of actions (e.g., following and liking posts). Second, we are planning to expand our research to include multiple LLMs, which will allow us to analyze potential biases across different models. Lastly, we emphasize the flexibility of our framework beyond social media simulations, highlighting potential cross-disciplinary applications in simulation fields (e.g., epidemiology or business planning).

Acknowledgments. This work was supported by the European Project DEUCE, Digitalising European Uncontested Claims Enforcement. Grant Number: 101138437. Call: JUST-2023-JCOO. Type of Action: JUST-LS.

References

1. Adamic, L.A., Glanc, N.: The political blogosphere and the 2004 U.S. election: divided they blog. In: LinkKDD, vol. 36–43 (2005)
2. Beskow, D.M., Carley, K.M.: Agent based simulation of bot disinformation maneuvers in twitter. In: 2019 Winter Simulation Conference, WSC 2019, National Harbor, MD, USA, 8-11 December 2019, pp. 750–761. IEEE (2019)
3. Bonabeau, E.: Agent-based modeling: methods and techniques for simulating human systems. In: Proceedings of the National Academy of Sciences (2002)
4. Bruns, A.: Echo chamber? what echo chamber? reviewing the evidence. In: School of Communication. Digital Media Research Centre, Cardiff (2017)
5. Butts, D.J., Bollman, S.A., Murillo, M.S.: Mathematical modeling of disinformation and effectiveness of mitigation policies. Sci. Rep. **13**(1), 18735 (2023)
6. Campos, R., Mangaravite, V., Pasquali, A., Jorge, A., Nunes, C., Jatowt, A.: YAKE! keyword extraction from single documents using multiple local features. Inf. Sci. J. 509, 257–289 (2018)
7. Chen, E., Deb, A., Ferrara, E.: #election2020: the first public twitter dataset on the 2020 us presidential election. J. Comput. Soc. Sci. **5**(1), 1–18 (2022). https://doi.org/10.1007/S42001-021-00117-9
8. E. Elliott, L.D.K.: Agent-based modeling in the social and behavioral sciences. In: Nonlinear Dynamics, Psychology, and Life Sciences, vol. 8, no. 2 (2004)
9. Gao, C., et al.: S3: Social-network simulation system with large language model-empowered agents. arXiv:2307.14984 (2023)
10. Garimella, K., Morales, G.D.F., Gionis, A., Mathioudakis, M.: Quantifying controversy on social media. arXiv:1507.05224v5 (2017)
11. Gausen, A., Luk, W., Guo, C.: Can we stop fake news? Using agent-based modelling to evaluate countermeasures for misinformation on social media. In: ICWSM Workshops (2021)
12. Ghaffarzadegan, N., Majumdar, A., Williams, R., Hosseinichimeh, N.: Epidemic modeling with generative agents. arXiv:2307.04986 (2023)
13. Ji, Z., et al.: Survey of hallucination in natural language generation. ACM Comput. Surv. **55**(12), 1–38 (2023). https://doi.org/10.1145/3571730
14. Kim, K., Altmann, J.: Effect of homophily on network formation. Commun. Nonlinear Sci. Numer. Simul. **44**, 482–494 (2017)
15. Kojima, T., Gu, S.S., Reid, M., Matsuo, Y., Iwasawa, Y.: Large language models are zero-shot reasoners. In: Annual Conference on Neural Information Processing Systems 2022, NeurIPS 2022, New Orleans, LA, USA, November 28 - December 9, 2022 (2022)
16. La Gatta, V., Luceri, L., Fabbri, F., Ferrara, E.: The interconnected nature of online harm and moderation: investigating the cross-platform spread of harmful content between Youtube and twitter. In: Proceedings ACM Human-Computer Interact (2023)
17. La Gatta, V., Moscato, V., Postiglione, M., Sperlí, G.: Covid-19 sentiment analysis based on tweets. IEEE Intell. Syst. **38**(3), 51–55 (2023). https://doi.org/10.1109/MIS.2023.3239180

18. Lewis, P.S.H., et al.: Retrieval-augmented generation for knowledge-intensive NLP tasks. In: Annual Conference on Neural Information Processing Systems 2020 (2020)
19. Liang, T.P., Lai, H.J., Ku, Y.C.: Personalized content recommendation and user satisfaction: theoretical synthesis and empirical findings. J. Manag. Inf. Syst. **23**, 45–70 (2014)
20. van Maanen, P.P., van der Vecht, B.: An agent-based approach to modeling online social influence. In: Proceedings of the 2013 IEEE/ACM International Conference on Advances in Social Networks Analysis and Mining, pp. 600–607 (2013)
21. Mastroeni, L., Vellucci, P., Naldi, M.: Agent-based models for opinion formation: a bibliographic survey. IEEE Access **7**, 58836–58848 (2019). https://doi.org/10.1109/ACCESS.2019.2913787
22. McPherson, M., Smith-Lovin, L., Cook, J.M.: Birds of a feather: homophily in social networks. Ann. Rev. Sociol. **27**, 415–444 (2001)
23. Murdock, I., Carley, K.M., Yağan, O.: An agent-based model of reddit interactions and moderation. In: Proceedings of the 2023 IEEE/ACM International Conference on Advances in Social Networks Analysis and Mining, pp. 195–202. Association for Computing Machinery, New York, NY, USA (2024). https://doi.org/10.1145/3625007.3627489
24. OpenAI: GPT-4 technical report. arXiv:2303.08774 (2023)
25. Park, J.S., O'Brien, J.C., Cai, C.J., Morris, M.R., Liang, P., Bernstein, M.S.: Generative agents: Interactive simulacra of human behavior. arXiv:2304.03442 (2023)
26. Sharma, P., Li, Y.: Self-supervised contextual keyword and keyphrase retrieval with self-labelling. Preprints (2019)
27. Troitzsch, K.G.: Social Science Microsimulation. Springer Science & Business Media, Heidelberg (1996). https://doi.org/10.1007/978-3-662-03261-9
28. Tseng, S., Nguyen, T.S.: Agent-based modeling of rumor propagation using expected integrated mean squared error optimal design. Appl. Syst. Innov. **3**, 48 (2020)
29. van Veen, D., Kudesia, R.S., Heinimann, H.R.: An agent-based model of collective decision-making: how information sharing strategies scale with information overload. IEEE Trans. Comput. Soc. Syst. **7**(3), 751–767 (2020). https://doi.org/10.1109/TCSS.2020.2986161
30. Wang, L., et al.: A survey on large language model based autonomous agents. arXiv:2308.11432 (2023)
31. Wu, Q., et al.: AutoGen: enabling next-Gen LLM applications via multi-agent conversation. arXiv:2308.08155 (2023)
32. Yang, S.Y., Liu, A., Mo, S.Y.K.: Twitter financial community modeling using agent based simulation. In: 2014 IEEE Conference on Computational Intelligence for Financial Engineering and Economics (CIFEr), pp. 63–70 (2014)

DeepFairRank: A Multi-objective Framework for Fair Top-k Node Ranking in Network Data

Francisco Santos$^{(\boxtimes)}$, Farzan Masrour, Pang-Ning Tan,
and Abdol-Hossein Esfahanian

Michigan State University, East Lansing, MI 48825, USA
{santosf3,masrours,ptan,esfahanian}@msu.edu

Abstract. The fair top-k node ranking problem aims to find the k most significant nodes in a network without discriminating against particular groups of nodes as defined by their protected attribute. However, unlike fair ranking problems for independent and identically distributed (i.i.d.) data, the rank assigned to a node may influence the perception of fairness among its neighbors with similar acceptability scores due to the interconnectivity among the nodes. Fairness perception, which is an individual-level fairness metric, has thus been proposed to measure the degree to which a node perceives its ranking outcome as fair. While existing fair node ranking algorithms can help maximize its fairness perception, they are susceptible to the oversmoothing effect due to their message passing mechanism. Thus, a key challenge in designing fair node ranking algorithms is to balance the trade-off between maximizing the acceptability of the highly ranked nodes while satisfying both individual-level and group-level fairness criteria. To address this challenge, this paper presents a novel framework called `DeepFairRank` that integrates the potentially diverging criteria in a unified, multi-objective optimization framework using neural networks. Experimental results demonstrate the effectiveness of the framework when applied to real-world data.

1 Introduction

The ranking of nodes in a network [23] plays a central role in a myriad of applications, from viral marketing to recommender systems. Unlike traditional ranking problems, the node ranking algorithm must consider the link structure of the network in addition to its node attribute information. Current node ranking algorithms can be broadly categorized into centrality-based, influence-based and score-based methods. Centrality-based methods [3] identify the top-k most significant nodes based on their relative positions in the network. Such methods would employ node-centric measures such as PageRank and eigenvalue centrality to determine the degree of importance of each node. In contrast, influence-based methods [11] evaluate the prominence of a node in terms of the expected number of nodes influenced given a diffusion model such as the linear threshold or independent cascade models [8]. Score-based ranking methods [24], which are the

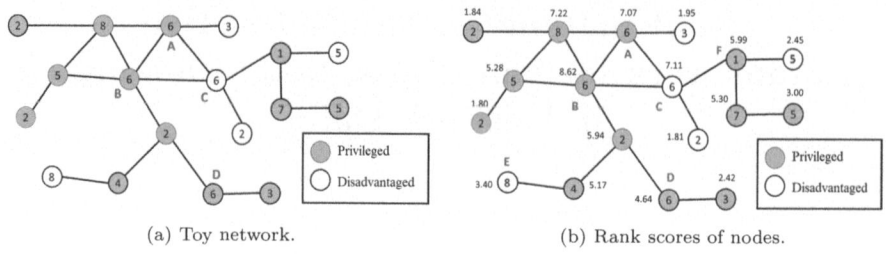

(a) Toy network.   (b) Rank scores of nodes.

Fig. 1. A network of student applicants for college admission. (a) The number associated with each node represents acceptability score while the node color denotes its protected group membership (privileged or disadvantaged). (b) Rank scores after applying a fairness-sensitive PageRank (FSPR) algorithm [20].

focus of this study, select the most "qualified" individuals given their acceptability scores.

To illustrate the score-based node ranking problem, consider the network shown in Fig. 1(a). Assume each node represents a high-school student applying for college admission. The acceptability score of a student could be based on a standardized test result or its combination with other criteria such as high school grades. Assume the acceptability score ranges from 1 (unacceptable) to 10 (must accept). Our goal is to select the top-k candidates for admission based on their acceptability and fairness criteria. We further assume that the candidates can be divided into groups based on some sensitive attribute, say, race or gender. For brevity, the groups are denoted as *privileged* and *disadvantaged* in the diagram. To ensure that the admission decision will not discriminate against the disadvantaged group, the ranking algorithm should consider the proportional representation of nodes from different groups in its top-k ranking [24] by incorporating group-level fairness criteria such as statistical parity [4] in its formulation.

Furthermore, due to their social ties, the decision to admit a student can significantly impact the perception of fairness within the student's social circle. For example, suppose node **C** in Fig. 1(a) observes both of its neighbors with similar acceptability scores, **A** and **B**, are accepted. This raises **C**'s expectation that it should also be accepted given their similar qualifications. However, if **C** is rejected, then the decision will likely be perceived as unfair since the outcome is below **C**'s expectation. In contrast, if **C** observes both **A** and **B** get rejected, this will lower its expectation, making **C** more amenable to the rejection decision. To ensure such expectation is met, individual-level criteria such as *fairness perception* [12] should be considered when ranking the nodes.

The notion of fairness perception is well-aligned with ideas from social equity theory [1], which suggests that the degree of satisfaction of individuals should be based on what they expect to receive. An individual will more likely perceive an outcome to be fair if the decision meets or exceeds the individual's expectation. However, since it is only an individual-level fairness metric, it does not ensure fairness for all groups of the protected attribute. In addition, the fairness perception measure introduced in [12] was designed for node classification instead of node ranking problems.

Another limitation of existing fairness-aware node ranking algorithms is their primary focus on enhancing fairness of the nodes' PageRank scores in a network [6,19,20]. While these approaches can indeed improve individual-level fairness, they are vulnerable to the *oversmoothing* problem due to their inherent message passing mechanism. For example, Fig. 1(b) shows the rank scores after applying the FSPR algorithm [20] to rank the nodes shown in Fig. 1(a):

$$\mathbf{h} = (1 - \gamma)\mathbf{P}^T\mathbf{h} + \gamma\mathbf{s},$$ (1)

where $\mathbf{h}$ is the rank vector of the nodes, $\mathbf{P}$ is the transition probability between nodes, $\mathbf{s}$ is the acceptability score vector of the nodes, and γ is a parameter associated with the jump probability, which is often set to 0.15. Observe that the score for the highly acceptable node $\mathbf{E}$ has reduced significantly from 8 to 3.4 while the score for node $\mathbf{F}$ increases from 1 to 5.99 due to the smoothing effect of their neighbors. As a result, node $\mathbf{F}$ is ranked higher than node $\mathbf{E}$ despite having a much lower acceptability score.

To overcome these challenges, this paper presents a novel score-based node ranking algorithm called `DeepFairRank`, designed to balance the trade-off between maximizing acceptability scores and fairness perception while minimizing the disparity among different groups of nodes using a neural network. Specifically, we first introduce an approach to compute fairness perception for node ranking problems. `DeepFairRank` also employs a multi-objective optimization framework to integrate the potentially diverging criteria. Experimental results on real-world datasets demonstrate the effectiveness of our proposed algorithm compared to other conventional fairness-aware ranking algorithms.

2   Related Work

There is a vast body of literature on the node ranking problem. Centrality-based measures [3] such as the Katz index [7] and PageRank [15] employ various metrics to determine the relative importance of a node in a network. Influence-based approaches [11] consider models of the diffusion process [8] to assess the degree of influence of the nodes. Both centrality-based and influence-based methods generally utilize the link information only to rank the nodes. In contrast, score-based models [24] consider other node attribute information to perform the ranking.

Algorithmic fairness is an important topic that has attracted considerable interest in recent years. Fairness can be defined at *individual level* as the absence of any prejudice towards an individual in a decision-making task [14]. In other words, similar people should be treated similarly [4]. For *group fairness*, a fair decision should not favor one group over another. Examples of group fairness metrics include *demographic parity* or *statistical parity* [4] and *equalized odds* [5]. Aside from defining fairness metrics, methods for debiasing decision outcomes from machine learning models have also been developed [13].

Recent years have also witnessed growing interest in ensuring fairness in ranking algorithms [16,24]. For example, the `FA*IR` algorithm [22] was designed to generate rankings that maximize utility while ensuring fairness across different groups. [2] presented an approach for designing fair ranking schemes by assisting users in selecting better weights for combining attributes when ranking items in a database. [18] considered exposure allocation between groups in designing a framework to achieve fairness in ranking. However, these approaches are mostly designed for i.i.d data. This has led to growing research focusing on imparting fairness into node ranking algorithms. Kang et al. [6] presented the InFoRM algorithm to mitigate individual biases when computing the PageRank scores of the nodes. However, since their approach did not consider group-level fairness, it may not be able to prevent discrimination against certain underprivileged groups of nodes. Other algorithms for debiasing the output of the PageRank algorithm include [9,20]. Nevertheless, such algorithms are susceptible to oversmoothing due to their message passing mechanism.

3   Preliminaries

3.1   Problem Statement

Let $G =< V, E, X >$ be an attributed network, where V is the set of nodes, $E \subseteq V \times V$ is the set of links, and X is the set of node attributes. The node attributes $X = (X^{(p)}, X^{(u)})$ are assumed to be a combination of the protected attribute, $X^{(p)}$, and other attributes, $X^{(u)}$. The protected attribute indicates whether the node belongs to a *privileged* ($X^{(p)} = 0$) or *disadvantaged* ($X^{(p)} = 1$) group, as shown by the example given in Fig. 1(a). We assume there exists a function $\phi : V \to [0, 1]$ such that $s_v = \phi(X_v)$ is the normalized **acceptability (utility) score** of node v. Note that $s_v = 1$ implies the node is highly acceptable whereas $s_v = 0$ implies the node is unacceptable. The choice of ϕ function is domain-dependent and is assumed to be given. We also consider a kernel function, $K : V \times V \to \mathbb{R}_+$, that measures the similarity between nodes in terms of their acceptability scores. We use the following Gaussian radial basis function as our kernel function for nodes u and v:

$$K(u, v) = \exp\left(-\frac{\|s_u - s_v\|^2}{2\sigma^2}\right), \qquad (2)$$

where σ is a hyperparameter.

Let $\Pi =< \pi_1, \pi_2, \cdots, \pi_{|V|} >$ be an ordering of the nodes in the network, where each $\pi_i \in V$. We consider node u to be ranked higher than node v, denoted as $u \succ v$, if $u = \pi_i$, $v = \pi_j$, and $i < j$. Furthermore, let $h : V \to [0, 1]$ be a ranking function that maps each node to a value between 0 and 1. We denote $\Pi(G, h)$ as the node ordering of G induced by the function h, where $\forall u, v : h(u) > h(v) \implies u \succ v$. Our goal is to learn the ranking function h that maximizes the acceptability scores $\frac{1}{k} \sum_{i=1}^{k} s_{\pi_i(G,h)}$ subject to the constraints imposed by both individual and group fairness criteria.

3.2   Fairness Measures

Fairness perception was introduced as an individual-level fairness criterion in [12] based on the premise that "the reaction of an individual to the outcome of a decision process is based on the expectation of the individual ... (which) not only depends on one's own outcome but also the outcomes of other individuals that belong to the same reference group." In [12], the reference group that shapes an individual's expectation is defined by the neighborhood of the node in a network.

Definition 1 (Fairness Perception [12]). *Given a prediction function h and a network $G =< V, E, X >$, the fairness perception of a node $v \in V$ is:*

$$f(v, h) = \begin{cases} 1 & \text{if } \mathbb{E}[h(v)] \leq h(v) \\ 0 & \text{otherwise} \end{cases} \tag{3}$$

where $\mathbb{E}[h(v)]$ is the neighborhood-based expectation of $h(v)$.

If $f(v, h) = 1$, then node v will perceive the decision as fair since its decision outcome $h(v)$ is no worse than its expectation. Otherwise, v will perceive the decision as unfair. The overall fairness perception for h on G, also known as fairness visibility [12], is given by the average fairness perception of its nodes:

$$f_G(h) = \frac{1}{|V|} \sum_{u \in V} f(v, h) \tag{4}$$

The formula below was used in [12] to calculate the neighborhood-based expectation of a node:

$$\mathbb{E}[h(v)] = \frac{y_v}{k_1} \left[\sum_{u \in N(v)} y_u h(u) \right] + \frac{1 - y_v}{k_0} \left[\sum_{u \in N(v)} (1 - y_u) h(u) \right] \tag{5}$$

where $N(v)$ is the set of nodes in the neighborhood of v, $k_0 = \sum_{u \in N(v)} (1 - y_u)$, $k_1 = \sum_{u \in N(v)} y_u$, and y_u is the true class label of node u. Note that the measure requires access to the ground truth labels y for computing the neighborhood-based expectation since it is designed for supervised classification task. It is therefore inapplicable to unsupervised node ranking problems, in which the ground truth ranking is unavailable when computing $\mathbb{E}[h(v)]$.

4   Methodology

This section presents our proposed `DeepFairRank` framework. We first introduce our neighborhood expectation function for computing fairness perception.

4.1   Neighborhood Expectation Function

Instead of using Eq. (5), we propose the following neighborhood-based conditional expectation $\mathbb{E}[h(v)]$ function:

$$\mathbb{E}[h(v)] = \frac{\sum_{u \in N(v)} h(u) K(v, u)}{\sum_{u \in N(v)} K(v, u)} \tag{6}$$

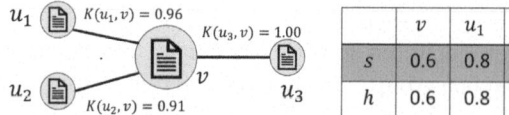

Fig. 2. An example illustration of fairness perception and calculation of the conditional expectation, $\mathbb{E}[h(v)]$.

Intuitively, the expected value of h for node v is computed based on the weighted average value of its neighbors, where the weights are given by similarity of their acceptability scores s (see Eq. (2)). Unlike the approach in [12], h is a continuous-valued ranking function instead of a binary decision function.

Example 1. Consider the example graph shown in Fig. 2. The central node v has three neighbors, u_1, u_2, and u_3. Let s_v denotes its normalized acceptability score and $K(v, u_i)$ denotes the similarity of its acceptability score to a neighboring node u_i. The table in the diagram shows the normalized acceptability scores s for all 4 nodes as well as the output of a ranking function h. Using Eq. (6), the conditional expectation for node v given its neighbors is:

$$\mathbb{E}[h(v)] = \frac{0.96 \times 0.8 + 0.91 \times 0.3 + 1 \times 0.7}{0.96 + 0.91 + 1} = 0.61.$$

Based on the definition in Eq. (3), since $h(v) = 0.6 < \mathbb{E}[h(v)]$, v will perceive the ranking function h as unfair.

4.2 Maximizing Fairness Perception

Let $\mathbb{C} = \{C_1, \ldots, C_k\}$ be the set of connected components in G. The theorem below shows a trivial approach for finding a ranking function h that maximizes fairness perception in G.

Theorem 1. *Given a network $G =< V, E, X >$, the ranking function h maximizes fairness perception, $\forall v \in V : f(v, h) = 1$, if and only if $\forall C \in \mathbb{C} : h(v) = h(v')$ for all $v, v' \in C$.*

Proof. If h yields the same value for all the nodes in a connected component C, then according to Eq. (6), $\forall v \in C : \mathbb{E}[h(v)] = h(v)$, and thus, $f(v, h) = 1$. Since this property holds for all $C \in \mathbb{C}$, therefore $\forall v \in V : f(v, h) = 1$. Conversely, if $\forall v \in V : f(v, h) = 1$, then all the nodes in the same connected component must have the same $h(v)$. This condition is trivially satisfied if the connected component has only one node. Thus, we consider the case when C has at least two nodes. By contradiction, assume that fairness perception is maximized but the nodes in C can have different values of h. Let $C_{min} = \{u \in C | h(u) = m\}$, where $m = min\{h(u) | u \in C\}$. Since the values of h are not uniform in C, there must exist a node $u_m \in C_{min}$ connected to another node $v \in C \setminus C_{min}$, in which $h(u_m) \neq h(v)$. Thus, $\forall v \in N(u_m) : h(u_m) \leq h(v)$ and there exists a neighboring node $v \in N(u_m)$ for which $h(v) > h(u_m)$. Using the definition of neighborhood-based expectation given in Eq. (6), we have

$$\mathbb{E}[h(u_m)] = \frac{\sum_{v \in N(u_m)} h(v)K(u_m, v)}{\sum_{v \in N(u_m)} K(u_m, v)} > \frac{\sum_{v \in N(u_m)} h(u_m)K(u_m, v)}{\sum_{v \in N(u_m)} K(u_m, v)} = h(u_m)$$

Since $h(u_m) < \mathbb{E}[h(u_m)]$, $f(u_m, h) = 0$, which contradicts the assumption that the fairness perception of all the nodes in C is maximized. Thus, the original assumption that C has nodes with different values of h must be wrong.

Theorem 1 suggests a trivial way to maximize fairness perception is to assign a constant h to every node in the same connected component. However, as will be shown in the next subsection, maximizing fairness perception alone is insufficient as it does not guarantee high utility of the ranking algorithm since the nodes in the same connected component or the same neighborhood may not have the same acceptability score s. Furthermore, since fairness perception is a local, individual-level fairness criterion instead of a global fairness criterion, it does not guarantee nodes from different groups will be treated equally.

4.3 Weighted Statistical Disparity

One way to ensure equity across the different groups of the protected attribute is to apply a group-level criterion such as statistical parity [4]. However, since the measure was designed for classification problems, we consider a **weighted statistical disparity (WSD)** metric to quantify the disparity in the average rank scores h of similarly qualified nodes in different groups of the protected attribute. Specifically, we discretize the normalized acceptability scores $\{s_v \in [0, 1]\}$ into K bins, denoted as $\hat{s}_1 < \hat{s}_2 < \cdots < \hat{s}_K$. For example, the normalized acceptability scores can be discretized into 5 bins, $[0, 0.2], (0.2, 0.4], (0.4, 0.6], (0.6, 0.8]$, and $(0.8, 1]$. Let $\hat{s}_1 = 0.1, \hat{s}_2 = 0.3, \cdots, \hat{s}_5 = 0.9$ be the centroids of the bins, $B_i^{(j)} = \{v \in V \mid \hat{s}_{i-1} < s_v \leq \hat{s}_i, X_v^{(p)} = j\}$ be the set of nodes from the protected group $X^{(p)} = j$ assigned to bin $\hat{s}_i$ and $\bar{h}_i^{(j)} = \frac{1}{|B_i^{(j)}|} \sum_{v \in B_i^{(j)}} h(v)$ be its average value of h. Assuming there are 2 groups, the weighted statistical disparity measure is given by:

$$\Gamma(h) = \sum_{i=1}^{K} \hat{s}_i \left| \bar{h}_i^{(0)} - \bar{h}_i^{(1)} \right| \tag{7}$$

Intuitively, $\Gamma(h)$ ensures that the disparity in rank scores of nodes with high acceptability scores be given higher emphasis than the disparity among nodes with low acceptability scores.

Example 2. Consider the plot shown in Fig. 3, where the horizontal axis denotes the normalized acceptability score (s) and the vertical axis denotes the rank score given by some function h. Each data point in the plot represents a node in the graph shown in Fig. 1(a). Assume the acceptability scores were discretized into 5 bins. The average value of h in bin i for each group j of the protected

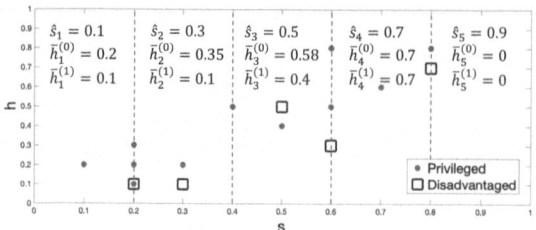

Fig. 3. An illustration of weighted statistical disparity.

attribute is denoted as $\bar{h}_i^{(j)}$, respectively. The weighted statistical disparity for the ranking function h can be computed using Eq. (7) as follows:

$$\Gamma(h) = 0.1 \times 0.1 + 0.3 \times 0.25 + 0.5 \times 0.18 + \ 0.7 \times 0 + 0.9 \times 0 = 0.175$$

The theorem below shows the relationship between fairness perception and the weighted statistical disparity measure.

Theorem 2. *Given a ranking function h that maximizes fairness perception, i.e., $\forall v \in V : f(v, h) = 1$, its weighted statistical disparity satisfies the following inequality:*

$$\kappa \times max\{\delta^*, 0\} \leq \Gamma(h) \leq \kappa \times \Delta^*,$$

where $\kappa := \sum_{i=1}^{K} \hat{s}_i$,

$$\delta^* = sign\left((h_{min}^{(0)} - h_{max}^{(1)})(h_{max}^{(0)} - h_{min}^{(1)})\right) \times min\left\{ \left| h_{min}^{(0)} - h_{max}^{(1)} \right|, \left| h_{max}^{(0)} - h_{min}^{(1)} \right| \right\},$$

$$\Delta^* = max\left\{ \left| h_{min}^{(0)} - h_{max}^{(1)} \right|, \left| h_{max}^{(0)} - h_{min}^{(1)} \right| \right\}.$$

Here, we denote $h_{min}^{(j)} = min\{h(v) \mid X_v^{(p)} = j\}$ and $h_{max}^{(j)} = max\{h(v) \mid X_v^{(p)} = j\}$.

Proof. First, note that $h_{min}^{(j)} \leq \bar{h}_i^{(j)} \leq h_{max}^{(j)}$ since

$$\bar{h}_i^{(j)} = \frac{1}{|B_i^{(j)}|} \sum_{v \in B_i^{(j)}} h(v) \leq \frac{1}{|B_i^{(j)}|} \sum_{v \in B_i^{(j)}} h_{max}^{(j)} = h_{max}^{(j)}$$

$$\bar{h}_i^{(j)} = \frac{1}{|B_i^{(j)}|} \sum_{v \in B_i^{(j)}} h(v) \geq \frac{1}{|B_i^{(j)}|} \sum_{v \in B_i^{(j)}} h_{min}^{(j)} = h_{min}^{(j)}$$

Using these inequalities, it can be easily shown that

$$h_{min}^{(0)} - h_{max}^{(1)} \leq \bar{h}_i^{(0)} - \bar{h}_i^{(1)} \leq h_{max}^{(0)} - h_{min}^{(1)}$$

$$\text{and } \left| \bar{h}_i^{(0)} - \bar{h}_i^{(1)} \right| \leq max\left\{ \left| h_{min}^{(0)} - h_{max}^{(1)} \right|, \left| h_{max}^{(0)} - h_{min}^{(1)} \right| \right\}$$

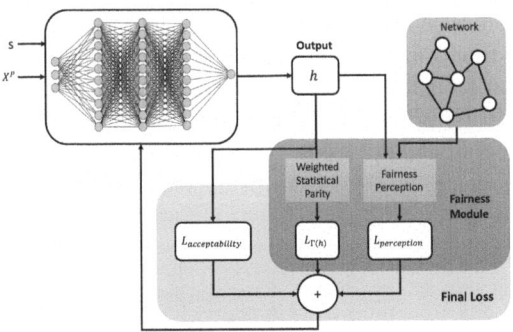

Fig. 4. Overall Framework of DeepFairRank.

Replacing the above inequality into Eq. (7) and using the definition of Δ^*, we obtain the following upper bound:

$$\Gamma(h) \leq \sum_i \hat{s}_i \max \left\{ \left| h_{\min}^{(0)} - h_{\max}^{(1)} \right|, \left| h_{\max}^{(0)} - h_{\min}^{(1)} \right| \right\} = \kappa \times \Delta^*$$

To obtain the lower bound for $\Gamma(h)$, we use $\left| \bar{h}_i^{(0)} - \bar{h}_i^{(1)} \right| \geq \delta^*$ where

$$\delta^* = \text{sign}\left((h_{\min}^{(0)} - h_{\max}^{(1)})(h_{\max}^{(0)} - h_{\min}^{(1)}) \right) \times \min \left\{ \left| h_{\min}^{(0)} - h_{\max}^{(1)} \right|, \left| h_{\max}^{(0)} - h_{\min}^{(1)} \right| \right\}$$

If $(h_{\min}^{(0)} - h_{\max}^{(1)})$ and $(h_{\max}^{(0)} - h_{\min}^{(1)})$ have opposite signs, then δ^* is negative. Since $\left| \bar{h}_i^{(0)} - \bar{h}_i^{(1)} \right|$ is non-negative, we can obtain a tighter bound as follows: $\left| \bar{h}_i^{(0)} - \bar{h}_i^{(1)} \right| \geq \max \left\{ \delta^*, 0 \right\}$. Replacing this into Eq. (7) yields

$$\Gamma(h) \geq \frac{1}{K} \sum_i \hat{s}_i \max \left\{ \delta^*, 0 \right\} = \kappa \times \max \left\{ \delta^*, 0 \right\}$$

Theorem 2 suggests that maximizing fairness perception alone may not guarantee group fairness since the upper bound of $\Gamma(h)$ can be small depending on how the nodes from different groups are distributed in the network. Thus, it would be desirable to find an algorithm that considers the trade-off between maximizing fairness perception and minimizing weighted statistical disparity.

4.4   DeepFairRank: **Proposed Framework**

This section presents DeepFairRank, our proposed neural network approach to learn a ranking function h that considers (1) the normalized acceptability scores of the nodes, s, and the trade-off between (2) maximizing fairness perception and (3) minimizing weighted statistical disparity measure, $\Gamma(h)$.

Figure 4 depicts the architecture of DeepFairRank, which consists of a stack of fully connected neural networks with an input layer, three hidden layers and an output layer. Each hidden layer contains 256 hidden neurons. The model employs a `tanh` activation function in all layers except for its output layer, which uses a `sigmoid` function to restrict the output to be a value between 0 and 1. The network output will be provided as input to a fairness perception module that computes the fairness perception loss associated with the ranking function h.

To illustrate how the fairness perception loss is computed, we first express the neighborhood-based expectation function (see Eq. (6)) in matrix notation as follows:

$$\mathbb{E}[h(v_i)] = \frac{\sum_{v_j \in N(v_i)} h(v_j) K(v_i, v_j)}{\sum_{v_j \in N(u_i)} K(v_i, v_j)} = \sum_j W_{ij} h(v_j), \tag{8}$$

where $W = D^{-1}(K \odot A)$, $\odot$ denotes a Hadamard product between the adjacency matrix A and kernel matrix K, while D is a diagonal matrix with $D_{ii} = \sum_j K_{ij} A_{ij}$. Given the output h of the fully connected network, the following fairness perception loss is computed:

$$\ell_{\mathrm{FP}}(h) = \sum_i \big(\mathbb{E}[h(v_i)] - h(v_i)\big) = \sum_i \big[\sum_j W_{ij} h(v_j) - h(v_i)\big]$$

which can be simplified as $\ell_{\mathrm{FP}} = \|(\mathbf{W} - \mathbf{I})h\|_1$, where $\mathbf{I}$ is the identity matrix. The fairness perception module will compute the loss, which will be combined with the weighted statistical disparity to learn the ranking function h. Additionally, h should be close to the acceptability score s to ensure that nodes with high acceptable scores are ranked higher than those with low scores. Putting everything together, the objective function to be optimized by our framework is:

$$\min_h \|s - h\|^2 + \alpha \Gamma(h) + \beta \ell_{\mathrm{FP}} \tag{9}$$

where the hyperparameters α and β control the trade-off between matching h to the node acceptability scores, ensuring group parity, and maximizing fairness perception. The model was trained end-to-end for 1000 epochs using Adam as its optimizer with a learning rate of 0.0001.

During training, the hyperparameter values for α and β are chosen to minimize the following power mean:

$$\alpha^*, \beta^* = \mathrm{argmax}_{\alpha, \beta} \left(\frac{r(h, s)^3 + (1 - \Gamma(h))^3 + f_G(h)^3}{3} \right)^{\frac{1}{3}}, \tag{10}$$

where $r(h, s)$ is the Spearman rank coefficient between the ranked output h and normalized acceptability score s, $\Gamma(h)$ is the weighted statistical disparity, and $f_G(h)$ is the average fairness perception. We randomly choose $\alpha \in [10^{-2}, 10^2]$ and $\beta \in [10^{-2}, 10^2]$ in logarithmic space for our experiments.

5   Experimental Evaluation

This section describes the experiments performed to evaluate the performance of `DeepFairRank`. Our code and data are available at https://github.com/frsantosp/DeepFairRank.

Table 1. Summary statistics of submitted and accepted papers for the ICLR conference from 2017 to 2020.

Year	# submitted papers	# accepted papers	% papers by famous authors	% accepted papers by famous authors
2017	488	243	21.7%	28.4%
2018	402	229	17.2%	20.1%
2019	1419	502	14.5%	19.7%
2020	2212	687	12.3%	15.6%

5.1 Data

We consider the following datasets for our experiments.

1. **Co-authorship network** (ICLR): This dataset [12] corresponds to the co-authorship network of papers submitted to the ICLR conference from 2017 to 2020. The nodes of the co-authorship network correspond to submitted papers, while an edge is created if two papers share a co-author. Table 1 shows the summary statistics of the submitted and accepted papers from 2017 to 2020. Similar to the approach used in [12], the submitted papers were categorized into 2 groups using *famous author* as the protected attribute. The acceptance decision of the conference is used as the true label during evaluation.

2. **COMPAS**: The COMPAS (Correctional Offender Management Profiling for Alternative Sanctions) [17], which is a widely used benchmark for evaluating fairness algorithms. Each node corresponds to a jailed offender. An edge is created between two nodes if the time in jail overlaps between two offenders. We use race as the protected attribute. The resulting network contains 8,946 nodes and 2,440,315 edges. We use the COMPAS decile score as acceptability score and the recidivism attribute as the true label during evaluation.

3. **Credit Default** [21]: The dataset contains 2,000 nodes using marital status as the protected attribute. The links in the network are established based on similarity of their spending and payment patterns. Each node (individual) has a binary class indicating whether the individual will default on their credit card payment in the next month.

4. **Facebook** [10]: The dataset is a social network from Facebook. Each node represents a facebook user and the link represents friendship. The protected attribute used is gender. The binary label is based if the user went to college. The dataset contains 1046 nodes with $53,000$ edges.

5.2 Experimental Setup

Baseline Algorithms We compare the performance of `DeepFairRank` (DFR) against the following baselines:

1. **No Calibration**, in which the nodes are ranked according to their normalized acceptability scores, i.e., $\mathbf{h} = \mathbf{s}$.
2. **Fa*ir** [22], a fair ranking algorithm that selects the top-k candidates to recommend based on a ranked group fairness criterion. We use the python implementation[1] provided by the authors for our experiment.
3. **InFoRM** [6], which debiases the normalized acceptability score $\mathbf{s}$ by minimizing the following loss:

$$\min_{\mathbf{h}} \|\mathbf{h} - \mathbf{s}\|^2 + \alpha \mathrm{tr}\left(\mathbf{h}^T \mathbf{L}_s \mathbf{h}\right),$$

   where $\mathbf{L}_s$ is the graph Laplacian matrix. As shown in [6], the optimization problem is equivalent to the PageRank formulation (Eq. (1)) if $\mathbf{s}$ corresponds to a uniform probability vector $\mathbf{e}$ (with $\mathbf{L}_s$ equals to the identity matrix). We use the code provided by the authors[2] for our experiment.
4. **FSPR** [20], which is a fairness-aware PageRank algorithm that considers the group fairness criterion. The algorithm is trained to solve the following constraint optimization problem:

$$\min_{\mathbf{h}} \ \|\mathbf{Q}^T \mathbf{h} - \mathbf{s}\|^2$$
$$\text{s.t.} \ \mathbf{Q}_p^T \mathbf{h} = \phi, \sum_i h_i = 1$$
$$0 \le h_i \le 1, \forall i \in \{1, 2, \cdots, n\}$$

   where $\mathbf{Q} = \gamma \left[1 - (1 - \gamma)\mathbf{P}\right]^{-1}$, $\mathbf{Q}_p$ is the PageRank mass allocated to the protected group, ϕ is the desired proportion for achieving equity, and $\gamma = 0.15$ is the parameter associated with the jump probability of the PageRank algorithm. The preceding convex optimization problem is solved using the CVXOPT software package.

Evaluation Metrics. We compared the performance of the fairness-aware algorithms according to 4 criteria. First, to demonstrate their utility, the *average precision* of each algorithm is computed from their respective ranked output as follows: $\frac{1}{m} \sum_k P@k \times I[r_k]$, where $P@k$ is the precision computed from the top-k highest ranked nodes, $I[r_k]$ is a binary indicator function whose value is 1 only if the i-th ranked node should be accepted according to the ground truth label, and m is the total number of "relevant" nodes. We used the python implementation of average precision from the scikit-learn library to compute the measure. Furthermore, we also compute the *average acceptability scores* of the selected top-k ranked nodes, to ensure that the highly ranked nodes are well-qualified to be selected. Third, to assess their individual-level fairness, we computed their *average fairness perception*, which is given by $\frac{1}{|V|} \sum_{v \in V} f(v, h)$. The metric ranges between 0 and 1, with larger values suggest a higher perception that the ranking is fair among the nodes. Finally, to ensure group fairness, we employed the weighted statistical disparity measure given in Eq. (7).

---

[1] https://github.com/fair-search/fairsearch-fair-python.
[2] https://github.com/jiank2/inform.

Table 2. Performance comparison of `DeepFairRank` against baseline algorithms. $\nearrow$ or $\searrow$ indicates larger or smaller values are better. The red color represents first place and blue represents second place. The last row in each table denotes average rank of each algorithm over all datasets.

Data	$ICLR_{17}$	$ICLR_{18}$	$ICLR_{19}$	$ICLR_{20}$	COMPAS	Credit	Facebook	Avg Rank
No Calib	0.9257	0.8480	0.9160	0.8917	0.4911	0.8706	0.8476	1.86
Fa*ir	0.8947	0.8212	0.8630	0.8449	0.3918	0.8581	0.7666	3.29
InFoRM	0.7044	0.6991	0.5175	0.3689	0.4097	0.4152	0.3258	4.85
FSPR	0.7760	0.7745	0.7270	0.7157	0.4648	0.8206	0.7279	3.86
DFR	0.9319	0.8641	0.9269	0.9111	0.5063	0.8668	0.8477	1.14

(a) Average Precision ($\nearrow$)

Data	$ICLR_{17}$	$ICLR_{18}$	$ICLR_{19}$	$ICLR_{20}$	COMPAS	Credit	Facebook	Avg Rank
No Calib	0.6790	0.6549	0.6674	0.6414	0.7923	0.8275	0.7850	1.00
Fa*ir	0.6746	0.6540	0.6656	0.6413	0.5638	0.8255	0.7849	3.00
InFoRM	0.6044	0.5878	0.5810	0.4727	0.6727	0.7770	0.3687	4.71
FSPR	0.6123	0.5878	0.6244	0.6030	0.7361	0.8100	0.6910	3.85
DFR	0.6773	0.6534	0.6673	0.6414	0.7923	0.8275	0.7849	1.71

(b) Average Acceptability Score ($\nearrow$)

Data	$ICLR_{17}$	$ICLR_{18}$	$ICLR_{19}$	$ICLR_{20}$	COMPAS	Credit	Facebook	Avg Rank
No Calib	0.7377	0.7537	0.6660	0.5967	0.5872	0.6046	0.4057	2.71
Fa*ir	0.7848	0.7587	0.6596	0.4910	0.4181	0.8140	0.3818	3.00
InFoRM	0.6885	0.6915	0.5441	0.3513	0.4315	0.3379	0.2842	4.71
FSPR	0.6803	0.7040	0.5849	0.6700	0.4735	0.8635	0.3722	3.14
DFR	0.7540	0.7611	0.6821	0.6103	0.5989	0.8170	0.4201	1.43

(c) Average Fairness Perception ($\nearrow$)

Data	$ICLR_{17}$	$ICLR_{18}$	$ICLR_{19}$	$ICLR_{20}$	COMPAS	Credit	Facebook	Avg Rank
No Calib	1.0397	0.8374	0.0610	0.0108	0.0198	0.00004	0.000093	3.86
Fa*ir	1.4751	1.1510	0.3580	0.1527	0.6084	0.0077	0.0002	4.86
InFoRM	0.0069	0.0055	0.0009	0.0002	0.00001	0.0000	0.0000008	1.14
FSPR	0.3528	0.0941	0.0027	0.0084	0.0023	0.00001	0.00064	3
DFR	0.0043	0.0055	0.0024	0.0028	0.0104	0.0075	0.000002	2.14

(d) Weighted Statistical Disparity ($\searrow$)

5.3 Experimental Results

We first compare the average precision of `DeepFairRank` against the ranking results generated by other baseline methods. The results are summarized in Table 2(a). Observe that `DeepFairRank` achieves the highest average precision on 6 out of the 7 datasets. The next best result is obtained by the **No Calibration** method, which is not surprising since there is strong correlation between acceptability score and the true class label of the nodes in the network. The **Fa*ir** algorithm [22] performs relatively well on the ICLR and credit datasets but worse on the COMPAS dataset whereas **FSPR** performs well on COMPAS but struggles with the other datasets. **InFoRM** has the lowest average precision values among all the methods on the ICLR datasets. This is likely due to the over-smoothing effect of the message passing schemes used by the algorithms.

Table 2(b) shows the average acceptability scores of the top-k nodes selected by each algorithm. As expected, the **No Calibration** method is the best performer since it is designed to rank the nodes according to the highest normalized acceptability scores. DeepFairRank is the next best performer, achieving acceptability scores that are comparable to **No Calibration** in all the datasets. As its average precision is also the highest, this shows that the nodes chosen by DeepFairRank not only have among the highest acceptability scores, their top-k rankings are also consistent with the true class labels despite the unsupervised learning nature of the algorithm.

In terms of their average fairness perception, the results shown in Table 2(c) suggest that DeepFairRank has the best average fairness perception in 4 out of the 7 datasets evaluated. This is not surprising as DeepFairRank is trained to optimize fairness perception as one of its criteria while accounting for its trade-off with other metrics. The next best performer is **No Calibration** followed by the **Fa*ir** and **FSPR** algorithms. Note that **FSPR** achieves the highest average fairness perception for the $ICLR_{20}$ and the credit dataset. DeepFairRank followed in those two datasets.

Finally we compare the performance of the various algorithms in terms of their group fairness criterion. The results in Table 2(d) suggest that **InFoRM** has the best average ranking with respect to the weighted statistical disparity, which is to be expected since the method is designed to explicitly optimize the group fairness criterion only. The next best performer is DeepFairRank, which has the best weighted statistical disparity measure in 2 datasets and second place in 3 datasets. While **InFoRM** excels at maintaining group fairness, it performs poorly on all the other metrics including average fairness perception and precision. These results suggest that **InFoRM** has difficulty managing the trade-off between utility and fairness compared to other metrics. Table 2(d) also shows that **Fa*ir** performs worse than **No Calibration** in terms of weighted statistical parity. This is due to the difficulty in tuning the hyperparameters of the algorithm (k, p, and α) using the code provided by the authors. In particular, the code has a tendency of breaking down and returning warning messages indicating the code library has not been tested outside the range of values used.

In summary, these results demonstrate the effectiveness of using the proposed multi-objective framework in DeepFairRank to achieve high average precision in its node rankings while balancing the trade-off between maximizing the individual-level fairness perception measure and minimizing the group-level weighted statistical disparity measure for the different groups. The framework gives flexibility for users to tune the algorithm towards achieving their desired utility and fairness goals for their application domain by choosing the appropriate values of the hyperparameters α and β. For example, increasing α will promote higher group-level fairness, as illustrated in Fig. 5a, since the hyperparameter promotes lower weighted statistical disparity in the loss function of

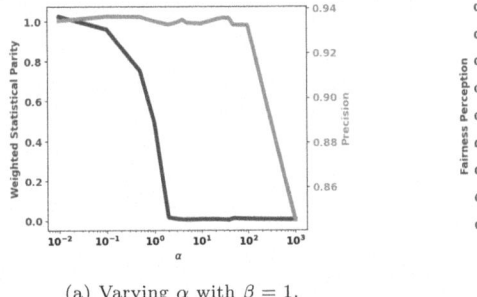
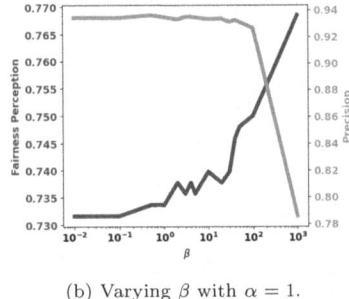

(a) Varying α with $\beta = 1$.   (b) Varying β with $\alpha = 1$.

Fig. 5. Effects of varying the hyperparameters α and β of the `DeepFairRank` algorithm on average precision and weighted statistical disparity (for varying α) or fairness perception (for varying β) when applied to the $ICLR_{17}$ dataset.

`DeepFairRank` when α increases. However, Fig. 5b shows that selecting a higher β can hurt both precision and fairness perception as it will unbalance the loss. Nevertheless, it is important to note that the algorithm maintains stable performance on a relatively wide range of hyperparameter values, showing that algorithm is relatively easy to tune.

6 Conclusions

This paper presents a fairness-aware node ranking algorithm called `DeepFairRank` that considers a multi-objective criteria based on acceptability score, fairness perception, and weighted statistical disparity between different groups of the protected attribute. We introduce a novel conditional expectation measure for fairness perception that is designed for node ranking problems. We also provide theoretical analysis to show that maximizing fairness perception alone is insufficient as it may lead to bias in terms of the group fairness criteria. Finally, we empirically demonstrate the effectiveness of `DeepFairRank` in terms of balancing the three competing requirements unlike other fairness-aware algorithms.

Acknowledgment. This material is based upon work supported by NSF under grant #IIS-1939368 and #IIS-2006633. Any opinion, findings, and conclusions or recommendations expressed in this material are those of the author(s) and do not necessarily reflect the views of the National Science Foundation.

References

1. Adams, J.S.: Inequity in social exchange. In: Advances in Experimental Social Psychology, vol. 2, pp. 267–299. Academic Press (1965)
2. Asudeh, A., Jagadish, H., Stoyanovich, J., Das, G.: Designing fair ranking schemes. In: Proceedings of the 2019 International Conference on Management of Data, pp. 1259–1276 (2019)

3. Das, K., Samanta, S., Pal, M.: Study on centrality measures in social networks: a survey. Soc. Netw. Anal. Min. **8**(1), 1–11 (2018). https://doi.org/10.1007/s13278-018-0493-2
4. Dwork, C., Hardt, M., Pitassi, T., Reingold, O., Zemel, R.: Fairness through awareness. In: Proceedings of the 3rd Innovations in Theoretical Computer Science Conference, pp. 214–226 (2012)
5. Hardt, M., Price, E., Srebro, N., et al.: Equality of opportunity in supervised learning. In: NeurIPS, pp. 3315–3323 (2016)
6. Kang, J., He, J., Maciejewski, R., Tong, H.: Inform: individual fairness on graph mining. In: Proceedings of the Ninth ACM SIGKDD International Conference on Knowledge Discovery and Data Mining, pp. 379–389 (2020)
7. Katz, L.: A new status index derived from sociometric analysis. Psychometrika **18**(1), 39–43 (1953)
8. Kempe, D., Kleinberg, J., Tardos, É.: Maximizing the spread of influence through a social network. In: Proceedings of the Ninth ACM SIGKDD International Conference on Knowledge Discovery and Data Mining, pp. 137–146 (2003)
9. Krasanakis, E., Papadopoulos, S., Kompatsiaris, I.: Applying fairness constraints on graph node ranks under personalization bias. In: Benito, R.M., Cherifi, C., Cherifi, H., Moro, E., Rocha, L.M., Sales-Pardo, M. (eds.) COMPLEX NETWORKS 2020 2020. SCI, vol. 944, pp. 610–622. Springer, Cham (2021). https://doi.org/10.1007/978-3-030-65351-4_49
10. Leskovec, J., Mcauley, J.J.: Learning to discover social circles in ego networks. In: NeurIPS, pp. 539–547 (2012)
11. Li, Y., Fan, J., Wang, Y., Tan, K.L.: Influence maximization on social graphs: a survey. IEEE Trans. Knowl. Data Eng. **30**(10), 1852–1872 (2018)
12. Masrour, F., Tan, P.N., Esfahanian, A.: Fairness perception from a network-centric perspective. In: 2020 IEEE International Conference on Data Mining (ICDM) (2020)
13. Masrour, F., Wilson, T., Yan, H., Tan, P.N., Esfahanian, A.: Bursting the filter bubble: Fairness-aware network link prediction. In: Proceedings of the AAAI Conference on Artificial Intelligence, vol. 34, pp. 841–848 (2020)
14. Mehrabi, N., Morstatter, F., Saxena, N., Lerman, K., Galstyan, A.: Survey on bias and fairness in machine learning. ACM Comp. Surv. **54**(6), 115:1–115:35 (2022)
15. Page, L., Brin, S., Motwani, R., Winograd, T.: The PageRank citation ranking: bringing order to the Web. Tech. rep, Stanford InfoLab (1999)
16. Patro, G.K., Porcaro, L., Mitchell, L., Zhang, Q., Zehlike, M., Garg, N.: Fair ranking: a critical review, challenges, and future directions, pp. 1929–1942 (2022)
17. ProPublica: Compas recidivism risk score data and analysis (2016), data retrieved from ProPublica Data Store (2016). https://www.propublica.org/datastore/dataset/compas-recidivism-risk-score-data-and-analysis
18. Singh, A., Joachims, T.: Fairness of exposure in rankings. In: Proceedings of the 24th ACM SIGKDD International Conference on Knowledge Discovery & Data Mining, pp. 2219–2228 (2018)
19. Tsioutsiouliklis, S., Pitoura, E., Semertzidis, K., Tsaparas, P.: Link recommendations for PageRank Fairness. In: Proceedings of the ACM Web Conference 2022, pp. 3541–3551 (2022)
20. Tsioutsiouliklis, S., Pitoura, E., Tsaparas, P., Kleftakis, I., Mamoulis, N.: Fairness-aware PageRank. In: Proceedings of the Web Conference 2021, pp. 3815–3826 (2021)

21. Yeh, I.C., Lien, C.: The comparisons of data mining techniques for the predictive accuracy of probability of default of credit card clients. Expert Sys. with Appl. **36**(2), 2473–2480 (2009)
22. Zehlike, M., Bonchi, F., Castillo, C., Hajian, S., Megahed, M., Baeza-Yates, R.: Fa* ir: A fair top-k ranking algorithm. In: Proceedings of the 2017 ACM on Conference on Information and Knowledge Management, pp. 1569–1578 (2017)
23. Zehlike, M., Sühr, T., Baeza-Yates, R., Bonchi, F., Castillo, C., Hajian, S.: Fair top-k ranking with multiple protected groups. Inf. Process. Manag. **59**, 102707 (2022)
24. Zehlike, M., Yang, K., Stoyanovich, J.: Fairness in ranking, part I: score-based ranking. ACM Comp. Surv. **55**(6), 1–36 (2022)

RUMs with Ties: A Discrete Choice Model Allowing Multiple Winners

Flavio Chierichetti[1], Ravi Kumar[2], Giuseppe Re[3(✉)], and Andrew Tomkins[2]

[1] Sapienza University, Rome, Italy
flavio@di.uniroma1.it
[2] Google Research, Mountain View, CA, USA
tomkins@google.com
[3] Algorand Labs, Rome, Italy
giuseppe.re@algorandlabs.com

Abstract. Discrete choice models are used to describe, explain, and predict choices made by people among a finite set of alternatives. However, standard discrete choice models come with an unrealistic assumption: that users are able to provide an *unequivocal* clear winner from any slate of alternatives. Often the user knows the winner but cannot report it, as when a UI does not allow a user to specify which of two movies they rated five stars is better. And often, among the myriad options available, the user is able to identify some good candidates, but finds it difficult to distinguish between the top contenders. In this paper, we study the problem of interacting with user choice data in which, sometimes, the user is unable to settle on a single compelling winner.

To address this issue, we introduce an extension to the well-known random utility models (RUMs), which we call *RUMs-with-Ties*, where comparisons can result in a tie. We begin with an axiomatic formulation of Luce dating to the 1950s, and provide algorithms and matching lower bounds for operating on data with ties. We also provide a comprehensive comparison of RUMs versus RUMs-with-Ties from different angles. We present theoretical results indicating that simple ways of incorporating ties into existing approaches are unlikely to perform well. We also prove in our setting that the presence of additional items, even if lower in quality, allows an algorithm to learn the highest ranked element with far fewer trials. Finally, we provide experimental evaluations of different approaches to handling indistinguishable items in choice settings and demonstrate the advantages of direct modeling of ties via our approach.

Keywords: random utility models · discrete choice · ties

1 Introduction

Users are increasingly faced with a so-called *Paradox of Choice* [46], in which more and more alternatives are available, paradoxically resulting in a decrease

F. Chierichetti—Partially supported by a Google Focused Research Award, by BiCi – Bertinoro international Center for informatics, and by the PRIN project 20229BCXNW.

L. M. Aiello et al. (Eds.): ASONAM 2024, LNCS 15211, pp. 188–206, 2025.
https://doi.org/10.1007/978-3-031-78541-2_12

in welfare due to the time and anxiety of understanding and weighing so many options [26]. Worse yet, as products are increasingly optimized, each generation improves by smaller amounts, causing the degree of difference between options to shrink towards zero, and exacerbating the difficulty of making a well-reasoned choice [15]. In this paper, we consider one important implication of increasingly complex decision spaces: machine learned models are frequently trained to predict a user's choice from a set of options, but users are sometimes unable to distinguish between the top options, resulting in a *tie*. A tie may come about because a user is restricted to a fixed scoring scale, such as five stars, so the user's preference may exist but the platform provider cannot observe it. It may similarly occur because the user takes actions on multiple items that do not indicate which is preferred, such as adding two items to a comparison set, or watching several short videos to completion. Or the user herself may not be able to distinguish between several top choices, such as when researching some options without coming to a final decision during a session [18].

Our models formalize the notion of a user being unable to distinguish the best option, and study approaches to accounting for these ties in modeling.

For clarity, there is a related problem that we do *not* consider: the user electing to select no item from the set of alternatives. This *non-consumption* provides relevant information [6] and helps predict future choices better [11,12]. Non-consumption may occur for many reasons [16]. A user asked whether they drive to work or take public transit may not respond because they telecommute or commute by bicycle [53]. Or a user offered several products may find that none of them provides compelling value for money. Generally, adding more alternatives will decrease the probability of non-consumption [43]. On the other hand, our topic of study—the phenomenon of indistinguishability of top alternatives—is different in part because the probability of being unable to distinguish the best choice may either increase or decrease with more options: an obvious winning option may be added, or a previous standout best option may become conflated with a new option of similar utility.

Formalizing Ties via Just Noticeable Differences. Early efforts to model indistinguishable items in utility theory assumed transitivity: if a and b were indistinguishable, and likewise b and c, then a must be indistinguishable from c.

This assumption was convenient but was recognized as unrealistic [5]. As a classic example, consider a sequence of weights ranging from 10g to 10kg, each 0.1% heavier than the previous one. Neighboring weights are indistinguishable by humans, and so under transitivity all weights in the sequence must be indistinguishable, but the lightest and heaviest are obviously different. For scalar quantities like weight, heat, or size, the psychometrics community has a long-standing notion of a *just noticeable difference* (JND), introduced by Weber and formalized by Fechner in the late 1800s [20]. Under this notion, weight w_1 might be indistinguishable from w_2, but eventually the difference between w_1 and w_i exceeds the JND and can then be sensed. Discrete choice relies on the scalar quantity of an item's overall utility, where a JND might reasonably be applied.

Thus, two items (e.g., cars, movies) might have a clear winner or, if within the JND, might be "too close to call" for a user trying to distinguish them.

In 1956, R. D. Luce developed a theory extending the preferences of a user to incorporate intransitive indistinguishability using a notion of JND [35]. He developed an axiomatic framework connecting a new set-theoretic concept he called a *semiorder* to a numerical realization based on utility values of items and the JNDs between them, and showed these two formulations were equivalent. In his numeric model, each item a has a utility w_a, and there is JND value δ such that a and b are indistinguishable if $|w_a - w_b| \leq \delta$.[1] Luce formalizes an important but subtle point about JNDs, which we introduce by example. Fix δ and consider the setting where $|w_a - w_b| \leq \delta$, so a and b are indistinguishable. Now the user considers c, and realizes that, while c is indistinguishable from a, it is clearly worse than b: $w_b - w_c > \delta$. We can perform an inference by contradiction: say a is no worse than b. Then $w_a \geq w_b > w_c + \delta$, contradicting our observation that a and c are indistinguishable. While the user cannot make a direct comparison between a and b, introducing a separate element c allowed an *indirect* comparison, and the user can now *infer* that b is preferred to a. This type of indirect reasoning follows directly from the definition of indistinguishability using JNDs, and also connects to the human intuition that comparing hard-to-distinguish options to some new alternatives can provide new insights and help us to reach a conclusion. We must therefore differentiate between *observed preferences*, which occur when $w_b > w_a + \delta$ and *inferred preferences*, which occur when $|w_a - w_b| < \delta$ but $\exists c :$ $w_c \in [\min(w_a, w_b) - \delta, \max(w_a, w_b) - \delta) \cup (\min(w_a, w_b) + \delta, \max(w_a, w_b) + \delta]$. We study below how powerful inferred preferences are compared to observed preferences.

Representing Inferred Preferences. Following Luce's definitions, we formalize the distinction between observed and inferred preferences and state the computational problem of deriving all possible inferred preferences. While Luce was not focused on computational questions (his work predates P vs NP!), his proof connecting semiorders with utilities and JNDs implicitly specifies an algorithm for determining inferred preferences;[2] we formalize and analyze it. We also develop an improved algorithm, which is implied by later work in the area, and we show that this second algorithm is more computationally efficient. Complementing, we also introduce a matching lower bound on the running time to conclude this is the optimal algorithm for determining all inferred preferences. Once all such preferences have been inferred, Luce showed that the final distinguishable pref-

---

[1] Luce's model is slightly more general, in that each element may have a more general interval of indistinguishability around it.

[2] Luce's formulation of a semiorder starts with a binary relation $<$. If $a \not> b$ and $b \not> a$ then a and b are indistinguishable, written $a \parallel b$. To establish the equivalence with the more actionable formulation based on utilities, Luce shows two additional axioms are required. First, if $a > b$, $b \parallel c$, and $c > d$, then $a > d$. Second, if $a > b$, $b > c$, and $b \parallel d$, then either $a \not\parallel d$ or $c \not\parallel d$. These additional axioms are used in his proof via an argument that implicitly gives the information necessary to compute all inferred preferences.

erences of the user may be perfectly represented as a *bucket order* in which the universe is partitioned into an ordered set of buckets. Items in distinct buckets are distinguishable according to the ordering of the buckets, while items within a bucket are indistinguishable. Interestingly, beginning with an intransitive distinguishability relation based on JNDs, and allowing all possible inferences, results in an (often fine-grained) *equivalence* relation in which the remaining indistinguishable elements are transitive, as the elements of a bucket must all be within δ of one another.

From User Models to Population Models. The upshot of these observations is a model grounded in JNDs to characterize a user's preferences, including a characterization of pairs of elements that are too close to distinguish, even by inferred preferences. This bucket order may be viewed as an augmentation (incorporating JNDs) of the total ordering implied by a user's utility vector. In practical settings, we are faced not with a single user but with many users, each with potentially different utilities. In the classical theory of discrete choice, the behaviors of the entire population are captured by a distribution over utility vectors, known as a *Random Utility Model (RUM)* [8,9,50,55,56]. We define an analogous notion of *RUMs-with-Ties*, characterized by a distribution over user bucket orders, to capture the behavior of a population of users making choices with the possibility of being unable to distinguish the best item. The rest of this paper explores the properties of RUMs-with-Ties, along with a series of experiments studying their representational power and learnability. Our results regarding bucket orders, inferred preferences, and the definition of RUMs-with-Ties are covered in Sects. 3 and 4.

RUMs Versus RUMs-with-Ties. The first question is whether the new model is truly different from RUMs. We address this from a few perspectives. (i) We observe that any RUM may be represented as a RUM-with-Ties whose buckets are all singletons. However, the converse is not true—there are RUMs-with-Ties that cannot be approximated by any RUM on the same vocabulary plus a "no-choice" element. (ii) Alternately, we might try to use a RUM to represent just the cases in which a RUM-with-Ties provides a winner; here also, there are RUMs-with-Ties such that the winner distribution for each slate, conditioned on some element winning, cannot be approximated by any RUM. Hence, RUMs-with-Ties appear to be strictly more representationally powerful. (iii) We ask next how succinctly a RUM-with-Ties can be written down, allowing a small approximation error to provide a more robust measure. Here, we are able to extend a result of [14] for RUMs, showing that both types of models can be approximated using $\Theta(n^2 \log n)$ bits. (iv) Finally, we ask whether a deployed system receives more or less information when interacting with a RUM versus a RUM-with-Ties. On the one hand, the RUM specifies the winner of every slate, which seems like a clear improvement. On the other hand, a RUM-with-Ties might give more information as the vocabulary is larger by a single element. We study the question of the information content of responses from these two models using a statistical testing framework, asking in which model the hypotheses can be distinguished more rapidly. We show that, surprisingly, the sample complexity may either increase or decrease in the presence of ties. (See Sect. 5.)

Can Ties be Useful? We raised the tantalizing prospect that an additional item might help a user to differentiate between two very close options. In Sect. 6, we study the theoretical justification for this question: could adding an additional (worse) item uncover the better option with only, say, 1% as many trials as direct comparison? We answer this question in the affirmative for the most popular discrete choice model, the multinomial logit (MNL), augmented with a JND; the resulting model is known as δ-*MNL* (or δ-*logit*) [33]. We consider two close elements, a and b. Under MNL, each trial between them will add Gumbel random noise to the utilities of a and b, and with some small probability, they will become distinguishable. However, many such trials will be required to determine the winner by direct comparison between a and b. We then consider introducing a third element, c, worse than either, and allow comparisons between c and each of a and b in turn. By analyzing the information available from this indirect comparison versus the direct comparison between a and b, we show that comparisons with a well-chosen third element c can be arbitrarily more efficient by any desired multiplicative factor in the number of trails.

2   Related Work

Discrete choice models have been extensively studied in disparate settings [4,21, 31,32]. RUMs lie at the core of choice theory [9,16,45,47,50], but they have also been used as components of larger models for voting [27] and text generation [48]. MNLs [23] are a special case of RUMs, but their simplicity makes them easier to learn from data and so are widely adopted [39,41] and used beyond choice theory [7]. The discrete choice literature is vast [24]; we only focus on works closest to ours.

Learning Choice Models from Preference Data. There has been recent work on learning ranking and choice models from preference data [1,47] and on learning RUMs from preferences [3,13,42]. Others have focused on the identifiability of RUMs [51,56].

Learning from Partial Orders. Learning from incomplete preferences inducing a partial order have been studied for MNLs [34,36,37,57] and RUMs [55]. These assume the underlying model is a RUM and only the observed data induces a partial order on items. However, we show that if the partial order comes from the indistinguishability of top alternatives, RUMs do not provide an accurate description of the choice distribution, while RUMs-with-Ties can.

Modeling Indifference. The seminal work by [35] spearheaded different research directions on including indifference threshold in discrete choice models. There have been attempts at axiomatizing choice models resulting from the indistinguishability of top alternatives for a single "user" [10,19]. In this regard, [40] analyze the properties of a stochastic choice function resulting from indecisiveness between close alternatives. In particular, they show that RUMs may or may not be induced by indistinguishability, validating the need for a more comprehensive model like ours. Another research direction, which we also pursue in this work, is to provide probabilistic models incorporating indistinguishability and learning algorithms for them. These include the MPD model [30] having an indifference threshold in a 2-MNL and the δ-MNL model [33] for more than two items. The latter has been used in several practical settings [11,12,54]. Recently, [38] provided an adaptive online learning algorithm for δ-MNL, but they only looked at pairwise preferences and not at the partial order induced by δ-MNLs. Our RUMs-with-Ties model is a generalization of δ-MNLs and accounts for more complex dynamics. Finally, some works have modeled indifference by adding a fictitious "no-choice" item to an MNL [22,25]. We show that, both in theory and in practice (MNL++), this is quite inaccurate at predicting choices.

3   Preliminaries

Order Relations. A *partial order* (or *poset*) $P = (X, \prec)$ is a binary ordering relation $\prec$ on X^2 that is transitive ($x \prec y$ and $y \prec z \implies x \prec z$) and antisymmetric ($x \prec y$ and $y \prec x \implies x = y$). If x, y are two elements of a poset such that $x \not\prec y$ and $y \not\prec x$, then we say that x and y are *incomparable* and write $x \parallel y$. An *antichain* of a poset is a set of pairwise incomparable elements. Given a poset P, we sometimes use $\prec_P$ to denote the ordering relation of P.

A *strict weak order* is a poset $P = (X, \prec)$ such that X can be partitioned into antichains $L_1, \ldots, L_k$, and for each $1 \leq i < j \leq k$, for each $x \in L_i$ and for each $y \in L_j$, it holds $x \prec y$, i.e., the antichains that make up the strict weak ordering are totally ordered. We will use *bucket order* as a synonym for strict weak order.

Let $\mathcal{S}_n$ be the set of all permutations (or *total orders*) of $[n]$. Also, let $\mathcal{B}_n$ denote the set of all the bucket orders of $[n]$; clearly, $\mathcal{B}_n \supseteq \mathcal{S}_n$. The number of bucket orders on $[n]$ is called the *Fubini number* (or *ordered Bell number*) and it satisfies $\log |\mathcal{B}_n| = \Theta(n \log n)$ [49].

Probability Distributions. For a discrete distribution D, let supp(D) denote its support. Let $x \sim D$ denote that $x \in$ supp(D) is sampled from D. For any element x, let $D(x)$ denote the probability that D assigns to x; clearly, $x \in$ supp(D) if and only if $D(x) > 0$. We generalize this to any subset $S \subseteq$ supp(D): $D(S) = \sum_{x \in S} D(x)$. When it is clear from the context, let $D(S)$ denote $D(S \cap$ supp(D)).

The *total variation distance* between distributions D and D' is

$$|D - D'|_{\mathrm{tv}} = \frac{1}{2} \sum_{\substack{x \in \mathrm{supp}(D) \\ \cup\, \mathrm{supp}(D')}} |D(x) - D'(x)| = \frac{1}{2} |D - D'|_1 .$$

This is also equal to the maximum, over all events S, of the absolute difference between the probabilities of S in D and in D', i.e.,

$$|D - D'|_{\mathrm{tv}} = \max_{S \subseteq \mathrm{supp}(D) \cup \mathrm{supp}(D')} |D(S) - D'(S)|.$$

Discrete Choice. Let a *slate* denote any non-empty subset of $[n]$. Given a bucket order $B \in \mathcal{B}_n$ and a slate $T \subseteq [n]$, let

$$B(T) = \begin{cases} i & \text{if } i \in T \text{ satisfies } i \succ_B j, \forall j \in T \smallsetminus \{i\}, \\ \bot & \text{otherwise.} \end{cases}$$

i.e., the unique maximum item in T according to B if it exists, and $\bot$ otherwise (i.e., in the case where the maximum item is not unique).

A *RUM* on $[n]$ is a probability distribution over $\mathcal{S}_n$.[3] As we will see later, a *RUM-with-Ties* on $[n]$ can be understood as a probability distribution D over $\mathcal{B}_n$ (although our first definition will be based on a user model). We drop the quantifier "on $[n]$" when it is obvious from the context. Given a slate $T \subseteq [n]$, we use D_T to denote the distribution of the random variable $B(T)$ for $B \sim D$, i.e., the distribution of the winner in the slate T with a random bucket order from D. Note that $\mathrm{supp}(D_T) \subseteq T \cup \{\bot\}$. A *Multinomial Logit (MNL)* is a special type of RUM where each item $i \in [n]$ is associated with *base value* w_i. A random permutation is generated by iteratively sampling without replacement, where at each step, an unsampled item is chosen w.p. proportional to the exponential of its base value. For a slate S, the probability $i \in S$ wins is $\frac{e^{w_i}}{\sum_{j \in S} e^{w_j}}$.

To define an approximation notion for RUM-with-Ties, we first define a *distance* between two RUM-with-Ties D, D', following [14]: $\mathrm{dist}(D, D') =$

$$\max_{\varnothing \subset S \subseteq [n]} |D_S - D'_S|_{\mathrm{tv}} = \max_{\substack{\varnothing \subset S \subseteq [n] \\ S' \subseteq S \cup \{\bot\}}} |D_S(S') - D'_S(S')|.$$

I.e., the distance is the maximum, over the slates S, of the total variation distance of the winner distributions of S with D and D'. Equivalently, it is the maximum over S and $S' \subseteq S \cup \{\bot\}$ of the absolute difference of the probabilities, with D and D', that the result of a random choice in S lies in S'. (Such a random choice could either return an element of S, or $\bot$ if no choice is made in S.)

We omit all proofs due to space constraints.

---

[3] RUMs are typically presented in terms of noisy item evaluations made by users. Each item is assumed to have a base value; each user samples utility for each item from a joint distribution. The user then chooses an item "rationally" as the one with the highest utility among the available ones (breaking ties, if any, u.a.r.). As the utilities are random, the family of resulting models is named "Random Utility Models," or RUMs. In an equivalent definition, the user first sorts all the items decreasingly according to their observed utilities (breaking ties, if any, u.a.r.), obtaining a permutation; given a slate, a rational user will choose the item with the highest rank in the permutation.

3.1   System 1 and System 2 Models

In our paper we follow Luce's pioneering work [35] and consider the following evaluations-with-noise models. We use the terminology popularized by the Nobel laureate Daniel Kahneman in his book [28] inspired by the way in which humans think: they can use their "System 1"—an instinctive and fast mode of thought— or their "System 2"—a logical and relatively slower way of thinking.

Definition 1 (System 1 Model and G_1). *Fix $\delta \geq 0$ and let $(U_1, \ldots, U_n)$ be the user's utilities sampled from a joint value distribution $\mathcal{D}$. Given $S \subseteq [n]$ and given $\{i, j\} \in \binom{S}{2}$, the user infers $i > j$ (i beats j) if $U_i > U_j + \delta$, the user infers $j > i$ (j beats i) if $U_j > U_i + \delta$, and the user infers $i \parallel j$ (i incomparable to j) if $|U_i - U_j| \leq \delta$. (Equivalently, given a slate $S \subseteq [n]$, the user sees a digraph G_1 with the elements of S as its nodes and for each $i, j \in S$, an arc from i to j if $i > j$, i.e., $U_i > U_j + \delta$.) Now, if there exists $i \in S$ such that $i > j$ for each $j \in S \setminus \{i\}$, then the user will choose i in S; otherwise, the user will choose nothing (i.e., a $\perp$ choice).*

Thus, if a user is told explicitly by System 1 that an object from the slate is better (by δ) than every other object in the slate, then they will choose that object. Note that if $(U_1, \ldots, U_n)$ is such that $\Pr[\exists \{i, j\} \in \binom{[n]}{2} \mid U_i = U_j] = 0$— e.g., if the distribution is continuous and has full support—then System 1 with $\delta = 0$ is the standard RUM model. For $\delta > 0$, instead, System 1 enables the user to choose "nothing". The relation $>$ in Definition 1 is also called a *semiorder*.

Sometimes, however, users can use inductive reasoning to improve their ability to choose the best option in a slate. For instance, if the user feels that $1 > 3$ but also feels that $1 \parallel 2$ and $2 \parallel 3$, then the user can infer that 1 is the item with the largest utility. Indeed, $1 > 3 \implies U_1 > U_3 + \delta$ and $2 \parallel 3 \implies |U_2 - U_3| \leq \delta \implies U_3 \geq U_2 - \delta$. Together we get $U_1 > U_3 + \delta \geq U_2$, letting the user infer 1 is better than 2. A similar argument shows that $2 > 3$. Thus, inductive reasoning allows the user to infer the full ordering $1 > 2 > 3$, whereas System 1 only allowed the user to observe that $1 > 3$. We codify this additional inference power below.

Definition 2 (System 2 Model and G_2). *Fix $\delta \geq 0$ and let $(U_1, \ldots, U_n)$ be the user's utilities sampled from a joint value distribution $\mathcal{D}$. Given a slate S and G_1 from System 1, the user produces a digraph G_2 on the same node set S containing each arc $i \to j$ for which the user can infer that $U_i > U_j$ from G_1. If $i \to j$ is in G_2, we say that $i \succ j$. Now, if there exists $i \in S$ such that $i \succ j$ for each $j \in S \setminus \{i\}$, then the user will choose i in S; otherwise, the user will choose nothing (i.e., a $\perp$ choice).*

Thus System 2 can add arcs to G_1 so to obtain a more informative rational graph G_2; the arc $i \to j$ is in G_2 only if the user can formally prove that the $U_i > U_j$. Such proofs could be direct (e.g., $i \to j$ is in G_1, thus $U_i > U_j + \delta$) or, as for the above example, might involve comparisons to other items. Note that for the example slate $S = \{1, 2, 3\}$, the user's System 1 chooses nothing from S whereas the user's System 2 is able to determine the best option in S.

Finally, we show a result on the structure of the optimal inferred G_2. This result is somewhat implicit in [35] but not in terms of G_2.

Theorem 1 ([35]). G_2 *induces a bucket order.*

3.2   RUMs-with-Ties

Finally, we recall a combinatorial model that is simpler to state yet equivalent to System 2, i.e., the choice systems representable with System 2 coincide with the choice systems representable by this combinatorial model. This model will allow us to find an optimal fit as well to succinctly represent a generic System 2 model.

Definition 3 (RUM-with-Ties). *A RUM-with-Ties is a probability distribution D over $\mathcal{B}_n$. Given $S \subseteq [n]$, the probability $D_S(i)$ that $i \in S$ gets chosen is equal to the probability, over bucket orders B from D, that i is the unique highest item of S in B, i.e., $D_S(i) = \Pr_{B \sim D}[B(S) = i]$. The probability $D_S(\perp)$ that no choice is made from S is defined as $D_S(\perp) = \Pr_{B \sim D}[B(S) = \perp]$.*

The following was proved in [35].

Theorem 2 ([35]). *Consider any joint value distribution $\mathcal{D}$, any $\delta \geq 0$, and any choice distribution that System 2 induces on $[n]$. Then, there is a RUM-with-Ties inducing the same choice distribution. Conversely, for any RUM-with-Ties, for each $\delta > 0$, there exists a joint value distribution $\mathcal{D}$ such that System 2 (as well as System 1) induces the same choice distribution as the RUM-with-Ties.*

3.3   System 1 Vs System 2

System 1 and System 2 can be different in the ability to discern the best item in a slate. Clearly, System 2 is never worse than System 1. We will show distributions/slates for which both are identical in power, and distributions/slates such that System 2 can determine the winner while System 1 cannot.

Let $n \geq 3$ and let item $i \in [n]$ have utility value sampled independently and u.a.r. from $[i - \epsilon, i + \epsilon]$ for some $\epsilon \in (0, 1/4)$. Then, System 1 will be able to determine the winner of $S = \{1, 2, 3\}$ w.p. 1 if $\delta < 1 - 2\epsilon$ and w.p. 0 if $\delta > 1 + 2\epsilon$. On the other hand, System 2 will be able to determine the winner of S w.p. 1 if $\delta < 2 - 2\epsilon$ and w.p. 0 if $\delta > 2 + 2\epsilon$. In particular, the two systems coincide in their discriminatory power on S if $\delta < 1 - 2\epsilon$ (in which case, they will both find the winner of S), or if $\delta > 2 + 2\epsilon$ (in which case, they will both result in a no-choice). If $\delta \in (1 + 2\epsilon, 2 - 2\epsilon)$, then System 2 will return the winner of S w.p. 1, while System 1 will return the winner of S w.p. 0. Letting $\epsilon = 0$, this construction can be made deterministic.

4 Inferring Preferences

In this section we present algorithms to construct G_2 from System 2 given G_1 from System 1. We obtain a quadratic-time algorithm that returns G_2 containing all and only the arcs $i \rightarrow j$ for which the user (who has only "rational" access to G_1) can prove $U_i > U_j$. We then show a quadratic lower bound on the running time of any algorithm that infers all the missing arcs by querying System 1.

4.1 Algorithm

Let $V(G)$ denote the set of nodes in the graph G and let $V = V(G_1)$. For $i \in V$, let $N_1^-(i)$ be the set of predecessors of i in G_1, i.e., nodes with a directed edge to i. Similarly, let $N_1^+(i)$ be set of successors of i in G_1, i.e., nodes with a directed edge from i. Let $\deg_1^-(i)$ (resp., $\deg_1^+(i)$) be the in-degree (resp., out-degree) of node i in the digraph G_1, i.e., $\deg_1^-(i) = |N_1^-(i)|$ and $\deg_1^+(i) = |N_1^+(i)|$.

Before introducing the algorithm, we state a relationship between the perceived utilities and the node degrees in G_1; a version of this statement also appears in [2]. Let $P_{i,j}$ be the predicate $P_{i,j} =$"$\deg_1^-(i) < \deg_1^-(j)$ or $\deg_1^+(i) > \deg_1^+(j)$".

Lemma 1 ([2]). *Let $\{i,j\} \in \binom{V}{2}$. Then, (i) $P_{i,j} \implies U_i > U_j$, (ii) $P_{j,i} \implies U_j > U_i$, and (iii) $\overline{P_{i,j} \lor P_{j,i}} \implies N_1^-(i) = N_1^-(j)$ and $N_1^+(i) = N_1^+(j)$. It is then impossible for $P_{i,j} \land P_{j,i}$ to hold.*

Let us define the *score* s_i of $i \in V$ to be $s_i = \deg_1^+(i) - \deg_1^-(i)$, i.e., the number of items that i beats minus the number of items that beat i. We point out an important consequence of Lemma 1.

Theorem 3 ([2]). *Let $\{i,j\} \in \binom{V}{2}$. Then, (i) $s_i > s_j \implies U_i > U_j$, (ii) $s_i < s_j \implies U_j > U_i$, and (iii) $s_i = s_j \implies N_1^-(i) = N_1^-(j)$ and $N_1^+(i) = N_1^+(j)$.*

We now present Algorithm 1 and prove that it produces a correct and optimal inference from G_1.

Algorithm 1. An algorithm for producing graph G_2, given G_1.

1: $V \leftarrow V(G_1)$
2: **for all** $i \in V$ **do**
3:     $s_i \leftarrow \deg_1^+(i) - \deg_1^-(i)$ {score}
4: $E \leftarrow \{i \rightarrow j \mid \forall i, j \in V$ s.t. $s_i > s_j\}$
5: **return** $G_2(V, E)$

Theorem 4. *The graph G_2 produced by Algorithm 1 contains all and only the arcs $i \rightarrow j$ for which one can prove, given access to G_1, that $U_i > U_j$. Moreover, in general, for no $\{i,j\} \in \binom{V}{2}$ it is possible to prove that $U_i = U_j$.*

We remark that the first part of Theorem 4 is implicit in [2]. Also note that Algorithm 1 can be implemented to run in quadratic time; a constructive version of the method in [35] needs cubic time.

Theorem 5. *Algorithm 1 runs in time $O(|V|^2)$.*

Note that once the scores of the nodes are computed, the bucket order corresponding to G_2 (see Theorem 1) can be constructed in time $O(n \log n)$ by sorting the nodes by scores.

4.2 Lower Bound

Next, we prove a lower bound on the query complexity of any algorithm that aims to infer all the missing arcs by querying System 1. In fact, we show the same lower bound on the number of calls to System 1, even to find the best item.

Theorem 6. *An algorithm that queries System 1 on pairs of items needs $\Omega(|V|^2)$ queries in expectation to determine the best item of V, even in settings where System 2 is guaranteed to be able to determine the best item of V.*

5 The Power of RUMs-with-Ties

RUMs-with-Ties are at least as powerful as classical RUMs by definition, but the precise relationship between these models is not obvious. In this section we explore different measures of the power and complexity of RUMs-with-Ties.

We begin in Sect. 5.1 by attempting to replicate the behavior of RUMs-with-Ties by *augmenting* regular RUMs with a "no-choice" element. In Sect. 5.2, we consider a related question, attempting to use RUMs to replicate the behavior of RUMs-with-Ties specifically in cases for which the RUM-with-Ties gives a winner (i.e., does not return "no-choice"). We show that RUMs-with-Ties are not approximable by RUMs in either setting. Next, in Sect. 5.3, we ask how many bits are required to specify almost perfectly the behavior of a RUM-with-Ties—the ability to introduce some small errors in the approximation provides a more robust measure that is not subject to corner cases of perfect fidelity. We give tight bounds on the representation complexity, essentially characterizing the compressibility of RUMs-with-Ties. These results show that RUMs-with-Ties can be written down in essentially the same space as RUMs. Finally, in Sect. 5.4, we ask whether the ability of RUMs-with-Ties to employ the additional "no-choice" tag is a blessing or a curse in terms of the information available to the model user.

5.1 RUMs-with-Ties Vs Augmented RUMs

A natural idea is to augment $[n]$ of the RUM with a special "no-choice" item $\perp$ [22]. This augmented RUM (with $n + 1$ items) is only applied to slates containing $\perp$: when $\perp$ is chosen from a slate, we say that a "no-choice" happened in the original slate.

We prove that RUMs-with-Ties are more expressive than augmented RUMs. The difference is due to the non-monotonicity of the "no-choice" event in RUMs-with-Ties. Indeed, in an augmented RUM R, if $S \ni \perp$ is one of its slates, then given any $T \supset S$ it must be that $D_S(\perp) \geq D_T(\perp)$, i.e., the probability of a tie is monotonically decreasing. In a RUM-with-Ties, instead, the probability of a tie satisfies no such property, e.g., even if no tie could have happened in S, it could be that the top-most element of $T \smallsetminus S$ is always tied with the top-most element of S. We formalize this next.

Lemma 2. *For $n = 3$, there exists a a RUM-with-Ties R on $[n]$ such that, for each augmented RUM R' on $[n] \cup \{\perp\}$, there exists at least one slate $S \subseteq [n]$, such that the ℓ_∞-distance between R's distribution on S and the distribution on $S \cap \{\perp\}$ given by R' is at least $1/2$.*

5.2 RUMs-with-Ties Vs Projected RUMs

A different natural reduction is to project a RUM-with-Ties T to eliminate the ties: given a set $\{S_1, \ldots, S_t\}$ of slates with their observed (no-)choice distributions $T_{S_1}, \ldots, T_{S_t}$, condition each distribution on not causing a tie; let $T'_{S_1}, \ldots, T'_{S_t}$ be the resulting distributions. Then, the hope is to fit a RUM to $T'_{S_1}, \ldots, T'_{S_t}$ obtaining a representation of all the winning events and represent the "no-choice" events separately. We prove that the error induced by projected RUMs can be large. The construction uses the fact that such a conditioning can cut off a large part of the probability space, so that the conditioning blows up the probability of some rare events.

Lemma 3. *For each $\epsilon > 0$, and for each large enough n, there exists a RUM-with-Ties T on $[n]$ whose no-ties conditioned winning distributions $T'_{S_1}, \ldots, T'_{S_{2^n}}$ have the property that, for each RUM R on $[n]$, there exists at least one slate $S_i \subseteq [n]$, such that the ℓ_∞-distance between S_i and R's distribution on S_i is at least $1 - \epsilon$.*

5.3 Succinctly Representing RUMs-with-Ties

We next consider the problem of approximately representing RUMs-with-Ties with few bits. In particular, we show that RUM-with-Ties on the ground set $[n]$ can be sketched to $O(n^2 \cdot \epsilon^{-2} \log n)$ bits. This generalizes the result in [14] for classical RUMs.

Theorem 7. *Let $0 < \epsilon, \delta < 1$. There is a polynomial-time algorithm that, given any distribution D on $\mathcal{B}_n$, produces a multiset B of $O\left(\epsilon^{-2} \cdot \left(n + \ln \delta^{-1}\right)\right)$ bucket orders such that, with probability $\geq 1 - \delta$, the uniform distribution $\tilde{D}$ on B guarantees that $\mathrm{dist}(D, \tilde{D}) \leq \epsilon$.*

The algorithm is to sample independent bucket orders $B_1, \ldots, B_t$ from D, where $t = O(\epsilon^{-2} \cdot (n + \ln \delta^{-1}))$, and to let $\tilde{D}$ be the uniform distribution on this multiset of samples.

An easy consequence of Theorem 7 is that any RUM-with-Ties can be approximately represented using $O(n^2 \log n)$ bits, since a bucket order can be represented using $O(n \log n)$ bits.

Corollary 1. *For each $0 < \epsilon < 1$, and for each RUM-with-Ties D, there is a data structure using $O(\epsilon^{-2} \cdot n^2 \log n)$ bits that can be used to return, for each slate $S \subseteq [n]$, a distribution $\tilde{D}_S$ satisfying $\left| D_S - \tilde{D}_S \right|_{tv} \leq \epsilon$.*

Using the lower bound construction in [14], Corollary 1 can be shown to be near-optimal.

Corollary 2. *Fix a constant $0 < \alpha \leq 1/2$. A data structure for a generic RUM-with-Ties D that can be used, for each slate $S \subseteq [n]$, to return a distribution $\tilde{D}_S$ satisfying $\left| D_S - \tilde{D}_S \right|_{tv} \leq \frac{1-\alpha}{4}$, requires at least $\frac{\alpha^3}{6} \cdot n^2 - 1$ bits.*

5.4 Comparing the Information Present in Responses from RUMs-with-Ties Vs RUMs

Our goal is to understand whether the addition of a "no-choice" response represents additional information returned by a RUM-with-Ties, or whether the loss of perfect information on the max-utility element of the slate represents a loss of information. To study this, we rely on the hypotheses testing framework and ask whether the responses from a RUM-with-Ties can more efficiently differentiate between two underlying data models than the responses from a RUM. For simplicity, instead of RUMs, we study MNLs vs MNLs-with-Ties in Systems 1 and 2. We show that, for some parameter choice, MNLs-with-Ties allow for more efficient predictions, while for some others, MNLs have more efficient predictions.

Lemma 4. *There exist settings where, using a single sample, (i) with MNLs, no algorithm can guess correctly with probability larger than $3/4$, while (ii) MNL-with-Ties can be used to guess the unknown hypothesis with probability $1 - o(1)$.*

Lemma 5. *There exist settings where, using a single sample, (i) MNLs can be used to guess the unknown hypothesis with probability at least $3/4 - o(1)$, while (ii) with MNLs-with-Ties, the probability of guessing correctly is at most $1/2 + o(1)$.*

6   Learning the Best Element of Pair

In this section we focus on algorithms to learn the highest utility member of a pair in a δ-MNL. Let the two objects have base values x, y (wlog, $x < y$) and we will have access to outcomes of 2-slates by different users in the MNL-with-ties model. Let $\delta > 0$ be the indifference threshold and $c := \frac{y-x}{\delta}$.

Let $X \sim \text{Gumbel}(x, 1)$ and $Y \sim \text{Gumbel}(y, 1)$ be two independent r.v.'s. Then, $X - Y \sim \text{Logistic}(x - y, 1)$. We can show that if δ is large, we have a tie whp., making it necessary to sample the outcome a large number of times in order to infer if $y > x$. Suppose, instead, we can leverage a third value $Z \sim$ Gumbel$(z, 1)$ where $x - \delta < z < y - \delta$. Then, if $\delta \gg 1$, we have that $|X - Y| \leq \delta$ (tie) whp.; conditioned on this, $|X - Z| \leq \delta$ (tie) whp.; conditioned on both, $Y - Z > \delta$ (win) whp. Thus we can infer $y > x$. E.g., if $x = \delta$, $y = 2\delta$, we can pick $z = \delta/2$ as a "lower quality" pivot value. By carefully choosing a pivot, we can significantly decrease the number of necessary slates for inference.

A simple PAIRS algorithm can check N many independent 2-slates of the original two objects and then output the final winner as the object that got strictly more wins than the other.

Lemma 6. *Given $\rho \in (0, 1)$, the PAIRS algorithm outputs "Y wins" with probability $\geq 1 - \rho$ if the number of slates is $N_{\text{PAIRS}} = 2\ln(1/\rho) \cdot \left(\frac{1}{1+e^{(1-c)\delta}} - \frac{1}{1+e^{(1+c)\delta}} \right)^{-2}$.*

Suppose we have access to a pivot object Z with base value $z := (x+y)/2 + \delta = x + \delta \cdot (1 + c/2) = y + \delta \cdot (1 - c/2)$ (it would be analogous for the "symmetric" case $z = \frac{x+y}{2} - \delta$, with lower base value than x, y if $c < 1$). The PIVOT algorithm checks N independent 2-slates: $N/2$ of the first object against the pivot and $N/2$ of the second object against the pivot. It tracks the pivot and outputs the final winner as the object who got the least losses against the pivot.

Lemma 7. *Given $\rho \in (0, 1)$, the PIVOT algorithm outputs "Y wins" with probability $\geq 1 - \rho$ if the number of slates is $N_{\text{PIVOT}} = 4\ln(2/\rho) \cdot \left(\frac{1}{1+e^{-c\delta/2}} - \frac{1}{1+e^{c\delta/2}} \right)^{-2}$.*

We can show that PIVOT can outperform PAIRS in terms of the slates used for inference by an unbounded factor.

Lemma 8. *Given $A > 1$, if $0 < c < 1$ and $\delta \geq \frac{4\ln(2)+\ln(A)}{1-c}$, then $N_{\text{PAIRS}} > A^2 N_{\text{PIVOT}}$.*

We can further generalize Lemma 8 as follows. Consider the following two settings. In the first (as in PAIRS), we can only see outcomes of 2-slates comparing the two original objects, and we denote with Ψ_{PAIRS} the minimum number of samples needed to correctly guess the winner with probability $\geq 2/3$. In the second (as in PIVOT), we are allowed to pick a pivot object to compare against any of the two original objects in a slate with size 2. We consider the experiment in which we fix a pivot and, each time, we compare it to a random object of the pair and see the outcome; again, let Ψ_{PIVOT} be the number of samples needed in this second setting for correctness with probability $\geq 2/3$.

Theorem 8. *Given an arbitrarily large $A > 1$, if $0 < c < 1$ and $\delta \geq \frac{4\ln(2)+\ln(G\cdot(A+12/c))}{1-c}$, it holds $\Psi_{\text{PAIRS}} \geq A \cdot N_{\text{PIVOT}} \geq A \cdot \Psi_{\text{PIVOT}}$, where G is an absolute constant.*

Theorem 8 shows that the number of samples needed for decision with a pivot can be arbitrarily smaller than that needed with only pairs of the original objects. Moreover, PIVOT algorithm can perform much better than *any* algorithm with access only to slates of the original objects.

7   Experiments with Modeling Ties in Practice

In this section we provide a summary of an experimental analysis of RUM-with-Ties in contexts where user preferences induce bucket orders over the set of available items. We provide experimental evidence in support of the use of RUMs-with-Ties in the setting of indistinguishability of close alternatives. We have made the code publicly available on GitHub[4] for reproducibility.

Our experiments are over several real-world "user ratings" datasets: SUSHI [29], containing ratings on 100 sushi ingredients; TRAVEL [44], containing TripAdvisor reviews scores from 980 users on 10 activities; *young people spending habits* (YPSH) [52], containing a student survey on spending habits for 7 different items (e.g., branded clothing, shopping); GOODBOOKS[5], containing ratings on 10 000 books; MOVIELENS[6], containing ratings from 270 000 users on 45 000 movies.

We interpret the ratings given by a user as utility values, with a fixed indifference threshold δ. We split each dataset randomly into training (80%) and test (20%).

We consider the RUM-with-Ties induced by the training dataset (train-RUMwt). For comparison, we consider the following algorithms. The first, MNL++, approximates a RUM-with-Ties using an augmented MNL, i.e., an MNL augmented with the "no-choice" item [22,25], learned from a training set using standard stochastic gradient descent. The second, δ-MNL, is an MNL with an indifference threshold δ [33]: the utilities are computed as output values of an MNL, but a winner is selected if and only if its value is at least δ more than all the other values; otherwise, the outcome is "no-choice." . The third, LP-RUMwt, solves the following LP to learn a RUM-with-Ties.[7]

$$\begin{cases} \min \frac{1}{2} \cdot \sum_{\varnothing \neq S \subseteq [n]} \sum_{x \in S \cup \{\perp\}} \delta_{S,x} \\ -\delta_{S,x} \leq D_S(x) - \sum_{\substack{B \in \mathcal{B}_n \\ x = B(S)}} p_B \leq \delta_{S,x}, \quad \forall S \subseteq [n], \forall x \in S \cup \{\perp\} \\ \sum_{B \in \mathcal{B}_n} p_B = 1 \text{ and } p_B \geq 0 \; \forall B \in \mathcal{B}_n \end{cases} \quad (1)$$

As ground-truth, we use the RUM-with-Ties induced by the whole dataset (OPT-RUMwt). We compare the algorithms against OPT-RUMwt on the slates test set. Each of these models induces a distribution over the outcomes of each slate in the slates test set, also known as stochastic choice [17]. We compare those distributions according to the average KL-divergence between the ground-truth given by OPT-RUMwt and a models' distribution.

From Table 1, we can see that train-RUMwt is by far the best model. This finding strongly supports the use of RUMs-with-Ties to model ties. However,

---

[4] https://github.com/re-gius/rums-with-ties.

[5] https://github.com/zygmuntz/goodbooks-10k.

[6] https://www.kaggle.com/datasets/rounakbanik/the-movies-dataset.

[7] We cannot compare directly to [38] since they provide an adaptive online learning algorithm, which assumes to have access to outcomes of any slate. Our datasets, instead, only contain a subset of all possible slates. Moreover, we focus on offline learning algorithms.

Table 1. Average KL-divergence ($\times 10^{-3}$) achieved by the models on the slates test set. The best overall metric is in bold (always train-RUMwt). The best metric for models learned on slates (i.e., excluding train-RUMwt) is in blue.

dataset	MNL++	δ-MNL	LP-RUMwt	train-RUMwt
SUSHI	7.9	7.3	**4.5**	**0.39**
YPSH	5.4	3.1	**2.3**	**0.026**
TRAVEL	5.5	8.9	**3.6**	**0.073**

in some cases we only have access to a fixed set of slates with their outcomes: in this setting, we cannot learn train-RUMwt, but only the other models. Thus, it is also useful to restrict the comparison to MNL++, δ-MNL, LP-RUMwt. LP-RUMwt consistently outperforms MNL++ and δ-MNL in terms of average KL-divergence, suggesting that it is better than all previous models for learning the choice distribution of a random slate. However, there is still a significant gap between LP-RUMwt and train-RUMwt. This highlights the opportunity to develop new algorithms to reduce the complexity of the previous LP so to make LP-RUMwt scalable to larger datasets but closer to train-RUMwt in terms of performance.

Finally, we use Algorithm 1 to compute the actual number of new inferences obtained in System 2. For each user u, we consider the average review scores $\{r_u^i\}_{1 \leq i \leq 10}$ as sampled utilities from its value distribution over the categories. For a fixed indifference threshold $\delta > 0$, we say that (i, j) is inferred by System 1 if $\|r_u^i - r_u^j\| > \delta$; we say that (i, j) is inferred by System 2 if $i \to j$ or $j \to i$ in the G_2 graph returned by Algorithm 1. For each user u, we can therefore define the number of inferences from System 1, denoted $S_1(u, \delta)$, and from System 2, denoted $S_2(u, \delta)$. We report the average over all users of the incremental percentage of new inferences of System 2 with respect to System 1 for different values of δ: $\text{avg}_u \frac{S_2(u,\delta) - S_1(u,\delta)}{S_1(u,\delta)} \%$.

Table 2. Average percentage of new inferences with respect to System 1 provided by Algorithm 1 for different values of δ.

dataset	$\delta = 0.25$	$\delta = 0.5$	$\delta = 1$	$\delta = 2$
TRAVEL	2.8	10.0	25.3	–
MOVIELENS	–	6.3	42.4	–
GOODBOOKS	–	–	26.4	44.7

Table 2 reports the results of the experiment for the datasets. Notice that, even for seemingly low and realistic values of δ, Algorithm 1 gives significantly more inferred preferences, on average, with respect to System 1. This gives experimental evidence of the importance of System 2 new inferences and the need for an optimally efficient algorithm to compute them like our Algorithm 1.

References

1. Agarwal, S.: On ranking and choice models. In: IJCAI, pp. 4050–4053 (2016)
2. Aleskerov, F., Bouyssou, D., Monjardet, B.: Utility Maximization, Choice and Preference. Springer, Heidelberg (2007)
3. Almanza, M., Chierichetti, F., Kumar, R., Panconesi, A., Tomkins, A.: RUMs from head-to-head contests. In: ICML, pp. 452–467 (2022)
4. Anderson, S.P., De Palma, A., Thisse, J.-F.: Discrete Choice Theory of Product Differentiation. MIT Press, Cambridge (1992)
5. Armstrong, W.E.: A note on the theory of consumer's behavior. Oxf. Econ. Pap. **2**(1), 119–122 (1950)
6. Bahamonde-Birke, F.J., Navarro, I., de Dios Ortúzar, J.: If you choose not to decide, you still have made a choice. J. Choice Model. **22**, 13–23 (2017)
7. Bang, D., et al.: Logit mixing training for more reliable and accurate prediction. In: IJCAI, pp. 2812–2819 (2022)
8. Benson, A.R., Kumar, R., Tomkins, A.: On the relevance of irrelevant alternatives. In: WWW, pp. 963–973 (2016)
9. Benson, A.R., Kumar, R., Tomkins, A.: A discrete choice model for subset selection. In: WSDM, pp. 37–45 (2018)
10. Block, H.D., Marschak, J.: Random orderings and stochastic theories of response. In: Economic Information, Decision, and Prediction: Selected Essays: Volume I Part I Economics of Decision, pp. 172–217 (1974)
11. Cantillo, V., Amaya, J., Ortúzar, J.D.D.: Thresholds and indifference in stated choice surveys. Trans. Res. Part B: Meth. **44**(6), 753–763 (2010)
12. Cantillo, V., Ortúzar, J.D.D.: Implications of thresholds in discrete choice modelling. Trans. Rev. **26**(6), 667–691 (2006)
13. Chierichetti, F., Giacchini, M., Kumar, R., Panconesi, A., Tomkins, A.: Approximating a rum from distributions on k-slates. In: AISTATS, pp. 4757–4767. PMLR (2023)
14. Chierichetti, F., Kumar, R., Tomkins, A.: Light RUMs. In: ICML, pp. 1888–1897 (2021)
15. Chierichetti, F., Kumar, R., Tomkins, A.: On the number of trials needed to distinguish similar alternatives. PNAS **119**(31) (2022)
16. Corbin, R., Marley, A.: Random utility models with equality: an apparent, but not actual, generalization of random utility models. J. Math. Psych. **11**(3), 274–293 (1974)
17. Debreu, G.: Stochastic choice and cardinal utility. Econometrica, 440–444 (1958)
18. Dhar, R.: Consumer preference for a no-choice option. J. Consum. Res. **24**(2), 215–231 (1997)
19. Falmagne, J.-C.: A representation theorem for finite random scale systems. J. Math. Psych. **18**(1), 52–72 (1978)
20. Fechner, G.T.: Vorschule der aesthetik. Breitkopf & Härtel, Leipzig (1876)
21. Feng, Z., Zhu, Y., Cao, J.: Modeling air travel choice behavior with mixed kernel density estimations. In: WSDM, pp. 611–620 (2017)
22. Gao, S., Frejinger, E., Ben-Akiva, M.: Cognitive cost in route choice with real-time information: an exploratory analysis. Procedia-Social Behav. Sci. **17**, 136–149 (2011)
23. Gensch, D.H., Recker, W.W.: The multinomial, multiattribute logit choice model. J. Mark. Res. **16**(1), 124–132 (1979)

24. Haghani, M., Bliemer, M.C., Hensher, D.A.: The landscape of econometric discrete choice modelling research. J. Choice Model. **40**, 100303 (2021)

25. Hess, S., Beck, M.J., Chorus, C.G.: Contrasts between utility maximisation and regret minimisation in the presence of opt out alternatives. Transp. Res. Rec. P. A: Policy Pract. **66**, 1–12 (2014)

26. Iyengar, S.S., Lepper, M.R.: When choice is demotivating: can one desire too much of a good thing? J. Pers. Soc. Psy. **79**(6), 995 (2000)

27. Jiang, A., Soriano Marcolino, L., Procaccia, A.D., Sandholm, T., Shah, N., Tambe, M.: Diverse randomized agents vote to win. In: NIPS (2014)

28. Kahneman, D.: Thinking, Fast and Slow. Farrar, Straus & Giroux (2011)

29. Kamishima, T.: Nantonac collaborative filtering: recommendation based on order responses. In: KDD, pp. 583–588 (2003)

30. Krishnan, K.: Incorporating thresholds of indifference in probabilistic choice models. Manag. Sci. **23**(11), 1224–1233 (1977)

31. Kumar, R., Mahdian, M., Pang, B., Tomkins, A., Vassilvitskii, S.: Driven by food: modeling geographic choice. In: WSDM, pp. 213–222 (2015)

32. Labutov, I., Schalekamp, F., Luu, K., Lipson, H., Studer, C.: Optimally discriminative choice sets in discrete choice models: application to data-driven test design. In: KDD, pp. 1665–1674 (2016)

33. Lioukas, S.: Thresholds and transitivity in stochastic consumer choice: a multinomial logit analysis. Manag. Sci. **30**(1), 110–122 (1984)

34. Liu, A., Zhao, Z., Liao, C., Lu, P., Xia, L.: Learning Plackett–Luce mixtures from partial preferences. In: AAAI, pp. 4328–4335 (2019)

35. Luce, R.D.: Semiorders and a theory of utility discrimination. Econometrica **24**(2), 178–191 (1956)

36. Ma, J., et al.: Learning-to-rank with partitioned preference: fast estimation for the Plackett–Luce model. In: AISTATS, pp. 928–936 (2021)

37. Ma, J., Zhang, X., Mei, Q.: Fast learning of mnl model from general partial rankings with application to network formation modeling. In: WSDM, pp. 715–725 (2022)

38. Nguyen, Q.P., Tay, S., Low, B.K.H., Jaillet, P.: Top-k ranking Bayesian optimization. In: AAAI, pp. 9135–9143 (2021)

39. Oh, M.-H., Iyengar, G.: Multinomial logit contextual bandits: provable optimality and practicality. In: AAAI, pp. 9205–9213 (2021)

40. Ok, E.A., Tserenjigmid, G.: Indifference, indecisiveness, experimentation, and stochastic choice. Theor. Econ. **17**(2), 651–686 (2022)

41. Ou, M., Li, N., Zhu, S., Jin, R.: Multinomial logit bandit with linear utility functions. In: IJCAI, pp. 2602–2608 (2018)

42. Parkes, D.C., Soufiani, H.A., Xia, L.: Random utility theory for social choice. In: NIPS (2012)

43. Patall, E.A., Cooper, H., Robinson, J.C.: The effects of choice on intrinsic motivation and related outcomes: a meta-analysis of research findings. Psych. Bull. **134**(2), 270 (2008)

44. Renjith, S., Sreekumar, A., Jathavedan, M.: Evaluation of partitioning clustering algorithms for processing social media data in tourism domain. In: RAICS, pp. 127–131 (2018)

45. Saha, A., Gopalan, A.: Best-item learning in random utility models with subset choices. In: AISTATS, pp. 4281–4291 (2020)

46. Schwartz, B.: The Paradox of Choice: Why More is Less. Harper Perennial (2005)

47. Seshadri, A., Ragain, S., Ugander, J.: Learning rich rankings. In: NeurIPS, vol. 33, pp. 9435–9446 (2020)

48. Sim, Y., Routledge, B.R., Smith, N.A.: The utility of text: the case of amicus briefs and the supreme court. In: AAAI (2015)
49. Sklar, A.: On the factorization of squarefree integers. Proc. AMS **3**(5), 701–705 (1952)
50. Train, K.E.: Discrete Choice Methods with Simulation. Cambridge University Press, Cambridge (2003)
51. Turansick, C.: Identification in the random utility model. J. Econ. Theo., 105489 (2022)
52. Vilčeková, L., Sabo, M.: The influence of demographic factors on attitudes toward brands and brand buying behavior of Slovak consumers. Intl. J. Ed. Res. **1**(11), 1–10 (2013)
53. Weis, C., et al.: Surveying and analysing mode and route choices in Switzerland 2010–2015. Travel Behav. Soc. **22**, 10–21 (2021)
54. Yu, Y., Han, K., Ochieng, W.: Day-to-day dynamic traffic assignment with imperfect information, bounded rationality and information sharing. Trans. Res. Part C: Emerg. Tech. **114**, 59–83 (2020)
55. Zhao, Z., Liu, A., Xia, L.: Learning mixtures of random utility models with features from incomplete preferences. In: IJCAI, pp. 3780–3786 (2022)
56. Zhao, Z., Villamil, T., Xia, L.: Learning mixtures of random utility models. In: AAAI (2018)
57. Zhao, Z., Xia, L.: Learning mixtures of Plackett–Luce models from structured partial orders. In: NeurIPS (2019)

Thresholds as Mechanisms for Weighting Influence in the Linear Threshold Rank

Maria J. Blesa[✉][iD], Alejandro Dominguez-Besserer, and Maria Serna[iD]

Universitat Politècnica de Catalunya – BarcelonaTech, 08034 Barcelona, Spain
{maria.j.blesa,maria.serna}@upc.edu

Abstract. Social networks are the natural space for the spreading of information and influence and have become a media themselves. Several models capturing that diffusion process have been proposed, most of them based on the Independent Cascade (IC) model or on the Linear Threshold (LT) model. The IC model is probabilistic while the LT model relies on the knowledge of an actor to be convinced, reflected in an associated individual threshold. Although the LT-based models contemplate an individual threshold for each actor in the network, the existing studies so far have always considered a threshold of 0.5 equal in all actors (i.e., a simple majority activation criterion).

Our main objective in this work is to start the study on how the dissemination of information on networks behaves when we consider other options for setting those thresholds and how many network actors end up being influenced by this dissemination. For doing so, we consider a recently introduced centrality measure based on the LT model, the Forward Linear Threshold Rank (FwLTR), which is the natural interpretation of the Linear Threshold Rank on directed networks.

We experimentally analyze the ranking properties for several networks in which the influence resistance threshold follows different schemes. Here we consider three different schemes: (1) uniform, in which all players have the same value; (2) random, where each player is assigned a threshold u.a.r. in a prescribed interval; and (3) determined by the value of another centrality measure on the actor. Our results show that the selection has a clear impact on the ranking, even quite significant and abrupt in some cases. We conclude that the social networks ranks that provide the best assignments for the individual thresholds are FwLTR and the well-known PageRank.

Keywords: Social Networks · Centrality measures · (Forward) Linear Threshold Rank · Social Media Analytics

Supported by MCIN/AEI/10.13039/501100011033 under grant PID2020-112581GB-C21 (MOTION).

1   Introduction

Nowadays the importance of social networks as a marketing tool is growing rapidly and spanning through diverse social networks and objectives. Think for example about viral marketing, that seeks to spread information about a product or service from person to person by word of mouth or sharing via the internet or email. The goal of viral marketing is to inspire individuals to share a marketing message (or opinion) to friends, family, and other individuals to create exponential growth in the number of its recipients. The so-called *spread of influence* process is used here with a very specific goal and needs to be maximized until it stabilizes and no additional convictions are possible. How to select the initial group of individuals that start the spreading process is a key issue.

Traditional social networks are labelled graphs in which the nodes represent individuals and the edges represent weighted relations between them. There have been several proposal for modeling the spread of influence in a network. Among them, the *Independent Cascade* (IC) model is a stochastic model proposed in [7]. It is based on the assumption that whenever a node is activated, it will do (stochastically) attempt to activate a neighbor. The Linear Threshold (LT) model is a deterministic model for influence spread based on some ideas of collective behavior [9]. In the LT model the strength of the tie between every pair of actors quantifies the capacity of one to influence the other and, additionally, each actor opposes a resistance to be influenced. A node gets influenced when their active predecessors can exert enough influence to overpass its resistance. The node resistance is quantified by a threshold that quantifies the minimum required percentage of the total income weight needed to convince the actor. Several computational problems have been proposed that attempt to solve the problem of selecting the initial set of participants to spread the product in the most effective. The two most relevant being the *influence maximization* [9,17] and the *target set selection problems* [4]. Another line of research uses the spread of influence process to define centrality measures [2,14].

Trying to experimentally evaluate centrality measures based on the linear threshold model, we came with a lack of suitable data sets. Most of the social networks you can find in repositories do not include any kind on information about the resistance of a participant to be influenced. Therefore, as no information about suitable threshold values to run the linear threshold model was collected, a single majority rule was assumed. However this rule might not be a realistic one in many scenarios. In this paper, we attempt to shed some light on the effect that the selection of threshold has on one of the centrality measures based on the LT model. We focus our analysis in the FwLTR rank which measures the influence that a node together with its immediate successors can exert [2]. FwLTR is the natural interpretation of the Linear Threshold Rank (LTR) on directed networks.

We performed a series of experiments in order to assess how the threshold values affect the FwLTR. We considered a subset of the social networks used in [14]. To evaluate the effect of the threshold selection, we fix the method to assign the influence thresholds and compute the FwLTR ranking. Over the ranking, we compute the traditional statistics and then we select some relevant sets of participants, the ten in the top of the ranking, the 10% in the top, and the participants whose rank is in the 10% of the highest values, and compute the number of nodes influenced by those sets.

We run three kind of experiments. In the first one we assign to each node a common threshold θ. This type of uniform assignment was the usual scheme used in the literature until now, with $\theta = 0.5$ to implement a simple majority rule. However, we try different values for θ others than 0.5. In the second experiment we assign random thresholds in different ranges. In the third experiment we follow a completely different schema and use the value of another centrality measure on the node to determine its threshold.

Besides more specific comments given later, we comment here on the main discovered trends. Our first set of experiments clearly show that, in most of the treated networks, the range of threshold values more relevant to the maximization of the influence is within $[0.2, 0.5]$, where it is possible to accumulate greater proportion of the total influence diffused in groups of actors of small size. Interestingly enough, we found that a random distribution of the thresholds behaves practically the same with respect to our measures for the diffusion of influence whether they are in $(0, 0.5]$ or $(0, 1]$. This suggests that the fact that a minority fraction of the actors have high thresholds affects little in the actor's ability to influence in general. Finally, experiments also suggest that the FwLTR centrality and the well-known PageRank provide the best thresholds, in the sense that they generate favorable conditions for a great influence expansion even when starting with reduced sets. In nearly all cases, it is enough to consider the first 10 top-ranked actors to reach the entire network, or the vast majority of it.

2   Centrality Measures

We outline the centrality measures used in different parts of this paper. We have focused only on some of the most significant ones, and refer the reader to existing surveys (e.g., [16]) for a wider overview on other measures. As we will see, some of them are based on the relevance and the topological properties of the nodes, e.g., PageRank, while others focus on their influence over the network, e.g., the Linear Threshold Rank and the Independent Cascade Rank. We start describing the topology-based centrality measures, which were the ones used traditionally as centrality measures, and continue describing the influence-based ones that have been proposed more recently. As usual, we assume that the social network is represented by a directed graph $G = (V, E)$.

Betweenness. In the betweenness centrality, a node is more important if it belongs to the shortest path between any pair of nodes in the graph [5]. For every node $i \in V$, define

$$\mathsf{Btwn}(i) = \sum_{s,t \in V - \{i\}} \frac{\sigma_{st}(i)}{\sigma_{st}}$$

where σ_{st} is the number of shortest paths between s y t, and $\sigma_{st}(i)$ the number of such shortest paths that pass by i.

PageRank. One of the most popular centrality measures is the PageRank [13], that Google uses to assign relevance to web pages. A web page is more relevant if other important web pages point to it. It uses a parameter $\alpha \in (0,1]$, that represents the probability that a user keeps jumping from a web page to another through the links that are between them (and thus, $1 - \alpha$ represents the probability that the user goes to a random web page). Let A be the adjacency matrix of G (i.e., $a_{ij} = 1$ if $(i,j) \in E$, and 0 otherwise), the PageRank (PR) of i is given by

$$\mathsf{PgR}(i) = (1 - \alpha) + \alpha \sum_{j \in V} \frac{a_{ji}\ \mathsf{PgR}(j)}{\delta^+(j)}$$

where $\delta^+(i)$ is the out degree of $i \in V$.

Perhaps the two most prevalent diffusion models in computer science are the *Independent Cascade* model [7] and the *Linear Threshold* model [9] (see also [15]). We define their corresponding influence-based centrality measures:

Independent Cascade Rank. It is an influence-based centrality measure [10] based on the Independent Cascade (IC) Model [7], which is a stochastic model. It is based on the assumption that whenever a node is activated, it will (stochastically) do attempt to activate each actor he targets. Given an activated node $i \in V$, any neighbor j such that $(i,j) \in E$ will be activated with a probability p_{ij}. When a new actor is activated, the process is repeated for this actor. The whole process ends when there are no active nodes with a new chance to spread its influence.

Given an initial core $X \subseteq V$ and a probability $p \in [0,1]$ (where $\forall (i,j) \in E : p_{ij} = p$), the influence spread of X is denoted by $F'(X,p)$. The Independent Cascade Rank of a node $u \in V$ is then defined as:

$$\mathsf{ICR}(u,p) = \frac{|F'(u,p)|}{\max_{v \in V}\{|F'(v,p)|\}}.$$

Linear Threshold Rank. It is based on the Linear Threshold (LT) Model [9]. Every node has an influence threshold, which represents the resistance of this node to be influenced by others. In the model, every edge (u,v) also has a weight representing the influence that node u has over node v. In practice, that weight is set to one since it is very different to quantify that concept.

The influence algorithm starts with an initial predefined set of activated nodes. At every iteration, the active nodes will influence their neighbors. When the total influence that a node receives exceeds its influence threshold, then this node will become active and join the set of active nodes. The algorithm stops when the set of active nodes converges, i.e., when no new nodes are influenced.

Given an initial set of active nodes $X \subseteq V$ and a threshold assignment $\theta : V \to \mathbb{N}$, let $F_t(X) \subseteq V$ denote the set of activated nodes at the t-th iteration of the spreading process. At the first step ($t = 0$) only the nodes in X are active, which means that $F_0(X) = X$. At the $t + 1$ iteration, a node i will be activated if, and only if, the sum of all the weights of the edges $\{i, j\}$, where j is already active, is higher than the resistance (or influence threshold) of i, i.e., $\sum_{j \subseteq F_t(X)} w_{ij} \geq \theta(i)$, In practice, it is usual to consider that all the edge weights are equal to one and thus a node i will be activated if, and only if,

$$\frac{|F_t(X) \cap \mathcal{N}(i)|}{|\mathcal{N}(i)|} \geq \theta(i),$$

where, $\mathcal{N}(u) = \{v \mid (u, v) \in E \vee (v, u) \in E\}$. Observe, that the process is monotonic, therefore it stops after at most $n = |V|$ steps. Thus can define the spread of X as $F(X) = F_n(X)$. The *Linear Threshold Rank* [14] of a node $i \in V$ is given by

$$\mathsf{LTR}(i) = \frac{|F(\{i\} \cup \mathcal{N}(i))|}{n}.$$

Forward Linear Threshold Rank. It is a centrality measure very similar to the LTR, but with a different initial activation set [2]. Formally, let $\mathcal{N}^+(i) = \{j \in V \mid (i, j) \in E\}$, then the Forward Linear Threshold Rank of a node $i \in V$ is given by

$$\mathsf{FwLTR}(i) = \frac{|F(\{i\} \cup \mathcal{N}^+(i))|}{n}.$$

3   Statistical and Top Measures

For analysing the results of a centrality measure on its own, three statistical metrics will be used later in the experimental part: the number of different ranks assigned ($\#$), their standard deviations (σ) from the mean and, Gini coefficient of the ranks.

The Gini coefficient [3,6] comes originally from the field of sociology as a measure of the inequality of populations with respect to different criteria (e.g., wealth spread). It is often represented graphically through the Lorenz curve [11], which shows wealth distribution by plotting the population percentile by income on the horizontal axis and cumulative income on the vertical axis. The Gini coefficient is equal to the area below the line of perfect equality (i.e., 0.5) minus the area below the Lorenz curve, divided by the area below the line of perfect equality.

Definition 1. *Given a list of values X of size n, the Gini coefficient of X is calculated as follows:*

$$Gini = \frac{\sum_{i=1}^{n} \sum_{j=1}^{n} |x_i - x_j|}{2n \sum_{i=1}^{n} x_i}$$

The Gini coefficient is a value in $[0, 1]$, where 0 indicates a completely equitable distribution of values, and 1 represents the most radical inequality. This coefficient is lately being very much used in data science as a measure for quantifying the fairness of data distributions coming from other scientific areas.

In [8] the case of selecting sets of actors with high centrality values was studied as initial trigger sets. Specifically, they used the Degree, Betweenness and Eigenvector centralities, but a mixed influence diffusion model was used instead of the usual IC model or LT model, and only two networks of a relatively small size (of the order of a thousand nodes), both of academic collaboration (similar to the ArXiv network that we use here). However, we were inspired by the idea in that article for the introduction of three metrics related to the maximization of influence, and we study some variants that use measures of centrality other than the FwLTR.

The following variables are conceived as natural measures that can allow us to see if the individual actors with the highest FwLTR in the network are able to exert an amount of influence meaningfully together. We define:

Top10 - This parameter is the proportion of actors that are influenced if we execute the influence expansion algorithm taking as initial activating set the 10 actors with the highest FwLTR, i.e.,

$$\text{Top10}(G) = \frac{|F(X_{10})|}{n}$$

where X_{10} is the 10 actors of G with the highest FwLTR.

Top10%Actors - From the FwLTR ranking of actors, let Y_{10} be the 10% with the greatest capacity to disseminate direct influence. Then,

$$\text{Top10\%Actors}(G) = \frac{|F(Y_{10})|}{n}$$

Top10%Values - Analogously, let Z_{10} be the set of actors whose influence make up the first 10% of different ranking values (ties included). Then,

$$\text{Top10\%Values}(G) = \frac{|F(Z_{10})|}{n}$$

4   Networks

Table 1 summarizes the characteristics of the networks considered for our experiments. The structural characteristics of the networks are described by seven common attributes: the number of vertices, the number of edges, whether the graph is weighted, whether the graph is directed, the average clustering coefficient, and the size of the main core.

The *average clustering coefficient* (ACC) is the average of the local clustering coefficients in the graph. The local clustering coefficient C_i of a node i is the number of triangles T_i in which the node participates normalized by the maximum number of triangles that the node could participate in.

$$\text{ACC} = \frac{1}{n} \sum_{i=1}^{n} C_i, \quad \text{where } C_i = \frac{T_i}{\delta_i(\delta_i - 1)}$$

where δ_i is the degree of the node i, and $n = |V|$. Given a graph G and $k \in \mathbb{Z}^+$, a *k-core* is the maximal induced subgraph of G where every node has at least degree k. The *main core* is a k-core of G with the highest k.

Table 1. Characteristics of the real networks under consideration (in alphabetical order). ACC = Average Clustering Coefficient, MC = size of the main core. When the diameter is ∞, the diameter of the biggest connected component is provided.

Data set	n	m	Directed?	Edge-weighted?	ACC	Diameter	MC
Amazon	334863	925872	✗	✗	0.3967	44	497
ArXiv	5242	14496	✗	✗	0.5296	∞ (17)	44
Caida	26475	106762	✓	✓	0.2082	17	50
ENRON	36692	183831	✗	✗	0.4970	11	275
Epinions	75879	508,837	✓	✗	0.1378	14	422
Gnutella	62586	147892	✓	✗	0.0055	11	1004
Higgs	256491	328132	✓	✓	0.0156	19	10
Wikipedia	7115	103689	✓	✗	0.1409	7	336

5 Experiments

Although the LT model contemplates an individual threshold for each actor in the network, the existing studies so far have always considered a threshold of 0.5 equal in all actors. In practice, this means that a majority activation criterion is assumed: a node is activated (or influenced) when at least half of its neighbors have been activated. Our main objective in this work is to start the study on how the dissemination of information on networks is affected when we consider other options for setting those thresholds, how many network actors end up being influenced by that dissemination process, and to which extend the top-ranked actors guarantee that dissemination.

Among the experiments performed, we would like to point out here the most interesting ones. The first experiment still considers the same threshold for all the actors but, apart from the standard 0.5, we consider other values in $[0,1]$. The second experiment sets random thresholds to the actors following a random uniform distribution. Finally, the third experiment sets to each node a threshold that depends on its centrality value according to several different measures. All the algorithms were implemented in ANSI C++, using GCC 7.5.0 for compiling the code. The experimental evaluation was performed on a cluster of computers with Intel® Xeon® CPU 5670 CPUs of 12 nuclei of 2933 MHz and (in total) 32 Gigabytes of RAM.[1]

---

[1] Infrastructure available at the Research & Development Lab (\\RDlab) of the Universitat Politècnica de Catalunya - BarcelonaTech.

5.1   Uniform Thresholds

The first experiment carried out consists in calculating the FwLTR of the network actors for the cases in which the threshold is the same for each of them and is defined as $1/4$, $1/2$, $3/4$ and 1. The results of this experiments are summarized in Table 2. For this experiment we obtain six metrics of the distribution of the FwLTR on the actors.

Table 2. Experimental results when considering uniform thresholds for $\theta \in \{1/4, 1/2, 3/4, 1.00\}$.

	θ	σ	#	Gini	Top10	Top10%Actors		Top10%Values	
Amazon	0.25	0.001358	1146	0.697121	0.00658478	0.754688	(33486)	0.628117	(518)
	0.50	0.000037	224	0.411186	0.000979505	0.301711	=	0.00142745	(25)
	0.75	0.000031	203	0.372209	0.00078241	0.210316	=	0.000958601	(24)
	1.00	0.000017	153	0.326807	2.9863e-05	0.0999991	=	5.0767e-05	(17)
ArXiv	0.25	0.149279	435	0.854073	0.648417	0.670355	(524)	0.656429	(104)
	0.50	0.005950	169	0.540466	0.00267074	0.263068	=	0.00839374	(20)
	0.75	0.004339	137	0.519565	0.00209844	0.167684	=	0.0034338	(14)
	1.00	0.003578	118	0.513171	0.00190767	0.0999618	=	0.0022892	(12)
Caida	0.25	0.234984	2033	0.827517	0.890954	0.979339	(2647)	0.979339	(2426)
	0.50	0.014930	1234	0.704623	0.114145	0.555694	=	0.362606	(137)
	0.75	0.012896	1174	0.693649	0.0957129	0.47169	=	0.266931	(130)
	1.00	0.001261	158	0.503259	0.000377715	0.0999811	=	0.000642115	(17)
ENRON	0.25	0.268885	2786	0.833348	0.110406	0.845307	(3669)	0.841655	(426)
	0.50	0.034267	2322	0.752884	0.0292707	0.588139	=	0.442222	(236)
	0.75	0.013211	1852	0.715277	0.0168429	0.39202	=	0.148643	(190)
	1.00	0.009538	1448	0.726360	0.000272539	0.0999945	=	0.00411534	(151)
Epinions	0.25	0.034836	1395	0.961041	0.0462051	0.612976	(7587)	0.609418	(375)
	0.50	0.001259	828	0.830114	0.014879	0.484587	=	0.0422778	(85)
	0.75	0.001185	814	0.828013	0.0144045	0.370326	=	0.0377838	(84)
	1.00	0.000343	327	0.707198	0.000131789	0.0999881	=	0.000487618	(37)
Gnutella	0.25	0.299963	593	0.892894	0.974148	0.979692	(6258)	0.979788	(6716)
	0.50	0.000173	111	0.689648	0.00615154	0.97402	=	0.00672674	(13)
	0.75	0.000172	111	0.689222	0.00594382	0.956076	=	0.00640718	(12)
	1.00	0.00070	52	0.545161	0.00015978	0.0999904	=	0.000111846	(7)
Higgs	0.25	0.000010	77	0.279112	0.000557524	0.147467	(25649)	0.000557524	(10)
	0.50	0.000006	52	0.218244	0.00035089	0.13094	=	0.000323598	(8)
	0.75	0.000006	53	0.214459	0.000362586	0.125685	=	0.000304104	(6)
	1.00	0.000005	46	0.192700	3.89877e-05	0.0999996	=	2.33926e-05	(6)
Wikipedia	0.25	0.026782	309	0.875415	0.32298	0.366268	(711)	0.324666	(37)
	0.50	0.006443	234	0.766315	0.00196767	0.359663	=	0.00477864	(25)
	0.75	0.006125	238	0.763096	0.00196767	0.288686	=	0.00463809	(26)
	1.00	0.005943	238	0.761196	0.00140548	0.0999297	=	0.00337316	(24)

Observing the sharp differences obtained in the results of this first experiment, we decided to do more experiments for the Gnutella and ArXiv networks with fine-grained threshold values between 0.2 and 0.5, and for Amazon, Caida, ENRON, Epinions and Wikipedia between 0.2 and 0.4. On the contrast, the remaining networks showed a considerable degree of monotony in metrics despite threshold variation. Table 3 shows the refined results for the Caida and Gnutella networks.[2] Here we only expose and comment on the most relevant results.

The extreme variation of Top10%Values due to widespread ties in the allocation of the FwLTR questions its usefulness as a metric for maximizing influence, since its behavior is not justified by the variation in the other measures.

Note that there is an obvious correlation between Top10%Actors and the Gini, since the more unequal a distribution is, the greater the proportion of the value accumulated by a small group of actors will be. However, this does not necessarily occur in Top10, where the set of actors has a fixed size of 10, regardless of the size of the network considered. The value of the Gini decreases when the influence threshold is increased, but the behavior of Top10 seems independent to this trend (see, e.g., Wikipedia or Caida; see Table 3).

We see a curious rise and fall of the cardinality of the set of Top10%Values in the Gnutella network between the thresholds 0.26 and 0.34 (see Table 3), which is not reflected in the variation of the other metrics for the same range of thresholds. There is also a sudden drop after 0.48, reflected in all metrics except Top10%Actors, which suffers only a small decrease. This suggests that, despite been the distribution of the actors' FwLTR radically affected and its influence severely limited, 10% of the most influential actors is a robust and large enough set for expanding its influence to most of the Gnutella network at those threshold magnitudes. This behavior is not observed in ArXiv, where Top10%Actors follow an online behavior with the other metrics. Despite Gnutella's dimensions being larger than ArXiv's, the most significant differences between these two networks, in terms of the parameters that we have obtained from them, is that Gnutella is directed and ArXiv is not, and ArXiv has greater diameter and ACC (around 100 times greater), and a much smaller main core. These data suggest that ArXiv is a much more interconnected and distributed network than Gnutella, whose parameters indicate that most nodes have few interconnections, but there are a subset of nodes with very high degrees, which would be very capable of diffusing the influence effectively, and therefore would be in the Top10%Actors. Gnutella's Gini is especially high compared to these networks, a fact that could validate this explanation.

The findings thus far indicate that within the majority of the analyzed networks, resistance thresholds falling within the range of approximately $[0.2, 0.5]$ emerge as particularly significant. Within this range, small actor groups are able to induce a significant proportion of the total network influence.

5.2   Random Thresholds

In this experiment, we set random thresholds to the actors according to a random uniform distribution in $[0, 1]$, and later on two different intervals.

---

[2] Results for the remaining ones can be found in the Appendix.

Table 3. Experimental results when considering uniform thresholds θ: refined zoom in [0.2, 0.4] and [0.2, 0.5].

	θ	σ	#	Gini	Top10	Top10%Actors		Top10%Values	
Caida	0.20	0.365273	1455	0.747064	0.907347	0.995392	(2647)	0.995392	(6562)
	0.22	0.338180	1615	0.775445	0.906327	0.995392	=	0.995392	(5223)
	0.24	0.322551	1740	0.790862	0.904363	0.98916	=	0.994334	(4594)
	0.26	0.218601	2077	0.825349	0.871124	0.978357	=	0.978357	(2225)
	0.28	0.197560	2173	0.822601	0.860057	0.933975	=	0.930727	(1818)
	0.30	0.142933	2291	0.792907	0.787762	0.92729	=	0.918829	(1399)
	0.32	0.130483	2355	0.783452	0.765703	0.913957	=	0.899452	(1302)
	0.34	0.062324	2366	0.720697	0.187875	0.87101	=	0.795958	(408)
	0.36	0.057867	2395	0.714390	0.18085	0.859754	=	0.791917	(386)
	0.38	0.049777	2424	0.704256	0.184816	0.849745	=	0.769481	(334)
	0.40	0.044526	2402	0.697489	0.171709	0.838414	=	0.745647	(277)
Gnutella	0.20, 0.22	0.373341	280	0.819895	0.974547	0.988176	(6258)	0.987873	(212)
	0.24	0.373303	277	0.819943	0.974547	0.988176	=	0.987361	(186)
	0.26	0.299963	593	0.892894	0.974148	0.979692	=	0.979788	(6716)
	0.28	0.299944	594	0.892910	0.974148	0.979692	=	0.979788	(6715)
	0.30, 0.32	0.297771	615	0.894629	0.974148	0.979596	=	0.979596	(6606)
	0.34, 0.38	0.090475	581	0.986210	0.971863	0.981753	=	0.971863	(610)
	0.40, 0.42	0.080219	666	0.986708	0.971783	0.981657	=	0.971783	(495)
	0.44	0.080126	663	0.986709	0.971783	0.981657	=	0.971783	(497)
	0.46, 0.48	0.080126	662	0.986709	0.971783	0.981657	=	0.971783	(495)
	0.50	0.000173	111	0.689648	0.00615154	0.97402	=	0.00672674	(13)

Table 4. Experimental results (for 100 executions) for random thresholds under a random uniform distribution.

Network	σ	#	Gini	Top10	Top10%Actors		Top10%Values	
Amazon	0.007479	2317	0.002218	1	1	(33486)	1	(1492)
ArXiv	0.316806	1478	0.208256	0.793209	0.793209	(524)	0.793209	(234)
Caida	0.007981	3248	0.002557	1	1	(2647)	1	(396)
ENRON	0.248287	13117	0.083801	0.918347	0.918347	(3669)	0.918347	(1390)
Epinions	0.269755	5816	0.259223	0.628316	0.637106	(7587)	0.628316	(636)
Gnutella	0.405467	2288	0.768062	0.972805	0.992394	(6258)	0.978238	(264)
Higgs	0.003500	9404	0.709183	0.00924399	0.139252	(25649)	0.0245662	(1472)
Wikipedia	0.142819	1048	0.277731	0.327056	0.383696	(711)	0.333942	(109)

Random Uniform Distribution - For this experiment we assigned thresholds in $[0, 1]$ following a uniform probability distribution. In order to determine centrality of the actors, we simulate the influence expansion several times, each with new randomly assigned values, and finally we get the average FwLTR of the actors for those executions. Table 4 shows the results for this experiment when performing 100 executions.

We can see that there are no large differences for any of the variables between the two experiments and that, for the vast majority of networks, a uniform random distribution is very conducive to maximizing influence, achieving very satisfactory results with only the top 10 actors of the ranking. However, we also see that we rarely achieve more expansion if we increase the set to the first 10% of actors. We also see, for the first time, that the Top10 sets have managed to expand the influence over all the actors in the Amazon and Caida networks.

Table 5. Experimental results (for 20 executions) for random thresholds under a random uniform distribution within $(0, 0.5]$.

Network	σ	#	Gini	Top10	Top10%Actors		Top10%Values	
Amazon	0.021248	519	0.008737	1	1	(33486)	1	(1192)
ArXiv	0.316359	646	0.213341	0.793209	0.793209	(524)	0.793209	(151)
Caida	0.031134	526	0.015298	1	1	(2647)	1	(117)
ENRON	0.249355	3265	0.089727	0.918347	0.918347	(3669)	0.918347	(380)
Epinions	0.266832	702	0.269977	0.629397	0.7452	(7587)	0.763597	(9994)
Gnutella	0.407975	424	0.768950	0.974579	0.991276	(6258)	0.991628	(12851)
Higgs	0.003177	4318	0.714451	0.00971964	0.139479	(25649)	0.0190182	(1027)
Wikipedia	0.139828	571	0.285285	0.327056	0.374982	(711)	0.334505	(77)

Table 6. Experimental results (for 100 executions) for random thresholds under a random uniform distribution within $[0.5, 1]$.

Network	σ	#	Gini	Top10	Top10%Actors		Top10%Values	
Amazon	0.163244	35500	0.712485	0.717867	0.719664	(33486)	0.717935	(9784)
ArXiv	0.228914	2007	0.240972	0.791683	0.791683	(524)	0.791683	(240)
Caida	0.246121	7448	0.609839	0.926308	0.926308	(2647)	0.926308	(818)
ENRON	0.208083	13455	0.152003	0.918075	0.918075	(3669)	0.918075	(1572)
Epinions	0.124726	11641	0.759954	0.537092	0.542825	(7587)	0.537092	(1365)
Gnutella	0.000176	1059	0.691910	0.00560828	0.973988	(6258)	0.856725	(145)
Higgs	0.000043	2344	0.528186	0.002234	0.141759	(25649)	0.00643687	(358)
Wikipedia	0.091907	1960	0.678521	0.319606	0.364301	(711)	0.331553	(214)

We observe that the number of different values produced in the ranking is much higher in all cases than those of the FwLTR with $\theta = 0.5$ (see Table 2) and they are more dispersed also, as evidenced by the standard deviation σ. The Gini also indicates that, in general, the ranking values are significantly more equally distributed when $\theta = 0.5$. We see a significant reduction in Gini in almost all networks (e.g., in the case of ENRON, it drops approximately from 0.75 to 0.08). The exceptions are Gnutella and Higgs, where there is an ascent with respect to the majority criterion.

Random Uniform Distribution in Two Different Intervals - As a result of the previous experiment, we wondered what would happen if we assigned the thresholds uniformly at random but separately for domains (0, 0.5] and [0.5, 1]. By selecting these two intervals we can see what happens when the thresholds are in the more permissive or restrictive half for the transmission of influence. In Table 5 and Table 6 we can check the results.

The values of the variable Top10 in Table 6 suggest that for most of networks it is possible to achieve a good expansion of influence even when the thresholds of resistance are distributed in the most restrictive half of the interval. In some cases, we obtain practically the same expansion as when assigning the thresholds in (0, 0.5], as is the case for the networks ArXiv, Wikipedia or ENRON. As expected, a smaller value is reached in general. The values of $\#$ are notably different between the two experiments, being the values of the second significantly larger than those of the first in almost all cases. This may indicate that, when the thresholds tend to be more restrictive, there is a more diverse ranking. Despite this, there does not appear to be any trend in the value of the Gini between the two experiments, and the standard deviation increases for some networks and decreases for others.

If we compare Table 5 with Table 4 in the previous section, we observe little difference in the variables that measure the expansion of influence. In fact, the only relevant improvement occurs on the Epinions network. We also see a general reduction in the number of rankings assigned to the actors ($\#$), but very similar standard deviation and Gini values. These results indicate that a random distribution of the thresholds behaves practically the same with respect to the diffusion of influence whether they are in $(0, 0.5]$ or $(0, 1]$, suggesting that perhaps the fact that a minority fraction of the actors have high thresholds affects little to the ability to influence of the actors in general.

5.3   Thresholds Fixed by Other Centrality Measures

In this experiment we will use different centrality criteria to set the thresholds of the actors of a network. We will denote $FwLTR_m$ to the FwLTR where the threshold θ is set to the value given by the centrality measure m; for example $FwLTR_{ICR}$ is the FwLTR ranking of the actors in a network when we set the resistance threshold of each actor to its ICR centrality value. We introduce these metrics to study the possibility that an actor's resistance to being influenced in a network is related to its centrality value, perhaps by a topological criterion or by own ability to spread influence, and in turn, we study the case that this related with the lack of these qualities.[3]

---

[3] How the thresholds are assigned according to the different centrality measures in each network can be seen in the Appendix.

Table 7. Experimental results for node thresholds assigned according to Btwn.

Network	σ	#	Gini	Top10	Top10%Actors		Top10%Values	
Amazon	0.144358	185	0.021906	0.986335	0.996363	(33486)	0.987983	(572)
ArXiv	0.321546	84	0.207490	0.793209	0.793209	(524)	0.793209	(11)
Caida	0.097087	10	0.009517	1	1	(2647)	1	(26228)
ENRON	0.269095	431	0.104976	0.915458	0.918347	(3669)	0.915458	(1392)
Epinions	0.277439	39	0.265431	0.628725	0.757126	(7587)	0.629173	(28)
Gnutella	0.410393	73	0.767687	0.974978	0.996421	(6258)	0.974579	(9)
Higgs	0.003131	198	0.817138	0.00899057	0.148863	(25649)	0.0103551	(109)
Wikipedia	0.145380	14	0.275292	0.32818	0.431061	(711)	0.339002	(55)

Table 8. Experimental results for node thresholds assigned according to their ICR value.

Network	σ	#	Gini	Top10	Top10%Actors		Top10%Values	
Amazon	0.048237	159	0.002346	0.997814	1	(33486)	0.998719	(67)
ArXiv	0.016946	389	0.626108	0.0721099	0.603396	(524)	0.183518	(43)
Caida	0.058035	3025	0.664515	0.355354	0.95966	(2647)	0.913692	(364)
ENRON	0.016987	2322	0.691726	0.0317235	0.526927	(3669)	0.251199	(237)
Epinions	0.003255	1614	0.862370	0.0452035	0.599573	(7587)	0.152493	(169)
Gnutella	0.387287	267	0.801691	0.974643	0.990701	(6258)	0.983383	(106)
Higgs	0.000028	144	0.448542	0.00132948	0.15022	(25649)	0.0015907	(18)
Wikipedia	0.028498	781	0.744640	0.113563	0.355727	(711)	0.195643	(92)

Table 9. Experimental results for node thresholds assigned according to PgR.

Network	σ	#	Gini	Top10	Top10%Actors		Top10%Values	
Amazon	0.000000	1	0.000000	1	1	(33486)	1	(334863)
ArXiv	0.321327	84	0.207110	0.793209	0.793209	(524)	0.793209	(11)
Caida	0.066887	7	0.004494	1	1	(2647)	1	(26356)
ENRON	0.256195	360	0.085229	0.918347	0.918347	(3669)	0.918347	(37)
Epinions	0.275013	39	0.258302	0.628725	0.756837	(7587)	0.629173	(28)
Gnutella	0.410393	73	0.767687	0.974978	0.996421	(6258)	0.974579	(9)
Higgs	0.003319	189	0.785266	0.00894768	0.151744	(25649)	0.0101797	(83)
Wikipedia	0.145312	14	0.274871	0.32832	0.431061	(711)	0.339002	(55)

Table 10. Experimental results when the thresholds of the nodes are assigned according to their value of FwLTR.

Network	σ	#	Gini	Top10	Top10%Actors		Top10%Values	
Amazon	0.004888	4	0.000024	1	1	(33486)	1	(334855)
ArXiv	0.322199	84	0.208631	0.793209	0.793209	(524)	0.793209	(11)
Caida	0.151072	20	0.023374	1	1	(2647)	1	(25895)
ENRON	0.296371	385	0.118261	0.918347	0.918347	(3669)	0.918347	(39)
Epinions	0.277610	39	0.265945	0.628725	0.757087	(7587)	0.629173	(28)
Gnutella	0.410393	73	0.767687	0.974978	0.996421	(6258)	0.974579	(9)
Higgs	0.003712	183	0.706886	0.00899837	0.157916	(25649)	0.0103473	(101)
Wikipedia	0.146984	14	0.285386	0.32832	0.430921	(711)	0.338721	(54)

FwLTR$_{Btwn}$ – When fixing the threshold according to Btwn (see Table 7), we can observe similar results to those of the experiment detailed in Table 4 for the thresholds distributed in a uniform random way. The standard deviation and the Gini are also very similar. Where the results of these two experiments really differ is in the number of distinct values in the ranking, which are significantly smaller now. These results suggest that we can expect achieve a very good expansion with very small initial sets, but note that we gain little or nothing by increasing the pool of the top 10 to the first 10%.

FwLTR$_{ICR}$ – When fixing the threshold according to ICR (see Table 8) the values of Top10, Top10%Actors and Top10%Values are lower than when fixed according to Btwn (compare with Table 7) for all networks except Amazon and Higgs, where we see a slight increase. The values of σ are also lower, indicating higher concentration, and gives more variety of values, which are also more unequally distributed, as indicated the Gini, except in the Higgs network, which undergoes the opposite behavior.

FwLTR$_{PgR}$ and FwLTR$_{FwLTR}$ – When fixing the threshold according to PgR and FwLTR (see Tables 9 and 10, respectively) we see extremely similar values in the variables related to the expansion of influence. We expected similarities, and we observe that they give very comparable threshold assignments. However, it is surprising to what extent the diffusion of the influence on the Top10 variables between both methods. The values obtained for the metrics σ, # and Gini are not disparate, which show similar results under both centralities. From these tables we deduce that assigning the thresholds according to the PgR and FwLTR centralities of the actors generates favorable conditions for a great influence expansion from reduced sets. In nearly all cases, it is enough to have the first 10 actors of the ranking to reach the entire network or its vast majority, except for two networks (Higgs and Wikipedia) whose individual characteristics seem to imply difficulties for the diffusion of influence, as by this and previous experiments suggest.

5.4   Complementaries

We have also performed an additional experiment that complements the previous one. We will denote by FwLTR$_{1-m}$ the case in which the thresholds are set by the complement of the value obtained from the centrality m, that is, if an actor obtains a centrality value c in the measure m, we will assign a threshold equivalent to $1 - c$. In this way we can observe the case in which a greater centrality corresponds to a greater resistance to influence, and the opposite.

The FwLTR$_{1-Btwn}$ (see Appendix) behaves in the opposite way in the Top10 variables, where we barely managed to expand the influence. The values of the Gini they have also been reversed: the high values in Table 7 are here low, and vice versa. The standard deviation σ is very low in all cases, indicating that ranking values are produced very clustered around the mean.

In the FwLTR$_{1-ICR}$ (see Appendix) we see that we hardly managed to diffuse the influence in none of the Top10 variables. We also have low values of the standard deviation, but generally similar to those of the FwLTR$_{ICR}$. The Gini values are less high, indicating a more equitable distribution of ranks. In the same way, we see practically the same values for FwLTR$_{1-PgR}$ and FwLTR$_{1-FwLTR}$ (see Appendix) in all metrics. The expansion of influence is predictably restrictive unlike the cases described in the previous paragraph. The values of σ are very low and those of the Gini higher in most cases, compared to the previous ones obtained in the non-complementary version. The exceptions are in Higgs and Gnutella, where they previously had high values but are now diminished.

6   Conclusions

We have considered the spread of influence in the LT model and the absence of potential values for the threshold defining the individual resistance that is the basis for the model. We have selected to study the behaviour of the FwLTR rank in a popular selection of social networks, and studied different ways to fix the individual threshold of the actors. Our results show that the selection has a clear impact on the ranking. Surprisingly enough, we have found a abrupt variation of the ranking when setting the thresholds uniformly, being thresholds in the interval $[0.2, 0.5]$ the best selection. For the case of random thresholds, a similar phenomenon appears as there are almost no differences on the ranking when the individual thresholds are selected in $(0, 0.5]$ than in $(0, 1]$. Finally, we have been able to asses that the social networks ranks that provide the best assignments for the individual thresholds are PgR and FwLTR.

There are many ways to complement this study, for example by including additional social networks in the analysis which might provide insight on the best models to study influence spread in the LT model. Another extension will be to study how the threshold selection impacts on other influence spread problems like the influence maximization or the target set selection problems. We would also like to compare FwLTR to recently-proposed centrality measures, e.g. [1,12].

Appendix

Additional data supporting this work can be found at https://cs.upc.edu/~mjblesa/ASONAM.2024.

References

1. Arhachoui, N., Bautista, E., Danisch, M., Giovanidis, A.: A fast algorithm for ranking users by their influence in online social platforms. In: Proceedings of the 2022 IEEE/ACM International Conference on Advances in Social Networks Analysis and Mining (ASONAM), pp. 526–533. Association for Computing Machinery (2022). https://doi.org/10.1109/ASONAM55673.2022.10068673

2. Blesa, M., García-Rodríguez, P., Serna, M.: Forward and backward linear threshold ranks. In: Proceedings of the 2021 IEEE/ACM International Conference on Advances in Social Networks Analysis and Mining (ASONAM), pp. 265–269. Association for Computing Machinery (2021). https://doi.org/10.1145/3487351. 3488355

3. Ceriani, L., Verme, P.: The origins of the Gini index: extracts from Variabilità e Mutabilità (1912) by Corrado Gini. J. Econ. Inequal. **10**(3), 421–443 (2012)

4. Chen, N.: On the approximability of influence in social networks. SIAM J. Discret. Math. **23**(3), 1400–1415 (2009). https://doi.org/10.1137/08073617X

5. Freeman, L.: A set of measures of centrality based on betweenness. Sociometry **40**(1), 35–41 (1977)

6. Gini, C.: Variabilitàe Mutabilità. Contributo allo Studio delle Distribuzioni e delle Relazioni Statistiche. C, Cuppini (1912)

7. Goldenberg, J., Libai, B., Muller, E.: Using complex systems analysis to advance marketing theory development: modeling heterogeneity effects on new product growth through stochastic cellular automata. Acad. Market. Sci. Rev. **9**(1) (2001)

8. Hasson, S.T., Akeel, E.: Influence maximization problem approach to model social networks. In: 2019 International Conference on Advanced Science and Engineering (ICOASE), pp. 135–140 (2019). https://doi.org/10.1109/ICOASE.2019.8723703

9. Kempe, D., Kleinberg, J., Tardos, E.: Maximizing the spread of influence through a social network. In: Proceedings of the 9th ACM SIGKDD Conference on Knowledge Discovery and Data Mining, pp. 137–146 (2003)

10. Kempe, D., Kleinberg, J., Tardos, É.: Influential nodes in a diffusion model for social networks. In: Caires, L., Italiano, G.F., Monteiro, L., Palamidessi, C., Yung, M. (eds.) ICALP 2005. LNCS, vol. 3580, pp. 1127–1138. Springer, Heidelberg (2005). https://doi.org/10.1007/11523468_91

11. Lorenz, M.O.: Methods of measuring the concentration of wealth. Publ. Am. Stat. Assoc. **9**(70), 209–219 (1905). https://doi.org/10.2307/2276207

12. Namtirtha, A., Dutta, A., Dutta, B., Sundararajan, A., Simmhan, Y.: Best influential spreaders identification using network global structural properties. Nat. - Sci. Rep. **11**(2254) (2021). https://doi.org/10.1038/s41598-021-81614-9

13. Page, L., Brin, S., Motwani, R., Winograd, T.: The PageRank citation ranking: bringing order to the web. In: The Web Conference. Stanford Digital Library (1999)

14. Riquelme, F., Gonzalez-Cantergiani, P., Molinero, X., Serna, M.: Centrality measures in social networks based on linear threshold model. Knowl.-Based Syst. **40**, 92–102 (2017). https://doi.org/10.1016/j.knosys.2017.10.029

15. Shakarian, P., Bhatnagar, A., Aleali, A., Shaabani, E., Guo, R.: The independent cascade and linear threshold models. In: Diffusion in Social Networks. SCS, pp. 35–48. Springer, Cham (2015). https://doi.org/10.1007/978-3-319-23105-1_4

16. Wan, Z., Mahajan, Y., Kang, B.W., Moore, T.J., Cho, J.H.: A survey on centrality metrics and their network resilience analysis. IEEE Access **9**, 104773–104819 (2021). https://doi.org/10.1109/ACCESS.2021.3094196

17. Zareie, A., Sakellariou, R.: Influence maximization in social networks: a survey of behaviour-aware methods. Soc. Netw. Anal. Min. **13**(78) (2023). https://doi.org/10.1007/s13278-023-01078-9

PARFAITE: PageRank-Matrix Factorization for Interpretable Graph Embeddings

Gabriel Damay$^{(\boxtimes)}$ and Mauro Sozio

Institut Polytechnique de Paris, Télécom Paris, Palaiseau, France
{gabriel.damay,sozio}@telecom-paris.fr

Abstract. There have been increasing efforts in recent years to develop so-called graph embeddings, which among other things allow to employ standard machine learning techniques to solve urgent real-world problems. However, developing *interpretable* graph embeddings has received much less attention. In our work, we develop PARFAITE, an algorithm for finding an interpretable and effective graph embedding, based on the factorization of the PageRank matrix of the input graph. We evaluate the interpretability of our method against popular graph embedding techniques, such as node2vec, showing that PARFAITE boasts significantly higher interpretability scores. Another contribution of our work is the release of a novel dataset constructed from all pages of the French version of Wikipedia, which we release for reproducibility and benchmarking.

Keywords: network embedding · interpretability · personalized PageRank · graph mining

1 Introduction

Graph embeddings consist of low-dimensional vector representations reflecting the relationships between vertices as well as some other properties of the underlying graph. They allow to employ standard machine learning techniques to perform vertex classification, link prediction and many other tasks. To compute such embeddings, one often aims at optimising some objective functions reflecting proximity relationships between the vertices or edges of the input graph.

There have been significant efforts in recent years to compute effective graph embeddings. They can be classified in three main categories [5]. The first category regroups the approaches based on matrix factorization, such as [1] and the more recent approach HOPE [14]. The second category regroups the approaches based on sampling random walks from the input graph, with the most prominent work being perhaps node2vec [6]. Finally, the last category regroups Graph Neural Network-based approaches, with GCN [10] being one of the best-known approaches.

This work was partially supported by the French National Agency (ANR) under project APY (ANR-20-CE38-0011).

L. M. Aiello et al. (Eds.): ASONAM 2024, LNCS 15211, pp. 223–238, 2025.
https://doi.org/10.1007/978-3-031-78541-2_14

In our work, we focus on matrix-factorization based approaches which typically consist of factorizing a matrix representing some proximity relationships between the vertices of the input graph. In particular, approaches based on the Personalized PageRank (PPR) matrix, which we focus on in our work, have been shown to provide very good results in a wide range of tasks, such as graph reconstruction and nodes classification [20,23], link prediction [20,21,23], as well as nodes recommendation [21]. There are other methods based on the factorization of the PPR matrix, such as APP [23], STRAP [21], and NRP [20]. HOPE [14] is considered to be one of the state-of-the-art approaches based on matrix factorization. One of its variants also focuses on the SVD of the PPR matrix. However, most of the analysis and the experiments in [14] focus on the Katz proximity matrix.

The interest for explainable algorithms is growing recently among the broad research community and in the general public alike. Many data-processing algorithms are fed with embeddings of complex data. If the embedding methods are not interpretable, there is little hope to provide satisfying explanations of the result of the algorithm. As a result, many recent papers focus on developing embedding methods that are interpretable, such as [13] and [18], for image and video processing, respectively. In the field of graph embeddings, most of the efforts have focused on defining measures to assess the interpretability of a graph embedding algorithm, with community-based metrics emerging as one of the most popular metrics. In particular, [4,9] develop three different community-based interpretability metrics and evaluate node2vec and HOPE in terms of those metrics. Those works suggest that a satisfactory solution for an interpretable graph embedding is still missing. Our work aims at filling this gap, focusing on developing an effective and interpretable graph embedding.

We develop PARFAITE, a novel approach based on the SVD of the PPR matrix. Our experimental evaluation against the state-of-the-art shows that our method provides higher interpretability scores, while boasting similar results in link prediction. To improve the interpretability of our method, we depart from the related work in a number of ways. In particular, in contrast with previous work, our method focuses on the *centered* PPR matrix. It seems natural to center the PPR matrix, in that, the uncentered matrix is positive and therefore its first singular vectors are also positive. Moreover, an embedding based on such a positive matrix would use only half of the available space. Observe that centering is also performed in principal component analysis (PCA). As we show in Sect. 3.2, centering helps us both reducing biases in the PPR matrix and to interpret the left part of the decomposition of the SVD. We also depart from the related work in the way our embedding is derived from the singular values of the SVD. In particular, virtually all methods based on the SVD of the PPR matrix, use the square root of the singular values to build the vertex embeddings $U\Sigma^{1/2}$ and $V\Sigma^{1/2}$. However, $U\Sigma$ and $V\Sigma$ contain some relevant information because they represent the lines and columns of the original matrix projected onto the eigenspaces. Such information is not leveraged in previous work, to the best of our knowledge. Another contribution of our work is a new metric measuring

Table 1. Notation

Symbol	Meaning	Definition
A	Adjacency matrix of the graph	
D	Diagonal matrix of out-degrees	
P	Stochastic matrix of random walk in the graph	$D^{-1}A$
U, V, Σ	Result matrices of the SVD, U and V are unit matrices, and Σ is diagonal	
Π	PPR matrix, each line is the PPR vector of a node	$\alpha \sum_{i=0}^{+\infty} (1-\alpha)^i P^i$
Π_l	PPR matrix, approximated using only $l+1$ iterations	$\alpha \sum_{i=0}^{l} (1-\alpha)^i P^i$
G	Number of known ground-truth communities	
K	Number of dimensions of an embedding	
D	Number of dimension of the SVD	

the interpretability of a graph embedding addressing some of the limitations of previous metrics. Finally, we release a novel dataset constructed from all pages of the French version of Wikipedia, which we release for reproducibility and benchmarking.

The rest of our work is organized as follows. In Sect. 2 we present the main preliminary concepts needed for our work. In Sect. 3, we present our approach, first through an overview and an explanation of its interpretation and then through the technical details of its use and implementation. In Sect. 5 we conduct an experimental evaluation showing that our method boasts higher interpretability scores than node2vec. Finally, Sect. 6 summarizes our work and discussed interesting directions for future work.

2 Preliminaries

2.1 Notations

Vectors and matrices are represented respectively by bold lower- and uppercase letters. Indices and exponents after a vector or matrix name are part of the name if they are bold, but are respectively entries in the object or power operator if they are not bold. Each time the cardinality of a set is noted with an uppercase letter (e.g. K is the number of dimensions of an embedding), we use the corresponding lowercase letter as a variable for the index of an element of the set (e.g. k is the index of a dimension of the embedding) (Table 1).

2.2 Personalized PageRank (PPR)

The Personalized PageRank (PPR) score is an extension of the PageRank score introduced in [15] in which the starting vertices are in a defined subset. We call *Personalized PageRank of a node u* the PPR score in which the starting subset contains only the node u, and we note $\boldsymbol{\pi_u}$ the vector containing the scores.

Let A be the adjacency matrix of a graph (directed or not) and D the diagonal matrix of its out-degrees. The matrix $P = D^{-1}A$ is the stochastic matrix of a random walk in the graph.

Let α be the probability at each step that our random walk terminates given that it didn't terminate before, and X_0 and X_f the vertex on which the walker is respectively before the first step and after the last step.

$$\boldsymbol{\pi}_{\boldsymbol{uv}} = \mathbb{P}\left(X_f = v | X_0 = u\right) \tag{1}$$

$$\boldsymbol{\pi_u}^\top = \alpha \sum_{i=0}^{\infty} (1-\alpha)^i \boldsymbol{e_u}^\top \boldsymbol{P}^i \tag{2}$$

where $\boldsymbol{e_u}$ is the uth vector from the canonical basis of $\mathbb{R}^n$, i.e. a vector of which u^{th} entry is 1 and all other entries are 0.

We call *PPR matrix of the graph*, and we note $\boldsymbol{\Pi}$ the matrix of which each line is the PPR vector of the related node. We can see from (2) that

$$\boldsymbol{\Pi} = \alpha \sum_{i=0}^{\infty} (1-\alpha)^i \boldsymbol{P}^i \tag{3}$$

We call *Reversed PPR vector of a node* the vector $\boldsymbol{\gamma_u}$ that contains, for each dimension v, the PPR score when the node v is the source node and the node u is the target node, i.e. $\boldsymbol{\gamma_{uv}} = \boldsymbol{\pi_{vu}}$.

2.3   SVD

The Singular Values Decomposition (SVD) is a well-known method of matrix factorization. Each matrix $\boldsymbol{M} \in \mathbb{R}^{m \times n}$ is factorized into two unit matrices $\boldsymbol{U} \in \mathbb{R}^{m \times m}$ and $\boldsymbol{V} \in \mathbb{R}^{n \times n}$, and a diagonal matrix $\boldsymbol{\Sigma} \in \mathbb{R}^{m \times n}$ so that $\boldsymbol{M} = \boldsymbol{U\Sigma V}^\top$. The *truncated SVD* of $\boldsymbol{M}$, approximates $\boldsymbol{M}$ with the matrices $\boldsymbol{\Sigma_d} \in \mathbb{R}^{d \times d}$, $\boldsymbol{U_d} \in \mathbb{R}^{m \times d}$ and $\boldsymbol{V_d} \in \mathbb{R}^{n \times d}$ that are the matrices made respectively of the d highest values of $\boldsymbol{\Sigma}$ and the related columns of $\boldsymbol{U}$ and $\boldsymbol{V}$. The acronym SVD is usually used as a metonymy for the Truncated SVD and we will do so in the rest of this paper.

One of the main advantages of the SVD, that make it especially attractive for embedding, is that it is known to filter out the noise in the data. More specifically, the SVD removes the small variations between data so that only the main patterns in the matrix are kept in the final result [12].

3   Our Approaches

3.1   Overview

We introduce the new PAgeRank FActorization-based InTerpretable Embedding (PARFAITE) method. It produces two embeddings, PARFAITE_L and PARFAITE_R .

Our method consists of three main parts. First, a truncated Singular Values Decomposition is performed. To overcome the issues of representing the PPR matrix exactly for very large graphs, we employ a function m that, given any

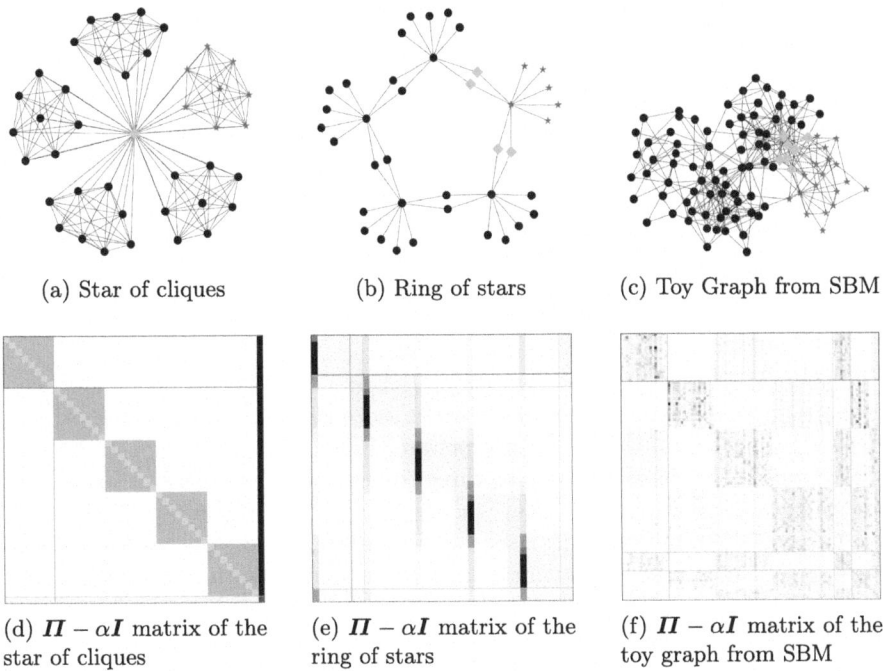

(a) Star of cliques   (b) Ring of stars   (c) Toy Graph from SBM

(d) $\Pi - \alpha I$ matrix of the   (e) $\Pi - \alpha I$ matrix of the   (f) $\Pi - \alpha I$ matrix of the
star of cliques   ring of stars   toy graph from SBM

Fig. 1. Toy graphs and their respective $\Pi - \alpha I$ matrices. On each graph, a community is highlighted, the red stars vertices belonging exclusively to the community, and the green diamonds ones belonging both to the highlighted community and to at least another one. On each matrix, the rows and columns relative to the red star vertices are highlighted through red and orange boxes, and the rows and columns relative to the green diamond vertices are highlighted through green and cyan boxes. We use $\Pi - \alpha I$ instead of Π for better readability. (Color figure online)

vector v, approximates the multiplication of the matrix with the vector. This results in two representations of the vertices as $U\Sigma$ and $V\Sigma$. Then, vertices are clustered, with each vertex being represented by a concatenation of its left and right representation. This provides the central points of the communities in the space of these representations, stored in the matrix C. Finally, our left and right embeddings, respectively PARFAITE_L and PARFAITE_R are computed by projecting back these central points onto the original spaces. A pseudocode of our algorithm is shown in Algorithm 1.

3.2   Interpretation of the Steps

The PARFAITE method relies on the well-established fact that the PPR vector of each vertex often contains large scores at dimensions corresponding to vertices it shares at least one community with, while it contains small scores at other dimensions [8,11].

Algorithm 1. PARFAITE

1: **procedure** PARFAITE(P: stochastic random walk matrix, m_P: function that approximate $\bar{\Pi}v$ for any v)
2:     $U, \Sigma, V \leftarrow$ SVD(m_P)
3:     clustering $\leftarrow$ kmeans(concat(normalize($U\Sigma$), normalize($V\Sigma$)))
4:     $C \leftarrow$ clustering.clusters_centers
5:     PARFAITE_L $\leftarrow UC^{\top}_{:,d:}$
6:     PARFAITE_R $\leftarrow VC^{\top}_{:,:d}$
7:     **return** PARFAITE_L , PARFAITE_R
8: **end procedure**

This property strengthens the one stated by [22] that two nodes are likely to share a community not only if the PPR score from one node to the other is high, but especially if their PPR vectors are similar.

We illustrate this property with three toy graphs represented in Fig. 1. The first two graphs provide ideal cases where we can find communities according to two natural community definitions: a) a clique and b) a star where all nodes but one are connected to a single node (e.g. they are connected to a same influencer in social media). In particular, Fig. 1a represents a star of cliques, while Fig. 1b represents a ring of stars. Observe that the communities in the graphs of Fig. 1 do *overlap*, with red star vertices belonging exclusively to one community, while green diamond vertices belonging to multiple communities. The graph in Fig. 1c is built using the Stochastic Block Model (SBM) with 100 vertices, 300 edges, 4 communities, with each pair of vertices of a same community being 100 times more likely to be connected than two vertices from different communities. As a result, in Fig. 1c, there are 19 vertices belonging to exactly two communities, 7 of which are depicted in green.

Figures 1d, 1e and 1f show the corresponding PPR matrices. Recall that each row corresponds to some vertex v and it represents the PPR vector of v, that is when v is the source vertex of the random walk. Indeed, we can see that PPR vectors contain larger scores to vertices of a same community as the source node, generating "square" patterns in the matrices.

Similarly, we observe that the reversed PPR vectors (columns in the matrix) exhibit the same pattern of large scores inside the community and low scores outside.

However, there might be relatively few vertices that have very large PPR scores even if they do not share communities with other nodes. This is apparent in Fig. 1e, where we observe that the central nodes in the neighboring cliques have higher scores in the PPR vector of the studied community than other nodes in the same community. The reversed PPR is not affected by that issue.

To avoid this bias, we define the $\bar{\Pi}$ matrix obtained by centering the columns of Π.

Property 1.

$$\bar{\varPi}_{i,j} = \frac{1}{n}\mathbb{P}\left(X_f = j\right)\left(\mathbb{P}\left(X_0 = i|X_f = j\right) - \mathbb{P}\left(X_0 = i\right)\right)$$

Proof.

$$\bar{\varPi}_{i,j} = \mathbb{P}\left(X_f = j|X_0 = i\right) - \mathbb{P}\left(X_f = j\right)$$
$$= \frac{\mathbb{P}\left(X_0 = i|X_f = j\right)\mathbb{P}\left(X_f = j\right)}{\mathbb{P}\left(X_0 = i\right)} - \mathbb{P}\left(X_f = j\right)$$

If we look at the rows of this new matrix, each entry is then the excess of probability to go to each vertex from the reference vertex, compared to the agnostic probability. If we look at the columns and because $\mathbb{P}\left(X_f = j\right)$ is a constant along a column, each entry is proportional to the excess of probability to come from each vertex given that the walk arrived at the reference vertex.

Because the SVD only keeps the main patterns of the matrix it decomposes, we expect the result of the SVD to exhibit the typical rows and columns for each community. We could then interpret these vectors as respectively the belonging of each node to the community and its relevance wrt. the community.

Our last problem is that, although the truncated SVD should make the community patterns in the matrix apparent, each dimension of the decomposition does usually not match a community, hindering the interpretation. To tackle this issue we perform a clustering on the vertices represented by the SVD, and we use the central points to obtain the desired representative vectors for each community.

3.3   Decomposition of the PPR Matrix

The PPR matrix is a dense matrix belonging to $\mathbb{R}^{n\times n}$. In most cases of big graphs, this matrix is too big to be explicitly represented. As we saw in Sect. 3.1, most modern SVD algorithms don't require an explicit representation of the matrix M but only a function $m(v) = Mv$. We use a reduced form of (3) to approximate the PPR matrix.

$$m(v) = \bar{\varPi}_l v = \left(\alpha \sum_{i=0}^{l}(1 - \alpha)^i P^i v\right) - \left(\pi \cdot v\right)\mathbf{1} \tag{4}$$

where $\mathbf{1}$ is the vector of which all entries equal 1 and π is the (not-personalized) PageRank vector of the matrix.

We know from [11] that a few steps are usually enough to compute an approximation of the PageRank vector that outlines the community of the node. We fix $l = 10$ for the rest of this paper.

3.4   Finding the Communities

SVD provides representations that should, if used to reconstruct the matrix, contain only the main patterns of the matrix which are the communities. There is no reason however to think that the dimensions of the representations correspond themselves to communities. That is why we perform a clustering of the vertices using their SVD representations to find the communities. Since both matrices of the embedding provide relevant and distinct information, there is no reason to exclude one and therefore we use a concatenation of both sides as the vectors for the clustering. Choosing the best clustering algorithm for this step is out of the scope of this paper, we simply use the well-known k-means++ algorithm.

As we saw before, the PPR and reversed PPR vectors for vertices of a same community are expected to have similar patterns of high and low entries, which correspond to similar direction of the vectors. They are however not supposed to have similar norms, especially for the reversed PPR for which the norm typically follows the importance of the vertex. To make the clustering algorithm work on directions and not on euclidean distance both embeddings are normalized before concatenation and then the cosine similarity is used.

3.5   Reconstructing the Communities Rows and Columns

The $U\Sigma$ and $V\Sigma$ embeddings are the projection of the columns and rows of the centered PPR matrix onto the singular spaces, e.g. for a vertex of the graph w, we have $(U\Sigma)_{w,.} = \bar{\pi}_w^\top V$. Therefore the clusters centers are the central columns and rows of each community, projected on the singular spaces. To reconstruct the true central columns and rows of the communities, which are the typical centered reversed PPR and centered PPR vectors for the community, we multiply by the transposed of the projection matrices, which are U and V. Note that this reconstruction is not perfect because the dimensions of the singular spaces we use are smaller than the dimension of the original space.

4   Metrics

Several metrics have been proposed to evaluate the interpretability of a graph embedding, such as the Interpretability Score (IS) [4], the Betweenness Centrality Importance (BCI) and Closeness Centrality Importance (CCI) [9]. Those metrics all consider the vector representation of a given node interpretable if it encodes somehow whether that node belongs to some real-world communities.

We evaluate the results of our algorithms in terms of IS. We also study the BCI and CCI, however, due to the strong limitations of BCI and CCI, discussed in Sect. 4.2 and because of space limitations, we do not include them in our experimental evaluation. Finally, we propose the new Complete Interpretability Score Integrating Priority (CISIP) metric to tackle some weaknesses of the previous metrics.

4.1   Interpretability Score (IS)

Let k be the index of one of the embedding dimensions, and C_g one of the ground-truth communities. The IS for a (k, g) pair is decomposed into a top part

$IS_{top(k,g)}$ that equals the recall@$|C_g|$ of the top-scores vertices of the embedding and a bottom $IS_{bottom(k,g)}$ part defined similarly on the lowest scores. The article then proposes to aggregate the scores by dimension or ground-truth group using either average or maximum function. In order for us to obtain a single result score for the entire embedding, we will aggregate first by taking the maximum IS per embedding dimension over the ground-truth groups and then the mean over all embedding dimensions. The use of the mean allows us to obtain scores between 0 and 1. In contrast with [4], we do not multiply the result by 100, which only changes the results by this factor without any other impact.

$$IS = \frac{1}{K} \sum_{k=1}^{K} \max_{g \in [0,G]} \left(\max(IS_{top(k,g)}, IS_{bottom(k,g)}) \right) \qquad (5)$$

4.2   Betweenness Centrality Importance (BCI) and Closeness Centrality Importance (CCI)

These scores are defined based on the well-known Betweenness Centrality and Closeness Centrality scores, introduced in [2]. For a (k,g) pair, the top BCI score is defined as the normalized Betweenness Centrality in relation with the community, of the top nodes of the dimension that belong to the community. The CCI top score is defined similarly but using the Closeness Centrality instead of the Betweenness Centrality. However the normalization of these scores make them harder to use and interprete. For a (k,g) pair, the scores are normalized by the number of nodes with non-null contribution. Let's consider a community C_g and two embedding dimensions k_1 and k_2 so that the first one contains only one very central vertex among its top vertices, and the second one contains this same vertex plus another slightly less central vertex. Then, although arguably much more interpretable w.r.t the gth ground-truth community, the k_2 embedding dimension will have a lower score than the k_1 one.

4.3   CISIP Metric

A weak point of the metrics already proposed in the literature is that, although they evaluate the fitness of an embedding dimension to a ground-truth group, they don't evaluate how well the embedding separates the data and the redundancy between the dimensions. Let's take the IS as an example: if 50% of the $|C_1|$ highest scores for the first dimension belong to C_1, then $IS_{top(1,1)} = 0.5$. But then it is possible that 50% of the highest scores for the second dimension belong either to the exact same part of C_1, denoting strong redundancy for the interpretations of these dimensions, or the other points of C_1, denoting that the first group is halved between these dimensions hence reducing the interpretability of both dimensions. In both cases, $IS_{top(2,1)} = 0.5$.

On the top of that all these metrics only consider the nodes that receive top- or bottom-$|C_g|$ scores from the embedding, and all these vertices are consider with equal weight. It seems however natural to consider that the importance should be decreasing before reaching the $|C_g|$ threshold, e.g. the vertex with the highest score should have higher importance than the vertex with the second highest score. Similarly, it seems that the importance should not drop to 0 after the $|C_g|^{\text{th}}$ score: if two embedding dimensions have the exact same top-$|C_g|$ vertices, but one has its $(|C_g| + 1)^{\text{th}}$ vertex belonging to C_g while the other has not, it seems natural to consider that the first one is more interpretable wrt. C_g.

To tackle these weaknesses, we propose the new Complete Interpretability Score Integrating Priority (CISIP). First of all, we use a function f to smooth the binary belonging feature, providing a score of belonging or importance of each node in each community. The simplest of these functions is simply to use the belonging feature (identity function), but we can also use the Mean Belonging of the Neighbors, the PPR of the community or other functions.

A score is then attributed to each (k, g) pair. Using the Hungarian algorithm for optimal assignment, an exclusive mapping $S = \{(k_i, g_i), i \in [1, \min(K, G)]\}$ of the embedding dimensions to the ground-truth groups is performed. To achieve good performances, the scoring at this step for a couple (k, g) is computed by summing the embedding scores for the dimension k of the vertices that belong to C_g. To account for the possibility that the bottom part of the embedding is the part that matches the community, we score in the same way the opposite of the dimension and we take the max score of the two. If the score with the opposite is the max, we will then use the opposite of the dimension for the Weighted Kendall Tau scoring.

Then, for each pair $(k_i, g_i) \in S$, a score is computed using Vigna's weighted Kendall Tau score WKT [16]. This method provides a score between 0 and 1 of how much the two vectors (embedding score and smoothed ground-truth belonging) are ranked in the same way, with more importance given to the top-scores of each vector. Finally, our metric $CISIP$ is computed as the mean score over the dimensions and groups

$$CISIP = \frac{1}{\min(K, G)} \sum_{i=1}^{\min(K,G)} WKT(v_d, f(w_g)) \qquad (6)$$

5   Experiments

Our main goal is to show that our method provides better interpretability scores than state-of-the-art approaches, while boasting similar results for a popular machine learning task in graph analysis, such as link prediction.

Datasets. We use two datasets that are available on the SNAP website [19]. The first one named *Wikispeedia* contains 4592 vertices, and 120 000 edges. 105 overlapping ground-truth community are given. The second one named *Facebook* contains 10 graphs. We exclude 2 of them, numbered 698 and 3980, because they contain less than 128 nodes and we could therefore not compute the SVD for them using the same parameters used for the others. The remaining 8 graphs contain between 155 and 1035 vertices and between 3312 and 60050 edges. Between 7 and 46 overlapping ground-truth communities are given for each graph.

We also release a new dataset with ground-truth communities[1] (Wikipedia fr), constructed from all the pages of the French version of Wikipedia.[2] In such a graph, nodes represent Wikipedia pages while directed edges represent links between the corresponding Wikipedia pages. In the French version of Wikipedia, it is common to add links to so called "portals" at the end of the page which serve as reference pages for given topics and can be seen as ground-truth communities. Such novel graph contains 2.52 millions vertices, 102 million edges and 2 700 ground-truth communities. To keep the dimension of the embeddings manageable however we only study the 117 communities that contain more than 10 000 vertices.

Methods. We evaluate our method against two widely used algorithms for graph embedding:

- HOPE, a state-of-the-art approach based on SVD;
- node2vec, which is a state-of-the-art approach based on random walks.

We evaluate all methods in terms of the IS and CISIP metrics, discussed in the previous section. We include in our experiments embeddings produced by the state-of-the-art node2vec and HOPE algorithms and our 2 new embeddings PARFAITE_L and PARFAITE_R .

For node2vec, we use the most-used python3 implementation [3]. We keep the default parameters, i.e. the number of walks is 200 per vertex, the length of the walks is 30 and the window size is 10.

For HOPE, we use the official implementation in Matlab[3] provided by the authors of [14] that we reimplement in python3 (while using numpy and scipy), so as to provide a fair comparison with the other approaches.

Our algorithm is implemented in python3 using mainly the numpy [7] and scipy [17] packages. The jump parameter α for the PPR algorithm is set to $\alpha = 0.1$. The number l of iterations to approximate the PPR matrix is set to $l = 10$ and the dimension D of the intermediate SVD is set to $D = 128$.

For all embeddings the dimension K of the embedding is set to be the number G of known ground-truth communities.

Metrics. We measure the interpretability of the methods using each of the metrics mentioned in Sect. 4. The Interpretability Scores (IS) are aggregated along the ground-truth groups using the max function, and then along the embedding dimensions using the sum function. For CISIP, three smoothing functions f are considered. The first one, that we call "identity" is the direct use of the binary belonging feature, the second one is the Mean Neighbors Belonging (MNB), that is the mean of the belonging features of the neighbors, and the last one is the PPR of the community.

---

[1] https://gitlab.telecom-paris.fr/gabriel.damay/WikipediaFRNetwork.
[2] All the pages from the main space of Wikipedia, that is all the pages usually accessed by public, excluding discussions, user pages etc.
[3] https://github.com/ZW-ZHANG/HOPE.

5.1   Results

The results are given in Tables 2 and 3. Because of space limitation, only Facebook 107 and 1912, as parts with highest degree and order of Facebook, are presented, as well as Facebook 3437 because of the good performances of HOPE on this dataset with the CISIP-identity metric. The results for the other versions of Facebook are similar to what is presented.

As we see in these results, our algorithm's interpretability is much higher than node2vec's when evaluated using the Interpretability Score or any of the variants of CISIP. HOPE's interpretability is closer but still generally below PARFAITE's.

Our left embedding greatly outperforms the right one when using CISIP with the PPR smoothing, and this result was expected, as the left embedding is an approximation of the typical PPR vector of each clusters detected. The results are however equivalent between our two embeddings when compared using either the IS or CISIP with the identity smoothing, which both evaluate directly the matching between the embedding and the belonging features, or with the MNB smoothing. This is consistent with our interpretation that the left embedding represents which communities a vertex belongs to, while the right embedding measures somehow the "importance" of a vertex in a community.

5.2   Is the Clustering Needed?

To check if the last step of our algorithm of clustering the data and computing the final embeddings is really needed, we compare our results to what we would have without the clustering step. To achieve this, we take the $U\Sigma$ and $V\Sigma$ results from the SVD, and we keep only the G first dimensions to match the dimension of the PARFAITE embeddings.

The results for IS and for CISIP with the three smoothing functions already used are presented in Tables 4 and 5. We see that in most cases the results of our PARFAITE_L and PARFAITE_R outperform those before clustering, sometimes significantly (e.g. more than 0.1 points of difference for the IS). This is especially true for the IS and CISIP with PPR smoothing. We note however that the SVD provides better results in several occurrences when compared to our embeddings using CISIP with the identity or the MNB smoothing.

Overall we conclude that PARFAITE_L and PARFAITE_R do generally perform better than single SVD, i.e. without the clustering and reconstruction steps.

5.3   Is Our Embedding Efficient?

We check the efficiency of PARFAITE against HOPE and node2vec at the task of Link Prediction. We build test graphs by removing 0.1% of the edges of a real-world graph. We store the pairs of vertices of these edges as "positive" pairs, and build a set of "negative" pairs by drawing the same number of pairs of vertices that are not connected by an edge.

The embedding of the test graph is computed and a score is attributed to each positive or negative pair of vertices using this embedding.

For HOPE, the score is the dot product of the left embedding of the first vertex in the pair and the right embedding of the second vertex. The score for PARFAITE is similar but the left embedding of the first vertex is normalized to account for the entire use of the singular values on each side of the embedding. For node2vec, following [6], a Logistic Regression is trained on the test graph by representing the pairs with a concatenation of their vertice's embeddings.

We run this experiment on 10 test graph for both the Wikispeedia dataset and on the 1912 part of the Facebook dataset, as the part with the highest order. The mean results are given in Fig. 2. As we can see, node2vec is outperformed by both HOPE and PARFAITE. HOPE achieves similar results as PARFAITE on Facebook 1912 and slightly better on Wikispeedia but overall we can say that the greater interpretability of PARFAITE does not come at the cost of lower efficiency on the task of link prediction.

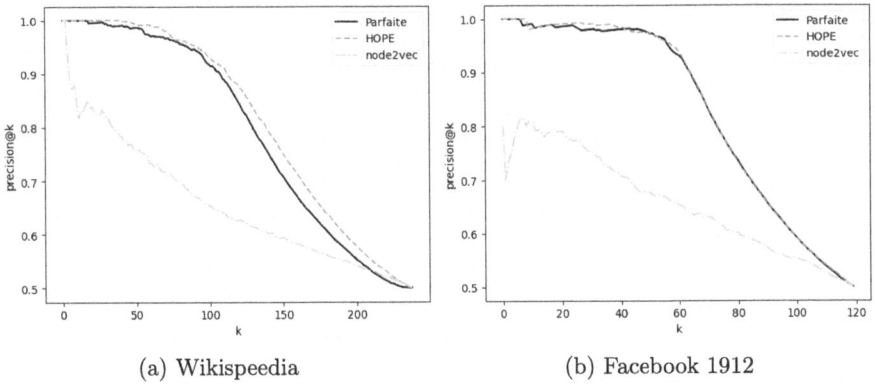

(a) Wikispeedia   (b) Facebook 1912

Fig. 2. Mean Precision@k of the Link Prediction on 20 test graphs built from the Wikispeedia and Facebook 1912 datasets

Table 2. Results for the Interpretability Score

Dataset	PARFAITE		HOPE		node2vec
	left	right	left	right	
Wikispeedia	**0.389**	0.375	0.200	0.266	0.139
Facebook 107	**0.626**	0.601	0.550	0.550	0.323
Facebook 1912	**0.766**	0.764	0.603	0.603	0.348
Facebook 3437	0.430	**0.759**	0.429	0.429	0.188
WikipediaFr	0.040	**0.083**	0.041	**0.083**	*

* The embedding for this graph with node2vec as been stopped after 36h of computation.

Table 3. Results for CISIP

Dataset	PARFAITE		HOPE		node2vec
	left	right	left	right	
No smoothing (identity function)					
Wikispeedia	**0.429**	0.414	0.339	0.318	0.255
Facebook 107	0.403	**0.408**	0.316	0.316	0.234
Facebook 1912	0.401	**0.425**	0.352	0.353	0.266
Facebook 3437	0.300	0.310	**0.331**	**0.331**	0.256
WikipediaFr	0.350	**0.376**	0.339	0.353	*
Smoothing with MNB					
Wikispeedia	0.431	**0.508**	0.282	0.369	0.105
Facebook 107	**0.545**	0.515	0.384	0.384	0.161
Facebook 1912	0.504	**0.511**	0.269	0.267	0.029
Facebook 3437	0.513	**0.524**	0.373	0.373	0.067
WikipediaFr	0.272	**0.499**	0.400	0.493	*
Smoothing with PPR					
Wikispeedia	0.449	**0.516**	0.189	0.174	0.012
Facebook 107	**0.585**	0.576	0.304	0.304	0.093
Facebook 1912	0.557	**0.633**	0.278	0.279	0.101
Facebook 3437	0.568	**0.688**	0.402	0.402	0.073
WikipediaFr	**0.123**	−0.067	0.049	−0.015	*

Table 4. SVD Results for the Interpretability Score

Dataset	SVD (left)	SVD (right)
Wikispeedia	0.234	0.204
Facebook 107	0.538	0.516
Facebook 1912	0.589	0.604
Facebook 3437	0.270	0.463
WikipediaFr	**0.043**	0.061

Table 5. SVD Results for CISIP

Dataset	No smoothing		MNB smoothing		PPR smoothing	
	SVD_L	SVD_R	SVD_L	SVD_R	SVD_L	SVD_R
Wikispeedia	0.363	0.362	0.280	0.369	0.244	0.202
Facebook 107	**0.493**	**0.464**	**0.596**	**0.571**	**0.609**	**0.604**
Facebook 1912	0.377	0.392	0.299	0.320	0.319	0.320
Facebook 3437	**0.304**	**0.337**	0.429	0.443	0.419	0.421
WikipediaFr	**0.350**	**0.381**	0.259	**0.502**	**0.150**	**−0.011**

6   Conclusion and Future Work

We have presented PARFAITE, a novel graph embedding algorithm and we have evaluated its interpretability against the state-of-the-art node2vec and HOPE algorithms. We also presented the new interpretability score CISIP to assure that interpretations are well-separated and non-redundant. Our approach relies on the clustering of the vertex embeddings for which we use the k-means algorithm. An interesting direction for future work would be to study whether other clustering methods provide better results. We have also seen that, although the clustering phase increases the interpretability of our embedding, in some rare cases clustering might have a negative impact on the interpretability. It would be interesting to have a better understanding of this phenomenon which could pave the way for a more effective approach. Finally, the CISIP metric contains a smoothing function as a parameter and it would be interesting to study the properties of various functions for this task and what they imply in terms of interpretability of the embedding at hand.

References

1. Belkin, M., Niyogi, P.: Laplacian eigenmaps for dimensionality reduction and data representation. Neural Comput. **15**(6), 1373–1396 (2003). https://doi.org/10.1162/089976603321780317
2. Bloch, F., Jackson, M.O., Tebaldi, P.: Centrality measures in networks. Soc. Choice Welfare **61**(2), 413–453 (2023)
3. Cohen, E.: node2vec 0.4.6. https://pypi.org/project/node2vec/
4. Gogoglou, A., Bruss, C.B., Hines, K.E.: On the interpretability and evaluation of graph representation learning. In: NeurIPS Workshop on Graph Representation Learning (2019)
5. Goyal, P., Ferrara, E.: Graph embedding techniques, applications, and performance: a survey. Knowl.-Based Syst. **151**, 78–94 (2018). https://doi.org/10.1016/j.knosys.2018.03.022
6. Grover, A., Leskovec, J.: node2vec: Scalable feature learning for networks. In: Proceedings of the 22nd ACM SIGKDD International Conference on Knowledge Discovery and Data Mining, pp. 855–864 (2016). https://doi.org/10.1145/2939672.2939754
7. Harris, C.R., et al.: Array programming with NumPy. Nature **585**(7825), 357–362 (2020). https://doi.org/10.1038/s41586-020-2649-2
8. Hollocou, A., Bonald, T., Lelarge, M.: Multiple local community detection. SIGMETRICS Perform. Eval. Rev. **45**(3), 76–83 (2018). https://doi.org/10.1145/3199524.3199537
9. Khoshraftar, S., Mahdavi, S., An, A.: Centrality-based interpretability measures for graph embeddings. In: 2021 IEEE 8th International Conference on Data Science and Advanced Analytics (DSAA), pp. 1–10 (2021). https://doi.org/10.1109/DSAA53316.2021.9564221
10. Kipf, T.N., Welling, M.: Semi-supervised classification with graph convolutional networks. In: International Conference on Learning Representations (2022)

11. Kloumann, I.M., Kleinberg, J.M.: Community membership identification from small seed sets. In: Proceedings of the 20th ACM SIGKDD International Conference on Knowledge Discovery and Data Mining, pp. 1366–1375 (2014). https://doi.org/10.1145/2623330.2623621

12. Konstantinides, K., Natarajan, B., Yovanof, G.: Noise estimation and filtering using block-based singular value decomposition. IEEE Trans. Image Process. **6**(3), 479–483 (1997). https://doi.org/10.1109/83.557359

13. Lee, S., Song, B.C.: Interpretable embedding procedure knowledge transfer via stacked principal component analysis and graph neural network. In: Proceedings of the AAAI Conference on Artificial Intelligence, vol. 35, no. 9, pp. 8297–8305 (2021). https://doi.org/10.1609/aaai.v35i9.17009

14. Ou, M., Cui, P., Pei, J., Zhang, Z., Zhu, W.: Asymmetric transitivity preserving graph embedding. In: Proceedings of the 22nd ACM SIGKDD International Conference on Knowledge Discovery and Data Mining, pp. 1105–1114 (2016). https://doi.org/10.1145/2939672.2939751

15. Page, L., Brin, S., Motwani, R., Winograd, T.: The pagerank citation ranking: bringing order to the web. Technical report, Stanford InfoLab (1999)

16. Vigna, S.: A weighted correlation index for rankings with ties. In: Proceedings of the 24th International Conference on World Wide Web, pp. 1166–1176 (2015). https://doi.org/10.1145/2736277.2741088

17. Virtanen, P., et al.: SciPy 1.0: fundamental algorithms for scientific computing in Python. Nat. Methods **17**, 261–272 (2020). https://doi.org/10.1038/s41592-019-0686-2

18. Wu, J., Ngo, C.W.: Interpretable embedding for ad-hoc video search. In: Proceedings of the 28th ACM International Conference on Multimedia, pp. 3357–3366 (2020). https://doi.org/10.1145/3394171.3413916

19. Yang, J., Leskovec, J.: Defining and evaluating network communities based on ground-truth. In: Proceedings of the ACM SIGKDD Workshop on Mining Data Semantics (2012). https://doi.org/10.1145/2350190.2350193

20. Yang, R., Shi, J., Xiao, X., Yang, Y., Bhowmick, S.S.: Homogeneous network embedding for massive graphs via reweighted personalized PageRank. Proc. VLDB Endow. **13**(5), 670–683 (2020). https://doi.org/10.14778/3377369.3377376

21. Yin, Y., Wei, Z.: Scalable graph embeddings via sparse transpose proximities. In: Proceedings of the 25th ACM SIGKDD International Conference on Knowledge Discovery and Data Mining, p. 1429–1437 (2019). https://doi.org/10.1145/3292500.3330860

22. Zhang, Y., et al.: Robust hierarchical overlapping community detection with personalized PageRank. IEEE Access **8**, 102867–102882 (2020). https://doi.org/10.1109/ACCESS.2020.2998860

23. Zhou, C., Liu, Y., Liu, X., Liu, Z., Gao, J.: Scalable graph embedding for asymmetric proximity. In: Proceedings of the AAAI Conference on Artificial Intelligence, vol. 31, no. 1 (2017). https://doi.org/10.1609/aaai.v31i1.10878

Categorising Corruption in the Vaccine Discourse: A General Taxonomy, Data Set, and Evaluation of LLMs for Classifying Corruption Dialogue in Social Media

Vitor Gaboardi dos Santos[1]([⊠]) [ID], Guto Leoni Santos[1] [ID], Antonia Egli[1],
Estatira Kahvazadeh[2], Bill Doolin[3], Patricia Takako Endo[4] [ID],
and Theo Lynn[1] [ID]

[1] Dublin City University, Dublin, Ireland
{vitorgaboardidos.santos,guto.santos,antonia.egli,theo.lynn}@dcu.ie
[2] Georgia Institute of Technology, Atlanta, USA
ekahvaz@gmail.com
[3] Trinity College Dublin, Dublin, Ireland
doolinbill@gmail.com
[4] Universidade de Pernambuco, Caruaru, Brazil
patricia.endo@upe.br

Abstract. Real or perceived corruption can have a damaging effect on health care services and outcomes. In particular, research suggests perceived corruption had a significant impact on COVID-19 vaccination. Given the role of social media in health communications, identifying and understanding perceived corruption related to vaccines and vaccination is critical to build societal cohesion and public trust in health institutions and strategies, manage and combat misinformation and disinformation, and design more effective policies, interventions, and communications strategies. There is a dearth of research on binary and multi-class classification of corruption dialogues in health or otherwise. We address this gap by introducing a general hierarchical corruption dialogue taxonomy (HCDT) and formulating binary and multi-class classification tasks based on the HCDT. We also create a vaccine-specific labelled dataset for each task, and fine-tune three large language models (BERT, RoBERTa, and BERTweet) based on these datasets. We evaluate the performance of these models in the binary and multi-class classification tasks. While all models performed similarly for the binary task, RoBERTa performed best for multi-class classification of corruption dialogue.

Keywords: Corruption · Large Language Models · BERT · Twitter · Multi-class Classification · Vaccine · COVID-19

1 Introduction

As of April 2024, over 775 million cases and 7 million deaths resulting from COVID-19 were reported to the World Health Organisation [1]. While a wide

© The Author(s), under exclusive license to Springer Nature Switzerland AG 2025
L. M. Aiello et al. (Eds.): ASONAM 2024, LNCS 15211, pp. 239–254, 2025.
https://doi.org/10.1007/978-3-031-78541-2_15

range of countermeasures were implemented to mitigate the spread of the disease, immunisation is the primary response against severe acute respiratory syndrome coronavirus type 2 (SARS-CoV-2), particularly as it continues to evolve [2]. Unfortunately, there still remains a significant population who are either vaccine hesitant or opposed to vaccination for COVID-19 or in general. A variety of factors contribute to such beliefs and behaviours including perceptions of trust in the vaccine approval process and vaccine effectiveness in protecting individuals, vaccine conspiracy beliefs, perceived side effects, perceived availability, and free-riding based on herd immunity [3]. The consequences are significant. Research suggests that between 2021 and 2022, over 232,000 deaths could have been prevented among unvaccinated adults in the United States alone [4,5].

Corruption is commonly defined as "the abuse of entrusted power for private gain" [6,7]. Corruption, real or perceived, can have a damaging impact on health outcomes and the quality of health care services and result in higher healthcare costs, erosion of trust in the health system, and reduced service utilisation [8]. In the context of COVID-19, a study of 90 countries worldwide found that public corruption was one significant causes of cross-country variation in immunisation progress [9]. The speed, scale, and complexity of approving, allocating, distributing, and rolling out COVID-19 vaccines worldwide was unprecedented and as a result introduced new corruption risks [10–12]. These risks include corruption in the vaccine development and approval process, vaccine deployment and distribution systems, vaccine procurement, emergency funding for vaccines, preferential access to vaccines, and vaccine policy decisions [10]. While social media has many benefits, it can interfere with public health communication by spreading health misinformation and disinformation, creating a false sense of uniformity and validity, and legitimising questionable or false information [13,14]. This is particularly true in the context of the COVID-19 pandemic, where numerous studies highlight the adverse impact of social media on vaccine uptake [15–17]. Consequently, identifying and understanding perceived corruption related to vaccines and vaccination is critical to build social cohesion and public trust in health institutions and strategies, manage and combat misinformation and disinformation, and design more effective policies, interventions, and communications strategies.

In this paper, we propose a pipeline for automatically detecting and classifying perceived corruption in the vaccine discourse on social media platforms using Large Language Models (LLMs). While there is an extensive literature on corruption in health contexts, and specifically COVID-19, there are few studies on the use of Machine Learning (ML) to classify content for corruption-related dialogue on social media and vaccination. This can be explained by a number of significant challenges, not least a lack of corruption dictionaries and taxonomies, access to data both for testing and training models and empirical analysis, and availability of classification tools to identify posts relating to corruption.

In summary, we make four contributions in this paper to advance research on corruption dialogue detection and classification. First, we create a hierarchical corruption dialogue taxonomy (HCDT) that provides a structured framework for categorizing various forms of corrupt practices. Second, we create a labelled corruption dialogue dataset comprising tweets related to corruption within the COVID-19 vaccine discourse using a combined approach with humans and GPT [18]. Third, we fine-tune different LLMs using BERT-based architectures to initially detect corruption-related tweets and then categorize between 11 forms of corruption practices. Fourth, we evaluate the fine-tuned LLMs and report on their effectiveness in detecting and categorising binary and multi-class corruption-related tweets accurately. To the best of our knowledge, this is the first time an annotation scheme and computational model for identifying and classifying corruption-related discourse on social media has been designed and evaluated.

The remainder of this paper is organised as follows: Sect. 2 introduces related works on the detection of corruption using ML techniques. Section 3 describes our hierarchical taxonomy for corruption dialogue. Section 4 presents the data and methodology used to fine-tune and evaluate the LLM employed to detect and classify corruption dialogue. Section 5 discuss the results of our methodology. Section 6 describes limitations and avenues for future work. Finally, Sect. 7 presents the conclusion and final remarks of the paper.

2   Related Work

Several studies have been published on the use of ML techniques to detect corruption or related indicators in different domains. For example, Lima & Delen [19] conducted a study to predict levels of corruption perception indexes across 132 nations. They employed different ML models and used data collected from various website sources associated with indexes. The results revealed that the Random Forest (RF) model performed better, achieving an accuracy of 85.77%.

Rabuzin & Modrušan [20] compared different models to identify suspicious one bid tenders, which may raise suspicions of favouritism, collusion, or lack of transparency. They found that the Logistic Regression (LR) model produced the best accuracy overall, while Naïve Bayes (NB) exhibited the best performance in identifying potential signs of corruption in the public procurement process. While they found a lack of data within the field, they noted that the models show promise for public procurement corruption detection.

Denisova-Schmidt et al. [21] examined anti-corruption education effects on students' perceptions of academic integrity and corruption in Russia by employing a two-step ML regression process analysis using around 2,000 surveys. They found out that students who plagiarise frequently seem to have more negative opinions regarding corruption, suggesting that policymakers should consider unwanted diverse impacts across student groups before implementing anti-corruption education on a greater scale.

Ash et al. [22] used ML techniques to identify instances of corruption in local governments by analysing budgetary data from Brazilian municipalities. They employed annual budget data spanning from 2001 to 2012 to train models aimed at predicting the occurrence of corruption within these municipalities, achieving an accuracy rate of 76%. Their findings suggest that even in areas where corruption is not prevalent, there may be a tendency to manipulate records to mitigate potential indicators of corruption. Additionally, audits seem to play a significant role in disciplining and reducing corrupt behaviour.

We found one study that apply ML to classify social media posts for corruption. Li et al. [23] employed an Natural Language Processing (NLP) approach to gather and assess Twitter data to identify instances where users self-reported experiences of corruption, mainly within the healthcare sector. Following the analysis of tweets filtered using corruption-related keywords and using NLP techniques and manual annotation, they found 2,383 tweets. The authors clustered these tweets into actor-based topics, resulting in two main themes - police bribery and corruption in healthcare.

While existing works contribute to corruption detection, they primarily focus on identifying general corruption. In contrast, our study extends beyond this scope by considering a broader range of corruption topics, such as bribery, collusion, abuse of power, fraud, and obstruction of justice. Additionally, our pipeline incorporates fine-tuning state-of-the-art LLMs instead of relying on standard ML models. Furthermore, we explore the detection of corruption on Twitter and within the health context, offering a more comprehensive approach to addressing corruption detection.

3   Hierarchical Corruption Dialogue Taxonomy (HCDT)

As discussed in Sect. 2, there is a dearth of research on corruption dialogue classification using ML techniques; only one related work was identified, i.e., Li et al. [23]. Furthermore, establishing whether content pertains to corruption generally and/or a specific type of corrupt practice without a taxonomy is a significant challenge. There is little agreement on the definition of corruption let alone sub-categories of corruption [24,25]. While Li et al. [23] clustered tweets by actor-based themes, the clustering was not based on a comprehensive approach to corruption-related categories. Other categorisations proposed in the wider corruption literature are not sufficiently granular or comprehensive for use in classifying corruption dialogue. For example, Jancsics [26] categorises corruption into four types - market corruption, social bribes, corrupt organizations, and state capture. Similarly, Bussell [27] again categorises corruption into four types - legislative corruption, contracting, employment, and services. In both cases, these taxonomies are conceptual and were not applied empirically.

We organise the taxonomy on a two-level hierarchical structure. The first level is binary, i.e., *corruption* and *non-corruption*. As per Zhang et al. [28], we do not detail the non-corruption category as it is not our focus. The second level is based on sources from across all disciplines, including those from academia and practice. It contains ten specific categories and a miscellaneous or general corruption category. Table 1 presents the HCDT with definitions from Corruption Watch [29], UNODC [30] and LexisNexis [31]. In Sect. 4, we present terms, words, and word stems based on corrupt practices for each second level category. Again, these were sourced from commonly cited glossaries of corruption including those from Corruption Watch [29], UNODC [30], and Transparency International [32]. It should be noted that we excluded terms and practices that were not necessarily related to a corrupt practice *per se* e.g., whistleblowers, transparency, fiduciary risk, avoiding tax, etc.

Table 1. Hierarchical corruption categories with definitions

First Level	Second level	Definition
Corruption	Abuse of power/authority	"The use of one's position or authority to commit an unlawful act for the purpose of obtaining a personal advantage or an advantage for another person or entity, out of which one can derive personal gain. Abuse of power can also refer to the refusal to perform an act or function which forms part of prescribed duties." [29]
	Bribery	"The act of offering someone money, services or other valuables, in order to persuade him or her to do something in return." [29]
	Collusion/conspiracy	"Collusion is secret agreement between parties, in the public and/or private sector, to conspire to commit actions aimed to deceive or commit fraud with the objective of illicit financial gain. Conspiracy is an agreement between two or more persons to commit an offence, or which necessarily involves committing an offence." [29,31]
	Conflict of interest	"This arises when an individual with a formal responsibility to serve the public participates in an activity that jeopardises his or her professional judgment, objectivity and independence. Often this activity primarily serves personal interests and can potentially influence the objective exercise of the individual's official duties." [29]
	Embezzlement	"When a person holding office in an institution, organisation or company dishonestly and illegally appropriates, uses or traffics the funds and goods they have been entrusted with for personal enrichment or other activities." [29]
	Extortion	"The act of utilising, either directly or indirectly, one's access to a position of power or knowledge to demand unmerited cooperation or compensation as a result of coercive threats." [29]
	Fraud	"The unlawful and intentional making of a misrepresentation which causes actual prejudice or which is potentially prejudicial to another." [29]
	Money laundering	"Any act or attempted act to disguise the source of money or assets derived from criminal activity." [29]
	Nepotism/favouritism	"Favouritism refers to the normal human inclination to prefer acquaintances, friends and family over strangers. Nepotism is a form of favouritism based on acquaintances and familiar relationships whereby someone in an official position exploits his or her power and authority to provide a job or favour to a family member or friend, even though he or she may not be qualified or deserving." [29]
	Obstruction of justice	"The use of physical force, threats or intimidation, or the promise, offering of an undue advantage to induce false testimony or to interfere in the giving of testimony or the production of evidence in a proceeding in relation to the commission of offences established in accordance with the United Nations Convention against Corruption." [30]
	General corruption	"Other types of corruption not covered in the previous categories."

4   Data and Methods

Figure 1 illustrates our approach. Initially, we leverage the Twitter API to gather tweets about the COVID-19 vaccine discourse by employing specific keywords. Subsequently, we search for corruption- and non-corruption-related tweets and label them using both human and GPT-based coders. This annotation process leads to the creation of two distinct datasets: (1) a binary dataset with only corruption-related and non-corruption-related tweets (HCDT Level 1), and (2) a multi-class dataset consisting of 11 corruption-related topics (HCDT Level 2). Next, we fine-tune different BERT-based models using the annotated datasets, with the dual objective of first discerning whether tweets are related to corruption or not, and then identifying the specific type of corruption if present. Finally, we evaluate the performance of all models using a separate test dataset.

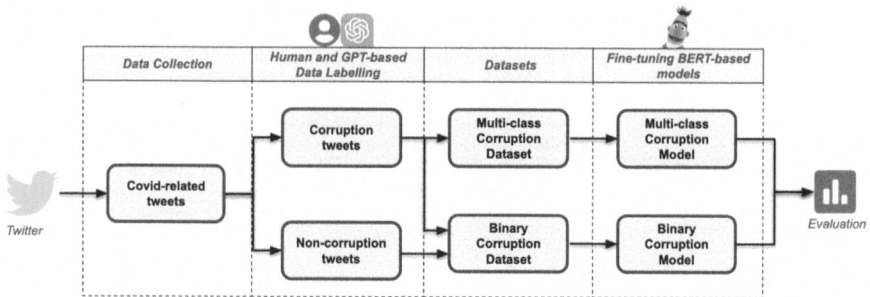

Fig. 1. Approach overview for automatically classifying perceived corruption in the vaccine discourse on Twitter using LLMs.

4.1 Data Collection

Data were collected on English language tweets associated with vaccines, vaccination and the COVID-19 pandemic posted on Twitter during a 12-month period between December 2020, when the COVID-19 vaccine was first released for public use in the United States, and November 2021. The following terms were used to generate the dataset: ("COVID-19" and "vaccine") or ("coronavirus" and "vaccine") or ("covid" and "vaccine"). Data was accessed through the Twitter enterprise API. This resulted in a dataset of 379,066,249 tweets.

Next, we generated a list of words, phrases and word stems related to corrupt practices associated with each second level category of the HCDT (see Table 2). This "bag of words" was used to filter potential tweets featuring corruption dialogue from the initial data set. Furthermore, only original tweets with at least 40 characters were considered; replies and retweets were excluded. We then pre-processed tweets by removing all URLs. After pre-processing, the dataset comprised 53,702,090 tweets.

4.2 Data Labelling

High-quality labelled data is essential for achieving good performance and generalisation when fine-tuning LLMs [33], especially in the case of text classification. However, manually labelling a large dataset can be costly, time-consuming, and prone to errors [34]. False positives are particularly prevalent when using a "bag of words" approach for filtering. To address these challenges, we employed a three-stage labelling process to identify and label tweets against both levels of HCDT.

In the first stage, two independent human coders coded tweets against HCDT Level 1 (i.e., corruption or not). If the tweets are related to corruption, the two independent coders then classified each tweet against one HCDT Level 2 category. This resulted in a dataset of 20,804 tweets.

GPT [18] has been effectively used as an annotator in natural language tasks. For example, studies suggest that the model performance trained on the GPT-3 annotated data is often comparable to or even better than trained on human-annotated data [34,35]. Therefore, in the second stage, we used GPT as an additional annotator to conduct the same exercise performed in the first stage. Specifically, we employed the following prompt using GPT-3.5-turbo: *Check whether the following tweet delimited by triple backticks is corruption-related (including topics such as: abuse of power; bribery, conspiracy, conflict of interest, embezzlement, extortion, fraud, money laundering, nepotism, and obstruction of justice) and in the coronavirus context (discussing COVID-19, vaccine, pandemic). Answer in a JSON file format according to the examples shown.*

This instruction was followed by two examples to improve the model's contextual understanding of the task. Once the GPT completed annotating the dataset, we compared the human and GPT classifications and only retained those tweets where both classifications agreed. This resulted in a dataset of 11,945 tweets.

Fine-tuning language models with balanced datasets enhances performance in downstream tasks [36,37]. As such, in the third stage, we randomly select tweets from the entire dataset and used GPT once again to annotate a dataset of non-corruption tweets until we get the same number of tweets as the corruption dataset generated in the second stage, i.e., 11,945 tweets.

4.3 Labelled Datasets

For training and testing the proposed models, two balanced labelled datasets are required - (1) a binary corruption dataset (HCDT Level 1) and (2) a multi-class corruption dataset (HCDT Level 2). In effect, the former is a subset of the latter, and thus we discuss in this order.

Multi-Class Corruption Dataset (HCDT Level 2). HCDT Level 2 comprises 11 classes - abuse of power/authority; bribery; collusion and conspiracy; conflict of interest; embezzlement; extortion; fraud; money laundering; nepotism and favouritism; obstruction of justice; and general corruption. Figure 2 shows the amount of tweets per class in the annotated dataset using the labelling strategy detailed in Sect. 4.2.

The *General Corruption* class is the most prevalent one, comprising 1,506 samples, while the *Obstruction of Justice* class has the least representation, with only 632 samples. The remaining classes demonstrate a more balanced distribution of samples. To ensure a fair evaluation of the model's performance across

Table 2. Corruption terms and word stems used to filter the corruption-related tweets.

Class	Word, phrase, word stem
Abuse of Power/Authority	abuse*, misuse, exploit*, manipulate, oppress*, control, dominat*, tyranny, subjugate, usurp, maltreat*, rent seeking, rent-seeking, trade influence, trading influence, elite capture, undue advantage
Bribery	bribe*, kickback, gratuity, payola, hush money, grease, palm-greasing, backhander, inducement, incentive, graft, solicit*, scoral*, fakelaki, facilitation payment, baksheesh, gift giving, interest peddling, fcpa
Collusion/Conspiracy	collusion, conspiracy, plot*, scheme, cabal, confederacy, intrigue, secret, connivance, complicity, collud*, deep state, state capture, elite capture, bid rig*, bid rotat*
Conflict of Interest	conflict, interest, unethical, bias, influence, personal gain, self-serving, double-dealing, impropriety, trading influence, trade influence, revolving door, conflict of interest
Embezzlement	embezzl*, misappropriate, steal, pilfer, purloin, peculate, defalcate, swindle, loot, skim, siphon, divert, misappropriat*, enrich*
Extortion	extort*, blackmail, shakedown, ransom, threat, pressure, coercion, squeeze, intimidation, strong-arm, coerci*
Fraud	fraud, scam, con, swindle, deceive, hoodwink, dupe, trick, cheat, sham, counterfeit, impostor, racket, spoof, crim*, concealment, gouging, tax evasion, evade tax, evading tax, carbon cowboys, mispric*, transfer mispric*
Money Laundering	money laundering, clean, dirty money, shell company, offshore account, front company, smurfing, layering, launder*, tax haven, underground bank, secrecy jurisdiction, shell company
Nepotism/Favouritism	nepotism, favouritism, favouritism, patronage, bias, partiality, preferential treatment cronyism, crony*, old boy network, clientelis*, enrich, neopatrimon*
Obstruction of Justice	obstruction*, *justice, impede, hinder, cover-up, conceal, pervert, tamper, stonewall, spoliat*
General Corruption	corrupt*

different classes, the test dataset was built with the same number of samples for each class. However, due to the imbalance in class distribution, a simple percentage-based selection wouldn't suffice. Instead, 126 samples were randomly chosen from each class. This value represents 20% of the samples from the least represented class. The remaining samples were used as training data for the multi-class corruption model, comprising 10,559 samples across the 11 HCDT Level 2 categories.

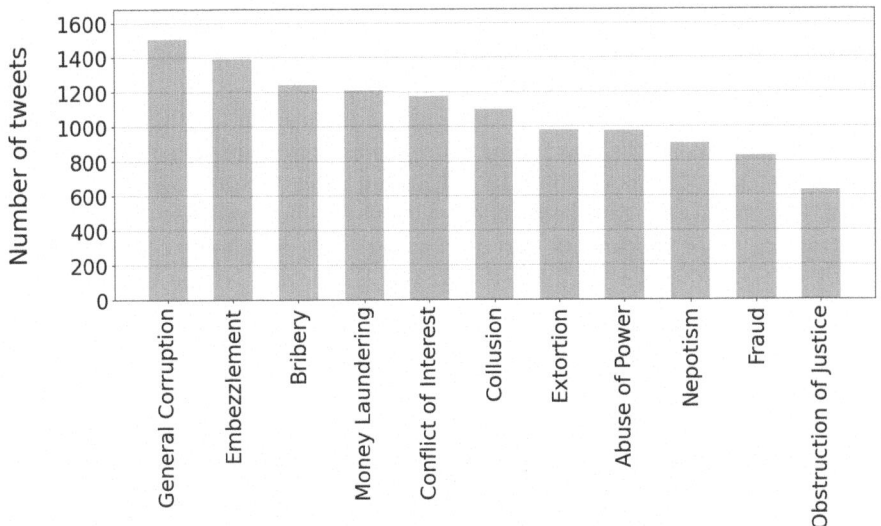

Fig. 2. Corruption-related class distribution.

Binary Corruption Dataset (HCDT Level 1). The HCDT Level 1 dataset comprised two classes - corruption-related and non-corruption-related tweets. To build the corruption-related test set, we selected the 126 testing samples from each HDCT 2 class as detailed above, resulting in a total of 1,386 instances. In this case, all samples were simply labelled as corrupted-related. The remaining 10,559 tweets were designated as training samples. Similarly, to construct the non-corruption test dataset portion, we also randomly sampled 1,386 samples from the pool of 11,945 non-corruption-related, creating a balanced test dataset. Again, the remaining 10,559 non-corruption related were considered as training samples.

4.4   Fine-Tuning LLMs

After preparing the training and testing datasets, we selected the following three pre-trained encoder-based LLMs to perform fine-tuning: BERT[1], RoBERTa[2], and BERTweet[3]. BERT [38] is a standard LLM that achieves outstanding performance in many text classification problems [39–41]. RoBERTa is a modified version of BERT with improvements in training methodology and performance across various NLP tasks [42]. On the other hand, BERTweet [43] is a BERT-based model specifically fine-tuned on a large corpus of English tweets related to COVID-19. It is optimized for the unique features of tweets, including informal grammar, short text length, and irregular vocabulary. We decided not to use decoder-based

---

[1] https://huggingface.co/bert-base-uncased.
[2] https://huggingface.co/FacebookAI/roberta-base.
[3] https://huggingface.co/vinai/bertweet-base.

LLMs, such as GPT [18] or Mistral [44], because they introduce challenges, such as the high costs associated with commercial APIs and computational resources and scalability issues in real-world scenarios [37].

These three models were fine-tuned for both binary and multi-class tasks discussed in this paper. The fine-tuning process was performed using the following hyperparameters: 20 epochs, 5×10^{-6} learning rate, AdamW optimizer, batch size of 8 samples, and saving the model with the highest accuracy on test data. The training and experiments were performed using a computer with Intel(R) Core(TM) i7-12700 CPU at 2.10GHz, 32 GB RAM, and Nvidia GeForce GTX 16660 SUPER.

5   Results

Table 3 presents the LLM classification performance for the binary classification task. All models achieved over 98% for all metrics, indicating strong performance for the binary classification of corruption or non-corruption.

Table 3. Binary classification results.

Model	Accuracy	Precision	Recall	F1-score
BERT	99.0260	99.0317	99.0260	99.0259
BERTweet	98.7734	98.7744	98.7734	98.7734
RoBERTa	98.6652	98.6665	98.6652	98.6652

RoBERTa performed worst compared with BERT and BERTweet, with all the metrics circa 98.66%. BERTweet performed slightly better, ranking as the second-best model in our experiments. BERT demonstrated the best performance compared with the other two models, achieving 99.02% for all the metrics. All models performed to a high level across all metrics, indicating that regardless of the model, they were effective in distinguishing corruption-related content from non-corruption-related content. This is also reflective of the binary and balanced nature of the classification task.

Table 4 presents the results for the multi-class classification problem. The traditional BERT model showed the performed worst amongst the models, with an accuracy of 88.09%. BERTweet followed with slightly better results, achieving 89.89% accuracy. RoBERTa outperformed both of the other models with an accuracy of 90.90%. All the models performed best in precision, albeit slight. A model exhibiting high precision, but low recall may be preferable in situations where false positives are undesirable. In our context, it's crucial to accurately identify corruption-related content in tweets, even at the risk of overlooking some instances (resulting in lower recall). Other metrics followed a similar trend in performance across all the models.

Table 4. Multi class classification results.

Model	Accuracy	Precision	Recall	F1-score
BERT	88.0952	88.7308	88.0952	88.2437
BERTweet	89.8989	90.0639	89.8989	89.9037
RoBERTa	90.9090	91.0716	90.9090	90.9350

In the multi-class classification task, model performance is lower across all metrics for all models and less consistent compared to the binary classification scenario. This outcome is expected since the classification task consists of 11 classes. Similarly, the best model for the multi-class classification task (RoBERTa) differed from the one for binary classification (BERT). Again, RoBERTa is an advancement on BERT and is better suited for the more complex task.

Table 5. Examples of models prediction (we replaced the original usernames by @user in order to keep the confidentiality).

Tweet	Label	BERT	BERTweet	RoBERTa
The FDA Cover-up that Led to the Approval of the Pfizer Vaccine	Corruption	Obstr. Just.	Obstr. Just.	Obstr. Just.
@user It's blatantly obvious anyone pushing vaccines this hard has had their palms greased	Bribery	Fraud	Embez.	Bribery
@user So state looted the covid funds thinking vaccines were for free	Embez.	Embez.	Embez.	Embez.

Figure 3 shows the confusion matrices for the multi-class classification task. The diagonals represent correct predictions. The models classified most tweets in the test dataset accurately. RoBERTa achieved higher values in the matrices diagonals for most classes. The other models outperformed RoBERTa in some classes. For instance, BERT outperformed other models for *nepotism and favouritism* and *obstruction of justice*. On the other hand, BERTweet outperformed other models for *collusion and conspiracy* and *extortion*.

Certain classes had a higher frequency of misclassifications. For instance, some tweets labelled as *conflict of interest* were misclassified as *collusion and conspiracy*, with BERT, BERTweet and RoBERTa misclassifying 13, 3, and 1 tweets, respectively. Similarly, *fraud* was mistaken with *collusion and conspiracy* by BERT (9 times), BERTweet (10 times), and RoBERTa (7 times). The highest number of incorrect classifications occurred when confusing *general corruption* with *obstruction of justice*, representing 17 misclassifications across all models. Due to the relatively small number of instances, a manual review was undertaken. The primary cause of the misclassification would seem to related to

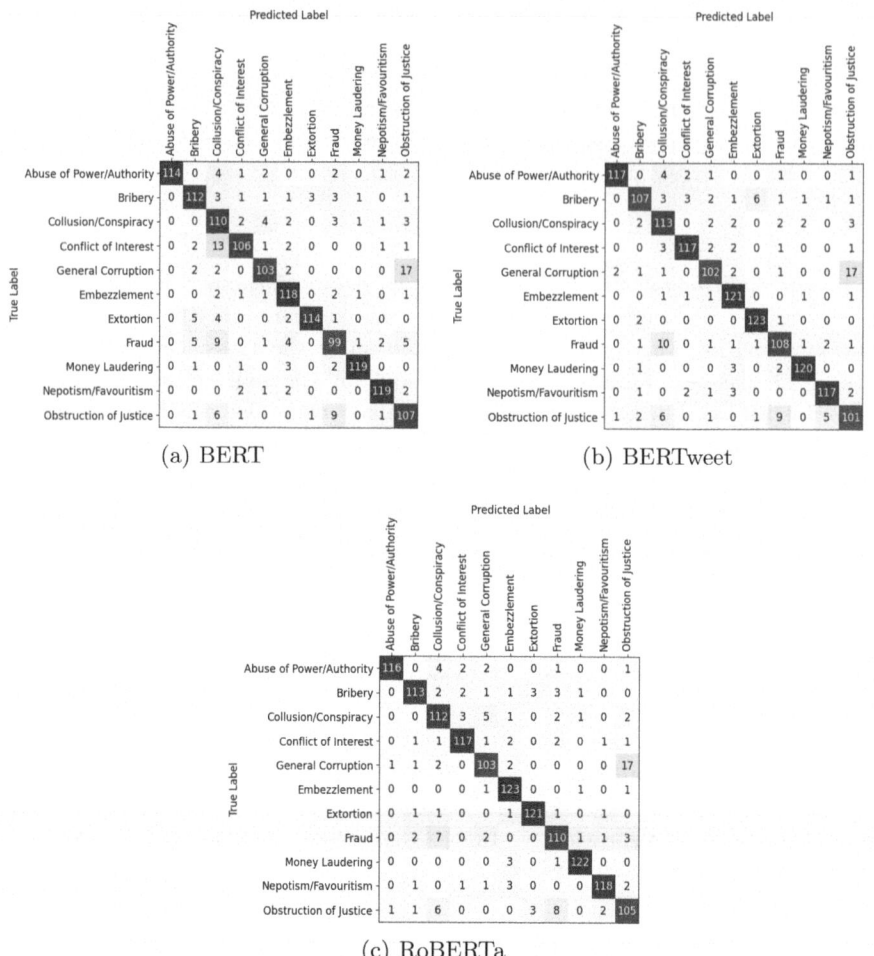

(a) BERT

(b) BERTweet

(c) RoBERTa

Fig. 3. Confusion matrices for the different BERT-based models.

a combination of overlapping vocabulary and semantic ambiguity. For example, terms and phrases in the context of obstruction of justice and general corruption were common in both, e.g., cover-up and tamper. Furthermore, corruption-related language can be vague and coded, e.g., making use of euphemisms or indirect language, thus making it hard for a model to understand the intent. This issue may be resolved using a larger and more diverse training dataset, and multi-label classification rather than merely multi-class classification.

Table 5 shows some instances of corruption-related tweets with the classification of each model. The first tweet implies a corrupt behaviour within the FDA approval process. Therefore, we labelled it as *general corruption* since it suggests that information was concealed or the approval process was manipulated. All models misclassified it as *obstruction of justice*, which would be understandable if the tweet's focus was on deliberate evidence concealment. We labelled the second tweet as *bribery* class since the phrase "palms greased" suggests the idea of someone receiving financial incentives or benefits in exchange for promoting vaccines. In this case, only the RoBERTa model correctly classified this tweet. Finally, we labelled the third tweet as *embezzlement* since the mention of "state looted the covid funds" implies that the funds allocated for COVID-related purposes were diverted for other uses, which aligns with the concept of embezzlement. All models correctly predicted this tweet.

6   Limitations and Future Work

This study is not without limitations, which in themselves provide future avenues for research. Firstly, the study was limited to one social media platform, Twitter, one language, English, and one health and vaccination context, COVID-19 and vaccination. Since the acquisition of Twitter, the owners of X have introduced new cost-based API access for researchers. This is a significant economic barrier to similar research moving forward and may encourage less ethical access to data through scraping without permission. Also, there exists significant potential for undertaking similar work on other social media platforms and specifically those with research APIs e.g., TikTok. Similarly, while English is widely spoken worldwide, it is not representative of many of the countries and regions most severely impacted by COVID-19. While the authors plan to localise the HCDT for Portuguese, Spanish, and Italian and develop associated annotated datasets and models, similar work is required for other major languages.

Secondly, the HCDT is limited to two levels and while a comprehensive set of indicative practices per second level category was identified to create a 'bag of words', the taxonomy could be extended further. For example, other corruption typologies could be integrated, and a third class could be introduced at the first level, i.e., anti-corruption, which would facilitate more nuanced analysis.

Thirdly, we identified some misclassification issues, which we believe are due to overlapping vocabulary and semantic ambiguity. While performance on binary and multi-class classification tasks was relatively high, even better performance may be achieved by increasing the size and diversity of the annotated datasets, applying multi-label classification or exploring ensemble solutions, which we plan to address in future works.

7   Conclusion

Perceived or real corruption can negatively impact trust in health services and associated outcomes. Such perceptions can be amplified and propagated on social media and interfere with public health communication. In the case of vaccination, this can result in lower levels of immunisation which can lead to illness and death. In this paper, we addressed the detection and classification of perceived corruption in the COVID-19 vaccine discourse on social media.

First, we proposed a hierarchical corruption dialogue taxonomy (HCDT), a two-level hierarchical taxonomy of corruption dialogue that can be used for categorising content by type of corruption. Second, to support future research in the identification and classification of corruption dialogue in the COVID-19 discourse on Twitter, we developed a labelled dataset for training and testing classification models. Third, we evaluated three different pre-trained BERT-based architectures (BERT, RoBERTa, and BERTweet) for (i) classifying tweets as corruption-related or non-corruption-related, and (ii) classifying tweets as one of 11 types of corruption according to the HCDT. For the binary classification, all the models obtained similar performance, with metrics around 98% and 99%; BERT marginally outperformed the other models. For the multi-class classification task, RoBERTa outperformed other models, with all the metrics around 91%.

Acknowledgment. This work was partially funded by the Irish Institute of Digital Business (IIDB) at Dublin City University, Ireland.

References

1. World Health Organisation. WHO COVID-19 dashboard (2024)
2. World Health Organisation (WHO): Global covid-19 vaccination strategy in a changing world july 2022 update (2022)
3. Burke, P.F., Masters, D., Massey, G.: Enablers and barriers to covid-19 vaccine uptake: an international study of perceptions and intentions. Vaccine **39**(36), 5116–5128 (2021)
4. Jia, K.M., et al.: Estimated preventable covid-19-associated deaths due to non-vaccination in the united states. Eur. J. Epidemiol. **38**(11), 1125–1128 (2023)
5. Zhong, M., et al.: Estimating vaccine-preventable covid-19 deaths under counterfactual vaccination scenarios in the united states. medRxiv, pp. 2022–05 (2022)
6. Eigen, P.: Measuring and combating corruption. J. Policy Reform **5**(4), 187–201 (2002)
7. Transparency International: What is corruption? (2024)
8. Naher, N., Hoque, R., Hassan, M.S., Balabanova, D., Adams, A.M., Ahmed, S.M.: The influence of corruption and governance in the delivery of frontline health care services in the public sector: a scoping review of current and future prospects in low and middle-income countries of south and south-east asia. BMC Public Health **20**, 1–16 (2020)
9. Farzanegan, M.R., Hofmann, H.P.: Effect of public corruption on the covid-19 immunization progress. Sci. Rep. **11**(1), 23423 (2021)
10. Kohler, J.C.: Covid-19 vaccines and corruption risks: preventing corruption in the manufacture, allocation and distribution of vaccines (2020)
11. Goel, R.K., Nelson, M.A., Goel, V.Y.: Covid-19 vaccine rollout-scale and speed carry different implications for corruption. Journal of Policy Modeling **43**(3), 503–520 (2021)
12. Spreco, A., Schön, T., Timpka, T.: Corruption should be taken into account when considering covid-19 vaccine allocation. Proc. Natl. Acad. Sci. **119**(19), e2122664119 (2022)

13. Egli, A., Rosati, P., Lynn, T., Sinclair, G.: Bad robot: a preliminary exploration of the prevalence of automated software programmes and social bots in the covid-19# antivaxx discourse on twitter. In: Proceedings of the The International Conference on Digital Society, Nice, France, pp. 18–22 (2021)
14. Broniatowski, D.A., et al.: Weaponized health communication: Twitter bots and russian trolls amplify the vaccine debate. Am. J. Public Health **108**(10), 1378–1384 (2018)
15. Rodrigues, F., Ziade, N., Jatuworapruk, K., Caballero-Uribe, C.V., Khursheed, T., Gupta, L.: The impact of social media on vaccination: a narrative review. J. Korean Med. Sci. 38(40) (2023)
16. Wilson, S.L., Wiysonge, C.: Social media and vaccine hesitancy. BMJ Glob. Health **5**(10), e004206 (2020)
17. Skafle, I., Nordahl-Hansen, A., Quintana, D.S., Wynn, R., Gabarron, E.: Misinformation about covid-19 vaccines on social media: rapid review. J. Med. Internet Res. **24**(8), e37367 (2022)
18. Brown, T., et al.: Language models are few-shot learners. Adv. Neural. Inf. Process. Syst. **33**, 1877–1901 (2020)
19. Lima, M.S.M., Delen, D.: Predicting and explaining corruption across countries: a machine learning approach. Gov. Inf. Q. **37**(1), 101407 (2020)
20. Rabuzin, K., Modrusan, N.: Prediction of public procurement corruption indices using machine learning methods. In: KMIS, 333–340 (2019)
21. Denisova-Schmidt, E., Huber, M., Leontyeva, E., Solovyeva, A.: Combining experimental evidence with machine learning to assess anti-corruption educational campaigns among russian university students. Empirical Econ. **60**, 1661–1684 (2021)
22. Ash, E., Galletta, S., Giommoni, T.: A machine learning approach to analyzing corruption in local public finances. Center for Law & Economics Working Paper Series, vol. 6 (2020)
23. Li, J., Chen, W.-H., Xu, Q., Shah, N., Kohler, J.C., Mackey, T.K.: Detection of self-reported experiences with corruption on twitter using unsupervised machine learning. Soc. Sci. Humanities Open **2**(1), 100060 (2020)
24. Graycar, A.: Corruption: classification and analysis. Policy Soc. **34**(2), 87–96 (2015)
25. Rose, J.: The meaning of corruption: testing the coherence and adequacy of corruption definitions. Public Integrity **20**(3), 220–233 (2018)
26. Jancsics, D.: Corruption as resource transfer: an interdisciplinary synthesis. Public Adm. Rev. **79**(4), 523–537 (2019)
27. Bussell, J.: Typologies of corruption: a pragmatic approach. In: Greed, Corruption, and the Modern State. Edward Elgar Publishing, pp. 21–45 (2015)
28. Zhang, Y., Ren, P., de Rijke, M.: A taxonomy, data set, and benchmark for detecting and classifying malevolent dialogue responses. J. Am. Soc. Inf. Sci. **72**(12), 1477–1497 (2021)
29. Corruption Watch: Glossary of corruption-related terms (2022)
30. UNODC: Glossary of corruption-related terms (2019)
31. LexisNexis: Glossary (2024)
32. Transparency International: Corruption A-Z (2024)
33. Zhao, W.X., et al.: A survey of large language models. arXiv preprint arXiv:2303.18223 (2023)
34. Wang, S., Liu, Y., Xu, Y., Zhu, C., Zeng, M.: Want to reduce labeling cost? gpt-3 can help. arXiv preprint arXiv:2108.13487 (2021)
35. Ding, B., et al.: Is gpt-3 a good data annotator? arXiv preprint arXiv:2212.10450 (2022)

36. Zevallos, R., Farrús, M., Bel, N.: Frequency balanced datasets lead to better language models. Find. Assoc. Comput. Linguist. EMNLP **2023**, 7859–7872 (2023)
37. dos Santos, V.G., Santos, G.L., Lynn, T., Benatallah, B.: Identifying citizen-related issues from social media using llm-based data augmentation. In: International Conference on Advanced Information Systems Engineering, pp. 531–546. Springer (2024)
38. Devlin, J., Chang, M.-W., Lee, K., Toutanova, K.: Bert: pre-training of deep bidirectional transformers for language understanding. arXiv preprint arXiv:1810.04805 (2018)
39. Lee, J.-S., Hsiang, J.: Patent classification by fine-tuning bert language model. World Patent Inf. **61**, 101965 (2020)
40. Santos, G.L., et al.: Kicking prejudice: large language models for racism classification in soccer discourse on social media. In: International Conference on Advanced Information Systems Engineering, pp. 547–562. Springer (2024)
41. Gupta, S., Bolden, S., Kachhadia, J., Korsunska, A., Stromer-Galley, J.: Polibert: classifying political social media messages with bert. In: Social, Cultural and Behavioral Modeling (SBP-BRIMS 2020) Conference, Washington, DC (2020)
42. Liu, Y., et al.: Roberta: a robustly optimized bert pretraining approach. arXiv preprint arXiv:1907.11692 (2019)
43. Nguyen, D.Q., Vu, T., Nguyen, A.T.: Bertweet: a pre-trained language model for english tweets. arXiv preprint arXiv:2005.10200 (2020)
44. Jiang, A.Q., et al.: Mistral 7b. arXiv preprint arXiv:2310.06825 (2023)

Evaluating and Improving Projects' Bus-Factor: A Network Analytical Framework

Sebastiano A. Piccolo[1]([✉]) [ID], Pasquale De Meo[2] [ID], and Giorgio Terracina[1] [ID]

[1] Department of DeMaCS, University of Calabria, Rende, (CS), Italy
sebastiano.piccolo@unical.it
[2] Department of DICAM, University of Messina, Messina, (ME), Italy

Abstract. When enough people leave a project, the project might stall due to lack of knowledgeable personnel. The minimum number of people who are required to disappear in order for a project to stall is referred to as bus-factor. The bus-factor has been found to be real and tangible and many approaches to measure it have been developed. These approaches are problematic: some of them do not scale to large projects, others rely on ad-hoc notions of primary and secondary developers, and others use arbitrary thresholds. None of them proposes a normalized measure of the bus-factor. Therefore, in this paper we propose a framework that, by modelling a project with a bipartite graph linking people to tasks, allows us to *1)* quantify the bus-factor of a project with a normalized measure which does not rely on thresholds; and *2)* increase the bus-factor of a project by reassigning people to tasks. We demonstrate our approach on a real case, discuss the advantages of our framework, and outline possibilities for future research.

Keywords: Bus-factor · Network Robustness · Graph Algorithms · Assignment Problem · Heuristics

1 Introduction

The *bus-factor* of a project – also known as *truck-factor* – is informally defined as the minimum number of people that have to disappear (as if they were hit by a bus) before the project stalls because nobody can complete certain tasks [5,14,19] or because of project fragmentation where integration between different modules, teams, and parts of the project becomes difficult [19]. Key people might become unavailable due to promotions, getting new jobs, going on parental leaves, or any other unforeseeable external event. The risk is tangible and empirical studies have shown that many projects suffer of low bus-factor. For instance, Yamashita et al. [25] analyzed a set of 2496 Github projects and found that, in more than 88% of the repositories, the number of core developers is less than 16. Similarly, Avelino et al. [5] considered a sample of 133 Github projects and estimated that 118 projects had a bus-factor lower than 9. There are even projects which rest on just one key person [5,22].

© The Author(s), under exclusive license to Springer Nature Switzerland AG 2025
L. M. Aiello et al. (Eds.): ASONAM 2024, LNCS 15211, pp. 255–270, 2025.
https://doi.org/10.1007/978-3-031-78541-2_16

Measuring the bus-factor of a project is not a trivial task, as establishing when a project stalls would require to consider the exact role of each person in the project, their contribution, and other project-specific factors as well as some subjective ones. For instance, the first estimation procedure proposed in literature [26], required to solve an NP-Hard problem and did not scale to projects with more than 30 people [5,11]. Another measure [10] requires to define primary and secondary developers and relies on two thresholds as inputs. The current state-of-the-art measure [5], assumes that a project stalls when more than 50% of files in the project are abandoned.

These assumptions and thresholds are arbitrary, targeted to computer science projects, and might even be project-specific. Furthermore, quantifying the bus-factor by simply counting the number of people makes the measure dependent on project size, preventing comparisons across different projects. Finally, there is still a lack of methods to improve the bus-factor of a project by reassigning people to tasks. Therefore, in this paper we develop a general graph theoretical framework that allows us to *1)* quantify the bus-factor of a project with a normalized measure that does not rely on arbitrary thresholds and enables meaningful comparisons across projects; and *2)* improve the bus-factor of a project by reassigning people to tasks.

We model a project with a bipartite graph $\mathcal{G}$ where vertices are people and tasks, and edges specify who works on which task. On $\mathcal{G}$, people's unavailability can be simulated by removing the corresponding nodes. As such, the problem of quantifying the bus-factor finds its natural position within Network Science, through the theory of *network robustness* [2,19,23]. Consequently, we consider the connected component of $\mathcal{G}$ with the largest number of tasks, and measure the relative fraction of tasks in it as people are removed from $\mathcal{G}$. We quantify the bus-factor as the area under this decay curve normalizing it by its theoretical maximum. To this end, we devise an algorithm which computes the bus-factor in linear time, with respect to the number of people, by using a union-find data structure.

Finally, we take on a crucial issue in project management: increasing the bus factor with the purpose of augmenting the project's chances of success. In this regards, we propose two algorithms, based on hill-climbing and simulated annealing respectively, that optimize our measure of bus-factor by reassigning people to tasks. We demonstrate our framework on a real case and find that we are able to improve the project bus-factor by 40%, in a way that is statistically higher than what we would expect by randomly assigning people to tasks.

In sum, we make the following contributions: *1)* we develop a theoretical framework grounded in Network Science to quantify and increase projects' bus-factor; *2)* we define $\mathcal{B}(\mathcal{G})$, a normalized measure of the bus-factor of a project; *3)* we provide an $\mathcal{O}(n)$ algorithm to compute $\mathcal{B}(\mathcal{G})$; and *4)* we provide two efficient heuristics to optimize $\mathcal{B}(\mathcal{G})$ by reassigning people to tasks.

2   Our Framework

In this section, we discuss our framework to compute and increase a project bus-factor. We treat the assignment of people to tasks as a bipartite network and use the topology of such a network in order to compute a bus-factor index. The bus-factor we compute is normalized; therefore, our measure enables comparisons between the bus-factor of projects of different sizes – both in terms of number of people and number of tasks. We begin our exposition with some preliminaries on graph theory. In the following we will use the words graph and network interchangeably.

2.1   Preliminaries on Graphs

Networks are a convenient mathematical formalism to represent and analyze complex systems, focusing on the way the components of such systems are connected; components are represented as vertices and their connections are represented with edges between the vertices. Graph methods are flexible enough to enable the representation and analysis of heterogeneous systems [1], with multiple layers of connections [8,16], which evolve over time [12,13].

An undirected graph is a pair $\mathcal{G} = (V, E)$ where V is the set of vertices and $E \subseteq V \times V$ is the set of edges. A graph is represented through its adjacency matrix A; with $A_{ij} = 1$ if vertices i and j are connected and $A_{ij} = 0$ otherwise. The degree k_i of a vertex i is the number of vertices connected to it; that is, $k_i = \sum_j A_{ij}$. A *subgraph* of a graph G is the graph induced by a subset of vertices of G. A connected component of a graph is a connected subgraph which is not part of any other connected subgraph. The largest connected component is called giant connected component and we refer to it as GCC. If the set of vertices V can be divided in two sets V_1 and V_2 with $V_1 \cap V_2 = \emptyset$, such that $\forall (i,j) \in E \ i \in V_1 \wedge j \in V_2$ the graph is called *bipartite*. In the remainder of this paper, we denote a bipartite graph as a tuple $\mathcal{G} = (V_1, V_2, E)$.

We use bipartite graphs in order to represent the assignment of people to tasks; in the following section, we develop a measure to evaluate the bus-factor of a project from such an assignment.

2.2   Evaluating the Bus-Factor of a Project from the Assignment of People to Tasks

Let us recall that the informal definition of bus-factor is the minimum number of people who must disappear before the project stalls. Defining when a project stalls in a formal way is far from easy, because many factors can determine when the project stalls; including the role of each person, the extent of their contributions to the project, and other subjective factors [11,22,24]. In particular, prior literature has relied on ad-hoc notions of primary and secondary developers, arbitrary thresholds to establish when a project stalls, and has produced non-normalized measures. To overcome these limitations, we propose a measure for the bus-factor which, with a single number, summarizes the robustness of a project to people removal and enables comparisons between different projects.

We represent the assignment of people to tasks with a bipartite graph $\mathcal{G} = (P, T, E)$, where P is the set of vertices representing the people involved in the project, T is the set of vertices representing the tasks, and E is the set of edges. Let us denote with TGCC the *task-wise giant connected component*; that is, the connected component which connects the highest number of tasks (nodes in T). To evaluate the robustness of $\mathcal{G}$, we draw upon the concept of network robustness [2]. Starting from $\mathcal{G}$, we can remove in a given order – typically in decreasing order of degree – vertices from P and, each time we remove a person, we measure the fraction of nodes in T that are still connected to the TGCC. The choice of removing people by their degree is motivated by three facts: *1)* real-world networks are resistant against random vertex removal [2,19]; *2)* vertex degree is generally correlated with other centrality measures and topological properties [17]; and *3)* finding the ordering of nodes which maximizes network disruption is, in general, a problem that can be mapped into the set cover problem, which is NP-Hard. The use of people's degree in decreasing order represents a greedy approximation to the set cover problem.

The people removal procedure describes a decay curve: on the y-axis we have the fraction of tasks still connected to the TGCC and on the x-axis we have the fraction of people removed from the project. We compute the bus-factor as the area under this decay curve, divided by the theoretical maximum.

Definition 1 (Bus-factor). *Given a network $\mathcal{G} = (P, T, E)$, with P the set of n people and T the set of m tasks, the bus-factor of such an assignment is obtained by computing the area under the decay curve of the TGCC as people are removed from the network in decreasing order of their degree. We denote the non-normalized bus-factor of a network $\mathcal{G}$ as $B(\mathcal{G})$ and compute it as follows:*

$$B(\mathcal{G}) = \frac{1}{2n} \sum_{q=1}^{n} [S(\mathcal{G}, q-1) + S(\mathcal{G}, q)] \tag{1}$$

where $S(\mathcal{G}, q)$ is the relative number of tasks still connected to the TGCC, after the removal of q people, with $1 \leq q \leq n$. The normalized bus-factor is computed by dividing the non-normalized bus-factor for the theoretical maximum ($\mathcal{B}_n$), which is obtained if $\mathcal{G}$ is a fully connected network (that is, if every person is assigned to every task):

$$\mathcal{B}(\mathcal{G}) = \frac{B(\mathcal{G})}{\mathcal{B}_n} = \frac{1}{2n-1} \sum_{q=1}^{n} [S(\mathcal{G}, q-1) + S(\mathcal{G}, q)] \tag{2}$$

$\mathcal{B}(\mathcal{G})$ is equal to 1 for the fully connected bipartite graph, and equal to 0 for the fully disconnected graph.

Algorithm 1. Fast bus-factor computation

1: **function** BUS-FACTOR($\mathcal{G}$, removal_order)
2:     $\mathcal{U} \leftarrow$ UnionFind($\mathcal{G}$.nodes())
3:     $S \leftarrow [0]$
4:     **for all** $p \in$ removal_order.reverse() **do**
5:         neigs $= \Gamma(\mathcal{G}, p) \cup \{p\}$
6:         $\mathcal{U}$.union(neigs)
7:         S.append($\mathcal{U}$.tasks_in_TGCC())
8:     **end for**
9:     $S \leftarrow S/m$
10:     **return** S.reverse()
11: **end function**

2.3   A Fast Algorithm to Compute the Bus-Factor

Computing the bus-factor naively results in an algorithm which is quadratic in the number of people. In fact, in order to find the TGCC of a network we need to perform a depth-first search (DFS) to find all the connected components and then we need to find the component with the highest number of tasks. The DFS has a computational complexity equal to $\mathcal{O}(|V| + |E|)$. In order to compute the bus-factor of a project, we need to remove n people and perform the DFS n times. Since in our bipartite graph we have n people and m tasks, $|V| = n + m$. Assuming a sparse network, we have $|E| = \mathcal{O}(|V|) = \mathcal{O}(n + m)$. As such, in a sparse network, the naïve computation of the bus-factor has a computational complexity of $\mathcal{O}(n^2 + nm)$.

Here, we present an algorithm (Algorithm 1) which computes the bus-factor, exactly, in linear time with respect to the number of people. We use a union-find structure and start with n people and m tasks disconnected. We proceed in increasing order of degree from the last person to remove to the first one and we progressively add edges. We keep track of the set with the largest number of tasks in the union-find structure and store the number of tasks in it. Such a set corresponds to the TGCC of the network.[1] In this way, we only perform n union operations and after each union operation we keep track of the number of tasks in the TGCC. Our approach is described in Algorithm 1.

The function to compute the bus-factor takes in input the bipartite graph of the assignment of people to tasks $\mathcal{G}$ and a vector of people in decreasing order of degree (removal_order). $\mathcal{U}$ is a union-find structure initialized with all the nodes of $\mathcal{G}$ and S is a vector where we store the number of tasks in the TGCC. For each person p in the vector (removal_order), in reverse order, we perform a union operation between the person p and the set of tasks to which p is connected (denoted with $\Gamma(\mathcal{G}, p)$). S is finally divided by the total number of tasks and is returned in reverse order. The output of Algorithm 1 is the relative number of tasks in the TGCC after each removal. In order to compute the normalized bus-

---

[1] Our approach can be adapted, by keeping track of the largest set in the union-find structure, to compute the robustness of an arbitrary network.

factor from S, we need to compute $\mathcal{B}(G)$ by computing the area under S in $[0, 1]$ normalized by $\mathcal{B}_n$, using (2).

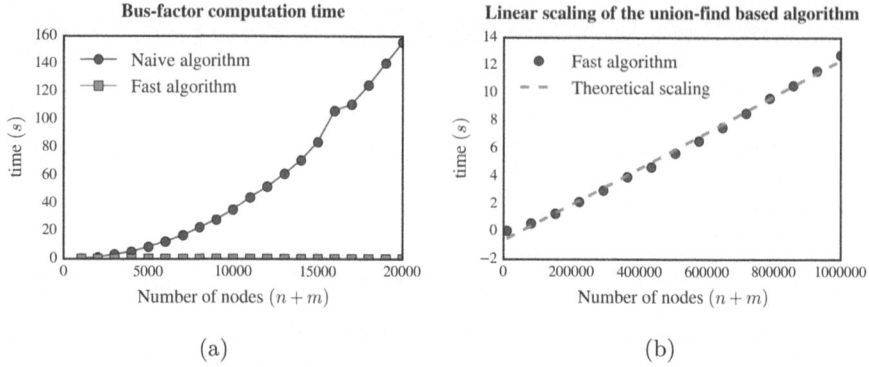

(a)   (b)

Fig. 1. (a) Time comparison between the naïve algorithm and our fast one, based on a union-find structure. **(b)** Empirical evaluation of the linear scaling of Algorithm 1. The execution time has been measured on 15 Erdos-Renyi bipartite graphs, with connection probability $p = 2\log(2n)/n$. The theoretical scaling line is obtained by fitting a line to the 15 points through ordinary least squares.

In Fig. 1(a) we contrast the execution time of our fast algorithm for the computation of the bus-factor against the naïve approach. It is possible to see that the naïve approach becomes impractical even for relatively small networks, and the speedup offered by our algorithm is significant. For our experiment we generated 20 Erdos-Renyi bipartite graphs, with $n = m$ and a total number of nodes $(n + m)$ varying in [1000, 20000] with increments of 1000. We set the connection probability $p = 2\log(2n)/n$; this value ensures that the resulting graphs are both sparse and connected. To better show the performance of our algorithm, we perform a second experiment in which we generate 15 Erdos-Renyi bipartite graphs, as before, with a total number of nodes $(n+m)$ equally spaced in [10000, 1000000], and measure the execution time to compute the bus-factor. We report our results in Fig. 1(b). It is possible to confirm that the union-find based algorithm for the bus-factor computation scales linearly with the number of people.

Now that we have both a measure to evaluate the bus-factor of a project and a fast algorithm to do so, we can turn our attention to methods for improving the bus-factor of a project by increasing $\mathcal{B}(\mathcal{G})$. In the next section, we are going to discuss some methods to improve the bus-factor of a project by reallocating people to tasks.

2.4   Increasing the Bus-Factor of a Project: Initial Models

It is possible to increase the bus factor of a project in many ways. One option is to assign people more responsibilities and, thus, more tasks. This option is

---

Algorithm 2. Bus-factor optimization with hill-climbing

1: **function** BUS-FACTOR-OPT-HC($\mathcal{G}$, n_iter)
2:     $\mathbf{x} \leftarrow \mathcal{G}$, $H_x \leftarrow H(\mathbf{x})$
3:     improved $\leftarrow$ **true**
4:     **while** improved **do**
5:         improved $\leftarrow$ **false**
6:         **for** $i \in 0 \ldots$ n_iter **do**
7:             $\mathbf{x}' \leftarrow \mathcal{P}(\mathbf{x})$, $H_{x'} \leftarrow H(\mathbf{x}')$
8:             **if** $H_{x'} < H_x$ **then**
9:                 $\mathbf{x} \leftarrow \mathbf{x}'$, $H_x \leftarrow H_{x'}$
10:                improved $\leftarrow$ **true**
11:            **end if**
12:        **end for**
13:    **end while**
14:    **return** x
15: **end function**

---

equivalent to increase the density of the bipartite network, which does increase the bus-factor of the project. However, the risk of such an option is to overload people with tasks and responsibilities making the project more prone to errors and rework [19,21].

Another option is hiring new people, assigning them strategically to tasks in order to increase the bus-factor of the project. This option is certainly interesting, particularly from the point of view of mathematical optimization, because it involves many variables, a hiring budget, managing the hiring process, and assigning people to tasks in order to optimize the bus-factor. However, this option is impractical and has important drawbacks: the hiring process is conditioned by a budget and can be slow, inefficient, and ineffective [6]. Additionally, adding people to a project might also be detrimental since it imposes higher coordination efforts [9,18,20].

Here, we focus on a third strategy. We focus on increasing the bus-factor of a project by reassigning tasks to people, improving the topology of the bipartite network against people removal. We do this by preserving the total number of tasks assigned to each person. In the following, for simplicity and without loss of generality, we make the simplifying assumption that a person can be assigned to any task. In those situations where this is not the case, the previous assumption can be relaxed by incorporating – in the objective function or in the edge swap procedure described in the following – the information about who can perform which task. Reassigning tasks to people while preserving the total number of tasks assigned to each person can be achieved by swapping pairs of edges as follows. Let (p_1, t_1) and (p_2, t_2), with $p_1, p_2 \in P, p_1 \neq p_2$ and $t_1, t_2 \in T, t_1 \neq t_2$ be two edges of an assignment graph $\mathcal{G} = (P, T, E)$; the edge-swap procedure takes (p_1, t_1) and (p_2, t_2) in input and returns the new edges (p_2, t_1) and (p_1, t_2). The swap is valid if it does not produce a double edge; that is, if none of new edges (p_2, t_1) and (p_1, t_2) is already present in $\mathcal{G}$. With such a procedure it is

---

Algorithm 3. Bus-factor optimization with simulated annealing

---

1: **function** Bus-Factor-Opt-SA($\mathcal{G}$, t_m, t_M, n_iter, α)
2:     $t \leftarrow t_M$
3:     $\mathbf{x} \leftarrow \mathcal{G}$, $H_x \leftarrow H(\mathbf{x})$
4:     **while** $t > t_m$ **do**
5:         **for** $i \in 0 \ldots$ n_iter **do**
6:             $\mathbf{x}' \leftarrow \mathcal{P}(\mathbf{x})$
7:             $H_{x'} \leftarrow H(\mathbf{x}')$
8:             $\Delta H \leftarrow H_{x'} - H_x$
9:             **if** $\Delta H < 0$ **or** $\mathrm{rand}(0,1) < e^{\frac{-\Delta H}{t}}$ **then**
10:                 $\mathbf{x} \leftarrow \mathbf{x}'$
11:                 $H_x \leftarrow H_{x'}$
12:             **end if**
13:         **end for**
14:         $t \leftarrow \alpha \times t$
15:     **end while**
16:     **return** x
17: **end function**

---

possible to rearrange the edges of $\mathcal{G}$ in order to increase the bus-factor measure $\mathcal{B}(\mathcal{G})$. We denote the swapping procedure as $\mathcal{P}(\mathcal{G})$.

The problem of assigning people to tasks in order to maximize $\mathcal{B}(\mathcal{G})$ reduces to the generalized assignment problem, which is NP-Hard [27]. Therefore, here we employ two heuristics that increase the bus-factor of a given project by minimizing $H(\mathcal{G}) = -\mathcal{B}(\mathcal{G})$. The first algorithm is based on an hill-climbing procedure, which is a greedy optimization approach. The hill-climbing algorithm works as follows: first, a solution proposal is generated as $\mathcal{G}' = \mathcal{P}(\mathcal{G})$. If $H(\mathcal{G}') < H(\mathcal{G})$, then the proposal is accepted and $\mathcal{G} = \mathcal{G}'$; otherwise, the proposal is rejected and $\mathcal{G}$ is unchanged. The process is iterated until convergence, when the algorithm reaches an optimum. Since hill-climbing is a greedy optimization strategy, there is no guarantee that the algorithm reaches the global optimum. The final, rewired, $\mathcal{G}$ is the solution; i.e. the optimized assignment network with an increased bus-factor. Our hill-climbing approach is described in Algorithm 2.

The second algorithm is based on simulated annealing, which is a global optimization algorithm [15]. In simulated annealing, a solution proposal is generated as $\mathcal{G}' = \mathcal{P}(\mathcal{G})$. Let $\Delta H = H(\mathcal{G}') - H(\mathcal{G})$ be the bus-factor variation between the new candidate solution and the current one, if $\Delta H < 0$, the solution is accepted – as in the hill-climbing algorithm. If $\Delta H \geq 0$, the proposal is accepted with probability $P(H) = e^{\frac{-\Delta H}{t}}$, where t is the temperature parameter. The initial temperature t_M is an input of the algorithm. After each iteration, a new temperature is computed as $t = t \times \alpha$, $\alpha \in (0,1)$. As such, simulated annealing can accept proposals that are worse than the current solution. This simple difference with the hill-climbing algorithm enables simulated annealing not to remain stuck into a local optimum [15]. It is worth noting that an iteration in simulated annealing can involve the generation of more than one solution proposal; thus, the algorithm can perform more than one edge swap at each iteration [15]. The number of solution proposals to generate at each iteration (n_iter) is an input of the algorithm. The algorithm terminates when the temperature t is lower than or equal to a minimum temperature t_m, provided as input. The pseudo-code of this algorithm is reported in Algorithm 3.

Tuning the parameters of the simulated annealing is not an easy task, since the right combination of parameters is problem dependent [7,15]. Nevertheless, there are some general indications and rules of thumb on how to set the parameters. In general, the higher the number of iterations, the better the final solution is [7]. However, there is a trade-off between the goodness of the solution and the computational time needed to reach it. The temperature should be cooled slowly, since, it has been shown that if temperature is cooled slowly enough, simulated annealing converges to the optimal solution [7]. The cooling factor α is typically chosen in $[0.99, 0.9999]$. The initial temperature, t_M should be chosen in such a way that around 80% of the proposals generated are accepted. The final temperature t_m should be very close to zero; typical values range in $[10^{-6}, 10^{-10}]$. The values of α, t_M, and t_m are modulated by the value of n_iter, which sets the number of proposals generated at each temperature t.

3   Experiments and Analysis

To demonstrate our framework, we use – as a case study – the bipartite network describing the allocation of people to tasks in the context of a real project of a biomass power plant, from Piccolo et al. [19]. The network consists of 111 people, 148 tasks, and 926 edges between people and tasks. The network exhibits the *small-world* property with density equal to 0.056, clustering coefficient equal to 0.3, and average shortest path length equal to 6.378. On average, a person is connected to 8.34 tasks. The network consists of 3 connected components (Fig. 2A) with a giant connected component containing 107 people and 145 tasks. Piccolo et al. [19], used this bipartite network to show the importance of people connected to a high number of tasks in determining both the assignment robustness, and the response to error propagation phenomena. In the following, we demonstrate our framework in two ways. First, we quantify the bus-factor of the project under examination and we assess its statistical significance with respect to an appropriate null model. Second, we improve the bus-factor of the project by reallocating people to tasks with the two algorithms previously discussed. We assess the statistical significance of the new assignments and evaluate the computational gains provided by our fast (union-find based) algorithm to compute the bus-factor.

3.1   Quantifying the Bus-Factor and Assessing Its Statistical Significance

Algorithm 1 provides us a fast way to compute the bus-factor decay curve. In Fig. 2B, we show the bus-factor decay curve of the project under examination

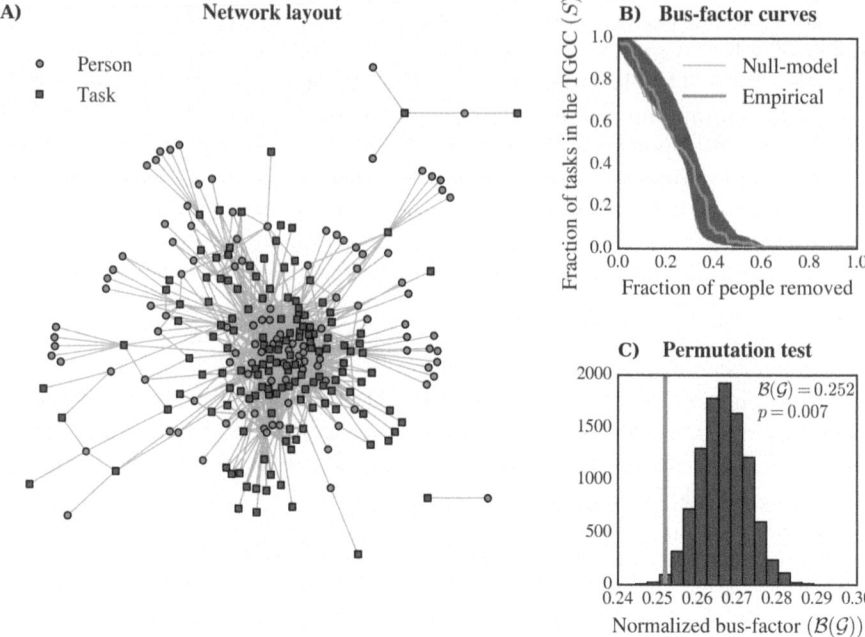

Fig. 2. A: Layout of the empirical network. The network consists of three connected components end exhibits modular organization. **B**: Decay curves of the TGCC as people are removed in decreasing order of their degree. The orange line shows the decay of the TGCC of the empirical network, while blue lines are computed on 10000 networks generated with a degree-preserving bipartite null model. **C**: Permutation test which assesses the statistical significance of the bus-factor of the empirical network against the degree-preserving null model. The bus-factor of the empirical network is statistically lower than what we would expect if people were randomly allocated to tasks (p-value = 0.007).

in orange. We compare this curve with 10000 random bipartite graphs obtained with a *degree-preserving* null model. The degree-preserving null model fixes the degree of each person and task and connects a random person with a random task, until all the connections have been assigned. In Fig. 2B, we show the bus-factor decay curve of the null model realizations with blue lines. It is evident that all the curves have similar shapes and that the curve of the empirical network is closer to the lower end of the range. Figure 2B shows the dependence of the bus-factor on the degree distributions of people and tasks. This behaviour is in line with other findings on network robustness [2].

Using formula (2) to compute the normalized bus-factor $\mathcal{B}(\mathcal{G})$, not only allows us to condense the information of the curve in a single number, but it also enables the assessment of the statistical significance of the bus-factor of a project. In fact, the 10000 random bipartite graphs realizations can be seen as random permutations of the edges of the original network, and can be used in a sort of permutation test. The bus-factor of the empirical network is $\mathcal{B}(\mathcal{G}) = 0.252$. A bus-factor which is 25% of the bus-factor of a fully connected bipartite assignment between 111 people and 148 tasks, with a connection density of 0.056, could appear as remarkable. However, when we compute the normalized bus-factor for all the 10000 random networks, we can see that the bus-factor of the project is actually lower than the bus-factor that we would expect at random. In Fig. 2C, we show the results of the permutation test. The vertical orange line shows the bus-factor of the project under examination and the blue histogram shows the distribution of the bus-factor under the null model. The p-value for the test $\mathcal{B}(\mathcal{G}) < \mathcal{B}(\text{null})$ is 0.007, indicating that the bus-factor of our project is statistically lower than what we would expect by randomly allocating people to tasks. This finding echoes those from prior research about the low bus-factor of many projects [5, 10, 14, 19, 25, 26].

3.2   Improving the Bus-Factor of the Case Study Project

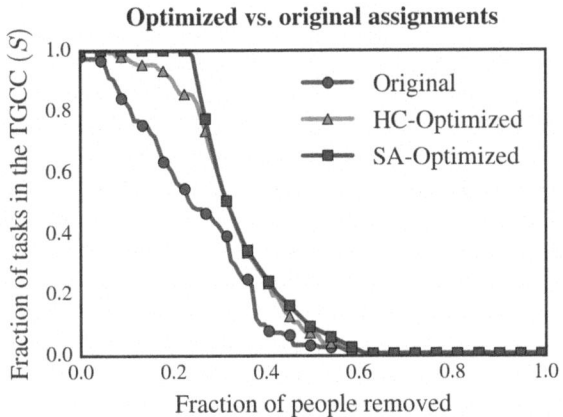

Fig. 3. Original vs. optimized assignments of people to tasks. For the hill climbing algorithm we set the parameter n_iter = 100. For the simulated annealing we set $t_M = 0.5$, $t_m = 10^{-8}$, n_iter = 100 and $\alpha = 0.99$. The bus-factor of the original network is 0.252; the bus-factor of the network optimized with the hill-climbing algorithm is 0.333; the bus-factor of the network optimized with simulated annealing is 0.352.

In this section, we focus on optimizing the bus-factor of the project under examination by rewiring the original assignment network so as to improve its bus factor. To this end, we employ the algorithms defined in Sect. 2.4. In Fig. 3, we plot the bus-factor decay curves for the original network $\mathcal{G}$ (in blue), the network $\mathcal{G}_{HC}$ obtained by rewiring $\mathcal{G}$ via the hill-climbing algorithm (in orange), and the network $\mathcal{G}_{SA}$ obtained by rewiring $\mathcal{G}$ via the simulated annealing algorithm (in gray). The improvement is evident. Numerically, $\mathcal{B}(\mathcal{G}_{HC}) = 0.333$ which is an improvement of 32% over $\mathcal{B}(\mathcal{G})$; and $\mathcal{B}(\mathcal{G}_{SA}) = 0.352$ which is an improvement of 40% over $\mathcal{B}(\mathcal{G})$. Compared with the statistical ensemble of networks from the degree-preserving null-model, both $\mathcal{B}(\mathcal{G}_{HC})$ and $\mathcal{B}(\mathcal{G}_{SA})$ are statistically higher than the bus-factor we would expect if people were randomly assigned to tasks, with p-value $p < 0.0001$. We note that optimizing the bus-factor as defined in (2), also leads to increasing the number of people that it is necessary to remove in order to disconnect the first task from the assignment network. This can clearly be seen by comparing, in Fig. 3, the point where the decay curves begin to decline. For the network rewired by simulated annealing, it is necessary to remove more than 20% of people (by their degree) in order to disconnect the first task from the TGCC. In contrast, after removing the same number of people from the original network, more than 50% of tasks result disconnected from the TGCC. The network found by simulated annealing is probably optimal. We experimented with different sets of parameters (see Table 1) and each instance of simulated annealing converged to the same final network. In contrast, the hill-climbing algorithm did not converge to the same network found by simulated annealing. However, when we set n_iter = 10000, hill-climbing finds a network only slightly sub-optimal, with a bus-factor ~ 0.352, very close to the one found by simulated annealing. In Table 1, we report the execution time for hill-climbing and simulated annealing under different parameters choices. We also contrast the execution time when we use the naïve quadratic algorithm to compute $H(\mathcal{G})$ with the execution time when we use Algorithm 1 to compute $H(\mathcal{G})$. The use of Algorithm 1 to compute the objective function, renders both hill-climbing and simulated annealing about 30 times faster than their counterparts which use the naïve algorithm. We also observe that simulated annealing with the linear algorithm to compute the objective function is faster or as fast as the hill-climbing algorithm which uses the quadratic algorithm to compute the objective function. Table 1 also shows the importance of tuning the parameters of simulated annealing in order to avoid wasted computation. Taken together, these results demonstrate the goodness of our framework both from a theoretical standpoint, in that they quantify the bus factor of a project and enable its optimization, and from an algorithmic standpoint, in that they provide efficient algorithms for measuring and improving the bus factor. Finally, our framework allows us to assess the statistical significance of the bus-factor of a project by comparing it with what we would expect if people were randomly allocated to tasks.

4   Related Works

The bus-factor is mainly covered in computer science literature by considering two related problems: *1)* finding the *core* developers – i.e. developers who contribute to the vast majority of the code base – and *2)* quantifying the bus-factor of a project. Core developers are detected by estimating the *degree of authorship* (DoA) of each file by combining factors extracted from repositories, code review, meetings, etc. [5,10,14,25]. These approaches are specific to computer science

Table 1. Speedup provided by the fast bus-factor computation over the naïve one, in the optimization algorithms.

Algorithm	parameters	T_n (sec)	T_f (sec)	speedup
Hill Climbing	n_iter = 100	42.3	2.4	17.6×
Hill Climbing	n_iter = 1000	260.8	8.4	31×
Hill Climbing	n_iter = 10000	1600.1	43.9	36×
Simulated Annealing	$t_M = 0.5$ $t_m = 10^{-8}$ n_iter = 1 $\alpha = 0.99975$	1523.5	53.5	28×
Simulated Annealing	$t_M = 1$ $t_m = 10^{-10}$ n_iter = 1 $\alpha = 0.99975$	2017.2	68.9	29×
Simulated Annealing	$t_M = 0.5$ $t_m = 10^{-8}$ n_iter = 100 $\alpha = 0.99$	4015.2	119.6	34×

T_n: computation time with naïve bus-factor computation.
T_f: computation time with fast bus-factor computation (Algorithm 1).

and, since in this paper we propose a general framework, we do not review them. We just point out that our framework can be paired with these approaches; for instance, by using the DoA to build the assignment graph $\mathcal{G}$.

Regarding the estimation of the bus-factor, Zazworka et al. [26] proposed the first approach, based of finding the minimal set X of people that would belong to more than t% (where t is an arbitrary threshold) of files; the bus-factor would be estimated as the size of X. Their approach was criticized for making too strong assumptions [5,11] and being non-scalable [11,22]. A second approach comes from Cosentino et al. [10]: a set of metrics to estimate the bus-factor based on the notions of primary and secondary developers, detected according to their contribution to the code base. When a certain number of files is abandoned, that is no primary or secondary developer is linked to it, the project is stalled. Avelino et al. [5] used the DoA to link developers to files and proposed the following heuristic to estimate the bus-factor: they iteratively remove the person associated with the highest number of files and increase the bus-factor by one, until more than 50% of the files are abandoned. A comparative analysis from Ferreira et al. [11] showed that the heuristic from Avelino et al. [5] is the best performing one. This heuristic is still the base for more recent approaches, such as the one from Jabrayildaze et al. [14]. A different angle is taken by CSDETECTOR [3,4]: a C4.5 algorithm, which detects so called *community smells*, including the bus-factor. However, CSDETECTOR makes binary predictions and does not estimate the bus-factor; it requires it as input.

Compared with existing heuristics, our measure $\mathcal{B}(\mathcal{G})$ offers several advantages: *1)* it does not use arbitrary thresholds; *2)* it is normalized; thus, it enables comparisons across different projects; *3)* it can be computed in linear time, scaling to very large networks. Finally, since our measure summarizes the decay of the TGCC, it quantifies nuances that prior approaches cannot take into account. For instance, as Fig. 3 shows, $\mathcal{G}_{HC}$ and $\mathcal{G}_{SA}$ remain with a TGCC lower than 50% after the same number of people have been removed. According to the heuristic from Avelino et al., these networks have the same bus-factor. However, this is not the case: the decay curve of $\mathcal{G}_{SA}$ Pareto-dominates the decay curve of $\mathcal{G}_{HC}$.

5   Conclusions and Future Work

The *bus-factor* evaluates the robustness of projects against people unavailability. Prior literature has shown that many projects have low bus-factor and that consequences of a low bus-factor can be quite severe. As such, a number of methods have been proposed to *1)* formally quantify the bus-factor of a given project; and *2)* detect the core people in a project. These approaches, however, rely on some assumptions and thresholds to define when a project stalls, which have been criticized as arbitrary. Furthermore, these approaches use non-normalized measures, making comparisons across projects difficult.

Therefore, in this paper, we proposed a framework grounded in network science to *1)* quantify the bus-factor with a normalized measure; and *2)* assign people to tasks in order to maximize the bus-factor of a project. We developed a linear algorithm to evaluate the robustness of a project and measure the bus factor. We showcased our approach using the network of allocation of people to tasks from a real design project. We have shown that our framework allows us to compare the bus-factor of different projects and we used this property to assess the statistical significance of the case project against a null model where people are randomly allocated to tasks. Finally, we developed two algorithms to reassign people to tasks in order to improve the bus-factor, obtaining an improvement of 40% over the case project.

We plan to extend this research in the following ways: *1)* with respect to the theory of bus-factor, we plan to characterize the hardness of measuring the bus-factor as well as assigning people to tasks in order to maximize it. *2)* with respect to the algorithms, we aim to expand our assignment algorithms by taking into account other factors such as the suitability of people to perform certain tasks. Furthermore, we will investigate other ways to remove people from a project in order to obtain better estimates of the bus-factor. *3)* with respect to practice, we plan to test our framework on a set of Github repositories comparing our measure with existing ones.

Acknowledgments. This research is partially supported by MUR under PNRR project PE0000013-FAIR, Spoke 9 - Greenaware AI – WP9.2.

Disclosure of Interests. The authors have no competing interests to declare that are relevant to the content of this article.

References

1. Agreste, S., De Meo, P., Ferrara, E., Piccolo, S., Provetti, A.: Analysis of a heterogeneous social network of humans and cultural objects. IEEE Trans. Syst. Man Cybernet. Syst. **45**(4), 559–570 (2015). https://doi.org/10.1109/TSMC.2014. 2378215
2. Albert, R., Jeong, H., Barabási, A.L.: Error and attack tolerance of complex networks. Nature **406**(6794), 378–382 (2000)
3. Almarimi, N., Ouni, A., Chouchen, M., Mkaouer, M.W.: Improving the detection of community smells through socio-technical and sentiment analysis. Journal of Software: Evolution and Process **35**(6), e2505 (2023). https://doi.org/10.1002/smr. 2505
4. Almarimi, N., Ouni, A., Mkaouer, M.W.: Learning to detect community smells in open source software projects. Knowl.-Based Syst. **204**, 106201 (2020). https:// doi.org/10.1016/j.knosys.2020.106201
5. Avelino, G., Passos, L., Hora, A., Valente, M.T.: A novel approach for estimating truck factors. In: 2016 IEEE 24th International Conference on Program Comprehension (ICPC), pp. 1–10 (2016). https://doi.org/10.1109/ICPC.2016.7503718
6. Behroozi, M., Shirolkar, S., Barik, T., Parnin, C.: Debugging hiring: what went right and what went wrong in the technical interview process. In: Proceedings of the ACM/IEEE 42nd International Conference on Software Engineering: Software Engineering in Society, ICSE-SEIS 2020, pp. 71–80. Association for Computing Machinery, New York (2020). https://doi.org/10.1145/3377815.3381372
7. Bertsimas, D., Tsitsiklis, J.: Simulated annealing. Stat. Sci. **8**(1), 10–15 (1993)
8. Boccaletti, S., et al.: The structure and dynamics of multilayer networks. Phys. Rep. **544**(1), 1–122 (2014)
9. Brooks, F.P., Jr.: The mythical man-month, anniversary Addison-Wesley Longman Publishing Co., Inc (1995)
10. Cosentino, V., Izquierdo, J.L.C., Cabot, J.: Assessing the bus factor of git repositories. In: 2015 IEEE 22nd International Conference on Software Analysis, Evolution, and Reengineering (SANER), pp. 499–503 (2015). https://doi.org/10.1109/ SANER.2015.7081864
11. Ferreira, M., Valente, M.T., Ferreira, K.: A comparison of three algorithms for computing truck factors. In: 2017 IEEE/ACM 25th International Conference on Program Comprehension (ICPC), pp. 207–217 (2017). https://doi.org/10.1109/ ICPC.2017.35
12. Holme, P., Saramäki, J.: Temporal networks. Phys. Rep. **519**(3), 97–125 (2012)
13. Holme, P., Saramäki, J.: Temporal network theory, vol. 2. Springer (2019)
14. Jabrayilzade, E., Evtikhiev, M., Tüzün, E., Kovalenko, V.: Bus factor in practice. In: Proceedings of the 44th International Conference on Software Engineering: Software Engineering in Practice, ICSE-SEIP 2022, pp. 97–106. Association for Computing Machinery, New York (2022). https://doi.org/10.1145/3510457.3513082
15. Kirkpatrick, S., Gelatt, C.D., Jr., Vecchi, M.P.: Optimization by simulated annealing. Science **220**(4598), 671–680 (1983)
16. Kivelä, M., Arenas, A., Barthelemy, M., Gleeson, J.P., Moreno, Y., Porter, M.A.: Multilayer networks. J. Complex Netw. **2**(3), 203–271 (2014)

17. Li, C., Li, Q., Van Mieghem, P., Stanley, H.E., Wang, H.: Correlation between centrality metrics and their application to the opinion model. Euro. Phys. J. B **88**(3), 1–13 (2015). https://doi.org/10.1140/epjb/e2015-50671-y
18. Parraguez, P., Piccolo, S.A., Perišić, M.M., Štorga, M., Maier, A.M.: Process modularity over time: modeling process execution as an evolving activity network. IEEE Trans. Eng. Manage. **68**(6), 1867–1879 (2019)
19. Piccolo, S.A., Lehmann, S., Maier, A.: Design process robustness: a bipartite network analysis reveals the central importance of people. Design Sci. **4**, e1 (2018)
20. Piccolo, S.A., Lehmann, S., Maier, A.M.: Different networks for different purposes: a network science perspective on collaboration and communication in an engineering design project. Comput. Ind. **142**, 103745 (2022)
21. Piccolo, S.A., Maier, A.M., Lehmann, S., McMahon, C.A.: Iterations as the result of social and technical factors: empirical evidence from a large-scale design project. Res. Eng. Design **30**(2), 251–270 (2019)
22. Ricca, F., Marchetto, A., Torchiano, M.: On the difficulty of computing the truck factor. In: Caivano, D., Oivo, M., Baldassarre, M.T., Visaggio, G. (eds.) PROFES 2011. LNCS, vol. 6759, pp. 337–351. Springer, Heidelberg (2011). https://doi.org/10.1007/978-3-642-21843-9_26
23. Schneider, C.M., Moreira, A.A., Andrade, J.S., Jr., Havlin, S., Herrmann, H.J.: Mitigation of malicious attacks on networks. Proc. Natl. Acad. Sci. **108**(10), 3838–3841 (2011)
24. Torchiano, M., Ricca, F., Marchetto, A.: Is my project's truck factor low? theoretical and empirical considerations about the truck factor threshold. In: Proceedings of the 2nd International Workshop on Emerging Trends in Software Metrics, WETSoM 2011, pp. 12–18. Association for Computing Machinery, New York(2011). 10.1145/1985374.1985379
25. Yamashita, K., McIntosh, S., Kamei, Y., Hassan, A.E., Ubayashi, N.: Revisiting the applicability of the pareto principle to core development teams in open source software projects. In: Proceedings of the 14th International Workshop on Principles of Software Evolution, IWPSE 2015, pp. 46–55. Association for Computing Machinery, New York (2015). https://doi.org/10.1145/2804360.2804366
26. Zazworka, N., Stapel, K., Knauss, E., Shull, F., Basili, V.R., Schneider, K.: Are developers complying with the process: an xp study. In: Proceedings of the 2010 ACM-IEEE International Symposium on Empirical Software Engineering and Measurement, ESEM 2010. Association for Computing Machinery, New York (2010). https://doi.org/10.1145/1852786.1852805
27. Özbakir, L., Baykasoğlu, A., Tapkan, P.: Bees algorithm for generalized assignment problem. Appl. Math. Comput. **215**(11), 3782–3795 (2010). https://doi.org/10.1016/j.amc.2009.11.018

Towards Generalized Offensive Language Identification

Alphaeus Dmonte[1](✉), Tejas Arya[2], Tharindu Ranasinghe[3],
and Marcos Zampieri[1]

[1] George Mason University, Fairfax, USA
admonte@gmu.edu
[2] Rochester Institute of Technology, Rochester, USA
[3] Lancaster University, Lancaster, UK

Abstract. The prevalence of offensive content on the internet, encompassing hate speech and cyberbullying, is a pervasive issue worldwide. Consequently, it has garnered significant attention from the machine learning (ML) and natural language processing (NLP) communities. As a result, numerous systems have been developed to automatically identify potentially harmful content and to mitigate its impact. These systems can follow two approaches; (i) Use publicly available models and application endpoints, including prompting large language models (LLMs) (ii) Annotate datasets and train ML models on them. However, both approaches lack an understanding of how generalizable they are. Furthermore, the applicability of these systems is often questioned in off-domain and practical environments. This paper empirically evaluates the generalizability of offensive language detection models and datasets across a novel generalized benchmark: *GenOffense*. We answer three research questions on generalizability. Our findings will be useful in creating robust real-world offensive language detection systems.

Keywords: Offensive Language · Large Language Models · Generalizability

1 Introduction

The presence of offensive posts on social media platforms leads to various negative consequences for users. Offensive posts have been linked to harmful outcomes such as increased suicide attempts [19,27] and mental health issues such as depression [3,8]. To address these serious repercussions, content moderation is typically employed on online platforms. Given the overwhelming volume of posts, however, human moderators alone cannot handle the task effectively, necessitating the development of automatic systems to assist them [41,48,51].

A highly effective method for constructing systems that can detect offensive language involves using publicly accessible application endpoints and models in an *unsupervised* fashion. Notably, the development of openly accessible services

L. M. Aiello et al. (Eds.): ASONAM 2024, LNCS 15211, pp. 271–286, 2025.
https://doi.org/10.1007/978-3-031-78541-2_17

such as perspective API [24] and models such as toxicBERT have greatly facilitated this approach. Furthermore, a more recent development involves the use of LLMs in a similar manner, employing specific prompts to identify offensive language [20]. The other most common method for offensive language identification is the *supervised* approach, where a dataset is annotated to serve as training material for ML systems. The datasets can be annotated with different goals in mind depending on the sub-task they address, such as aggression, cyberbullying and hate speech [43] as well as following a more general taxonomy [46].

While both the *unsupervised* and *supervised* approaches have provided excellent results in specific offensive language detection use cases, their generalizability [2,16] and the ability to perform in unseen use cases [1,38,44] have often been questioned. The ability to effectively generalize is consistently highlighted as a fundamental requirement for NLP models [14,26]. Particularly in a real-world application such as offensive language detection, generalization is crucial to ensure that the system exhibits robust, reliable, and fair behavior when making predictions on data that differs from their training data. However, to the best of our knowledge, no comprehensive evaluation of the generalizability of offensive language detection systems and datasets has been yet carried out. To fill this important gap in the literature, in this paper, we address the question of generalizability in offensive language identification.

Following [21], we define generalizability as the ability to perform consistently among different datasets. First, we construct a generalized offensive language detection benchmark; *GenOffense*, collecting eight datasets extracted from different social media platforms and mapping them to a general offensive language detection taxonomy. We evaluate publicly available APIs and models, including LLMs in *GenOffense*, and discuss the results. In the second part, we train various ML models on the training sets of these eight different datasets under different settings such as fully supervised, few-shot and zero-shot and evaluate the results. We answer three research questions as follows:

- **RQ1 - Generalizability:** How well do the publicly available systems and the models trained on different datasets generalize?
- **RQ2 - Dataset Size:** What is the impact of dataset size on generalizability? Does more data always result in better generalizability?
- **RQ3 - Domain Specificity:** What is the overlap and performance carryover between datasets collected from different platforms?

2   Related Work

Offensive Language Detection The problem of offensive language on social media has gained a lot of attention within the ML/NLP community. Researchers and organizations have developed systems to identify multiple types of offensive content such as *aggression, cyberbullying,* and *hate speech* [12,34]. Perspective API [24] is one such free API that was trained on the Toxic Comment Classification dataset [7]. More recently, with the rise of LLMs such as GPT, researchers have used LLMs to detect and identify various forms of offensive language [50]. [20] utilized ChatGPT for hateful speech detection and showed that ChatGPT provides satisfactory results for certain prompts. In a different study, [25] investigated the potential of using ChatGPT for annotating offensive comments and compared its results with those from crowdsourcing workers and the results show a high agreement. All these systems and APIs can be used in an *unsupervised* way to detect offensive content. However, these systems can induce bias to the task depending on the data they were used to train.

As discussed in the introduction, the most common approach to detect offensive content is the *supervised* approach, where the ML models are trained on annotated datasets. For this purpose, several datasets have been created for English [12,30,46]. The popular shared tasks such as OffensEval [47,49], HatEval [4] and HASOC [39] have also contributed to creating some of these popular English datasets. Researchers have trained various ML models ranging from SVMs [28] to neural transformers [35]. Recent studies have also fine-tuned transformer models on offensive language data and released domain-specific models such as HateBERT [9] and fBERT [37]. These supervised models have provided excellent results over several datasets.

Generalized Machine Learning. Good generalization, defined as the ability to successfully transfer representations, knowledge, and strategies from past experiences to new experiences, is a primary requisite for NLP/ ML models [21]. Generalization has been widely investigated on different NLP tasks, including machine translation [31], language modeling [10], and semantic parsing [22] and is crucial to ensure robustness, reliability, and fairness [40]. While the aforementioned offensive language detection methods have provided good results on the datasets they are evaluated, several studies have questioned their ability to perform on unseen use cases. [38] showed that hate speech classifiers often misclassify chess discussions as racist. [44] evaluate nine different offensive language detectors on political discussions and show that they have a low agreement. Furthermore, offensive language detection systems have been evaluated for geographic biases [17] and vulnerability to adversarial attacks [18]. Finally, [16] tested multiple intra- and cross-dataset offensive language identification scenarios. However, the study is limited to a few datasets and models. To the best of our knowledge, no work exists on a comprehensive evaluation of the generalizability of offensive language detection systems, which we address in this research.

3   GenOffense: A Generalized Offensive Language Detection Benchmark

The root cause for the lack of generalization research on offensive language detection is that no standard benchmark exists for the domain. While there are several popular datasets for offensive language identification, each of them has been annotated using different annotation guidelines and taxonomies. This, in theory, limits the possibility of combining existing datasets when training and evaluating robust offensive language identification models. To address this we construct the first Generalized Offensive Language Detection Benchmark; *GenOffense*.[1]

Table 1. The eight datasets used for *GenOffense*, including the number of instances (Inst.) in the training and testing sets, the OFF % in each set, the data source, and the reference.

| Dataset | Training | | Testing | | Data Sources | Reference |
	Inst.	OFF %	Inst.	OFF %		
AHSD	19,822	0.83	4,956	0.82	Twitter	[12]
HASOC	5,604	0.36	1,401	0.35	Twitter, Facebook	[29]
HatE	9,000	0.42	1,434	0.42	Twitter	[4]
HateX	11,535	0.59	3,844	0.58	Twitter, Gab	[30]
OHS	8,285	0.21	2,090	0.20	Reddit	[32]
OLID	13,240	0.33	860	0.27	Twitter	[46]
TCC	12,000	0.09	2,500	0.10	Wikipedia Talk	URL[1]
TRAC	4,263	0.20	1,200	0.42	Facebook, Twitter, YouTube	[5]

3.1   *GenOffense* Construction

We use eight popular publicly available datasets containing English data summarized in Table 1 to construct *GenOffense*. As the datasets were annotated using different guidelines and labels, following the methodology described in [33], we map all labels to OLID level A [46], which is offensive (OFF) and not offensive (NOT). We choose OLID due to the flexibility provided by its general three-level hierarchical taxonomy below, where the OFF class contains all types of offensive content, from general profanity to hate speech, while the NOT class contains non-offensive examples.

- **Level A:** Offensive (OFF) vs. Non-offensive (NOT).
- **Level B:** Classification of the type of offensive (OFF) tweet - Targeted (TIN) vs. Untargeted (UNT).
- **Level C:** Classification of the target of a targeted (TIN) tweet - Individual (IND) vs. Group (GRP) vs. Other (OTH).

In the OLID taxonomy, offensive (OFF) posts targeted (TIN) at an individual are often cyberbullying, whereas offensive (OFF) posts targeted (TIN) at a group are often hate speech.

AHSD is one of the most popular hate speech datasets available. The dataset contains data retrieved from Twitter, which was annotated using crowdsourcing. The annotation taxonomy contains three classes: Offensive, Hate, and Neither.

---

[1] https://github.com/TharinduDR/GeneralOffense.git.

We conflate Offensive and Hate under a class OFF while neither class corresponds to OLID's NOT class.

HASOC is the dataset used in the HASOC shared task 2020. It contains posts retrieved from Twitter and Facebook. The upper level of the annotation taxonomy used in HASOC is hate-offensive vs non Non hate-offensive, which is the same as OLID's. This allows us to directly map hate-offensive to OLID's OFF class and non hate-offensive to NOT class.

HatE is the official dataset at SemEval-2019 Task 5 (HatEval), which focuses on hate speech against migrants and women. The first level of annotation contains two classes, hate speech or not, which can be mapped directly to OLID's OFF and NOT categories.

HateX is a dataset collected for the explainability of hate speech. It contains both token- and post-level annotation of Twitter and Gab posts. Post-level annotations have three classes: Hateful, Offensive, and Normal. We map Hateful and Offensive classes to OFF class and Normal to NOT class.

OHS is a dataset collected from Reddit with the goal of studying interventions in conversations containing hate speech. Full conversations/threads have been retrieved and annotated at the post-level as hateful or not hateful, which we map to OFF and NOT classes correspondingly.

OLID is the official dataset of the SemEval-2019 Task 6 (OffensEval) [47]. It contains data from Twitter annotated with a three-level hierarchical annotation which we described before. We adopt the labels in OLID level A as our classification labels.

TCC is the Toxic Comment Classification dataset. TCC was created for the Kaggle competition with the same name. The dataset contains Wikipedia comments with various classes such as toxic, obscene, insult, and threat merged in the OLID OFF class. The rest of the instances were mapped to the NOT class.

TRAC is the dataset used in the TRAC shared task 2020 [23]. It focuses on aggression detection with three classes: overtly aggressive and covertly aggressive merged as OFF and non-aggressive which corresponds to the NOT class used in OLID. Finally, TRAC is the most heterogeneous dataset we used in terms of data sources containing posts from Facebook, Twitter, and YouTube.

3.2 *GenOffense* Properties

We highlight the following generalization types that *GenOffense* benchmark tests. These are shown as crucial generalization types by [21].

Platform Shift GenOffense benchmarks contains datasets from six different social media platforms. While most of the datasets are based on Twitter, *GenOffense* has datasets that are based on other social media platforms such as Facebook and Reddit. Therefore, *GenOffense* benchmark evaluates how the models can handle different platforms.

Language Shift. The datasets included in *GenOffense* range from 2017 to 2021. The language that was used to convey offense can be different from 2017 to 2021.

Therefore, *GenOffense* benchmark tests how the models can handle language shift.

Task Shift. As we mentioned before, these datasets contained different tasks such as aggression detection, hate speech detection and offensive language detection. As a result, *GenOffense* reflects these tasks and a model that can perform well in *GenOffense* will generalize well across different sub-tasks.

Topic Shift. Different datasets have been collected with different goals in mind depending on the 'offensive language detection sub-task' they address. Therefore, each dataset in *GenOffense* has different topics, and the models will be evaluated on how well they can handle different topics in the offensive language domain.

Finally, upon acceptance of this paper, *GenOffense* will be made available as an online platform where researchers can submit the model predictions and evaluate how the model generalizes over different datasets.

4   Unsupervised Offensive Language Detection Models

The following public models and APIs are evaluated in the test sets of *GenOffense* without any training or fine-tuning.

Table 2. Macro F1 score of the publicly available offensive language detection models. **Row I** shows public APIs/ models, **row II** shows the results of adapting transformers and **row III** shows the results for LLMs. The **Average** column shows the average score of all the experiments.

	Models	AHSD	HASOC	HatE	HateX	OHS	OLID	TCC	TRAC	Avg
I	Perspective	0.8603	0.6487	0.5340	0.6688	0.5578	0.7691	0.9228	0.6847	0.7058
	ToxicBERT	0.7430	0.6522	0.5283	0.6361	0.5416	0.7765	0.9606	0.6906	0.6911
II	BERT	0.1473	0.3951	0.4002	0.2986	0.4456	0.4328	0.3741	0.3961	0.3612
	fBERT	0.4589	0.3149	0.4075	0.3807	0.2403	0.3357	0.4178	0.3230	0.3599
	HateBERT	0.5335	0.4733	0.4968	0.5405	0.4466	0.4984	0.5945	0.3467	0.4913
III	Davinci-003	0.8152	0.5909	0.4881	0.6075	0.4780	0.7401	0.7617	0.7454	0.6534
	Falcon-7B	0.7406	0.6049	0.6033	0.6106	0.5291	0.7456	0.6178	0.7152	0.6458
	T0	0.6972	0.5005	0.4195	0.5631	0.5160	0.4907	0.6008	0.7126	0.5625
	MPT-7B	0.5313	0.3571	0.3621	0.5240	0.2832	0.3703	0.2998	0.7466	0.4343

4.1   Methods

Public APIs/ Models. We evaluate **Perspective API** [24] and **ToxicBERT** [11][2]. **Perspective API** is a free API developed by Google Jigsaw, that leverages machine learning to identify toxic comments. This API was first trained using a

---

[2] ToxicBERT is available at https://huggingface.co/unitary/toxic-bert.

BERT [13] model, which is then distilled into monolingual CNN based models. The model was mainly trained on the TCC dataset, which we also included in *GenOffense*. The model has six attributes, toxicity, severe toxicity, identity attack, insult, profanity, and threat. The model generates a score between 0 and 1 for each of these attributes. For each test dataset, we get all the attribute scores for each instance. If any of the attributes have a value greater than 0.5, we classify that instance as OFF, else it is classified as NOT.

We also evaluate **ToxicBERT** on *GenOffense*. ToxicBERT is a BERT model trained primarily on the TCC dataset. The model is a multi-label classification model with six labels similar to Perspective API. We follow a similar approach to Perspective API to convert the ToxicBERT outputs into OFF and NOT classes.

Adapting Transformers. We evaluate different general-purpose transformer models; BERT, and two domain-specific transformer models; fBERT [37] and Hate-BERT [9] on offensive language identification using an unsupervised approach. We classify a test sentence as positive or negative, where the positive label represents the NOT class and the negative represents the OFF class. We concatenate the last four hidden states returned by the model as the representative embeddings for the test sentence and the labels. We then find the cosine similarity between the representative embeddings of the labels and that of the test sentence. Finally, the sentence is assigned the label with the highest cosine similarity score.

Prompting LLMs. Finally, we evaluate how LLMs perform in *GenOffense* benchmark, a recent trend as we discussed before. We use the following prompt to get a response from LLMs.

> *Comments containing any form of non-acceptable language (profanity) or a targeted offense, which can be veiled or direct, are offensive comments. This includes insults, threats, and posts containing profane language or swear words. Comments that do not contain offense or profanity are not offensive. Is this comment offensive or not? Comment:*

We use several LLMs for prompting. We first use Davinci-003 through OpenAI API. Additionally, we use MPT-7B-Instruct, Falcon-7B-Instruct and T0-3B [36]. All of these models are available in HuggingFace[3] [45], and we use the LangChain implementation.

4.2   Results

The results of the aforementioned models are shown in Table 2. Public APIs/models generally performed well on *GenOffense* compared to the other two methods. However, LLMs also provide competitive results. From the LLMs, Davinci-003 performs best, closely followed by Falcon-7B. It is clear that recent

---

[3] MPT-7B-Instruct is available at https://huggingface.co/mosaicml/mpt-7b-instruct, Falcon-7B-Instruct is available at https://huggingface.co/tiiuae/falcon-7b-instruct and T0-3B is available at https://huggingface.co/bigscience/T0_3B.

LLMs produce better results on *GenOffense*. Overall, Perspective API performed best on the *GenOffense* benchmark. It provided the best results for six datasets out of eight and had the highest overall average.

Most of the models show inconsistent results on the datasets. Particularly, all the models do not perform well on HatE and OHS datasets which indicates that these models do not generalize well across different tasks and platforms.

5   Training Offensive Language Detection Models

In this section, we evaluate the *supervised* ML models on *GenOffense* benchmarks. We train the following ML models under different settings on the training sets in *GenOffense* benchmark and evaluate on the test sets.

LSTM. We experiment with a bidirectional Long Short-Term-Memory (BiLSTM) model, which we adapted from the baseline in OffensEval 2019 [47]. The model consists of *(i)* an input embedding layer with fasttext embedding [6], *(ii)* a bidirectional LSTM layer, and *(iii)* an average pooling layer of input features. The concatenation of the LSTM layer and the average pooling layer is further passed through a dense layer, whose output is ultimately passed through a *softmax* to produce the final prediction. We used updatable embeddings learned by the model during training as the input.

Transformers. We also use transformers as a classification model, which have achieved state-of-the-art on a variety of offensive language identification tasks. From an input sentence, transformers compute a feature vector $h \in \mathbb{R}^d$, upon which we build a classifier for the task. For this task, we implemented a softmax layer, i.e., the predicted probabilities are $y^{(B)} = \text{softmax}(Wh)$, where $W \in \mathbb{R}^{k \times d}$ is the softmax weight matrix and k is the number of labels. For the experiments, we use the bert-large-cased and domain-specific fBERT [37] and HateBERT [9] available in HuggingFace [45].

5.1   Model Configuration

For LSTM, we used a Nvidia Tesla k80 to train the models. We divided the dataset into a training set and a validation set using 0.8:0.2 split. We performed *early stopping* if the validation loss did not improve over 10 evaluation steps. For the LSTM model we used the same set of configurations mentioned in Table 3 in all the experiments. All the experiments were conducted for three times and the mean value is taken as the final reported result.

For transformers models, we used a GeForce RTX 3090 GPU to train the models. We divided the dataset into a training set and a validation set using a 0.8:0.2 split. For transformer models, we used the same set of configurations mentioned in Table 4 in all the experiments. We performed *early stopping* if the validation loss did not improve over 10 evaluation steps. All the experiments were conducted three times and the mean value is taken as the final reported result.

Table 3. LSTM Parameter Specifications.

Parameter	Value
batch size	64
epochs	3
first dense layer units	256
learning rate	1e-4
LSTM units	64
max seq. length	256

Table 4. BERT Parameter Specifications.

Parameter	Value
adam epsilon	1e-8
batch size	64
epochs	3
learning rate	1e-5
warmup ratio	0.1
warmup steps	0
max grad norm	1.0
max seq. length	256
gradient accumulation steps	1

5.2   Results

We use multiple strategies to answer the three **RQ**s considering generalizability with respect to training and testing data.

We address training set variation by training the three models in the following settings:

1 to 1 We train a separate machine learning model on each of the eight training sets. We then evaluate the trained model on each of the eight test sets in isolation.

All -1. We concatenate all training sets except one and train a single machine learning model. We then evaluate the model on the test set of that particular dataset that was left out.

All We concatenate the training sets of all the datasets and trained a single machine learning model. We then evaluate the model on each testing set of all eight datasets in *GenOffense*.

Few to 1. We also perform progress tests. We randomly selected 1000, 2000, 3000 etc. instances from each of the eight training sets and train separate machine learning models. We then evaluate the trained model on each of the eight test sets in isolation.

Table 5. Macro F1 score of the offensive language detection models. The **Training Dataset(s)** shows the training dataset while the subsequent columns show the results for each test set. The **Average** column shows the average score of all the experiments.

	Train Dataset(s)	AHSD	HASOC	HatE	HateX	OHS	OLID	TCC	TRAC	Avg
LSTM	AHSD	0.8872	0.5465	0.3735	0.4903	0.3757	0.4005	0.4598	0.5809	0.5143
	HASOC	0.4336	0.6539	0.5388	0.5339	0.5503	0.5832	0.5756	0.4056	0.5343
	HatEval	0.6605	0.5200	0.5825	0.5266	0.5479	0.5413	0.5212	0.4991	0.5498
	HateX	0.5531	0.4623	0.3976	0.7091	0.4943	0.5193	0.3710	0.4710	0.4927
	OHS	0.1487	0.3936	0.5234	0.5309	0.6984	0.4670	0.2604	0.8117	0.4793
	OLID	0.6391	0.6224	0.5283	0.5477	0.5636	0.7124	0.6366	0.7473	0.6247
	TCC	0.5432	0.4756	0.5581	0.5497	0.5841	0.5711	0.7930	0.5587	0.5791
	TRAC	0.1800	0.4058	0.5105	0.4868	0.5376	0.5457	0.5460	0.6853	0.4872
	All	0.8689	0.6134	0.4849	0.6775	0.6236	0.6754	0.6537	0.7490	0.6681
	All-1	0.8675	0.5745	0.4539	0.5842	0.4957	0.5569	0.6491	0.6231	0.6006
BERT	AHSD	0.9268	0.6300	0.5279	0.5867	0.5179	0.6991	0.8188	0.6278	0.6657
	HASOC	0.6203	0.7585	0.5850	0.5550	0.5798	0.4925	0.6541	0.5495	0.5993
	HatEval	0.6122	0.4418	0.5880	0.4966	0.5795	0.5795	0.6240	0.6884	0.6012
	HateX	0.5690	0.6049	0.6322	0.7829	0.6167	0.5049	0.7214	0.5382	0.6212
	OHS	0.1960	0.4100	0.4567	0.3875	0.7745	0.4225	0.5048	0.3920	0.4430
	OLID	0.6857	0.6366	0.5296	0.6206	0.5725	0.8074	0.8451	0.7402	0.6797
	TCC	0.7210	0.6448	0.5241	0.6297	0.5677	0.7453	0.8805	0.6678	0.6726
	TRAC	0.6225	0.6260	0.5757	0.6122	0.5579	0.6916	0.7692	0.8596	0.6643
	All	0.9257	0.7506	0.7412	0.7718	0.7263	0.7449	0.8578	0.7793	0.7872
	All-1	0.3805	0.5346	0.5557	0.5771	0.5652	0.6680	0.7829	0.6529	0.5896
HateBERT	AHSD	0.9299	0.6248	0.5367	0.6051	0.5311	0.6425	0.7313	0.5813	0.6478
	HASOC	0.5704	0.6529	0.5852	0.5873	0.5563	0.6666	0.6101	0.7247	0.6192
	HatEval	0.7033	0.4974	0.4748	0.5852	0.5392	0.4814	0.6336	0.5421	0.5571
	HateX	0.5276	0.5954	0.5765	0.7724	0.5805	0.4981	0.6547	0.5609	0.5958
	OHS	0.2149	0.4024	0.3785	0.3102	0.7591	0.4189	0.4982	0.3651	0.4184
	OLID	0.7610	0.6239	0.5465	0.5971	0.4855	0.7811	0.7822	0.6317	0.6511
	TCC	0.7885	0.6286	0.5376	0.6386	0.5502	0.7107	0.8408	0.6493	0.6680
	TRAC	0.2597	0.5083	0.5164	0.5614	0.5715	0.5838	0.6392	0.8239	0.5580
	All	0.9174	0.6180	0.6076	0.7803	0.7009	0.7278	0.7955	0.6789	0.7283
	All-1	0.4844	0.5487	0.5760	0.6073	0.5751	0.5257	0.7861	0.6625	0.5957
fBERT	AHSD	0.9241	0.6365	0.5318	0.6246	0.5096	0.6918	0.8032	0.5482	0.6587
	HASOC	0.6912	0.6753	0.5386	0.6343	0.5510	0.7778	0.8226	0.7443	0.6794
	HatEval	0.6810	0.5332	0.4917	0.5724	0.5693	0.5599	0.6893	0.6714	0.5960
	HateX	0.5276	0.5954	0.6263	0.7840	0.5784	0.5252	0.7156	0.5991	0.6189
	OHS	0.1615	0.3935	0.4649	0.5720	0.7558	0.5572	0.6905	0.5382	0.5167
	OLID	0.7239	0.6572	0.5474	0.6217	0.5217	0.7838	0.8234	0.7524	0.6789
	TCC	0.7497	0.6545	0.5243	0.6303	0.5452	0.7458	0.8486	0.6753	0.6717
	TRAC	0.5757	0.5975	0.5565	0.6277	0.5389	0.7049	0.8290	0.8416	0.6589
	All	0.9201	0.6338	0.5727	0.7768	0.7102	0.7350	0.8389	0.7677	0.7444
	All-1	0.3516	0.5590	0.5588	0.6369	0.5685	0.5854	0.7889	0.6601	0.5887

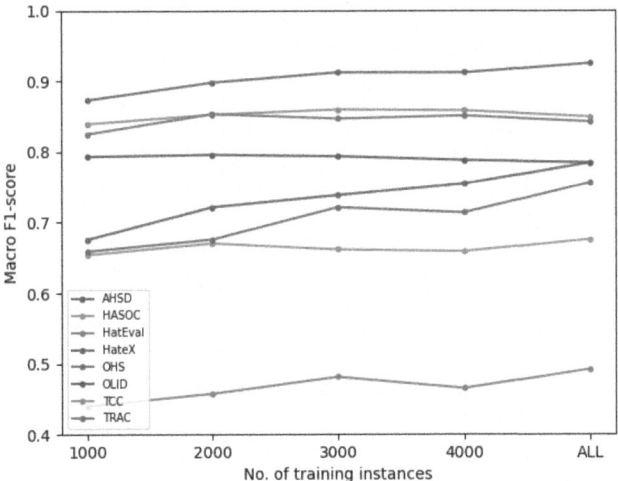

Fig. 1. Few-shot Learning Results for BERT

We present the results of the aforementioned strategies in Table 5 and Fig. 1 in terms of Macro F1 score. The transformer models outperform the LSTM for all tested dataset combinations. This is in line with the findings of popular competitions such as HatEval and OffensEval. However, domain-specific models such as fBERT and HateBERT did not outperform BERT in average scores of *GenOffense*. This can be because both of these models are fine-tuned on platform-specific data. Unsurprisingly the **all** strategy, achieves the best results in all four classifiers. However, few-shot results in Fig. 1 suggest and more training instances do not improve the average Macro F1 score of *GenOffense*. Furthermore, the **all -1** strategy was outperformed by many of the individual datasets suggesting that simply using a large dataset does not always result in better generalizability.

In terms of the individual dataset performance, models trained on OLID yielded the highest generalization followed by TCC. This is due to the general nature of these two datasets covering multiple types of offensive content rather than focusing on a particular type of offensive content (e.g. hate speech). AHSD also provided good generalization, likely due to the presence of both hate speech and general offensive language in the dataset. On the other hand, models trained on OHS yielded the worst performance. This can be explained by the platform-specificity of the dataset, as OHS is the only Reddit dataset in this collection.

5.3   Test Set Combination

We also look at the performance of the models on a single test set combining all individual test sets in *GenOffense*. We use a separate BERT model on each of the eight training sets and tested them on the concatenated test set. We present the results obtained on the consolidated test set in terms of Macro and Weighted F1 Table 6.

Table 6. BERT results for the combined test set in terms of Macro F1 and Weighted F1. Best results in bold.

Train Dataset	Macro F1	Weighted F1
AHSD	0.7348	0.7348
HASOC	0.6722	0.6743
HatE	0.6210	0.6239
HateX	0.6879	0.6899
OHS	0.4247	0.4348
OLID	**0.7543**	**0.7551**
TCC	0.7064	0.7060
TRAC	0.6467	0.6492

The results indicate that models trained on OLID offers the best performance on the combined test set, followed by AHSD, and TCC while OHS delivers the lowest performance by a very large margin. This is in line with the results obtained using individual test sets.

6   Conclusion

This paper introduced the first generalization benchmark for offensive language detection; *GenOffense*. We also presented a comprehensive evaluation of the generalizability of different computational models, including recently released LLMs. We hope that our findings motivate the community to further explore the question of generalizability as argued by other recent studies [15, 16].

We revisit the research questions posed in the introduction:

- **RQ1 - Generalizability:** Despite being popular, LLMs did not perform well in the *GenOffense* benchmark. APIs, such as Perspective, showed better generalizability. In the supervised setting, models trained on OLID, AHSD, and TCC provided the best generalizability to other datasets. This can be explained by their focus on general offensive language (in the case of OLID and TCC) and the presence of both hate speech and general offensive (in the case of AHSD), which is reflected in their annotation models. More specific datasets, such as HatEval, which focuses on women and migrants, displayed lower results. Finally, OHS, the only Reddit dataset, achieved the lowest performance, suggesting that the domain has a substantial impact on performance (see **RQ3**).
- **RQ2 - Dataset Size:** We observed that more data does not always result in better generalizability. The few-shot experiments showed that adding more training instances did not provide better generalizability. Even though the "All" strategy achieved the best performance for all datasets, the "All-1"

strategy achieved performance lower than most datasets in isolation. Therefore we have not found a direct correlation between generalizability and training dataset size in our experiments. The question of dataset size requires further investigation.

– **RQ3 - Domain Specificity:** Models trained on OHS, the only Reddit dataset in the collection, achieved the lowest performance of all datasets, suggesting that the domain plays an important role in generalizability. OHS is not the smallest dataset tested in our experiments, therefore we believe that the low performance is due to the specificity of their source material (Reddit) rather than its size. We would like to further investigate this by running more dataset ablation experiments.

In future work, we would like to extend *GenOffense* benchmark to adversarial test sets using popular augmentation techniques such as random insertion and random deletion. This will provide the opportunity for the researchers to explore probing in offensive language detection models. We believe this would provide us with even more insights into the generalizability of the datasets and the robustness of the models. Finally, we would like to extend *GenOffense* to support multilingual offensive language datasets and replicate these experiments for different languages. Such multilingual benchmarks will be useful for many real-world applications.

Acknowledgements. We would like to thank the anonymous reviewers for their positive and valuable feedback. We further thank the creators of the datasets used in this paper for making the datasets publicly available for our research.

The experiments in this paper were conducted on the High End Computing (HEC) Cluster at Lancaster University, which is funded through a combination of central funding and contributions from individual research grants. The experiments were designed in UCREL-HEX [42], which is a collection of GPU equipped hosts at the School of Computing and Communications, Lancaster University.

Marcos Zampieri is partially supported by a grant from the Virginia Commonwealth Cyber Initiative (CCI) award number N-4Q24-009 .

References

1. Aggarwal, P., et al.: HateProof: are hateful meme detection systems really robust? In: Proceedings of TheWebConf (2023)
2. Arango, A., Pérez, J., Poblete, B.: Hate speech detection is not as easy as you may think: a closer look at model validation. In: Proceedings of SIGIR, pp. 45–54 (2019)
3. Bannink, R., Broeren, S., van de Looij-Jansen, P.M., de Waart, F.G., Raat, H.: Cyber and traditional bullying victimization as a risk factor for mental health problems and suicidal ideation in adolescents. PloS one **9**(4) (2014)

4. Basile, V., et al.: Semeval-2019 task 5: Multilingual detection of hate speech against immigrants and women in twitter. In: Proceedings of SemEval (2019)
5. Bhattacharya, S., et al.: Developing a multilingual annotated corpus of misogyny and aggression. In: Proceedings of TRAC (2020)
6. Bojanowski, P., Grave, E., Joulin, A., Mikolov, T.: Enriching word vectors with subword information. Trans. Associat. Comput. Linguist. **5**, 135–146 (2017)
7. Borkan, D., Dixon, L., Sorensen, J., Thain, N., Vasserman, L.: Nuanced metrics for measuring unintended bias with real data for text classification. In: Companion Proceedings of WWW, pp. 491–500 (2019)
8. Bucur, A.M., Zampieri, M., Dinu, L.P.: An exploratory analysis of the relation between offensive language and mental health. In: Findings of the ACL (2021)
9. Caselli, T., Basile, V., Mitrović, J., Granitzer, M.: Hatebert: Retraining bert for abusive language detection in english. arXiv preprint arXiv:2010.12472 (2020)
10. Chronopoulou, A., Peters, M., Dodge, J.: Efficient hierarchical domain adaptation for pretrained language models. In: Proceedings of NAACL (2022)
11. Davidson, T., Bhattacharya, D., Weber, I.: Racial bias in hate speech and abusive language detection datasets. In: Proceedings of ALW (2019)
12. Davidson, T., Warmsley, D., Macy, M.W., Weber, I.: Automated hate speech detection and the problem of offensive language. In: Proceedings of ICWSM (2017)
13. Devlin, J., Chang, M.W., Lee, K., Toutanova, K.: BERT: pre-training of deep bidirectional transformers for language understanding. In: Proceedings of NAACL (2019)
14. Elangovan, A., He, J., Verspoor, K.: Memorization vs. generalization: quantifying data leakage in NLP performance evaluation. In: Proceedings of EACL (2021)
15. Fortuna, P., Soler, J., Wanner, L.: Toxic, hateful, offensive or abusive? what are we really classifying? an empirical analysis of hate speech datasets. In: Proceedings of the 12th Language Resources and Evaluation Conference, pp. 6786–6794 (2020)
16. Fortuna, P., Soler-Company, J., Wanner, L.: How well do hate speech, toxicity, abusive and offensive language classification models generalize across datasets? Inform. Process. Manag. **58**(3), 102524 (2021)
17. Ghosh, S., Baker, D., Jurgens, D., Prabhakaran, V.: Detecting cross-geographic biases in toxicity modeling on social media. In: Proceedings of W-NUT (2021)
18. Gröndahl, T., Pajola, L., Juuti, M., Conti, M., Asokan, N.: All you need is "love": evading hate speech detection. In: Proceedings of AISec (2018)
19. Hamm, M.P., et al.: Prevalence and effect of cyberbullying on children and young people: a scoping review of social media studies. JAMA Pediatrics **169**(8), 770–777 (2015)
20. Huang, F., Kwak, H., An, J.: Is chatgpt better than human annotators? potential and limitations of chatgpt in explaining implicit hate speech. In: Proceedings of WWW (2023)
21. Hupkes, D., et al.: State-of-the-art generalisation research in nlp: a taxonomy and review. arXiv preprint arXiv:2210.03050 (2022)
22. Jambor, D., Bahdanau, D.: LAGr: label aligned graphs for better systematic generalization in semantic parsing. In: Proceedings of ACL (2022)
23. Kumar, R., Ojha, A.K., Malmasi, S., Zampieri, M.: Evaluating aggression identification in social media. In: Proceedings of TRAC (2020)
24. Lees, A., et al.: A new generation of perspective api: efficient multilingual character-level transformers. In: Proceedings of KDD (2022)
25. Li, L., Fan, L., Atreja, S., Hemphill, L.: " hot" chatgpt: the promise of chatgpt in detecting and discriminating hateful, offensive, and toxic comments on social media. arXiv preprint arXiv:2304.10619 (2023)

26. Linzen, T.: How can we accelerate progress towards human-like linguistic generalization? In: Proceedings ACL (2020)
27. López-Meneses, E., Vázquez-Cano, E., González-Zamar, M.D., Abad-Segura, E.: Socioeconomic effects in cyberbullying: Global research trends in the educational context. Inter. J. Environmental Res. Public Health **17**(12) (2020)
28. Malmasi, S., Zampieri, M.: Detecting hate speech in social media. In: Proceedings of RANLP (2017)
29. Mandl, T., Modha, S., Kumar M, A., Chakravarthi, B.R.: Overview of the hasoc track at fire 2020: hate speech and offensive language identification in Tamil, Malayalam, Hindi, English and German. In: Proceedings of FIRE (2020)
30. Mathew, B., Saha, P., Yimam, S.M., Biemann, C., Goyal, P., Mukherjee, A.: HateXplain: a benchmark dataset for explainable hate speech detection. In: Proceedings of AAAI (2021)
31. Moisio, A., Creutz, M., Kurimo, M.: Evaluating morphological generalisation in machine translation by distribution-based compositionality assessment. In: Proceedings of NoDaLiDa (2023)
32. Qian, J., Bethke, A., Liu, Y., Belding, E., Wang, W.Y.: A benchmark dataset for learning to intervene in online hate speech. In: Proceedings of EMNLP (2019)
33. Ranasinghe, T., Zampieri, M.: Multilingual offensive language identification with cross-lingual embeddings. In: Proceedings of EMNLP (2020)
34. Ranasinghe, T., Zampieri, M.: Mudes: Multilingual detection of offensive spans. In: Proceedings of NAACL (2021)
35. Ranasinghe, T., Zampieri, M., Hettiarachchi, H.: BRUMS at HASOC 2019: deep learning models for multilingual hate speech and offensive language identification. In: Proceedings of FIRE (2019)
36. Sanh, V., et al.: Multitask prompted training enables zero-shot task generalization. In: Proceedings of ICLR (2022)
37. Sarkar, D., Zampieri, M., Ranasinghe, T., Ororbia, A.: fbert: a neural transformer for identifying offensive content. In: Findings of EMNLP (2021)
38. Sarkar, R., KhudaBukhsh, A.R.: Are chess discussions racist? an adversarial hate speech data set. In: Proceedings of AAAI (2021)
39. Satapara, S., et al.: Overview of the hasoc subtrack at fire 2022: hate speech and offensive content identification in english and indo-aryan languages. In: Proceedings of FIRE (2023)
40. Sharma, D., Buduru, A.B.: FAtNet: cost-effective approach towards mitigating the linguistic bias in speaker verification systems. In: Findings of ACL: NAACL (2022)
41. Vidgen, B., Nguyen, D., Margetts, H., Rossini, P., Tromble, R.: Introducing CAD: the contextual abuse dataset. In: Proceedings of NAACL (2021)
42. Vidler, J., Rayson, P.: UCREL - Hex; a shared, hybrid multiprocessor system. https://github.com/UCREL/hex, accessed: 2024
43. Waseem, Z., Davidson, T., Warmsley, D., Weber, I.: Understanding abuse: a typology of abusive language detection subtasks. In: Proceedings of ALW (2017)
44. Weerasooriya, T.C., Dutta, S., Ranasinghe, T., Zampieri, M., Homan, C.M., KhudaBukhsh, A.R.: Vicarious offense and noise audit of offensive speech classifiers (2023)
45. Wolf, T., et al.: Transformers: state-of-the-art natural language processing. In: Proceedings of EMNLP (2020)
46. Zampieri, M., Malmasi, S., Nakov, P., Rosenthal, S., Farra, N., Kumar, R.: Predicting the type and target of offensive posts in social media. In: Proceedings of NAACL (2019)

47. Zampieri, M., Malmasi, S., Nakov, P., Rosenthal, S., Farra, N., Kumar, R.: SemEval-2019 task 6: identifying and categorizing offensive language in social media (OffensEval). In: Proceedings of SemEval (2019)
48. Zampieri, M., et al.: Target-based offensive language identification. In: Proceedings of ACL (2023)
49. Zampieri, M., et al.: SemEval-2020 task 12: multilingual offensive language identification in social media (OffensEval 2020). In: Proceedings of SemEval (2020)
50. Zampieri, M., Rosenthal, S., Nakov, P., Dmonte, A., Ranasinghe, T.: Offenseval 2023: offensive language identification in the age of large language models. Nat. Lang. Eng. **29**(6), 1416–1435 (2023)
51. Zia, H.B., Castro, I., Zubiaga, A., Tyson, G.: Improving zero-shot cross-lingual hate speech detection with pseudo-label fine-tuning of transformer language models. In: Proceedings of ICWSM (2022)

Online Social Community Neighborhood Formation

Jiarui Wang[1]([✉]) [iD], George Barnett[1] [iD], Norman Matloff[1] [iD], and S. Felix Wu[2] [iD]

[1] University of California, Davis, CA 95616, USA
{jrwwang,gabarnett,nsmatloff}@ucdavis.edu
[2] National Cheng-Kung University, Tainan City, Taiwan
sfelixwu@gs.ncku.edu.tw

Abstract. Online social networks (OSNs) provide community platforms that engage users. The public page is a popular example. These pages are connected through "like" relationships, creating online community networks and neighborhoods. We investigated the pivotal features influencing link formation and neighborhood structuring within the page graph by exploring a series of potential features, both graph-based and content-based. Our methodology combines node similarity and Graph Neural Networks to perform link prediction. We identified the page state label as the single most accurate predictor in link prediction tasks, which also is the most efficient feature with the smallest number of classes. Moreover, we observe that augmenting the page state label feature with the page node degree and page city population features further enhances link prediction accuracy. Page location label shows a strong effect on pages connecting with their neighbors.

Keywords: Online social networks · Network formation · Location · Link prediction

1 Introduction

In the past decade, online social networks (OSNs) have witnessed exponential growth, attracting billions of users worldwide. These platforms empower individuals to create profiles, establish connections, and share content, offering unparalleled access without the traditional constraints of time and location associated with offline social groups. Users can effortlessly connect with others globally who share similar interests, fostering the rapid expansion of OSNs.

A prevalent activity on these online social platforms involves individual users setting up personal profiles, connecting with friends or strangers, and sharing content. These individuals form the basis of online social networks, with their numbers indicative of the platforms' business potential, as they represent prospective consumers for a wide array of products and services. This vast user base attracts a variety of entities, including businesses, non-profit organizations, and governmental bodies, all seeking to leverage these platforms for their respective interests. These entities, along with individual users, establish various online

L. M. Aiello et al. (Eds.): ASONAM 2024, LNCS 15211, pp. 287–304, 2025.
https://doi.org/10.1007/978-3-031-78541-2_18

social communities to cater to specific interests. These communities range from corporate and non-profit organization pages to user-created groups focusing on shared interests like neighborhood activities, workplace connections, and hobbies such as animal enthusiasts.

Numerous offline groups and communities have established their presence online through information pages or discussion forums. Additionally, the internet has seen the birth of myriad communities and groups that operate exclusively online, without any offline interactions. The rapid growth and sheer volume of these online social communities are remarkable, especially considering their relatively brief history. Unlike their offline counterparts, online communities face no constraints related to time or location, allowing for unlimited connections and interactions with other online entities. This paper delves into the dynamics of connections between various online social communities.

This study specifically focuses on public pages, which serve as a platform for disseminating information, facilitating user discussions, spreading news, and promoting businesses or public relations activities. Like individual users on social media, these pages can like or follow other pages, creating a network of connections among online social communities. This network, in turn, forms a vast graph of online social community interactions. Our research aims to uncover the pivotal factors that influence these connections and the development of neighborhoods within the online social community landscape.

In this study, we aim to contribute to the understanding of online social communities by investigating a range of page features to determine their impact on the formation of page neighborhoods. This is achieved through a methodology that applies link prediction techniques to each individual feature. We identified the page state label as the single most accurate predictor in link prediction tasks, which also is the most efficient feature with the smallest number of classes. Additionally, we find that a combination of features specifically the page state label, page node degree, and page city populationyields the best performance in link prediction accuracy.

2   Data Description

2.1   Data Acquisition

We use the same public page data as [6]. This dataset encompasses a broad spectrum of page metadata, including identifiers, names, descriptions, categories, as well as geographical data like country and city, alongside relational data such as liked pages. Notably, this collection process ensures the exclusion of any user-specific private information. The methodology for data acquisition relied on snowball sampling [11], initiating from a set of popular public pages and progressively encompassing pages liked by these initial nodes. This approach organically constructs a directed graph representation of the page network.

2.2   Data Cleaning

In this directed page-likes graph, each node represents a page, with outgoing edges indicating pages liked by this page. The graph comprises 61,263,729 pages connected by 789,494,545 edges. However, only 30.8% of these pages, totaling 18,895,994, have location information (country and city) specified by their page managers. We consider location information a key feature for predicting links between pages. Our analysis is centered on the subgraph comprising all U.S. pages, given that the U.S. encompasses the largest number of pages among all countries in our dataset.

The page-likes graph is constructed exclusively from ground truth data, comprising 6,194,277 pages with verified city locations within the United States and connections between them. We exclude 55,069,452 pages and their associated edges either located outside the United States or lacking city location information. Consequently, the resultant subgraph of U.S. pages exhibits disconnected components, primarily due to the exclusion of some connecting pages. The largest connected component encompasses 5,873,395 pages and 84,480,575 edges. Our analysis prioritizes this component due to its significant size relative to others.

All pages within our U.S. graph have their city locations in the United States, as listed by their managers. Among these, 36.6% of the pages are associated with cities that have unique names across all 50 states, making their city and state locations determinable. We refer to these as deterministic pages. Conversely, the remaining 63.4% of pages are linked to cities with names that duplicate across multiple states; we classify these as non-deterministic pages. Our study concentrates on the deterministic pages.

3   Page-Likes Link Prediction

3.1   Link Prediction

Link prediction spans various research fields, including statistics, network science, data mining, and machine learning, focusing on predicting the presence of links between nodes in a network. This task aligns with real-world applications such as predicting social connections in social networks or recommending products in user-product graphs.

From the social network perspective, Liben-Nowell and Kleinberg have developed link prediction techniques based on measures for analyzing the "proximity" of the nodes in a network [12]. The nodes within the "proximity" in the network are similar in some sense, leveraging the concept of homophily. Therefore, these nodes are more likely to interact with each other and be connected by edges. Thus, the most commonly used link prediction algorithms are similarity-based algorithms [14].

Given our data's graph structure, where edges represent "likes" between pages, graph-based algorithms are particularly suitable for link prediction. Graph-based representation learning effectively addresses this by encoding node features and graph topology into vector representations. These vectors are then used to calculate scores indicating the likelihood of edge formation between node pairs. Existing edges (positive edges) are labeled as 1, while non-existing edges (negative edges), introduced through uniform negative sampling, are labeled as 0 [13]. Our use of link prediction algorithms aims to identify key factors influencing the formation of page neighborhoods.

3.2 GraphSAINT

Graph Convolutional Networks (GCNs) face scalability challenges due to the necessity of updating all feature vectors within each iteration, making them less efficient for large graphs. To address these limitations, both GraphSAGE and GraphSAINT models adopt node sampling strategies, albeit through differing approaches. GraphSAGE employs uniform sampling to select a fixed number of neighboring nodes for each node in every layer and iteration. Conversely, GraphSAINT samples a sub-graph of the whole graph by nodes' importance as the mini-batch in each iteration, subsequently applying a GCN-like model on this sub-graph. This method effectively reduces the size of the original graph to a more manageable sub-graph, significantly enhancing training efficiency and time compared to GraphSAGE. Our prior research [6] has shown the GraphSAINT model to exhibit superior performance on the page graph, leading us to select GraphSAINT for encoding node representation vectors within the graph.

3.3 Feature Selection

In our study, we delve into the dynamics behind the "likes" relationships among public pages to unveil the mechanisms underlying online social community neighborhoods. This investigation is framed as a link prediction challenge, aiming to identify features that yield precise predictions within a directed page graph.

We introduce an array of candidate features for utilization within graph neural networks to forecast page-likes connections. By evaluating the predictive accuracy of these diverse features, we uncover the pivotal elements influencing the formation of page edges and neighborhoods. These features are categorized into two primary types:

1. **Topology-Related Features:** These features relate to the page's position and role within the graph's structure, such as its degree, or the network information for its neighborhood, such as state neighborhood distribution.
2. **Community-Specific Features:** These features relate to the intrinsic attributes of the page community, including the page's category, the population of the page's city, geographic coordinates of the page's city, and labels for both the city and state of the page.

By analyzing the effectiveness of these features in link prediction, we aim to elucidate the foundational factors that drive the establishment of online social community neighborhoods.

Constant Feature. Graph neural networks (GNNs) harness both node features and the graph's structural information to facilitate learning. The quality and informativeness of node features are crucial as they encapsulate the attributes of the nodes. Conversely, edge connections unveil the graph's structural intricacies. To enable a baseline comparison, we employ a constant value of 1 as the node feature across all nodes. This approach restricts the model to learning exclusively from the graph's topology and its connections, rendering all nodes indistinguishable based on their features.

The principle of homophily suggests that similar nodes tend to be closer or directly linked within a graph [15]. Our adoption of a uniform feature stems from the hypothesis that pages in proximity within the graph share certain similarities, thereby increasing their likelihood of forming connections. This method provides a foundational comparison, emphasizing the role of graph structure over individual node attributes in predicting linkages.

Page Degree. The degree of a page, defined as its number of neighbors, signifies its connectivity within the page-likes graph. The degree values range from a minimum of 1 to a maximum of 51,045. To visualize this distribution, we present the degree distribution across pages in Fig. 1a and 1b. Figure 1a uses linear scale and Fig. 1b uses logarithmic scale. The linear scale plots show an axis-aligned pattern, while the logarithmic scale plots show a heavy-tailed pattern. This pattern aligns with the degree distribution observed in other real-world networks, such as the MSN messaging network [16], indicating adherence to a log-normal distribution.

Node degree is an often used feature in network analysis. Therefore, we propose the degree of the page node as one candidate feature. The page graph is a directed graph. Hence, we use both normalized inward degree and normalized outward degree as features.

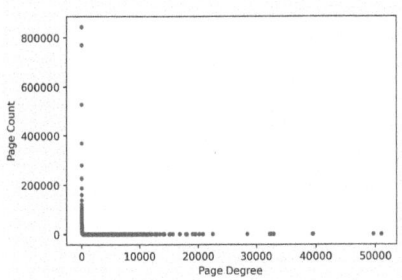

(a) Page Degree Distribution   (b) Page Degree Distribution on Log Scale

Fig. 1. xxx

Page's Category. Public pages categorize their topics as assigned by their managers, encompassing over a thousand distinct categories within the page-likes graph. For instance, the top 20 categories are enumerated in Table 1, extracted directly from the page metadata without modification. Despite the presence of duplicate categories, their impact on prediction accuracy is minimal. According to the theory of homophily [15], pages sharing the same category are more likely to form connections. Given the impracticality of employing one-hot vectors due to the extensive number of categories, binary encoding is utilized. This method efficiently compresses category data into eleven binary digits, significantly reducing memory usage while maintaining accuracy comparable to one-hot encoding [17].

Table 1. Top 20 Categories of Public Page

Page Category	number
Local Business	1,086,041
Non-Profit Organization	230,240
Professional Service	178,543
Restaurant	171,568
Real Estate	127,801
Company	124,187
Community	111,223
Education	109,761
Religious Organization	98,712
Shopping & Retail	95,284
Medical & Health	86,503
Shopping/Retail	84,897
Organization	79,744
Artist	79,668
Musician/Band	69,365
Arts & Entertainment	69,270
Public Figure	69,026
School	63,274
Community Organization	63,272
Nonprofit Organization	57,952

Page's State Label. In our previous study [6], neighborhood location information has emerged as a pivotal feature for classifying pages within the page-likes graph, aligning with the principles of homophily theory [15]. This theory suggests that pages within the same geographical state are more likely to establish connections than those across diverse states. Consequently, we advocate for the incorporation of a page's state label as a crucial feature for enhancing link prediction accuracy. Our analysis is confined to deterministic pages, whose state identities are verifiable, thereby ensuring the reliability of our predictions. To represent the geographical state of each page, we employ a 51-dimensional one-hot encoding scheme, accommodating the 50 states and Washington, D.C.

Page's State Neighborhood Distribution. In our previous study [6], we employed state neighborhood distribution (SND) vectors as node features for classifying the states of pages, yielding significant accuracy improvements. These vectors represent the distribution of a page's neighbors across different states, offering a nuanced perspective beyond mere state labels. While direct state labels provide definitive location information, neighborhood distribution vectors offer predictive insights based on the proximity and connections of pages within the graph. Although not as unequivocally accurate as state labels, these vectors serve as an informative feature, suggesting the potential state affiliation of a page based on the geographic distribution of its connections.

Page's City Label. Inspired by the insights gained from analyzing page state neighborhood distributions and aligned with the principles of homophily theory [15], our investigation extends into more granular location data of pages—their city locations. City-level data offer a finer granularity than state-level information, suggesting that pages within the same city may exhibit even tighter connections than those merely within the same state. However, the extensive variety of cities in our dataset, numbering in the tens of thousands, presents a much more complex challenge for classification compared to the 51 state-level categories. This analysis is confined to deterministic pages, as their city affiliations are unequivocally determined, in contrast to non-deterministic pages. Given the vast number of city categories, binary encoding serves as an efficient method to encode city information, mitigating the increase in feature dimensions associated with one-hot encoding methods.

Page's City Population. Observations from the dataset reveal a pattern where popular pages from major urban centers, such as New York and Los Angeles, exhibit higher connectivity, including links to pages from smaller municipalities. We posit that the population size of a city could serve as a pivotal feature in link prediction models. The underlying hypothesis is that larger cities, with their denser populations, host a broader array of activities and enterprises, casting a wider sphere of influence that captivates the attention of individuals from less populous areas. This dynamic is proposed to facilitate the formation of connections between pages representing large urban areas and those from smaller cities.

Page's City GPS Coordinates. In our previous study [6], our analysis revealed a notable trend among interstate pages, particularly those associated with cities situated along state borders. These pages demonstrated substantial connections to pages from proximate neighboring cities across state lines, suggesting a potential influence of geographical proximity on the establishment of page neighborhoods. Consequently, we propose incorporating the latitude and longitude of cities—specifically for deterministic pages—as features to explore the extent to which geographic location factors into the formation of these online community networks.

4   Evaluating Page Features

4.1   Experimental Setup

Link prediction inherently presents a binary classification challenge, necessitating a focus on accurately distinguishing between positive and negative edges. Consequently, the Area Under the Receiver Operating Characteristic Curve (AUC-ROC) serves as a critical metric for evaluating classifier performance, offering insights beyond mere accuracy by assessing the model's ability to differentiate each class effectively.

Table 2. Highest AUC-ROC on test set on different Positive/Negative edge ratio with page state label as feature

Ratio	1:1	1:5	1:10
AUC-ROC	0.8898	0.9175	0.9125

In the experiments, we employ a two-layer GraphSAINT model with random walk sampling to encode node features and topology. Each layer, implemented via the PyTorch Geometric (PyG) framework [19], contributes to a GNN layer [18]. The outputs from both layers are concatenated as the input to a linear layer, which outputs the node embedding vector. A dot product function, renowned for its efficacy in computing embedding similarities, acts as the decoder. Given the sparse nature of the page graph, actual edges are significantly outnumbered by potential non-existent edges. Because the total number of negative edges is enormous, we use negative sampling to sample a certain number of negative edges in the training [13]. We optimize the ratio of positive to negative edges at 1:5 for training and testing purposes. This specific ratio demonstrates superior AUC-ROC performance compared to alternative ratios, as evidenced in Table 2. The training process has 2000 epochs, necessitating approximately 20 h to complete.

4.2   Single Feature

In this section, each experiment isolates a proposed feature as the sole node attribute. Comparative analysis reveals the page state label as the superior node feature, distinguished by the highest scores in Table 3, the most stable training loss curve, and the most consistent testing AUC-ROC score curve in Fig. 3a.

Table 3. Summary of feature analysis across the entire dataset, ordered by average AUC-ROC

Single Page Feature	Avg. AUC-ROC	Avg. TPR	Avg. TNR
State Label	0.9308	0.8485	0.8832
SND	0.9306	0.8481	0.8729
City Label	0.9107	0.8113	0.8581
Constant Number 1(baseline)	0.9075	0.8109	0.8508
Page Category	0.9041	0.8040	0.8493
City GPS Coordinates	0.8964	0.7844	0.8609
Node Degree	0.7632	0.5832	0.9672
City Population	0.6849	0.5140	0.9043

Performance. The graph topology can affect the link formation between two nodes based on whether they are within their proximity. The Graph Neural Network algorithm automatically uses the graph topological information to learn the embeddings for the nodes in the graph. Inputting node features into Graph Neural Network adds node information to the graph topological information for generating node embeddings. It could be better or worse. Therefore, we need a baseline of the classifier performance, which is performed only on the graph topological information. We assign constant number 1 to all nodes as their features. Since all nodes have the same feature 1, the Graph Neural Networks only use the topological information in the training and testing.

Table 3 presents the prediction results for each feature. The results are averaged values of 3 runs. Column AUC-ROC represents how well the algorithm classifies the positive and negative edges. Columns TPR and TNR represent the true positive rate and true negative rate of the optimal threshold in the ROC curve for edge predictions. The table shows that the feature page state label has the best performance. Page state label, city label, and state neighborhood distribution features have better performance than the constant number 1 feature. These node features add useful node information to the graph topological information for the edge prediction. The rest of the features perform worse than the baseline feature constant number 1. Their node information interferes with the topological information, which causes the algorithm to perform worse on the prediction.

The marginal advantage of the page state label feature over the page state neighborhood distribution feature may stem from its direct and definitive representation of state labels. While the state neighborhood distribution offers insights into a page's state association, it does not achieve the exact correspondence of the actual state labels. Notably, the page state neighborhood distribution feature encompasses 306 dimensions, in contrast to the page state label feature's more concise 51 dimensions.

Table 4. Categorical feature comparison for positive edges

Feature	class#	intra-class edge		inter-class edge	
		ratio	accuracy	ratio	accuracy
State Label	51	0.7354	0.8512	0.2645	0.8321
City Label	12196	0.5080	0.8332	0.4919	0.7971
Constant baseline	1	–	0.8109	–	0.8109
Page Category	1412	0.1098	0.8155	0.8901	0.8034

Three categorical features, state label, city label, and category, demonstrate superior performance. We select features based on the homophily phenomenon in networks, which suggests that nodes are more likely to connect within the same class. Table 4 shows that the model performs better on intra-class edges for all features. Therefore, the higher the intra-class edge ratio, the better the model's performance. This explains why the state label exhibits the best performance. Page location label shows a strong effect on pages connecting with their neighbors.

Learning Curve. The learning curves in the training process offer insights into each feature's performance on the page graph data. Training loss curves and testing AUC-ROC score curves for all features are presented in Fig. 3, with each subplot applying a consistent log scale for training loss and a linear scale for testing AUC-ROC scores. Among these, the page state label feature, as illustrated in Fig. 3a, displays the most stable and conventional loss curve and AUC-ROC score, indicating its superior fit for the page graph data and effectiveness in link prediction. In contrast, features like the page state neighborhood distribution (Fig. 3b), page city label (Fig. 3c), constant number 1 (Fig. 3d), page category (Fig. 3e), and page city geographic coordinates (Fig. 3f) exhibit unstable training loss curves, particularly in their plateau phases. Both the page node degree and page city population features demonstrate atypical loss and AUC-ROC curves, further distinguishing the page state label feature's distinct advantage.

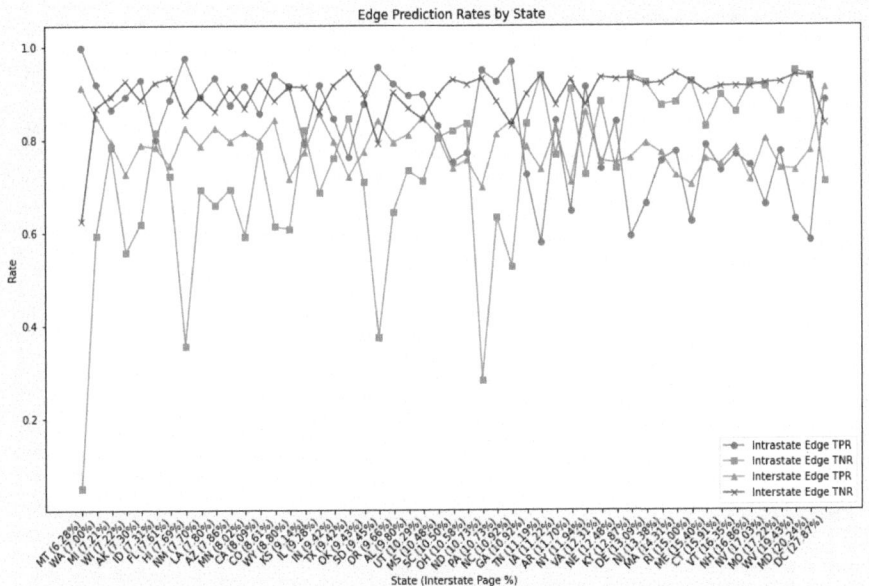

Fig. 2. Edge prediction rates by state

Table 5. Combine one feature with page state label feature analysis across the entire dataset

Feature	Feature	AUC-ROC	TPR	TNR
State Label	Node Degree	0.9317	0.8463	0.8850
State Label	City Population	0.9289	0.8436	0.8793
State Label	- (baseline)	0.9308	0.8485	0.8832
State Label	Category	0.9243	0.8358	0.8701
State Label	City GPS Coordinates	0.9241	0.8365	0.8623
State Label	Constant Number 1	0.9228	0.8298	0.8688
State Label	City Label	0.9131	0.8178	0.8595

4.3   Combined Feature

Performance. Initially, we explored the combination of two features, focusing on the page state label feature due to its superior performance. We paired it with other features to assess potential enhancements in accuracy. Table 5 reveals that combining the page state label feature with the page node degree feature improves performance beyond the baseline established by the sole use of the page state label feature. When the page state label feature is combined with the page city population feature, the performance is similar to the baseline. However,

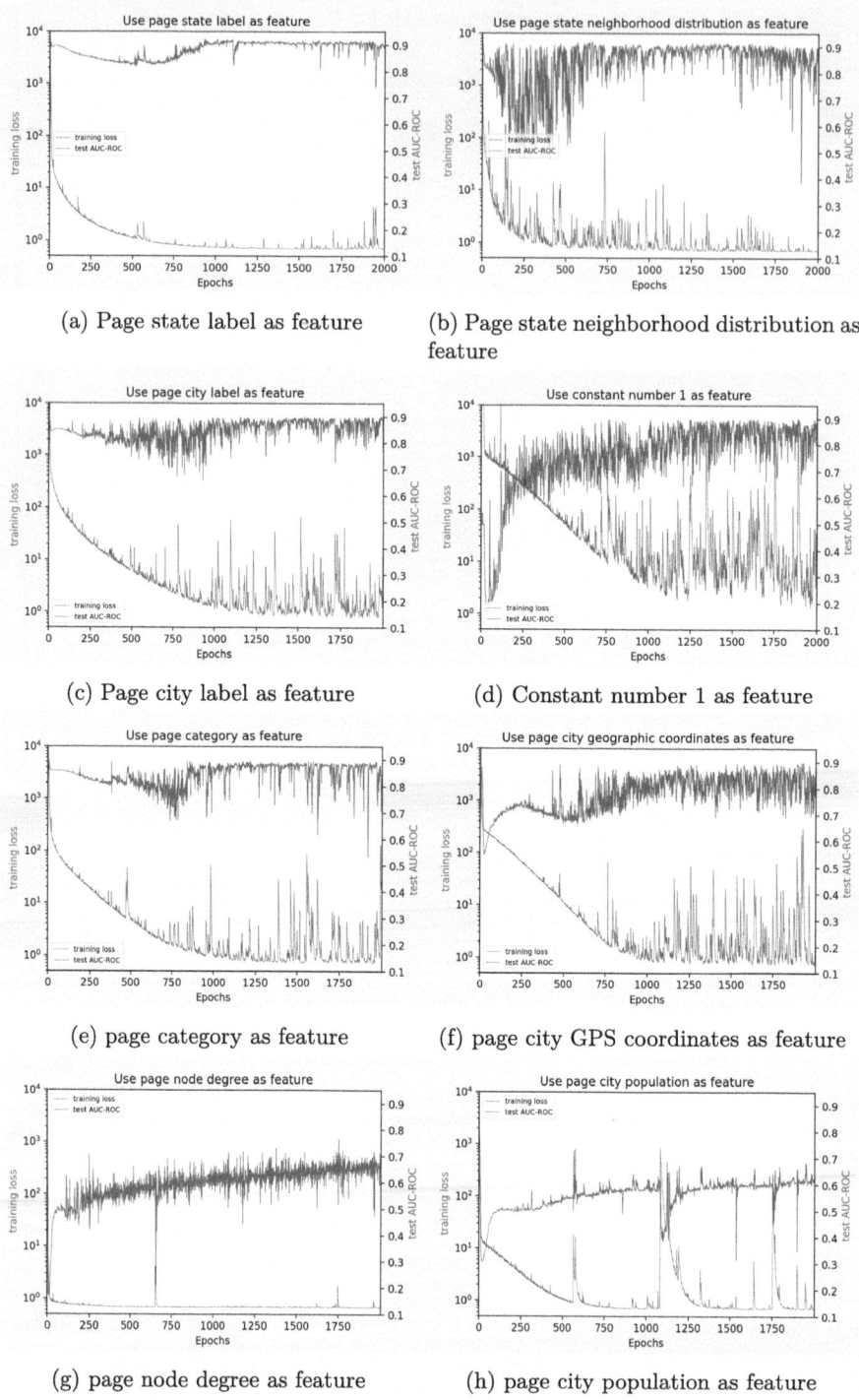

(a) Page state label as feature

(b) Page state neighborhood distribution as feature

(c) Page city label as feature

(d) Constant number 1 as feature

(e) page category as feature

(f) page city GPS coordinates as feature

(g) page node degree as feature

(h) page city population as feature

Fig. 3. Comparison of page features based on state and city

Table 6. Combine more features with page state label feature analysis across the entire dataset

Feature	Feature	Feature	Feature	Feature	AUC-ROC	TPR	TNR
State Label	Degree	Population	–	–	0.9394	0.8590	0.8857
State Label	Degree	–	–	–	0.9317	0.8463	0.
State Label	–	–	–	(baseline)	0.9308	0.8485	0.8832
State Label	–	Population	–	–	0.9289	0.8436	0.8798
State Label	Degree	Population	Category	Constant 1	0.9159	0.8168	0.8608
State Label	Degree	–	Category	–	0.9126	0.8141	0.8541

integrating other features with the page state label feature leads to a decrease in performance compared to the baseline.

Further experimentation led us to combine three features: page node degree, page city population, and page state label, which collectively exhibited the highest performance, as depicted in Table 6. The table also illustrates that merging the page category, page city GPS coordinates, constant number 1, and page city label features with the page state label feature resulted in suboptimal performance. While exhaustive combinations of these less effective features were not explored, a few examples are provided in Table 6 for illustrative purposes.

There are two types of edges in the page graph: interstate edges and intrastate edges [6]. Intrastate edges connect pages within the same state, while interstate edges link pages from different states. We evaluated the accuracy of predicting interstate and intrastate edges for each state, as well as for all states combined, using an algorithm that incorporates a feature combination of page state label, page node degree, and page city population. The results are presented in Table 7.

We detail the true positive rates (TPR) and true negative rates (TNR) for both intrastate and interstate edges across various states, as presented in Table 7. Given the uniform and random sampling of negative edges within the graph, intrastate and interstate negative edges constitute 4.85% and 95.15%, respectively, of all negative edges, as shown in Table 8. Conversely, intrastate and interstate positive edges represent 73.54% and 26.46%, respectively, of all positive edges. The distribution of intrastate edges, split into 75.22% positive and 24.78% negative, contrasts with interstate edges, which are divided into 5.27% positive and 94.73% negative, according to Table 9. This disparity in data distribution likely influences the observed discrepancies in TPR and TNR values for intrastate and interstate edges, underscoring the complexity of accurately predicting link formations within the page graph.

We visualize the data from Table 7 using a line chart in Fig. 2 for an intuitive understanding of the predictive statistics. The x-axis represents states ordered by their increasing interstate page percentages. Displayed are the true positive rates (TPR) and true negative rates (TNR) for both intrastate and interstate edges. Notably, the high TPR for intrastate edges (blue line) corresponds to states with lower interstate page percentages, whereas states with higher inter-

Table 7. Link Prediction Performance by State

State	Intrastate Edge		Interstate Edge		State	Intrastate Edge		Interstate Edge	
	TPR	TNR	TPR	TNR		TPR	TNR	TPR	TNR
AL	0.8939	0.7331	0.8107	0.8672	AK	0.9293	0.6189	0.7883	0.8856
AZ	0.8742	0.6932	0.7972	0.9101	AR	0.6455	0.9074	0.7083	0.9284
CA	0.8560	0.7862	0.7974	0.9266	CO	0.9402	0.6131	0.8423	0.8828
CT	0.7332	0.8960	0.7457	0.9125	DE	0.6622	0.9225	0.7927	0.9178
FL	0.8862	0.7226	0.7434	0.9322	GA	0.7234	0.8347	0.7847	0.8964
HI	0.9750	0.3560	0.8258	0.8542	ID	0.8014	0.8158	0.7839	0.9227
IL	0.9182	0.6853	0.8571	0.8606	IN	0.8454	0.7603	0.7961	0.9160
IA	0.8410	0.7661	0.8238	0.8747	KS	0.7926	0.8218	0.7730	0.9130
KY	0.5909	0.9390	0.7604	0.9304	LA	0.9322	0.6585	0.8258	0.8613
ME	0.7867	0.8281	0.7580	0.9023	MD	0.5835	0.9360	0.7749	0.9338
MA	0.7740	0.8788	0.7210	0.9413	MI	0.8666	0.7839	0.7960	0.8934
MN	0.9151	0.5924	0.8163	0.8683	MS	0.8294	0.8026	0.8104	0.8958
MO	0.7741	0.8600	0.7369	0.9215	MT	0.9986	0.0511	0.9133	0.6254
NE	0.8387	0.7380	0.7493	0.9298	NV	0.6582	0.9123	0.8005	0.9203
NH	0.7436	0.9211	0.7125	0.9121	NJ	0.7538	0.8722	0.7703	0.9204
NM	0.8925	0.6930	0.7865	0.8933	NY	0.9129	0.7251	0.8600	0.8739
NC	0.9677	0.5267	0.8389	0.8298	ND	0.9486	0.2824	0.6974	0.9308
OH	0.7705	0.8342	0.7552	0.9183	OK	0.8764	0.7080	0.7732	0.8985
OR	0.9206	0.6436	0.7933	0.9008	PA	0.9242	0.6318	0.8127	0.8808
RI	0.6227	0.9249	0.7019	0.9234	SC	0.7511	0.8189	0.7383	0.9288
SD	0.9551	0.3744	0.8402	0.7917	TN	0.5773	0.9375	0.7359	0.9354
TX	0.7614	0.8455	0.7194	0.9433	UT	0.8965	0.7098	0.8469	0.8433
VT	0.7668	0.8589	0.7828	0.9140	VA	0.7386	0.8817	0.7542	0.9333
WA	0.9186	0.5946	0.8503	0.8684	WV	0.6275	0.9462	0.7342	0.9368
WI	0.8921	0.5569	0.7257	0.9254	WY	0.9143	0.6074	0.7161	0.9134
DC	0.8838	0.7079	0.9104	0.8340	**Total**	0.8737	0.7392	0.8087	0.8976

state page percentages exhibit lower intrastate edge TPRs. This pattern suggests that pages with numerous out-of-state connections are more often involved in interstate edges, while intrastate pages, primarily linked within their own state, tend to form intrastate edges. Consequently, a lower interstate page percentage implies a higher number of intrastate pages and edges, resulting in increased intrastate edge TPRs. However, some states show anomalously low intrastate edge TNRs (orange line), attributed to a notably smaller number of intrastate negative edges than average, a byproduct of random sampling. This discrepancy likely contributes to the observed outliers.

Table 8. Positive/Negative edges percentage distribution

	Positive Edges	Negative Edges
Intrastate	73.54%	4.85%
Interstate	26.46%	95.15%
Total	100.00%	100.00%

Table 9. Intrastate/Interstate edges percentage distribution

	Positive Edges	Negative Edges	Total
Intrastate	75.22%	24.78%	100%
Interstate	5.27%	94.73%	100%

5 Related Work

5.1 User Social Network Analysis

The analysis of user social networks has received more focus than that of community networks in the fields of network science and social network analysis. Ugander et al. explored the global structure of the Facebook user network, identifying a range of network properties [1]. Barnett and Benefield [3] discovered that proximity and cultural homophily significantly influence Facebook friendship ties, noting that countries with international Facebook friendships often share borders, languages, and cultural traits [3].

5.2 Link Prediction

Link prediction has been a popular research area for the past decades. In social network link prediction, researchers typically employ three methodologies: similarity, probabilistic, and algorithmic approaches [4]. The similarity approach leverages graph-measures and content-measures (attributes of nodes or edges). Among algorithmic methods, deep learning has emerged as a particularly popular technique. In our study, we employ both similarity and algorithmic approaches to predict links.

5.3 Online Social Community Location Classification

Public pages represent a prominent platform for online communities, with each page embodying a distinct social community. While page managers have the option to label their pages with country and state/province locations, many pages lack this geographical information. Hong et al. [5] explored the page graph—a network where pages can "like" each other—and introduced a majority voting algorithm for inferring the missing country locations of pages. This method proved effective for country-level classification, leveraging shared cultural, linguistic, and social contexts among pages from the same country.

Nonetheless, the majority voting approach showed limitations in more granular subdivision location classifications, such as state labeling within the United States. We introduced the concept of neighborhood state distribution vectors and applied Graph Neural Networks for the classification of pages' subdivision locations, achieving notable accuracy [6]. This methodology offers insights into a page's influence across different states.

5.4   Graph Neural Networks

Graph neural networks (GNNs) are a subset of artificial neural networks designed for processing graph-structured data [7]. Graph convolutional networks (GCNs) are one type of GNN that are often used in graph representation learning. These representations aim to encapsulate the graph's topological structure in low-dimensional vectors, facilitating tasks such as node classification and link prediction. Nonetheless, GCNs' reliance on full graph adjacency matrices makes them computationally intensive, particularly for sizable graphs, leading to significant GPU memory demands and prolonged training durations [8] [9].

To mitigate these challenges, node sampling techniques have been developed to adapt GCNs for larger graphs. GraphSAINT, specifically, introduces an inductive learning strategy through graph sampling, enhancing both the efficiency and accuracy of training. It generates mini-batches by sampling sub-graphs from the entire graph for each iteration. This approach ensures that nodes influencing each other significantly are likely to be included in the same mini-batch, allowing for mutual support within the mini-batch and circumventing the need for broader graph traversal [10]. Such innovations significantly curtail the computational burden associated with GCNs, concurrently bolstering accuracy [10].

6   Conclusion

In this paper, we investigated the pivotal features that influence link formation and neighborhood structuring within the page graph. Initially, we explore a series of potential features, both graph-based and content-based, that may impact link connectivity. Subsequently, we present our methodology, combining the node similarity and GNN to perform the link prediction. Through meticulous experimentation with both individual and combined features, we ascertain that the page state label emerges as the most influential single feature for link formation. Moreover, we observe that augmenting the page state label feature with page node degree and page city population features further enhances link prediction accuracy. Ultimately, our analysis reveals a correlation between the true positive rate of intrastate positive edges and the interstate page percentage, underscoring the nuanced dynamics of link formation within the page graph. Page location label shows a strong effect on pages connecting with their neighbors.

References

1. Ugander, J., Karrer, B., Backstrom, L., Marlow, C.: The anatomy of the facebook social graph, arXiv preprint arXiv:1111.4503 (2011)
2. Wikipedia contributors, List of social platforms with at least 100 million active users — Wikipedia, the free encyclopedia (2023) (Accessed 30-May-2023)
3. Barnett, G.A., Benefield, G.A.: Predicting international Facebook ties through cultural homophily and other factors. New Media Soc. **19**(2), 217–239 (2017)
4. Daud, N.N., Ab Hamid, S.H., Saadoon, M., Sahran, F., Anuar, N.B.: Applications of link prediction in social networks: a review. J. Netw. Comput. Appli. **166**, 102716 (2020), https://www.sciencedirect.com/science/article/pii/S1084804520301909
5. Hong, Y., Lin, Y.-C., Lai, C.-M., Felix Wu, S., Barnett, G.A.: Profiling facebook public page graph. In: 2018 International Conference on Computing, Networking and Communications (ICNC), pp. 161–165 (2018)
6. Wang, J., Wang, X., Lai, C.-M., Felix Wu, S.: Online social community sub-location classification. In Proceedings of the 2023 IEEE/ACM International Conference on Advances in Social Networks Analysis and Mining (ASONAM 2023). Association for Computing Machinery, New York, pp. 276–280 (2024). https://doi.org/10.1145/3625007.3627504
7. Scarselli, F., Gori, M., Tsoi, A.C., Hagenbuchner, M., Monfardini, G.: The graph neural network model. IEEE Trans. Neural Netw. **20**(1), 61–80 (2009)
8. Bruna, J., Zaremba, W., Szlam, A., LeCun, Y.: Spectral networks and locally connected networks on graphs. In: Proceedings of International Conference on Learning Representations, pp. 1-14 (2014)
9. Yan, M., et al.: Characterizing and understanding GCNs on GPU. IEEE Comput. Archit. Lett. **19**(1), 22–25 (2020)
10. Zeng, H., Zhou, H., Srivastava, A., Kannan, R., Prasanna, V.: Graphsaint: Graph sampling based inductive learning method, arXiv preprint arXiv:1907.04931 (2019)
11. Lee, S.H., Kim, P.-J., Jeong, H.: Statistical properties of sampled networks. Phys. Rev. E **73**, 016102 (2006)
12. Liben-Nowell, D., Kleinberg, J.: The link-prediction problem for social networks. J. Am. Soc. Inform. Sci. Technol. **58**(7), 1019–1031 (2007). https://onlinelibrary.wiley.com/doi/abs/10.1002/asi.20591
13. Yang, Z., Ding, M., Zhou, C., Yang, H., Zhou, J., Tang, J.: Understanding negative sampling in graph representation learning. In: Proceedings of the 26th ACM SIGKDD International Conference on Knowledge Discovery & Data Mining, KDD 2020. Association for Computing Machinery, New York (2020), pp. 1666–1676. https://doi.org/10.1145/3394486.3403218
14. Yilmaz, E.A., Balcisoy, S., Bozkaya, B.: A link prediction-based recommendation system using transactional data. Sci. Rep. **13**(1), 6905 (2023)
15. McPherson, M., Smith-Lovin, L., Cook, J.M.: Birds of a feather: homophily in social networks. Annual Rev. Sociol. **27**, 415–444 (2001). http://www.jstor.org/stable/2678628
16. Leskovec, J., Horvitz, E.: Planetary-scale views on an instant-messaging network (2008)
17. Potdar, K., Pardawala, T.S., Pai, C.D.: A comparative study of categorical variable encoding techniques for neural network classifiers. Inter. J. Comput. Appli. **175**(4), 7–9 (2017)

18. Morris, C., et al.: Weisfeiler and leman go neural: Higher-order graph neural networks (2021)
19. Fey, M., Lenssen, J.E.: Fast graph representation learning with pytorch geometri, arXiv preprint arXiv:1903.02428 (2019)

Enhancing Stance Classification on Social Media Using Quantified Moral Foundations

Hong Zhang[1]([✉]), Quoc-Nam Nguyen[1], Prasanta Bhattacharya[2], Wei Gao[1], Liang Ze Wong[2], Brandon Siyuan Loh[2], Joseph J. P. Simons[2], and Jisun An[3]

[1] School of Computing and Information Systems, Singapore Management University, 80 Stamford Rd, Singapore178902, Singapore
hong.zhang.2022@phdcs.smu.edu.sg, {qnnguyen,weigao}@smu.edu.sg
[2] Institute of High Performance Computing (IHPC), Agency for Science, Technology and Research (A*STAR), 1 Fusionopolis Way, #16-16 Connexis, Singapore138632, Singapore
{prasanta_bhattacharya,wong_liang_ze,brandon_loh,simonsj}@ihpc.a-star.edu.sg
[3] Luddy School of Informatics, Computing, and Engineering, Indiana University Bloomington,, IN Bloomington, USA
jisunan@iu.edu

Abstract. This study enhances stance detection on social media by incorporating deeper psychological attributes, specifically individuals' moral foundations. These theoretically-derived dimensions aim to provide an interpretable profile of an individual's moral concerns which, in recent work, has been linked to behaviour in a range of domains including society, politics, health, and the environment. In this paper, we investigate how moral foundation dimensions can contribute to detecting an individual's stance on a given target. Specifically, we incorporate moral foundation features extracted from text, along with semantic features, to classify stances at both message- and user-levels using traditional machine learning and Large Language Models (LLMs). Our preliminary results suggest that encoding moral foundations can enhance the performance of stance detection tasks, but with notable heterogeneity across task type, models, and datasets. In addition, we illustrate meaningful associations between specific moral foundations and online stances on target topics. The findings from this study highlight the importance of considering deeper psychological attributes in stance classification tasks, and underscore the role of moral foundations in guiding online social behavior.

Keywords: Stance Detection · Moral Foundations · User Behaviour · LLM

H. Zhang and Q.-N. Nguyen—These authors contributed equally to this work.

L. M. Aiello et al. (Eds.): ASONAM 2024, LNCS 15211, pp. 305–319, 2025.
https://doi.org/10.1007/978-3-031-78541-2_19

1   Introduction

As a behaviour, a *stance* refers broadly to an expression of perspectives, attitudes, or judgments toward a given proposition. The study of stances is inherently interdisciplinary and traces its roots to psycho- and socio-linguistics. In their popular text, [6] define stance as the "lexical and grammatical expression of attitudes, feelings, judgments, or commitment concerning the propositional content of a message". Similarly, [12] refers to stance as "an articulated form of social action", which involves evaluation of an object, or an alignment with a given position.

Social media serves as a communicative platform that allows users to express their stance and views on a given topic of interest, providing researchers the opportunity to study this phenomenon at scale. Recent studies have proposed models for detecting or inferring the stance conveyed in social media posts on target topics [2,24]. While the vast majority of these studies have leveraged message-level characteristics such as language use and user interactions, the question of whether stance modeling can be improved through the incorporation of deeper user-level attributes, notably psychological characteristics, remains understudied, with only a few recent exceptions [36].

Our working hypothesis is that a user's broader value system carries relevant information for inferring their stance on a particular topic. In this study, we operationalize users' broader value system along the conceptual framework of Moral Foundations Theory (MFT) [19,22]. Notably, MFT proposes five distinct domains of moral concern rooted in universal evolutionary challenges: care/harm, fairness/cheating, authority/subversion, sanctity/degradation, and loyalty/betrayal. A key application case of this model is explaining political differences – liberals and conservatives endorse these moral foundations differently [18,21].

Table 1. Sample tweets for the stance target *Hillary Clinton* and their expressed bias towards each moral foundation.

Stance	Tweet	Care	Fairness	Loyalty	Authority	Sanctity
FAVOR	@HillaryClinton the @DalaiLama speaks of women in leadership roles bringing about a more compassionate world. #potus #SemST	+	+	+	+	+
FAVOR	Just met an awesome supporter on the CX bus! He said"Hillary is one strong woman and we need that for our country." #FellowsNV #SemST	+	-	+	+	+
AGAINST	I wish @OliviaPope could run @HillaryClinton 's campaign... #Scandal #livisreal #SemST	+	-	+	-	-
AGAINST	Why did you lie about the #Benghazi subpoena? @HillaryClinton No wonder no one trusts you. #SemST	-	-	-	+	+
NONE	@larryelder The more Republicans talk about social policy the better....for #SemST	-	-	+	+	+

Given the important role moral foundations play in shaping social behavior, we posit that they also enable the formation of human opinions and stances. In Table 1, we present a few tweets from the SemEval 2016 Task 6A dataset [28], which is widely used for stance detection tasks, and demonstrate that tweets across stance classes frequently express a positively- or negatively-valenced bias towards above-mentioned moral foundations. Hence, we hypothesize that the systematic extraction and incorporation of these moral foundations should enhance the detection of both, message- and user-level stances from social media-based content. In testing this hypothesis, we seek to augment online stance detection tasks on Twitter datasets by incorporating moral foundation features into message- and user-level stance detection models. Our findings reveal that the addition of moral foundation features significantly boosted the predictive performance of stance detection models. Furthermore, the insights generated from our association-based analyses highlight the prevalence and nature of moralized discourse surrounding key topics (e.g., wearing masks).

Through this study, we offer four key contributions. First, we perform a comprehensive analysis to assess the predictive utility of moral foundations for both message- and user-level stance detection tasks on social media. In doing so, we address the aforementioned limitations of current stance detection models, particularly their over-reliance on conventional textual and contextual features. Secondly, we highlight systematic variations in the predictive performance of moral foundation-based models across tasks (i.e., message-level vs. user-level stance detection), datasets, stance targets, and classifiers. Thirdly, we go beyond predictive performance to elucidate interesting associations between moral foundations and stances towards specific targets. Lastly, we show that the incorporation of moral foundations improves F1 scores on stance detection tasks by up to 24.86% points for LLM-based models, depending on the choice of model and dataset. This suggests that the addition of such psychological attributes might be particularly fruitful for LLM-based stance detection models. Further research can draw on these insights to explore the design of psychologically-rooted LLMs for related tasks.

2   Related Work

2.1   Stance Detection

As a natural language processing (NLP) task, stance detection has been widely studied in contexts spanning politics [39,44], climate change [41], and the COVID-19 pandemic [17]. When expressed in text, the stance of the message can typically be labeled into a number of constituent classes, e.g., as "Support", "Against" and "Neutral". Stance detection aims to automatically determine the position of a message or its author towards a given proposition or target [28] based on these classes. In target-specific stance detection tasks, the models are trained for a particular target [4]. However, more recent work has investigated the problem of zero-shot stance detection which is target-agnostic and aims to detect the stance towards new or unseen targets [3].

2.2   Moral Foundation Dictionary (MFD)

Moral Foundation Theory [20,22] posits five moral foundations that are prevalent across cultures and nations: Care/Harm, Fairness/Cheating, Loyalty/Betrayal, Authority/Subversion, and Sanctity/Degradation. To measure these five dimensions from text, [18] developed the first MFD using a two-phase approach. The first phase involved generating associations, synonyms and antonyms for the five moral foundations through thesauruses and conversations with peers. Next, words that were too distantly related to the foundations were removed. This resulted in a dictionary of 295 words related to five moral foundations, which is also used in linguistic tools such as the Linguistic Inquiry and Word Count program (LIWC) [18]. Based on the MFD, [35] expanded the morality lexicons to a dictionary of 4,636 words, and used this to measure *social effects* such as morality and stances on Twitter.

2.3 Extended MFD (eMFD)

Although the MFD offers an automatic and dictionary-based method to extract moral foundation cues from text, it has a few limitations [14,43]. For instance, the MFD was created by a small group of *experts* which limits its generalizability to a broader and *non-expert* population. Moreover, the MFD adopts a "winner takes all" strategy; it assigns a word to a single moral foundation. This precludes the possibility of a word being related to multiple moral foundations at once. To address these concerns, [23] proposed the *extended* MFD (eMFD) to help capture large-scale and intuitive moral judgments from text. Instead of relying on careful selection by a small group of experts, the eMFD lexicon was generated through a wider crowd-sourced task aimed at capturing a more comprehensive list of morally relevant content cues. Moreover, instead of a single moral dimension, the eMFD assigns each word a vector of scores, which reflects the probability of that word belonging to each moral foundation.

2.4 FrameAxis

The MFD is only effective when the corpus of interest contains words from the dictionary. However, moral information in text can be expressed using diverse linguistic cues and styles, which might include only few or none of the words present in the MFD. In such contexts, it is challenging to infer the underlying moral foundations effectively using just a dictionary-based approach. In their study, [25] proposed FrameAxis, a method for discovering the presence of framing and its associated biases from documents, by identifying the most relevant semantic facets. Using FrameAxis, any text can be projected to a high dimensional space using word embeddings to extract moral information, even when none of the words are present in the MFD.

3 Datasets

We use three public datasets in our study, which we will subsequently refer to as SemEval[1], Connected Behaviour (CB)[2], and P-Stance[3] in this paper.

---

[1] https://www.saifmohammad.com/WebPages/StanceDataset.htm.

[2] https://github.com/tommyzhanghong/connected_behavior.

[3] https://github.com/chuchun8/PStance.

The SemEval dataset was constructed for SemEval 2016 Task 6 [29], and contains 4,870 English tweets across six common targets, "Atheism" (AT), "Climate Change is a Real Concern" (CC), "Feminist Movement" (FM), "Hillary Clinton" (HC), "Legalization of Abortion" (LA), and "Donald Trump" (DT). The P-Stance [26] dataset is a popular stance detection dataset in the political domain, and contains 21,574 labeled tweets across 3 targets, namely "Donald Trump" (DT), "Joe Biden" (JB), and "Bernie Sanders" (BS). The CB dataset is a large Twitter dataset for *user-level* stance detection comprising over 100 million tweets [44]. This dataset was used for user-level stance detection towards three targets, namely "Donald Trump" (DT), "Wearing Mask" (WM), and "Racial Equality" (RE).

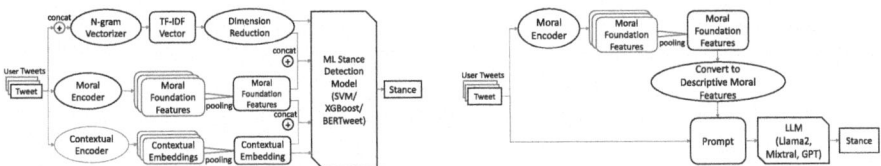

(a) With traditional stance detection model.

(b) With LLM-based stance detection model.

Fig. 1. Our method for enhancing stance classification with quantified moral foundations based on traditional ML models and LLMs. Components in red were only used for user-level stance detection, where tweets posted by the same user were concatenated before passing through TF-IDF vectorizer, and pooling was applied for moral and contextual embeddings.

4   Our Method

In this section, we discuss how moral features were encoded and incorporated into both traditional ML models and LLMs. The performance gain in stance detection tasks with the addition of moral features was studied using the design shown in Fig. 1.

4.1   Feature Encoding

To incorporate moral features with stance detection models based on distinct learning methods, we used TF-IDF vectors, contextual embeddings, and moral foundation features. As shown in Fig. 1, tweets were passed through separate channels to generate these embeddings. Next, the dimensions of these embeddings were reduced using PCA and UMAP. These embeddings were then combined in different ways in the stance detection models. For the LLM-based models, in particular, the embeddings were converted to descriptive moral features, as shown in Fig. 1b and explained in Sect. 5.

TF-IDF. The term frequency-inverse document frequency (TF-IDF) vectorizer represents n-gram features by using both character- or word-level TF-IDF values. This has been applied in many studies spanning document retrieval, text classification and stance detection [13,16,33,38].

Contextual Embedding. Sentence-BERT (SBERT) [34] is a variant of Transformer-based models [42] that uses siamese and triplet network structure in the training stage, and has been shown to outperform BERT and RoBERTa for sentence embedding tasks [27]. Here we encoded tweets with the SBERT all-mpnet-base-v2 version to generate contextual embeddings.

Moral Foundation Features. Two separate techniques were used to generate moral foundation features. The first was **eMFD** [23], a dictionary-based tool for detecting moral information in text. Using this method, *probability* and *sentiment* scores for a text along a moral foundation dimension were computed based on the frequency of occurrence of eMFD keywords in the text. This produced a 10-dimensional vector for each text (5 probabilities and 5 sentiment scores). The second was the Moral Foundation **FrameAxis** features [30], which combined the eMFD with the FrameAxis method described in [25]. In this method, the *bias* and *intensity* of a text along a moral foundation dimension are functions of the cosine similarity of word embeddings from the text and word embeddings of eMFD keywords for that moral dimension. This produced a 10-dimensional vector for each text (5 bias and 5 intensity scores).

User-level Representation. The CB dataset addresses the task of user-level stance detection. For this dataset, feature encodings were aggregated to form user-level representations. As TF-IDF is a statistical method relying on word and document frequencies, tweets posted by the same user were first concatenated to obtain a document for each of the targets. These user documents were then used to calculate TF-IDF embeddings. Conversely, other feature encodings have a fixed dimension. Hence we applied mean-pooling to obtain user-level embeddings from tweet embeddings.

4.2   Stance Detection Models

We compared three broad classes of models for the stance detection task.

1) Traditional Machine Learning Models In this study, we trained SVM and XGBoost classifiers using both n-gram and SBERT embeddings as basic features, and augmented these models by incorporating morality features, as described in the previous section. Such models have been widely used in stance detection studies [8,28].

2) Fine-Tuned Language Models Fine-tuned language models (FLMs) such as BERT-based models have been identified as state-of-the-art (SoTA) in stance detection tasks [5,26]. In this study, we fine-tuned a pre-trained BERTweet model

[31] on SemEval, CB and P-Stance datasets. BERTweet is the first public, large-scale and pre-trained language model for English tweets, having the same architecture as BERT-base [11]. Prior studies have shown that BERTweet outperforms strong pre-trained language models, and is SoTA on tweet-based stance tasks [10].

3) Large Language Models We implemented zero-shot and few-shot stance classification using three prominent LLMs: Llama2-70-chat[4], Mixtral-8x7B[5], and GPT-3.5-turbo[6]. Llama2-70-chat is based on the Llama 2 family of LLMs [40], and is fine-tuned for dialogue. Mixtral-8x7B is an LLM with a novel sparse mixture of experts (SMoE) architecture [37]. GPT-3 [7] is an autoregressive language model with 175 billion parameters, 10× more than any previous non-sparse language model. We chose these LLMs as they have performed well on many various NLP benchmarks, while also remaining cost-effective. Their pre-training data also includes social media posts, making them well-suited to process our tweet dataset. The LLMs were accessed via the Replicate API. Our prompting methods are presented in Sect. 5.

5   Prompting LLMs with Moral Features

5.1   Generating Descriptive Moral Features

As moral foundation features are represented as vectors of numbers, it is challenging for LLMs to interpret them. To address this challenge, we employed a two-step methodology: we first clustered these scores, and then converted the scores into textual expressions to be included in LLM prompts. Our method is described in Fig. 1b.

Discretization by Clustering: For each dimension of the eMFD or FrameAxis embeddings, we applied K-Means clustering with $K = 2$ on all scores for that dimension to determine the threshold between a *high* and a *low* score for that dimension. This allowed us to categorize each numerical score as either "High" or "Low", which makes it easier for LLMs to understand the prompt.

Converting to Text: Based on whether each tweet was "High" or "Low" on each moral dimension, we generated textual moral descriptions for that tweet. These descriptions were then included in the stance detection prompts.

5.2   Prompting for Stance Detection

For stance detection with LLMs, we implemented three distinct prompting schemes. These schemes were designed hierarchically, with the second and third schemes built upon the information integrated in the first one.

---

[4] https://replicate.com/meta/Llama2-70b-chat.
[5] https://replicate.com/mistralai/mixtral-8x7b-instruct-v0.1.
[6] https://platform.openai.com/docs/models/gpt-3-5-turbo.

Task + Context. The prompt included the task description and necessary contextual information, the stance target, and a clear definition of stance labels. This is consistent with previous studies leveraging ChatGPT for stance labeling [1].

Task + Context + FrameAxis. We further augmented the Task + Context prompt by including the moral descriptions generated from the FrameAxis embeddings.

Task + Context + eMFD. Alternatively, we augmented the Task + Context prompt by including the moral descriptions generated from the eMFD embeddings.

For the latter two schemes, we included the moral descriptions along with explanations of each moral dimension. Our hypothesis was that an LLM could make better stance predictions by considering these moral features. Our prompting strategies were adopted from the template samples available in HuggingFace's resources for LLaMa2[7] and Mixtral[8]. We implemented zero-shot, 1-shot, and 5-shot classification scenarios for the three schemes mentioned above.

6   Experiments and Results

6.1   Experimental Setup and Data Sampling

We tuned the hyperparameters for SVM and XGBoost following [15]. We performed grid search on SVM with two kernels. For the Radial Basis Function (RBF) kernel, we used parameter values {1e-3, 1e-4} for *gamma* and {1, 10, 100, 1000} for *C*. For the linear kernel, we used parameter values {1, 10, 100, 1000} for *C*. Similarly, we performed hyperparameter tuning for XGBoost using grid search in {100, 500} for *n_estimators*, {5, 10, 15, 20} for *max_depth* and {0.01, 0.1} for *learning_rate*. We applied a stratified 2-fold cross validation in the search process.

To correct for class imbalance in the SemEval dataset, we applied oversampling to balance training examples from different classes. In the CB dataset, we sampled 5 tweets from each of the top 500 active users, based on the number of posted tweets for each target and stance class, to avoid having an extremely large and sparse matrix due to the size of the vocabulary for TF-IDF. We did not observe a strong class imbalance for the P-Stance dataset, and hence, sampled 1,000 instances from the training data of each target for the same purpose.

We followed [31] to fine-tune BERTweet for each dataset and task over 30 epochs. We used AdamW with a learning rate of 1.e-5 and a batch size of 32, and assessed performance after each epoch with early stopping applied if no improvement occurred over 5 epochs. We selected the best checkpoint for test set evaluation. For prompting-based experiments, we set the models to only provide the most probable output (e.g., by setting `temperature=0`) and a maximum length of 5 tokens [9,32]. We repeated each experiment 10 times with random seeds, and reported the average F1 score as the final score from this exercise.

---

[7] https://huggingface.co/blog/llama2.

[8] https://huggingface.co/mistralai/Mixtral-8x7B-Instruct-v0.1.

6.2   Results and Analysis

Table 2. Average % F1-scores across multiple datasets and models. Llama, Mixtral, and GPT denote Llama2-70b-chat, Mixtral-7x8B, and GPT-3.5-turbo, respectively. CB denotes the Connected Behavior dataset.

Features/Schemes	Models	Datasets		
		SemEval	**CB**	**P-Stance**
n-gram	SVM/XGB (+PCA)	50.21/46.52	81.99/78.88	67.14/66.00
n-gram + eMFD		51.93/48.09	82.55/80.05	67.85/67.74
n-gram + FrameAxis		52.49/47.53	82.19/80.35	68.98/68.79
n-gram	SVM/XGB (+UMAP)	42.03/44.87	79.65/76.33	66.02/66.02
n-gram + eMFD		44.20/47.70	80.57/77.32	66.81/66.81
n-gram + FrameAxis		45.21/47.80	80.60/78.09	67.37/67.37
SBERT	SVM/XGB	62.15/60.31	77.38/77.44	75.13/73.79
SBERT + eMFD		62.53/61.12	77.76/78.00	75.73/75.43
SBERT + FrameAxis		62.96/61.38	77.82/79.51	76.00/76.98
Tweet only	FLM (BERTweet)	71.26	65.50	79.29
Tweet + eMFD		71.69	66.31	80.68
Tweet + FrameAxis		75.30	70.07	79.88
Task + Context	Llama/Mixtral/GPTZero-shot	54.14/37.19/66.46	43.48/22.01/72.28	61.94/22.57/74.80
Task + Context + eMFD		58.96/42.66/66.65	59.39/40.45/74.32	69.35/27.87/75.44
Task + Context + FrameAxis		58.61/39.35/66.82	63.70/34.37/74.00	80.58/34.65/75.48
Task + Context	Llama-/Mixtral/GPT1-shot	55.48/39.18/68.57	44.03/24.77/73.71	62.82/23.90/77.88
Task + Context + eMFD		58.98/44.75/70.05	61.50/42.89/76.73	67.76/29.16/78.68
Task + Context + FrameAxis		59.50/47.52/71.41	67.71/44.83/78.14	79.95/39.30/80.91
Task + Context	Llama/Mixtral/GPT5-shot	59.04/41.41/71.41	46.75/27.92/75.65	62.36/28.07/80.82
Task + Context + eMFD		63.99/47.86/74.39	68.14/50.37/79.25	68.31/31.97/82.61
Task + Context + FrameAxis		66.93/49.37/75.88	69.56/52.78/81.02	80.82/42.50/84.75

It is important to highlight that the goal of this study is not to produce a state-of-the-art stance detection model that can outperform existing stance detection benchmarks. Instead, we aim to study the effectiveness of encoding moral foundations in improving performance on stance detection tasks across a variety of models and datasets. We illustrate the experiment results in Table 2.

N-Gram Baseline. Using n-gram features as a baseline, the addition of moral features led to an improvement in the F1 scores on all datasets, with UMAP as a dimension reduction method showing greater improvement than PCA. The largest improvement with morality features was observed on the SemEval dataset, which has the smallest size. Specifically, the eMFD and FrameAxis features increased F1 scores by an average of 2.07% and 2.35% points, respectively, across models and dimension reduction methods.

SBERT Baseline. To validate the effectiveness of moral foundation features with contextual embedding, the same experiments were repeated using SBERT embeddings [34] as the baseline model. Although contextual embeddings have been shown to be a stronger baseline than n-gram based models, we still observed improvements in stance detection performance on adding moral features. Notably, the addition of eMFD and FrameAxis increased F1 score by up to 0.59% and 0.94% points, respectively, for tweet-level stance detection on the SemEval dataset, and by up to 0.47% and 1.25% points for user-level stance detection on the CB datset.

Fine-Tuned Language Models (FLMs). We conducted further experiments with FLMs, as they have achieved SoTA results in stance detection tasks [5,26]. However, in our analyses, their performance was mixed, with significant variance across datasets. Specifically, we found that the FLM underperformed compared to our baseline model, SBERT, on the CB dataset, but showed superior performance on the SemEval and P-Stance datasets. Similar to the results observed with traditional machine learning methods, we observed a significant improvement in stance detection performance upon integrating moral information into the FLM. This integration led to improvements in F1 scores by up to 4.04%, 4.57%, and 1.39% points for SemEval, CB, and P-Stance datasets, respectively.

LLMs Prompting. Our analysis revealed insightful trends and outcomes when evaluating the performance of Llama2-70b-chat, Mixtral-8x7B, and GPT-3.5-turbo models across three distinct learning scenarios: zero-shot, 1-shot, and 5-shot, and on the three datasets.

Zero-shot: As shown in Table 2, in the SemEval dataset, the inclusion of eMFD and FrameAxis led to performance improvements of up to 4.82% and 4.47% points, respectively, for Llama2-70b-chat, and up to 5.47% and 2.16% points for Mixtral-7x8B. In the CB dataset, enhancements were even more pronounced, with eMFD and FrameAxis boosting F1 scores by as much as 15.91% and 20.22% points for the Llama2-70b-chat model. For the P-Stance Dataset, improvements reached up to 7.41% points with eMFD and an impressive 18.64% points with FrameAxis, highlighting the substantial benefits of integrating moral foundation features.

1-shot: On the SemEval dataset, inclusion of eMFD and FrameAxis contributed to increases in F1 score by up to 3.50% and 4.02% points for Llama2-70b-chat, 5.57% and 8.34% points for Mixtral-7x8B, and 1.48% and 2.84% points for GPT-3.5-turbo. Similarly, for the CB dataset, including these moral features raised F1 scores by up to 17.47% and 23.68% points using the Llama2-70b-chat model, underscoring the value of incorporating moral foundation features. In the P-Stance dataset, the eMFD and FrameAxis moral features boosted F1 scores by 4.94% and 17.13% points using the same model, respectively, highlighting the effectiveness of FrameAxis in capturing moral narratives.

5-shot: On the SemEval dataset, we observed improvements of up to 4.95% and 7.89% points with the addition of eMFD and FrameAxis, respectively, for the Llama2-7b-chat model. We noted similar improvements on the CB dataset using the Llama2-7b-chat model, with eMFD and FrameAxis leading to increases of up to 21.39% and 22.81% points respectively. Highest improvement was observed

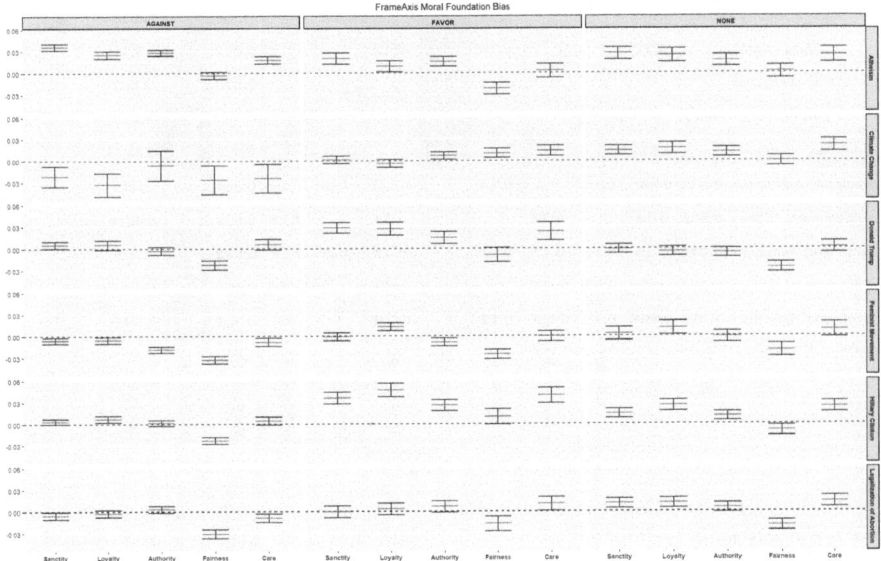

Fig. 2. Target- and stance-level heterogeneity in FrameAxis bias of moral foundations from the SemEval dataset.

using Mixtral model, where the inclusion of FrameAxis led to performance improvement of up to 24.86%. In the P-Stance dataset, integrating eMFD and FrameAxis resulted in an increase of 5.95% and 18.46% points respectively, using the Llama2-7b-chat model. Taken together, these performance improvements highlight the predictive importance of morality features.

7   How Do Moral Foundations Affect Stance Conveyance?

In the previous section, we highlighted the predictive value of incorporating moral foundations in stance detection tasks. In this section, we take a closer look at the associations between stances towards specific targets, and the expression of specific moral foundations.

In Fig. 2, we illustrate the prevalence of various FrameAxis bias features for our focal moral foundations, across targets and stance classes in the SemEval dataset. We note that moral bias generated from FrameAxis varies significantly across targets, as well as between stance classes within targets. For example, most users against the target *Climate Change is a Real Concern* scored lower on the Care, Fairness, Authority and Sanctity foundations than users supporting the target. This was a result of users making greater use of language relating to moral violation and/or lesser use of language relating to moral virtue, in these domains.

To further analyse how moral values are correlated with stances, in the SemEval and CB datasets, we encoded the "Favor" or "Support" stance as 1,

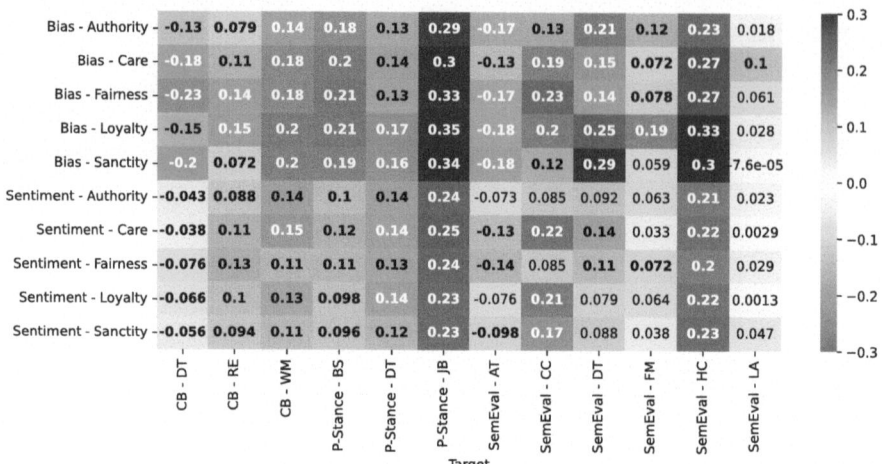

Fig. 3. Biserial correlations between moral foundation features and stance. Correlations that are statistically significant at $p < 0.05$ significance level are indicated in bold.

and "Against" stance as 0, and subsequently measured the biserial correlation between this binary stance, and the sentiment and bias features for each moral foundation. The statistical significance based on a two-tailed t-test was calculated for each moral foundation feature and target pair.

As evident from Fig. 3, the eMFD Sentiment and FrameAxis Bias features exhibited the same direction of correlations for most targets and moral foundations. This reflects the consistency between the moral information extracted using eMFD and FrameAxis techniques. The direction of correlation provides an indication of the moral polarity between individuals supporting a target and those against it. The correlational results help explain why the elicitation of moral values from tweets might have contributed to the observed performance enhancements in our stance detection models.

8 Conclusion and Future Work

Moral foundations play a key role in shaping our social behavior in a variety of contexts. Although past studies have explored the prevalence and associations of moral foundations in online discourse, the predictive value of moral foundation representations across a diverse range of tasks, targets, and datasets have remained understudied. In this paper, we investigate if the inclusion of moral foundations can improve the detection of online stances on social media. While existing stance detection models primarily use text- and interaction-based features, our proposed models highlight the importance of incorporating deeper user-level attributes, such as their moral foundations. Our models show improved performance in both message- and user-level stance detection tasks using traditional machine learning models, FLMs and more recent LLMs. Additionally, we highlight insightful associations between stances and each of the five moral foundations, which can provide useful inputs for researchers studying online discourse around societal and political events.

The moral encoders used in this paper can also be improved using recent modeling innovations, notably the incorporation of LLMs. For instance, future work can consider using LLMs to generate moral foundation features, as well as other related attributes (e.g., personalities and beliefs) to further improve performance on stance detection and related tasks.

Acknowledgments. This research is supported by the SMU-A*STAR Joint Lab in Social and Human-Centered Computing (SMU grant no.: SAJL-2022-CSS02, SAJL-2022-CSS003). This research is supported by A*STAR (C232918004, C232918005).

References

1. Aiyappa, R., An, J., Kwak, H., Ahn, Y.-Y.: Can we trust the evaluation on chatgpt? In: TrustNLP 2023, pp. 47–54 (2023)
2. AlDayel, A., Magdy, W.: Stance detection on social media: state of the art and trends. Inf. Process. Manag. **58**(4), 102597 (2021)
3. Allaway, E., Mckeown, K.: Zero-shot stance detection: a dataset and model using generalized topic representations. In: EMNLP 2020, pp. 8913–8931 (2020)
4. Alturayeif, N., Luqman, H., Ahmed, M.: A systematic review of machine learning techniques for stance detection and its applications. Neural Comput. Appl. **35**(7), 5113–5144 (2023)
5. Barbieri, F., Camacho-Collados, J., Espinosa Anke, L., Neves, L.: TweetEval: unified benchmark and comparative evaluation for tweet classification. In: Cohn, T., He, Y., Liu, Y., (eds.) ACL findings: EMNLP 2020, pp. 1644–1650. ACL (2020)
6. Biber, D., Finegan, E.: Adverbial stance types in English. Discourse Process. **11**(1), 1–34 (1988)
7. Brown, T., et al.: Language models are few-shot learners. Adv. Neural Inf. Process. Syst. **33**, 1877–1901 (2020)
8. Chen, T., Guestrin, C.: Xgboost: a scalable tree boosting system. In: Proceedings of the 22nd ACM SIGKDD International Conference on Knowledge Discovery and Data Mining, pp. 785–794 (2016)
9. Cruickshank, I.J., Ng, L.H.: Use of large language models for stance classification. arXiv preprint arXiv:2309.13734 (2023)
10. DeLucia, A., Wu, S., Mueller, A., Aguirre, C., Resnik, P., Dredze, M.: Bernice: a multilingual pre-trained encoder for Twitter. In: Goldberg, Y., Kozareva, Z., Zhang, Y., (eds.) EMNLP 2022, pp. 6191–6205. ACL (2022)
11. Devlin, J., Chang, M.W., Lee, K., Toutanova, K.: BERT: pre-training of deep bidirectional transformers for language understanding. In: Burstein, J., Doran, C., Solorio, T., (eds.) NAACL 2019: Human Language Technologies, Volume 1 (Long and Short Papers), pp. 4171–4186. ACL (June 2019)
12. Du Bois, J.W.: The stance triangle. Stancetaking Discourse: Subjectivity, Eval. Interact. **164**(3), 139–182 (2007)
13. Fürnkranz, J.: A study using n-gram features for text categorization. Austrian Res. Inst. Artif. Intell. **3**(1998), 1–10 (1998)

14. Garten, J., Hoover, J., Johnson, K.M., Boghrati, R., Iskiwitch, C., Dehghani, M.: Dictionaries and distributions: combining expert knowledge and large scale textual data content analysis: distributed dictionary representation. Behav. Res. Methods **50**, 344–361 (2018)
15. Gera, P., Neal, T.: A comparative analysis of stance detection approaches and datasets. In: Proceedings of Workshop on Eval4NLP 2022, pp. 58–69 (2022)
16. Ghanem, B., Rosso, P., Rangel, F.: Stance detection in fake news a combined feature representation. In: Proceedings of Workshop on FEVER 2018, pp. 66–71 (2018)
17. Teo, T.W., Choy, B.H.: STEM education in Singapore. In: Tan, O.S., Low, E.L., Tay, E.G., Yan, Y.K. (eds.) Singapore Math and Science Education Innovation. ETLPPSIP, vol. 1, pp. 43–59. Springer, Singapore (2021). https://doi.org/10.1007/978-981-16-1357-9_3
18. Graham, J., Haidt, J., Nosek, B.A.: Liberals and conservatives rely on different sets of moral foundations. J. Pers. Soc. Psychol. **96**(5), 1029 (2009)
19. Graham, J., Nosek, B.A., Haidt, J., Iyer, R., Koleva, S., Ditto, P.H.: Mapping the moral domain. **101**, 366 (2011). American Psychological Association (2011)
20. Haidt, J.: The new synthesis in moral psychology. Science **316**(5827), 998–1002 (2007)
21. Haidt, J., Graham, J.: When morality opposes justice: conservatives have moral intuitions that liberals may not recognize. Soc. Justice Res. **20**(1), 98–116 (2007)
22. Haidt, J., Joseph, C.: Intuitive ethics: how innately prepared intuitions generate culturally variable virtues. Daedalus **133**(4), 55–66 (2004)
23. Hopp, F.R., Fisher, J.T., Cornell, D., Huskey, R., Weber, R.: The extended moral foundations dictionary (eMFD): development and applications of a crowd-sourced approach to extracting moral intuitions from text. Behav. Res. Methods **53**, 232–246 (2021)
24. Küçük, D., Can, F.: Stance detection: a survey. ACM Comput. Surv. (CSUR) **53**(1), 1–37 (2020)
25. Kwak, H., An, J., Jing, E., Ahn, Y.-Y.: Frameaxis: characterizing microframe bias and intensity with word embedding. PeerJ Comput. Sci. **7**, e644 (2021)
26. Li, Y., Sosea, T., Sawant, A., Nair, A.J., Inkpen, D., Caragea, C.: P-stance: a large dataset for stance detection in political domain. In: ACL findings: ACL-IJCNLP 2021, pp. 2355–2365 (2021)
27. Liu, Y., et al.: Roberta: A robustly optimized BERT pretraining approach. arXiv preprint arXiv:1907.11692 (2019)
28. Saif Mohammad, Svetlana Kiritchenko, Parinaz Sobhani, Xiaodan Zhu, and Colin Cherry. Semeval-2016 task 6: Detecting stance in tweets. In *Proceedings of the 10th International Workshop on SemEval 2016*, pages 31–41
29. Mohammad, S.M., Sobhani, P., Kiritchenko, S.: Stance and sentiment in tweets. ACM TOIT, **17**(3), 1–23 (2017)
30. Mokhberian, N., Abeliuk, A., Cummings, P., Lerman, K.: Moral framing and ideological bias of news. In: Social Informatics: 12th International Conference, SocInfo 2020, pp. 206–219. Springer (2020)
31. Nguyen, D.Q., Vu, T., Tuan Nguyen, A.: BERTweet: a pre-trained language model for English tweets. In: Liu, Q., Schlangen, D., (eds.) EMNLP 2020: System Demonstrations, pp. 9–14. ACL (2020)
32. Nguyen, D.V., Nguyen, Q.N.: Evaluating the symbol binding ability of large language models for multiple-choice questions in vietnamese general education. In: SOICT 2023, pp. 379–386 (2023)

33. Popović, M.: chrF: character n-gram F-score for automatic MT evaluation. In: Proc. WMT 2015, pp. 392–395 (2015)
34. Reimers, N., Gurevych, I.: Sentence-BERT: sentence embeddings using siamese BERT-networks. In: EMNLP-IJCNLP 2019, pp. 3982–3992 (2019)
35. Rezapour, R., Shah, S.H., Diesner, J.: Enhancing the measurement of social effects by capturing morality. In: WASSA 2019, pp. 35–45 (2019)
36. Rezapour, R., Dinh, L., Diesner, J.: Incorporating the measurement of moral foundations theory into analyzing stances on controversial topics. In: Proc. ACM Hypertext 2021, pp. 177–188 (2021)
37. Sanseviero, O., Tunstall, L., Schmid, P., Mangrulkar, S., Belkada, Y., Cuenca, P.: Mixture of experts explained. Hugging Face Blog (2023)
38. Shannon, C.E.: A mathematical theory of communication. Bell Syst. Tech. J. **27**(3), 379–423 (1948)
39. Taulé, M., Pardo, F.M.R., Martí, M.A., Rosso, P.: Overview of the task on multimodal stance detection in tweets on catalan# 1oct referendum. In: IberEval@ SEPLN, pp. 149–166 (2018)
40. Touvron, H., et al.: Llama 2: Open foundation and fine-tuned chat models. arXiv preprint arXiv:2307.09288
41. Upadhyaya, A., Fisichella, M., Nejdl, W.: A multi-task model for sentiment aided stance detection of climate change tweets. In: AAAI ICWSM vol. 17, pp. 854–865 (2023)
42. Vaswani, A., et al.: Attention is all you need. NIPS (2017) 30
43. Weber, R., et al.: Extracting latent moral information from text narratives: relevance, challenges, and solutions. In: Computational Methods for Communication Science, pp. 39–59. Routledge, (2021)
44. Zhang, H., Kwak, H., Gao, W., An, J.: Wearing masks implies refuting trump?: towards target-specific user stance prediction across events in COVID-19 and us election. In: ACM WEBSCI 2023, pp. 23–32 (2020)

On Mining Dynamic Graphs for k Shortest Paths

Andrea D'Ascenzo[1]($\boxtimes$)(iD) and Mattia D'Emidio[2](iD)

[1] Luiss University, Rome, Italy
adascenzo@luiss.it
[2] University of L'Aquila, L'Aquila, Italy
mattia.demidio@univaq.it

Abstract. Mining graphs, upon query, for k shortest paths between vertex pairs is a prominent primitive to support several analytics tasks on complex networked datasets. The state-of-the-art method to implement this primitive is kPLL, a framework that provides very fast query answering, even for large inputs and volumes of queries, by pre-computing and exploiting an appropriate *index* of the graph. However, if the graph's topology undergoes changes over time, such index might become obsolete and thus yield incorrect query results. Re-building the index from scratch, upon every modification, induces unsustainable time overheads, incompatible with applications using k shortest paths for analytics purposes. Motivated by this limitation, in this paper, we introduce DECKPLL, the first dynamic algorithm to maintain a kPLL index under *decremental* modifications. We assess the effectiveness and scalability of our algorithm through extensive experimentation and show it updates kPLL indices orders of magnitude faster than the re-computation from scratch, while preserving its compactness and query performance. We also combine DECKPLL with INCKPLL, the only known dynamic algorithm to maintain a kPLL index under *incremental* modifications, and hence showcase, on real-world datasets, the first method to support fast extraction of k shortest paths from graphs that evolve by arbitrary topological changes.

Keywords: Graph Algorithms · Dynamic Networks · Algorithm Engineering · Experimental Algorithmics

1 Introduction

Computing, upon a *query*, a set of k *shortest paths* for a pair of vertices of a graph is considered a primitive of high importance in the context of graph analytics. Such paths, in fact, are often used to determine relations of closeness or similarity between vertices, based on the graph topological structure [4,9,29], and hence to support network classification tasks where graphs can be compared by sampling pairs of vertices and by analyzing their topological relationships [1, 28]. In particular, k shortest paths (ranked by their respective lengths, called

L. M. Aiello et al. (Eds.): ASONAM 2024, LNCS 15211, pp. 320–336, 2025.
https://doi.org/10.1007/978-3-031-78541-2_20

k shortest distances), are known to represent a natural, informative and robust measure of distance/similarity between vertices (and hence graphs) [1,12,20].

Methods to answer to queries on k shortest paths can be broadly divided into two groups, namely *without indexing* and *with indexing* [1]. In the former group, we find algorithms that directly solve, upon query, the single pair k shortest paths problem, i.e. that compute k shortest paths between a pair of vertices in a graph in non-descending order of their costs. Chiefly, reference algorithms in this group are: (i) the KBFS algorithm, a variant of the Breadth First Search (BFS) algorithm that visits a same vertex up to k times [1]; (ii) the Eppstein's algorithm which builds a shortest path tree, rooted at a vertex, and form paths to the other vertex by selecting edges that are outside the tree and connect different sub-trees [12]. For an n-vertex m-edge graph, the former runs in $\mathcal{O}((n+m)k)$ time while the latter reduces the time complexity to $\mathcal{O}(n+m+k)$. In the latter group, instead, we have strategies that divide the computational effort in two steps: in an offline phase, a one-time *pre-processing* of the input graph is performed to construct an *index* data structure which is then, at run-time, exploited to dramatically reduce the time necessary to answer queries on k shortest paths [1,19]. Methods without indexing are considered efficient in terms of space occupancy, and preferred in practical scenarios where memory is a constrained resource (e.g. embedded systems), while methods with indexing are superior in terms of query times and therefore are adopted in all real-world applications where query performance becomes is crucial, e.g. when large graphs are managed or several (possibly millions) queries have to be answered [1].

Among indexing-based methods, the KPLL framework, proposed in [1], has emerged as the one performing best in practice. Such framework consists of: (i) a pre-processing algorithm that computes, once, a so-called *k-2-Hop-Cover index* of the input graph storing, for each vertex, in the form of *labels*, selected information about paths and cycles passing through said vertex in the graph; (ii) a query algorithm, that retrieves, upon request, the k shortest paths by only accessing the labels of two queried vertices. Despite having high time and space complexities in the worst case, extensive experimental work has shown that carefully engineered implementations of KPLL, incorporating effective pruning strategies, outperform all other methods in the literature for queries on k shortest paths, since: (i) the time (space, resp.) overhead for pre-computing (storing, resp.) the index, is affordable and compatible with the requirements of applications of interest (at most few thousands of seconds to run the pre-processing and few GBs to store the index); (ii) using the index ensures extremely small query times, orders of magnitude smaller ($\approx \mu s$ per query) than any other known solution [1,9].

The main drawback of KPLL, and in general of indexing-based methods for graph mining, is that pre-computed information can easily become *obsolete* and yield incorrect query results when the managed network changes over time, i.e. if the underlying graph is *dynamic* [9,10]. In fact, most methods for fast query answering through pre-processing store topological data in the index that can be heavily altered even by very limited modifications occurring to the input (e.g. an edge deletion) [3,8,16,18]. One possibility to use indexing-based techniques with dynamic graphs is to re-compute the index from scratch after an change occurs on the network. However, this is considered impractical, since the pre-computation phase, though effective, induces large time overheads if repeated too often, e.g. with the frequency typically networks undergo updates in practice [3,17]. For this reason, and since most real-world graphs of interest for analytical purposes are inherently dynamic (e.g. social/road graphs) [3,17,24] researchers have worked on designing effective *dynamic algorithms* that are able to identify and update only the part of the pre-computed data structure that is altered by a graph change, faster than the pre-processing routine, while preserving the correctness of queries and the compactness of the index [3,15]. Investigations of this kind have concerned most indexing-based frameworks for retrieving non-trivial graph structures [3,8,13,14,16]. For KPLL indices, currently, the only known solution of this kind is algorithm INCKPLL [9], which however works only for graphs that change by *incremental updates* (vertex/edge insertions). To the best of our knowledge, no dynamic algorithm is known to update a KPLL index when the graph is *fully dynamic*, i.e. changes in an unrestricted, arbitrary fashion (including *decremental updates*, i.e. vertex/edge deletions).

Our Contribution. In this paper, we remove such limitation and present DECKPLL, the first algorithm to efficiently maintain a KPLL index under *decremental* changes. We prove the correctness of our method and assess its effectiveness through extensive experimental work. In particular, we provide empirical evidences on DECKPLL being able to update a KPLL index, on average, orders of magnitude faster than the re-computation from scratch, while at the same time preserving its compactness and query performance. We also experimentally show that DECKPLL can be used in combination with INCKPLL, the only known dynamic algorithm to maintain a KPLL index under *incremental* modifications hence showcasing the first method to support very fast, interactive extraction of k shortest paths from graphs that evolve by arbitrary topological changes.

Related Works. The problem of computing k shortest paths and distances has been widely investigated in the last decade, in several variants. The k *Shortest Simple Paths (*K-SSP*) problem*, for instance, requires that the sought k paths have to be *simple*, i.e. must not self-intersect. This version is generally considered computationally harder and, indeed, no asymptotic improvement is known to Yen's algorithm [27], which runs in $O(kn(m+n\log n))$ time in an n-vertex, m-edge graph. Some heuristics have been proposed to empirically accelerate Yen's algorithm (see, e.g. [30]), however none of them shows query performance comparable to that achieved by indexing-based methods for similar graph analytics problems (see, e.g., [1,2,10]). In this sense, designing an algorithmic method to solve the K-SSP problem with small query times is an open problem and represents an active area of research [30]. Among frameworks known in the literature to support graph mining processes at scale, those relying on computing compact indices via preprocessing have been observed to exhibit the best performance and scalability properties [10,13]. In particular, the hub-labeling technique, introduced in [6], has been successfully modified to handle, effectively, several graph analytics tasks, such as, e.g., path counting [29], best connections in transport systems [11,26] or constrained connectivity [23]. Due to the inherent time-evolving nature of real-world networks, most studies on pre-processing-based frameworks have been followed by investigations on corresponding *dynamic* algorithms to efficiently reflect topological changes, occurring on the underlying graph, on pre-computed data structures. Examples of such investigations include the design and experimental evaluation of dynamic algorithms for maintaining, over time: shortest path trees [7], transitive closure [16], centrality measures [21], or timetable graphs [5].

2 Preliminaries

We are given a graph $G = (V, E)$, with a vertex set V and a edge set E. We call $n = |V|$ ($m = |E|$, resp.) its number of vertices (edges, resp.). We denote by $(x, y) \in E$ an edge connecting two vertices $x, y \in V$ in G. For the sake of simplicity, we describe all methods for undirected, unweighted graphs. All presented techniques and algorithms can be extended to weighted digraphs.

A *path* $p = (v_1, v_2, \ldots, v_r)$ in G, connecting two vertices $s, t \in V$, is a sequence of r vertices such that $(v_i, v_{i+1}) \in E$ for all $i \in [1, r-1]$. We say edge (v_i, v_{i+1}) belongs to the path or $(v_i, v_{i+1}) \in p$ for each pair of vertices in p. We call: $v_1 = s$ and $v_r = t$ the *endpoints* of the path; and any $v_i, 1 < i < r$ an *internal vertex* of the path. Any path whose endpoints coincide is called a *cycle*, while a path is *simple* if the sequence has no vertex repetitions. The *length* $\ell(p)$ of a path p is the number of edges in p. Multiplicities of edge repetitions are counted if the path is not simple. A *shortest path* $p(s, t)$, for two vertices $s, t \in V$, is a path having minimum length among all those in G connecting s and t. The *distance* $d(s, t)$ for a pair $s, t \in V$ is the length $\ell(p(s, t))$ of a shortest path $p(s, t)$. Given a path $p = (v_1, \ldots, v_r)$, we use $p[v_i, v_j]$ to identify the *subpath* of p having v_i and v_j as endpoints, and whose internal vertices are those between v_i and v_j, for $1 \leq i < j \leq r$. If vertices v_i and v_j appear multiple times, the first (last, resp.) occurrence of v_i (v_j, resp.) is considered. Subpath $p[v_1, v_j]$ ($p[v_i, v_r]$, resp.) is called *prefix* (*suffix*, resp.) path of p. Additionally, given two paths $p = (v_1, \ldots, v_r)$ and $q = (w_1, \ldots, w_l)$ such that $v_r = w_1$, we denote by $p \oplus q = (v_1, \ldots, v_r = w_1, \ldots, w_l)$ the *concatenation* of p and q.

We assume vertices are associated to unique integer identifiers to enable natural comparisons for any pair $u, v \in V$ by expressions such as $u < v$ or $u \leq v$. Given two vertices $s, t \in V$, we call: (i) $\mathcal{P}_{st}$ the set of paths in G connecting s and t; (ii) $\mathcal{P}_{st}^{>v} \subseteq \mathcal{P}_{st}$ the subset of $\mathcal{P}_{st}$ containing paths whose internal vertices are all larger than some $v \in V$; (iii) $\mathcal{P}_{st}^{\not> v}$ the subset of $\mathcal{P}_{st}$ containing paths such that at least one internal vertex is smaller than or equal to v. Furthermore, we call $p_i(s, t)$ the *i-th shortest path* between s and t, i.e. the i-th element in $\mathcal{P}_{st}$, sorted in non-decreasing order of path lengths, and use $d_i(s, t) = \ell(p_i(s, t))$ to refer to the *i-th shortest distance* for pair s, t, i.e. the length of $p_i(s, t)$. Similarly, we use:

(i) $d_i^{>v}(s,t)$ ($d_i^{\geq v}(s,t)$ and $d_i^{\not> v}(s,t)$, resp.) to refer to the i-th shortest distance when paths are restricted to have internal vertices that all are larger (larger than or equal to and not greater, resp.) than some $v \in V$; (ii) $p_i^{>v}(s,t)$ ($p_i^{\geq v}(s,t)$, resp.) to identify the corresponding i-th shortest path; (iii) $d^{>v}(s,t) = d_1^{>v}(s,t)$ ($p^{>v}(s,t) = p_1^{>v}(s,t)$, resp.) to denote the distance (a shortest path inducing the distance, resp.) subject to the same restrictions on vertices. Finally, we call $\mathcal{P}_{st}^k$ a set of k shortest paths between s and t in G, i.e. a set containing an i-th shortest path for any $i \in [0, k-1]$, and by $\mathcal{P}_{st}^{k,>u}$ a set of k shortest paths between s and t in G having internal vertices larger than some $u \in V$.

Given a graph $G = (V, E)$, an integer $k > 1$, and a pair of vertices $s, t \in V$, the k *shortest paths* (K-SP) problem asks to compute a set $\mathcal{P}_{st}^k$ of k shortest paths between s and t in G, i.e. to answer a *query* on the k shortest paths for s and t. The KPLL framework [1] is the reference approach to address the K-SP problem in terms of computational time per query. The method is based on the pre-computation, for a given graph, of a data structure called k-*2-Hop-Cover* (k2HC, for short) *index*, a generalization of the *2-Hop-Cover index*, introduced in [6]. Such index can be used to solve both the K-SP problem and its natural specialization, named k *shortest distances* problem, that asks to retrieve only the lengths of k shortest paths, and is defined as follows. Given a graph $G = (V, E)$, define, for each vertex $v \in V$: (i) a *length label* $L(v)$, containing pairs in the form $(u, \mathcal{P}_{u,v}^{k,>u})$ where $u \in V$; (ii) a *loop label* $C(v)$, storing a sequence of k cycles $(c_1(v), c_2(v), \ldots, c_k(v))$ in G that include vertex v. Then, pair $I = (L, C)$, where $L = \{L(v)_{v \in V}\}$ and $C = \{C(v)\}_{v \in V}$, is called a k-2-Hop Cover index of G.

Note that the k2HC index is often referred to as k2HC labeling or simply k2HC. We use these notations interchangeably. Elements in the labels are called *entries*. A k2HC index is said to *cover* a graph G, or equivalently to satisfy the k-*cover property* for G, whenever it can be exploited to correctly solve the K-SP problem on G. Specifically, given a k2HC index $I = (L, C)$ of a graph $G = (V, E)$, let QUERY(I, s, t) identify the result of executing a *query* on I for a pair of vertices $s, t \in V$ and let such result consist of the shortest k paths, in terms of length, in multi-set $\Delta(I, s, t) = \{p_{vs} \oplus c_i(v) \oplus p_{vt} | (v, \mathcal{P}_{vs}) \in L(s), p_{vs} \in \mathcal{P}_{vs}, c_i(v) \in C(v), (v, \mathcal{P}_{vt}) \in L(t), p_{vt} \in \mathcal{P}_{vt}\}$. Then, I covers G, or equivalently satisfies the k-cover property for G, if and only if, for any pair $(s, t) \in V \times V$, we have QUERY$(I, s, t) = \mathcal{P}_{st}^k$. In other words, a k2HC covering a graph G allows to retrieve the k shortest paths in G for any pair of vertices $s, t \in V$, by a query on the index that selects shortest combinations in $\Delta(I, s, t)$, obtained by appending a cycle, originating at some vertex $v \in V$, to a path from s to v, and by concatenating to it a path from v to t. Any vertex v that forms one of the k smallest combinations of $\Delta(I, s, t)$ is called a *hub vertex* for pair s, t. Whenever a pair s, t is disconnected in G then $d_i(s, t) = \infty \, \forall i \in [1, k]$. Correspondingly, QUERY$(I, s, t)$ returns a single, default null value whenever there is no vertex $v \in V$ such that $(v, \mathcal{P}_{sv}) \in L(s), (v, \mathcal{P}_{tv}) \in L(t)$.

We call *size* of the index the total number of entries in length and loop labels. A naive k2HC of size $\Omega(kn^2)$ can be computed by $\mathcal{O}(n^2)$ executions of Eppstein's algorithm. However, to obtain superior query performance, a k2HC of minimum size is desirable. Unfortunately, by a reduction to the computation of a minimum 2-Hop-Cover index, the problem of computing a minimum-sized k2HC is $\mathcal{NP}$-hard [6]. Nevertheless, the heuristic of [1] has been shown to compute reasonably compact k2HC indices, which results in superior performance in terms of trade-off between preprocessing time, index size and query time. Specifically, the construction guarantees that, for any $v \in V$: (i) paths in $L(v)$, associated with a vertex $u \in V$, belong to $\mathcal{P}_{uv}^{k,>u}$; (ii) cycles in $C(v)$ belong to $\mathcal{P}_{vv}^{k,>v}$. Properties (i) and (ii), combined, guarantee that the resulting k2HC satisfies the k-cover property. Note that, in the reminder of the paper, for the sake of brevity, we use acronym KPLL also to refer to the preprocessing routine of the framework.

Dynamic Scenario. In a *dynamic* scenario we assume we are given an *initial* graph, say $G = (V, E)$, that can undergo either *incremental* (i.e. vertex/edge insertions) or *decremental* modifications (i.e. vertex/edge removals) at arbitrary times. In such scenario, pre-computed data structures, exploited to address the K-SP problem (e.g. the k2HC), might not reflect properly the graph structure as the topology changes over time, and thus the task of maintaining stored information updated emerges. Given a decremental modification x occurring on a graph G (e.g. the deletion of an edge $e \in E$), the *decremental* K-SP *problem* asks to compute the set $\mathcal{P}_{st}^{k} = \{p_1'(s,t), p_2'(s,t), \ldots, p_k'(s,t)\}$ of the k shortest paths between s and t in G', for some $s, t \in V'$, where $G' = (V', E')$ is the graph obtained by applying x to G (e.g. by removing e from E). If one relies on a k2HC to solve the K-SP problem, the decremental K-SP problem translates to a corresponding *decremental* k2HC *problem* which asks, given a graph $G = (V, E)$, a k2HC index I covering G, and a decremental modification x on G, to compute a k2HC I' that covers G' where G' is obtained by applying x to G. Note that, the incremental counterpart of the K-SP problem has been defined and addressed in [9]. In said work, the authors introduce INCKPLL, a *partially dynamic* (specifically *incremental*) algorithm that, given a graph G, a k2HC index $I = (L, C)$ covering G, and an incremental update x on G, updates I to an index $I' = (L', C')$ which covers graph G', obtained by applying x to G.

3   Decremental Algorithm

In this section we introduce DECKPLL, a dynamic algorithm to handle the decremental k2HC problem. Similarly to the incremental problem, if the change to be managed is a vertex update (insertion or deletion), this can be modeled as an appropriate sequence of edge operations (insertions or deletions) incident to the interested vertex [8]. Therefore, in what follows, for the sake of simplicity, we focus on handling edge deletions only. As in [1], we consider vertices are sorted by an easy-to-compute ordering, e.g. non-increasingly by degree, and indices reflect such ordering, i.e. $v_i \leq v_j \implies |N(v_i)| \leq |N(v_j)|$.

Let $G = (V, E)$ be a graph and let $I = (L, C)$ be a k2HC covering G. Let us be given a decremental update for G, namely the removal of an edge $e = (x, y) \in E$.

Algorithm 1: Algorithm DECKPLL.

Input: A k2HC $I = (L, C)$ covering graph $G = (V, E)$, edge $(x, y) \in E$
Output: A k2HC $I' = (L', C')$ covering graph $G' = (V, E \setminus \{(x, y)\})$
1 AFF-SET $\leftarrow \emptyset$;
2 **foreach** $u \in \{x, y\}$ **do** FINDAFF(u, AFF-SET);
3 AFF-HUBS $\leftarrow$ REMOVAL(I, AFF-SET);
4 $I' \leftarrow$ RESTORE(I, AFF-HUBS);
5 **return** I';

Differently from the incremental problem, in the decremental problem entries of the index might become *obsolete* (i.e. might not correspond to a path or a cycle in the graph) as a consequence of a decremental update. Such entries, therefore, must be removed to guarantee the correctness of queries. To this aim, our update procedure (see Algorithm 1), works in three phases: (i) detection of vertices *affected* by the edge removal, i.e. whose length label contains at least one obsolete entry; (ii) removal of obsolete entries, i.e. an entry associated to a path that traverses the removed edge; (iii) restoration of k-cover property, which might be broken by the removal of obsolete entries. In the detection of affected vertices (see Procedure 2), we search for vertices whose length label contains at least one obsolete entry. To this aim, we execute a BFS-like visit, on the new graph G', rooted at the two endpoints of the removed edge x and y. For each visited vertex v, the procedure executes QUERY to retrieve the k shortest paths from one of the endpoints to v encoded in the index: if none of such paths contains the removed edge, then we prune (i.e. stop) the visit at v. Vice versa, we add v to the set AFF-SET of affected vertices (initially empty). The algorithm is symmetric for x and y hence it is executed twice.

Procedure 2: Sub-routine FINDAFF of Algorithm 1.

Input: Endpoint u of removed edge, set AFF-SET
1 $E' \leftarrow E \setminus \{e\}$;
2 $G' \leftarrow (V, E')$;
3 **foreach** $t \in V$ **do** $visited[t] \leftarrow false$;
4 $visited[u] \leftarrow true$;
5 $Q \leftarrow \{u\}$;
6 **while** $Q \neq \emptyset$ **do**
7 $\quad$ Dequeue v from Q;
8 $\quad$ $visited[v] \leftarrow true$;
9 $\quad$ **foreach** $path \in$ QUERY(I, v, y) **do**
10 $\quad\quad$ **if** $(x, y) \in path$ **then**
11 $\quad\quad\quad$ AFF-SET $\leftarrow$ AFF-SET $\cup \{v\}$
12 $\quad$ **foreach** $(v, w) \in E' : \neg visited[w]$ **do**
13 $\quad\quad$ Enqueue w into Q;

After the detection of affected vertices, the algorithm removes obsolete entries by scanning length labels of each $v \in$ AFF-SET and by removing entries such that associated paths contain edge (x, y). This is done by sub-routine REMOVAL (see Procedure 3). During the removal, the algorithm traces and stores in a suited set, named AFF-HUBS, the hub vertices associated to removed entries. Finally, in the third and last phase (sub-routine RESTORE, see Procedure 4), the algorithm restores the k-cover property. To this purpose, both loop labels and length labels are updated. Specifically, for the former, we compute a set D_{cy} that contains any vertex v for which at least one loop label may be obsolete due to the change. In particular, such set, for a given loop labeling C, contains any vertex v which is: (i) connected either to x or y by a shortest path whose length is at most k and whose vertices are all larger than v; (ii) larger than or equal to the minimum between x and y (according to the vertex ordering); (iii) such that there exists a cycle in $C(v)$ containing the removed edge. These conditions can be summarized by Eq. 1:

$$D_{cy}(C) = \{v \in V : (d^{>v}(v,x) \le k \vee d^{>v}(v,y) \le k) \wedge v \le \min(x,y) \wedge \exists c_i \in C(v) \ : \ (x,y) \in c_i\} \quad (1)$$

Procedure 3: Sub-routine REMOVAL of Algorithm 1.

Input: A k2HC $I = (L, C)$ covering graph G, set AFF-SET
Output: Set AFF-HUBS.
1 AFF-HUBS $\leftarrow \emptyset$;
2 **foreach** $v \in$ AFF-SET **do**
3   **foreach** $(h, \mathcal{P}) \in L'(v)$ **do**
4   **foreach** $path \in \mathcal{P}$ **do**
5   **if** $(x,y) \in path$ **then**
6   Remove $path$ from $\mathcal{P}$;
7   AFF-HUBS $\leftarrow$ AFF-HUBS $\cup \{h\}$;
8 **return**AFF-HUBS;

For each vertex w of $D_{cy}(C)$, loop label entries are removed from $C(w)$ and new ones are computed in lines 1-13 of Algorithm 4. This part of the algorithm mimics the method in [1] for computing a k2HC from scratch, but only for a selected subset of vertices. Specifically, it starts a visit of the graph from each $v_i \in D_{cy}$, following the vertex ordering and traversing only vertices not smaller than v_i, and storing newly discovered cycles in the loop labels.

Finally, the algorithm searches the graph for new length label entries that must be added to the index to ensure that the k-cover property holds for the modified graph G'. By definition, set AFF-HUBS contains any vertex h such that the construction of the KPLL index, which performs a shortest-path like visit rooted at each vertex h [1], traversed the removed edge (x, y). Thus, to compute a k2HC index covering the modified graph, we execute lines 22-22 of Procedure 4 where we perform a shortest-path like visit, rooted at each vertex $h \in$ AFF-HUBS, that traverses only vertices larger than h and exploits information stored in the index to prune the visit when newly discovered paths are longer than those encoded in the current index. When new paths are discovered, a corresponding length label entry is added to the k2HC. We now state the correctness and provide a complexity analysis of DECKPLL. Due to space limitations, all proofs are deferred to a full version of this paper.

Lemma 1. *Let* $I = (L, C)$ *be a* k2HC *covering a graph* $G = (V, E)$. *Assume* $e = (x, y) \in E$ *is removed from* G. *Then, set* AFF-SET, *computed by Algorithm 2,*

Procedure 4: Sub-routine RESTORE of Algorithm 1.

Input: A k2HC $I = (L, C)$ not guaranteed to cover G', set AFF-HUBS
Output: A k2HC $I' = (L', C')$ covering G'

1 Compute $D_{cy}(C)$ as in Eq. 1;
2 **foreach** $w \in D_{cy}(C)$ **do**
3 | $C(v) \leftarrow \emptyset$;
4 | **foreach** $t \in V$ **do** $visited[t] \leftarrow 0$;
5 | $visited[w] \leftarrow 1$;
6 | $Q \leftarrow \{(w, \{w\})\}$;
7 | **while** $Q \neq \emptyset$ **do**
8 | | Dequeue (x, p) from Q;
9 | | $visited[x] \leftarrow visited[x] + 1$;
10 | | **if** $x = w$ **then** Add p to $C(w)$;
11 | | **if** $visited[x] < k$ **then**
12 | | | **foreach** $(x, v) \in E : v \geq w$ **do**
13 | | | | Enqueue $(v, p \oplus \{v\})$ into Q;
14 **foreach** $h \in$ AFF-HUBS **do**
15 | $L(h) \leftarrow \emptyset$;
16 | $Q \leftarrow \{(h, 0)\}$;
17 | **while** $Q \neq \emptyset$ **do**
18 | | Dequeue (x, p) from Q;
19 | | **if** $\ell(p) < \max\{\ell(p) : p \in$ QUERY$(I, h, x)\}$ **then**
20 | | | Add (h, p) to $L(x)$;
21 | | | **foreach** $(x, w) \in E : w > v$ **do**
22 | | | | Enqueue $(w, p \oplus \{w\})$ into Q;
23 **return** $I' = (\{C(v)_{v \in V}\}, \{L(v)_{v \in V}\})$;

---

contains vertices affected by the removal, i.e. any vertex $v \in V$ whose length label $L(v)$ contains an entry $(h, \mathcal{P}_{hv}^{k,>h})$ such that edge e belongs to a path in $\mathcal{P}_{hv}^{k,>h}$.

By Lemma 1, it follows that, after the execution of REMOVAL, we obtain: (i) an index that does not contain any obsolete length entry; (ii) set AFF-HUBS containing any vertex associated to at least one removal of a length entry. We can then prove the following.

Theorem 1. *Let $I = (L, C)$ be a k2HC covering a graph $G = (V, E)$ and let $G' = (V, E \setminus \{(x, y)\})$ for some $e = (x, y) \in E$. Call k2HC $I' = (L', C')$ the k2HC updated by Algorithm 4. Then, I' satisfies the k-cover property for G'.*

We now give the time complexity of DECKPLL in an output bounded sense [3,8].

Theorem 2. *Let $I = (L, C)$ be a k2HC covering a graph $G = (V, E)$. Let $l = \max_{v \in V} |L(v)|$, and ω be the maximum length of any path in G. Given an edge $e = \{x, y\} \in E$, let $G' = (V, E' = E \setminus \{e\})$. Then, algorithm DECKPLL takes $\mathcal{O}(n(nk\omega l + m))$ time to update I to a k2HC I' covering G'.*

Note that the time complexity of DECKPLL is larger than that of the KPLL pre-processing since: $l = \mathcal{O}(kn)$, $r = \mathcal{O}(n)$, $s = \mathcal{O}(m)$ and $c = \mathcal{O}(m)$. However, our experiments show that, on average, such values are smaller than the worst case.

4 Experimental Evaluation

In this section, we describe the experimental study we conducted to evaluate the performance of DECKPLL. To this aim, we implemented: (i) the pre-processing routine of KPLL; (ii) algorithm KBFS, the modified version of the BFS algorithm that is considered the state-of-the-art to retrieve k shortest paths when no index is available [1]; (iii) our new decremental algorithm DECKPLL. Moreover, for the sake of completeness, we implemented incremental algorithm INCKPLL and combined it with DECKPLL to form algorithm FULKPLL, a fully dynamic algorithm that maintains a k2HC index under any type of update, by applying INCKPLL or DECKPLL, resp., depending on whether the update to be handled is, resp., incremental or decremental. This is done with the purpose of assessing the performance of DECKPLL and INCKPLL when applied in combination on graphs that change by arbitrary modifications. In fact, the behavior of the resulting combination, referred to FULKPLL in what follows, might be influenced by the lazy strategy of INCKPLL, since the latter does not remove obsolete entries after incremental updates and tends to increase the index size, which in turn might increment the computational effort of DECKPLL to update it. Nonetheless, further in this section we provide empirical evidences confirming that the two dynamic algorithms are very effective also when combined.

As inputs to experiments we consider a collection of real-world graphs representing networks of various application domains of interest (e.g. web or social graphs) with heterogeneous topologies [22, 25] (see Table 1, where graphs are sorted top-to-bottom by size $|V| + |E|$). Concerning parameter k, we use values of k in $\{2, 4, 8, 16\}$, as other studies on the K-SP problem [1, 9].

Table 1: Overview of Input Graphs.

Graph	Short	$|V|$	$|E|$	avg. deg.	diameter
EatRS	ERS	23 219	328 105	28	5
Arxiv	ARX	34 401	420 828	24	14
PPIHuman	PPH	16 820	449 036	53	7
PPIMouse	PPM	17 349	733 037	84	6
Gowalla	GOW	196 591	950 327	10	16
NotreDame	NDM	325 729	1 090 108	6	46
MarkerCafe	MRC	69 413	1 644 801	47	9
Arabic	ARB	163 598	1 747 269	21	76
Stanford	SFD	255 265	1 941 926	15	164
FlyHemiBrain	FHB	21 739	2 897 925	266	5

For each input graph G and value of k, we perform two types of experiments, named DECR and FULL, resp., depending on the types of selected graph changes. In experiment DECR, we first execute KPLL to compute a k2HC index I covering G. Then, we select uniformly at random, for $\sigma > 0$ times, an edge (x, y) in the graph, remove it to obtain a graph G', and run DECKPLL to update index I to I' covering G'. We repeat the process for G' and, at the end, we compute from scratch a k2HC index I'' covering the last snapshot of the graph via KPLL. The purpose of this setting is to evaluate scenarios where deletions occur with uniform probability. In experiment FULL, instead, we uniformly select at random $\sigma > 0$ edges from the input graph G and remove them to obtain a graph G_{init}. We compute a k2HC index I covering G_{init} via KPLL and then re-insert the σ sampled edges, one after the other, until all removed edges are added back. In between such insertions, at fixed intervals (every 5 insertions), we apply randomly selected edge removals to the graph, for a total of $\frac{\sigma}{5}$ decremental update operations. Such fraction is selected following common distributions of updates in real-world graphs [7]. After each modification, we execute FULKPLL to update index I to I'. Eventually, we run KPLL to compute a k2HC index I'' covering the last snapshot of graph. This second experiment is designed with a twofold purpose, namely evaluating update times of DECKPLL and INCKPLL, when their executions are interleaved into FULKPLL, and assessing whether the space occupancy to store the k2HC changes when FULKPLL algorithm is applied, and specifically if it is impacted by the lazy strategy of INCKPLL which does not remove obsolete entries.

In both experiments, after each execution of the dynamic algorithms (after the final execution of KPLL, resp.) we perform 10^5 queries on I' (on I'', resp.) and measure: (i) median query times; (ii) size of index I' (I'', resp.); (iii) running time to update index I' via the dynamic algorithms after each change (to recompute I'' from scratch via KPLL, resp.). In all trials, vertex ordering follows non-increasing values of vertex degree, as in [1,9], while σ is set to 500. All our code is written in C++ and compiled with GCC 9.4.0 with opt. level $O3$. All tests have been executed on a workstation equipped with an Intel Xeon© CPU E5-2643 3.40 GHz and 128 GB of RAM, running Ubuntu Linux.

Analysis. A excerpt of our experimental results, for $k = 8$ and for experiment DECR and FULL is given in Table 2. Results for other values of k lead to similar interpretation, hence are omitted due to space limitations and will be given in a full version of this paper. For each input graph, we report: (i) the running time of KPLL to build the index and the median running time of DECKPLL or FULKPLL

Table 2: Results of the DECR (left) and FULL (right) experiment for $k = 8$.

Graph	CT (s) kPLL	CT (s) DEC KPLL	SPEED-UP	IS (MB) kPLL	IS (MB) DEC KPLL	QT (μs) kPLL	QT (μs) DEC KPLL	Graph	CT (s) kPLL	CT (s) FUL KPLL	SPEED-UP	IS (MB) kPLL	IS (MB) FUL KPLL	QT (μs) kPLL	QT (μs) FUL KPLL
ERS	3699	368.31	$1.0 \cdot 10^1$	554	554	2775	2776	ERS	3520	0.31	$1.1 \cdot 10^4$	554	555	2597	2598
ARX	3186	108.65	$2.9 \cdot 10^1$	1130	1130	2020	2023	ARX	3235	0.32	$1.0 \cdot 10^4$	1130	1132	1958	1963
PPH	1092	19.16	$5.7 \cdot 10^1$	264	264	1315	1316	PPH	1228	0.16	$7.2 \cdot 10^3$	264	264	1390	1391
PPM	1402	25.57	$5.4 \cdot 10^1$	360	360	1604	1605	PPM	1429	0.10	$1.4 \cdot 10^4$	360	360	2113	2114
GOW	2681	193.58	$1.3 \cdot 10^1$	2052	2052	580	580	GOW	2735	0.06	$4.0 \cdot 10^4$	2052	2053	592	594
NDM	1120	15.27	$7.3 \cdot 10^1$	2945	2945	324	324	NDM	1119	0.01	$9.4 \cdot 10^4$	2945	2949	453	455
MRC	3474	90.44	$3.8 \cdot 10^1$	1035	1035	1204	1204	MRC	3568	0.05	$6.2 \cdot 10^4$	1035	1035	1196	1199
ARB	1668	0.43	$3.8 \cdot 10^3$	2713	2713	1495	1495	ARB	1688	0.02	$7.4 \cdot 10^4$	2713	2713	1463	1463
SFD	1240	4.69	$2.6 \cdot 10^2$	1925	1925	272	272	SFD	1322	0.01	$1.1 \cdot 10^5$	1922	1936	360	366
FHB	13247	28.97	$4.5 \cdot 10^2$	772	772	3782	3782	FHB	13268	1.27	$1.0 \cdot 10^4$	772	772	4303	4303

to update it after each modification, resp. (see column *computational time* –
CT, in seconds); (ii) the median of the ratios of the time to re-build the index
by KPLL to that to update it by DECKPLL or FULKPLL after each modification
(see column SPEED-UP); (iii) the size of the index recomputed via KPLL and the
size of the index updated via DECKPLL or FULKPLL, after all modifications (see
column *index size* – IS, in MBs)); (iv) the median execution time to perform 10^5
queries on the index recomputed via KPLL and on that updated by DECKPLL or
FULKPLL, resp., after each modification (see column *query time* – QT, in μs).

The first conclusion that can be drawn from our experimental data is that the
analysis of Thm. 2 is pessimistic, since the measured running time of DECKPLL is
always at least an order of magnitude smaller than the time to re-compute from
scratch the index, regardless of graph size, diameter, and k. On top of that, we
notice that DECKPLL is always significantly faster than the recomputation from
scratch, with a speed-up against KPLL that ranges from one order of magnitude
up to 5 orders of magnitude (see Fig. 1, where we report the measured speed-
up, for all values of k and graphs, as a function of the graph size). Even more
interestingly, we observe that the speed-up increases as $|V|+|E|$ increases, which
suggests that DECKPLL scales well with the input size (see, again, Fig. 1, left).
Concerning experiment FULL, the combination of INCKPLL and DECKPLL yields
very small update times for any type of modification, with a speed-up against
KPLL that spans from 3 to 5 orders of magnitude. Remarkably, the median
execution time of FULKPLL is less than 2 seconds even in the largest input. In
terms of space occupancy, our experimentation highlights that the size of indices,
updated via DECKPLL and FULKPLL, is always comparable to that obtained
by the re-computation from scratch through KPLL (see column IS in Table 2).
This is a highly desirable observed behavior for our dynamic algorithms, since
it suggests they are able to preserve the compactness of the index, which is
known to be directly related to small average query time [1,9]. The latter aspect
is confirmed by our measures of such query times which do not change, even
after many graph updates, and remain comparable (few microseconds to few
milliseconds per query) to those obtained by querying indices recomputed via
KPLL, even for the largest graphs and values of k (see column QT in Table 2).

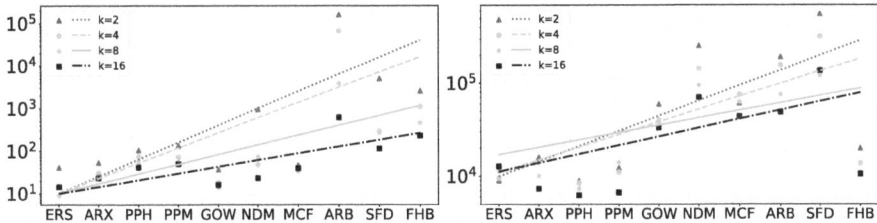

Fig. 1: Speed-up of DECKPLL and FULKPLL against KPLL as a function of graph
size in experiments DECR (left) and FULL (right). Lines show linear regressions.

Concerning the impact of parameter k on the performance of considered algorithms, our data show that the speed-up tends to decrease as k increases. This is expected, since the number of affected vertices depends on the number of paths encoded in the index. In particular, the value of k directly impacts on the size of sub-graph that must be visited by Procedure 2 to identify vertices whose loop label entries need to be recomputed (see Eq. 1). Nevertheless, machine-independent measures we collected in our experiments showcase that the cardinality of both sets D_{cy} and AFF-HUBS is always around an order of magnitude smaller than $|V|$, which explains the large speed-ups shown by both DECKPLL and FULKPLL in all tests (see Table 3 for an excerpt of such measures). Finally, we note that in experiment FULL, which is the closest to real-world applications (real networks tend to undergo many incremental updates interleaved by a fraction of decremental ones) algorithm FULKPLL is extremely fast at updating the k2HC while query times on the k2HC remain very low.

Experiment AMORTIZED. To consolidate the above observations on the effectiveness of DECKPLL and FULKPLL, and for the sake of fairness in the comparison, in what follows we present the results of an experiment, called AMORTIZED, whose aim is evaluate the average running time to perform k shortest paths queries by available solutions for dynamic graphs, regardless of whether they rely on indexing or not, in an amortized sense. In this direction, a common methodology, adopted in the literature for approaches that support queries on the graph with or without indexing [16], is to consider performance indicator *cumulative time* to execute a set of q queries. Such indicator includes preprocessing, query and update times for methods that rely on an initialization phase and pre-computed data (such as the KPLL index with DECKPLL or FULKPLL) while it includes only the running time of the query algorithm for methods that do not use any auxiliary data for query acceleration (e.g. KBFS or Eppstein's). To apply such methodology to our context, we executed the tests described in what follows. For each input graph, we randomly generate a number $\eta = 110$ of graph updates and a set of $q = 10^3$ vertex pairs. We pre-process the graph to compute a k2HC by KPLL and measure the time spent by the pre-processing routine. Then, for η times, we repeat the following process: (i) we apply one of the η graph updates; (ii) on the one hand, we execute and measure the total running time to determine a set of k shortest paths, for the q selected vertex pairs, by the KBFS algorithm (identified as the best option in practice to extract

Table 3: Excerpt of machine-independent measures.

Graph	$|V|$	$|E|$	k	D_{cy}	AFF-HUBS	Graph	$|V|$	$|E|$	k	D_{cy}	AFF-HUBS
PPH	16 820	449 036	2	25	153	ARB	163 598	1 747 269	2	96	10 491
			4	47	425				4	567	11 165
			8	63	1 123				8	808	36 151
			16	99	799				16	3 341	43 628

k shortest paths from real-world graphs without indexing [1]); (ii) on the other hand, we update the k2HC by dynamic algorithms, and compute answers to the same set of q queries, while measuring both the time for updating the k2HC and for q executions of the query routine on the k2HC. We execute twice per graph the above experiment, by fixing the η operations to be all decremental or of any type, and by therefore applying either DECKPLL or FULKPLL, accordingly (again, distribution of updates is inspired to real-world scenarios [7,9]).

An excerpt of the results of experiment AMORTIZED is shown in Fig. 2. We report, for one of the largest instances considered in this study, namely graph ARB: (i) the cumulative running time of DECKPLL or FULKPLL, i.e. the time taken to build the initial k2HC index, summed to the time spent (by DECKPLL or FULKPLL) to update it, for all η graph modifications (decremental or of arbitrary type), and to the time to answer the q queries, after each graph change, via the index; (ii) the cumulative running time of KBFS which is simply the total execution time of KBFS for extracting the same set of k shortest paths. Results for other graphs are similar, lead to equivalent conclusions, and hence are omitted due to space limitations. Our experiments show how dynamic indexing-based approaches outperform the KBFS method by up to more than an order of magnitude (even for the largest inputs and values of k), and regardless of updates being decremental only or arbitrary. In fact, we observe (see Fig. 2) the cumulative running time of DECKPLL in graph ARB to be more than an order of magnitude smaller than that of KBFS. Similarly, FULKPLL is from around 40 (80, resp.) times faster than KBFS to answer queries when $k = 16$ ($k = 8$, resp.). To summarize, our experimental work represents a strong empirical evidence of the fact that maintaining a k2HC index via FULKPLL is the most effective strategy to support the retrieval of k shortest paths in dynamic, possibly large graphs.

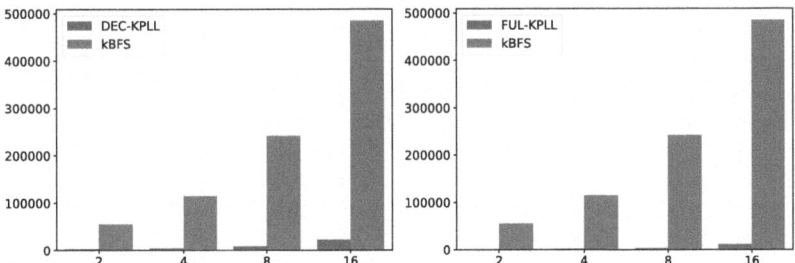

Fig. 2: Cumulative running time of DECKPLL (left) and FULKPLL (right) compared to that of KBFS on graph ARB (right), for all values of k.

5   Conclusion

In this paper, we have advanced the state-of-the-art in methods for mining graphs for k shortest paths in dynamic graphs. We have provided a full dynamization of framework KPLL by introducing DECKPLL, the first algorithm to efficiently maintain a k2HC under *decremental* modifications. We have assessed its practical effectiveness through extensive experimental work and have shown that, on average, it updates KPLL indices orders of magnitude faster than the recomputation from scratch while preserving its compactness and competitive query performance. We have experimentally shown that DECKPLL can be used in combination with INCKPLL, the only known algorithm to maintain a k2HC under *incremental* modifications hence showcasing the first method that supports very fast, interactive retrieval of k shortest paths from fully dynamic graphs.

Acknowledgments. Work partially supported by: Italian Ministry of University and Research through Project "EXPAND: scalable algorithms for EXPloratory Analyses of heterogeneous and dynamic Networked Data" (PRIN grant n. 2022TS4Y3N), funded by the European Union - Next Generation EU; Gruppo Nazionale Calcolo Scientifico-Istituto Nazionale di Alta Matematica (GNCS-INdAM); Project EMERGE, innovation agreement between MiSE, Abruzzo Region, Radiolabs, Elital, Leonardo, Telespazio, University of L'Aquila and Centre of EXcellence EX-Emerge (Grant n. 70/2017); PNRR MUR Project SCHEDULE "Smart interseCtions witH connEcteD and aUtonomous vehicLEs" (PE00000001, CUP J33C22002880001).

References

1. Akiba, T., Hayashi, T., Nori, N., Iwata, Y., Yoshida, Y.: Efficient top-k shortest-path distance queries on large networks by pruned landmark labeling. In: Proceedings of 29th International Conference on Artificial Intelligence, pp. 2–8. AAAI Press (2015)
2. Akiba, T., Iwata, Y., Yoshida, Y.: Fast exact shortest-path distance queries on large networks by pruned landmark labeling. In: Proceedings of 2013 ACM International Conference on Management of Data (SIGMOD 2013), pp. 349–360. ACM (2013)
3. Akiba, T., Iwata, Y., Yoshida, Y.: Dynamic and historical shortest-path distance queries on large evolving networks by pruned landmark labeling. In: Proceedings of 23rd Int'l World Wide Web Conference (WWW 2014), pp. 237–248. ACM (2014)
4. Chondrogiannis, T., Bouros, P., Gamper, J., Leser, U., Blumenthal, D.B.: Finding k-shortest paths with limited overlap. VLDB J. **29**(5), 1023–1047 (2020)
5. Cionini, A., et al.: Engineering graph-based models for dynamic timetable information systems. J. Discrete Algorithms **46–47**, 40–58 (2017)
6. Cohen, E., Halperin, E., Kaplan, H., Zwick, U.: Reachability and distance queries via 2-hop labels. SIAM J. Comput. **32**(5), 1338–1355 (2003)
7. D'Andrea, A., D'Emidio, M., Frigioni, D., Leucci, S., Proietti, G.: Dynamic maintenance of a shortest-path tree on homogeneous batches of updates: new algorithms and experiments. ACM J. Exp. Algorithmics **20**, 1.5:1.1–1.5:1.33 (2015)
8. D'Angelo, G., D'Emidio, M., Frigioni, D.: Fully dynamic 2-hop cover labeling. ACM J. Exp. Algorithmics **24** (2019)
9. D'Ascenzo, A., D'Emidio, M.: Top-k distance queries on large time-evolving graphs. IEEE Access **11**, 102228–102242 (2023)
10. Delling, D., Goldberg, A.V., Pajor, T., Werneck, R.F.: Robust distance queries on massive networks. In: Schulz, A.S., Wagner, D. (eds.) ESA 2014. LNCS, vol. 8737, pp. 321–333. Springer, Heidelberg (2014). https://doi.org/10.1007/978-3-662-44777-2_27

11. D'Emidio, M., Khan, I.: Dynamic public transit labeling. In: Misra, S. (ed.) ICCSA 2019. LNCS, vol. 11619, pp. 103–117. Springer, Cham (2019). https://doi.org/10.1007/978-3-030-24289-3_9

12. Eppstein, D.: k-best enumeration. In: Encyclopedia of Algorithms, pp. 1003–1006. Springer (2016)

13. Farhan, M., Koehler, H., Wang, Q.: BatchHL+: batch dynamic labelling for distance queries on large-scale networks. VLDB J. **33**, 101–129 (2024)

14. Farhan, M., Wang, Q.: Efficient maintenance of highway cover labelling for distance queries on large dynamic graphs. World Wide Web (WWW) **26**(5), 2427–2452 (2023)

15. Feng, Q., Peng, Y., Zhang, W., Lin, X., Zhang, Y.: DSPC: efficiently answering shortest path counting on dynamic graphs. In: Proceedings of 27th International Conference on Extending Database Technology (EDBT 2024), pp. 116–128. OpenProceedings.org (2024)

16. Hanauer, K., Henzinger, M., Schulz, C.: Faster fully dynamic transitive closure in practice. In: Proceedings of 18th International Symposium on Experimental Algorithms (SEA 2020), LIPIcs, vol. 160, pp. 14:1–14:14. Schloss Dagstuhl - Leibniz-Zentrum für Informatik (2020)

17. Hao, X., Lian, T., Wang, L.: Dynamic link prediction by integrating node vector evolution and local neighborhood representation. In: Proceedings of 43rd International ACM SIGIR Conference on Research and Development in Information Retrieval, p. 1717–1720. SIGIR 2020, ACM, New York, NY, USA (2020)

18. Hassan, M.S., Aref, W.G., Aly, A.M.: Graph indexing for shortest-path finding over dynamic sub-graphs. In: Proceedings of 2016 ACM Internatoional Conference on Management of Data (SIGMOD 2016), pp. 1183–1197. ACM (2016)

19. Jamil, H.M.: Efficient top-k shortest path query processing in sparse graph databases. In: Proceedings of 7th International Conference on Web Intelligence, Mining and Semantics (WIMS 2017). ACM (2017)

20. Lebedev, A., Lee, J.Y., Rivera, V., Mazzara, M.: Link prediction using top-k shortest distances. In: Calì, A., Wood, P., Martin, N., Poulovassilis, A. (eds.) BICOD 2017. LNCS, vol. 10365, pp. 101–105. Springer, Cham (2017). https://doi.org/10.1007/978-3-319-60795-5_10

21. Ni, P., Hanai, M., Tan, W.J., Cai, W.: Efficient closeness centrality computation in time-evolving graphs. In: Proceedings of 2019 IEEE/ACM International Conference on Advances in Social Networks Analysis and Mining (ASONAM), pp. 378–385. ACM (2019)

22. Peixoto, T.P.: The netzschleuder network catalogue and repository (2020). https://networks.skewed.de/

23. Peng, Y., Lin, X., Zhang, Y., Zhang, W., Qin, L.: Answering reachability and K-reach queries on large graphs with label constraints. VLDB J. **31**(1), 101–127 (2021). https://doi.org/10.1007/s00778-021-00695-0

24. Qiu, K., Zhao, J., Wang, X., Fu, X., Secci, S.: Efficient recovery path computation for fast reroute in large-scale software-defined networks. IEEE J. Sel. Areas Commun. **37**(8), 1755–1768 (2019)

25. Rossi, R.A., Ahmed, N.K.: The network data repository with interactive graph analytics and visualization. In: Proceedings of 29th AAAI Conference on Artificial Intelligence (AAAI 2015), p. 4292–4293. AAAI Press (2015)

26. Wang, S., Lin, W., Yang, Y., Xiao, X., Zhou, S.: Efficient route planning on public transportation networks: A labelling approach. In: Proceedings of 2015 ACM International Conference on Management of Data (SIGMOD 2015), pp. 967–982. ACM (2015)

27. Yen, J.Y.: An algorithm for finding shortest routes from all source nodes to a given destination in general networks. Q. Appl. Math. **27**(4), 526–530 (1970)
28. Yue, H., Guan, Q., Pan, Y., Chen, L., Lv, J., Yao, Y.: Detecting clusters over intercity transportation networks using k-shortest paths and hierarchical clustering: a case study of mainland china. Int. J. Geogr. Inf. Sci. **33**(5), 1082–1105 (2019)
29. Zhang, Y., Yu, J.X.: Hub labeling for shortest path counting. In: Proceedings of 2020 ACM International Conference on Management of Data (SIGMOD 2020), pp. 1813–1828. Association for Computing Machinery (2020)
30. Zoobi, A.A., Coudert, D., Nisse, N.: Space and time trade-off for the k shortest simple paths problem. In: Proceedings of 18th International Symposium on Experimental Algorithms (SEA 2020), Leibniz International Proceedings in Informatics (LIPIcs), vol. 160, pp. 18:1–18:13. Schloss Dagstuhl–Leibniz-Zentrum für Informatik (2020)

WIBA: What Is Being Argued? A Comprehensive Approach to Argument Mining

Arman Irani[1][(✉)], Ju Yeon Park[2], Kevin Esterling[1], and Michalis Faloutsos[1]

[1] University of California, Riverside, Riverside, USA
{airan002,kevin.esterling}@ucr.edu, michalis@cs.ucr.edu
[2] The Ohio State University, Columbus, USA
park.3509@osu.edu

Abstract. How can we effectively model arguments communicated in diverse environments? On the one hand, there is a great opportunity with the abundance of digitized speech across different contexts including online forums, official proceedings, or transcripts of spoken debates. On the other hand, there is a great challenge in correctly detecting arguments, especially since each medium has its own set of conventions, lingo, affordances, and styles of argumentative engagement. We propose WIBA, a novel framework and suite of methods that enable the comprehensive understanding of "**W**hat **I**s **B**eing **A**rgued" across contexts. Our approach develops a comprehensive framework that detects: (a) the existence, (b) the topic, and (c) the stance of an argument, correctly accounting for the logical dependence among the three tasks. Our algorithm leverages the fine-tuning and prompt-engineering of Large Language Models. We evaluate our approach and show that it performs well in all the three capabilities. First, we develop and release an Argument Detection model that can classify a piece of text as an argument with an F_1 score between 79% and 86% on three different benchmark datasets. Second, we release a language model that can identify the topic being argued in a sentence, be it implicit or explicit, with an average similarity score of 71%, outperforming current naive methods by nearly 40%. Finally, we develop a method for Argument Stance Classification, and evaluate the capability of our approach, showing it achieves a classification F_1 score between 71% and 78% across three diverse benchmark datasets. Our evaluation demonstrates that WIBA allows the comprehensive understanding of **W**hat **I**s **B**eing **A**rgued in large corpora across diverse contexts, which is of core interest to many applications in linguistics, communication, and social and computer science. To facilitate accessibility to the advancements outlined in this work, we release WIBA as a free open access platform (wiba.dev) and API.

Keywords: Argument Mining · Large Language Models · Natural Language Processing

L. M. Aiello et al. (Eds.): ASONAM 2024, LNCS 15211, pp. 337–354, 2025.
https://doi.org/10.1007/978-3-031-78541-2_21

1   Introduction

Arguments are fundamental building blocks of effective discourse. They are critical to decision-making and necessary for forming or altering one's opinion in a variety of communication contexts [9]. This extends from lively conversations among friends and loved ones to heated political debates in formal political forums. As a result, argumentation underwrites democratic legitimacy in many settings [5,8]. Despite their linguistic significance, arguments present considerable challenges for detection and quantification through text analytics at scale.

Opportunity and Challenges: The emergence of online text-based discourse provides an unprecedented opportunity: there are vast quantities of written opinions, including arguments, that can be mined to better understand what people think and believe, and understanding the reasoning that underwrites their opinions on any topic. These readily available data are any communication scholar's dream. But there is a catch: there are several theoretical and technical challenges which make argument mining difficult. Arguments can be extremely complex in their structure, while at the same time can be extremely sensitive in their semantic appearance, as specific features of different environments may significantly influence how argumentation is expressed. Different environments present different kinds of affordances for arguers [10]. An argument made in a legal procedure will look and read vastly different compared to an argument within a Reddit post. Handling these contextual variabilities, along with the structural and semantic nuances of arguments, presents a significant challenge we address in this work (Fig. 1).

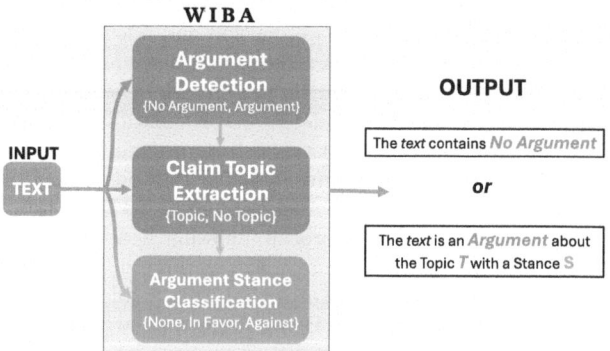

Fig. 1. This figure illustrates the proposed task and methods. The process begins with a given text input, where the first step is to determine the presence of an argument. If an argument exists, the subsequent steps involve extracting the topic being argued, and classifying the stance towards that topic. While each method can operate independently, out of order, Argument Stance Classification requires that a specific topic be provided along with the text.

Problem: How can we model argumentation in text-mediated discourse? To automatically and accurately identify argumentative characteristics of texts at scale across various digital environments introduces a set of complex challenges to overcome. For example, argumentation in online public forums is often informal and requires a level of contextual understanding in order to determine whether a post meets the formal conditions to be classified as an argument (stated below), and then to understand the content of the argument. In addition, it is challenging to develop a formalization of argumentation that can even be applied to an online public communication that is generalizable, yet specific enough to distinguish the fine line between a non-argument and a low-quality argument. Finally, developing tools that can meet these requirements to accurately understand the necessary argumentative characteristics poses several design challenges.

Previous Work: There has been relatively few efforts that have combined: (a) a comprehensive treatment of arguments, and (b) extensive computational and algorithmic development, especially with open-source tools. We can group previous works in a few large categories: (a) communications, linguistics and philosophy-oriented efforts, (b) algorithmic and implementation-oriented approaches, but typically with a niche focus, and (c) Large Language Model (LLM) approaches. We review related work and discuss how our work differs from them in Sect. 6.

Contribution: We propose WIBA, a systematic approach to enable the comprehensive understanding of **W**hat **I**s **B**eing **A**rgued. At a high level, our approach has two distinguishing features. First, it is comprehensive as we address all three questions regarding arguments: existence, topic, and stance toward the topic. Second, we develop algorithmic solutions that outperform previous techniques, with an ability to handle both informal (e.g., Reddit or Twitter) and formal (e.g., legal or political proceedings) types of arguments. Our approaches leverage LLMs, which we fine-tune and prompt-engineer appropriately. This combination of technical comprehensiveness and a theory-driven methodology sets WIBA apart as a robust tool for nuanced argument analysis across diverse contexts.

a. Framework Contributions. We develop a comprehensive framework, which introduces a theoretical foundation for computational argument mining. The key contributions of our framework include: (a) a proper formalization of arguments, shown to improve task performance when integrated with LLM prompt engineering; and (b) a "less is more" novel approach to task fine-tuning, which improves performance, increases consistency across new and unseen datasets while lowering the barrier of entry for future research which no longer requires massive datasets for task training; (c) we demonstrate that WIBA can identify arguments effectively, regardless of the argument type used or whether the syntax is informal or formal. Types of arguments include inductive, deductive, abductive, analogical, and fallacious. Informal and formal syntax refer to the syntactical style of communication, such as those found in casual online forums or formal legal proceedings. Further elaboration on the definitions of argument type and syntactical structure can be found in the WIBA Framework section.

b. Algorithmic Contributions. From an algorithmic point of view, we develop three methods. *Method 1:* we present a fine-tuned argument detection language model that takes a text input, and outputs whether the text is an argument or not, with an F_1 score between 80% and 86%, across three benchmark datasets, a 20–45pp improvement over existing state-of-the-art methods [18]. *Method 2:* We create a language model that has been trained to identify, for any text that meets the formal definition of an argument, the subject of what is being argued. This model is effective at extracting both implicitly and explicitly mentioned topics and aspects, with an efficacy advantage of nearly 40% over current keyword and topic modeling methods. *Method 3:* We develop a method that takes a topic and text as input and determines, if the text is an argument, the argumentative stance directed towards the topic provided. This method outperforms existing state-of-the-art methods [18] by 12 to 20pp with an F_1 score ranging between 71% and 78%.

c. Infrastructure Contributions. To maximize the practical impact of our research, we developed and launched the WIBA framework as an accessible online platform and API, depicted in Fig. 3 and further discussed in Sect. 5 (https://wiba.dev). This platform enables researchers to programmatically access our API endpoint to access the methods, or upload files for analysis, enabling them to apply any of the three WIBA methods discussed in this paper with a single click. The results are then made available for download. In addition, to support ongoing research in this field we: (a) open-source our code, (b) share our fine-tuned models, and (c) provide our datasets to the community.

2   WIBA Framework

What Is an Argument? The definitions provided in this work are based on the widely used philosophical theories of what an argument is and is made up of. An argument is defined as a piece of text or speech that contains a claim asserting something in favor or against a subject, supported by premises. In order for the argument to meet this formal definition, it must contain at least one claim AND at least one premise to support that claim. A claim is a statement that depends on a premise. Premises are statements that provide evidence, reasons, and support for the claim. In this work, we operate under the assumption that arguments are limited to 1–3 sentences, as that is the range of our data. In our training data, a text instance has an average of 27 words and an average of 2 sentences.

Our methods identify arguments and their contents, but do not make an assessment of the validity or truth of the arguments. Such an assessment is not necessary for our methodological purposes, nor is it normatively necessary;

for example, democracies are designed to enable true and false arguments to compete rather than to have some third party determine their validity [12].

Types of Arguments. Arguments can take many forms depending on their linguistic structure. Based on discourse theory, every argument conforms to one of the following five types: (a) deductive, (b) inductive, (c) abductive, (d) analogical, or (e) fallacious [6]. Below are brief definitions and examples for each type.

Deductive arguments require the claim of the argument to necessarily follow from in the premises, in that if one takes the premises to be true, the claim also must be true. For example, "Glyphosate is a chemical in GMOs and Glyphosate is bad for you, therefore GMOs are bad for you."

Inductive arguments require past observations or common knowledge to provide the necessary justification for the claim. For example, "Every time I recycle my waste, I reduce my carbon footprint, so recycling must help protect the environment."

Abductive arguments require that observations made as evidence provide a plausible explanation for the claim. For example,"Given the proliferation of misinformation and social media algorithms, it is plausible to suspect the manipulation of public opinion and polarization by foreign actors."

Analogical arguments require that if two claims are considered to be similar, then the truth of the first means the second is true. For example, "Enforcing dress codes in schools is like forcing everyone to wear uniforms to stop crime - they both make unnecessary rules in the name of control."

Fallacious arguments contain both a premise and a claim, but the latter does not follow from the former, such as when the premise is unjust, incorrect, or scientifically invalid. E.g., "Environmental activists exaggerate the threat of climate change to advance their political agenda, ignoring scientific dissent and manipulating data to justify costly and ineffective policies."

What Is an Argument Topic? An argument topic refers to the topic implicitly or explicitly central to the claim being made. An argument necessitates an assertion to be made, and this assertion must be made regarding *something*.

What Is an Argument Stance? An argument stance is a multi-class categorization of sentences into one of three classes: No Argument, Argument in Favor, and Argument Against. Stance classification is a difficult problem in computational linguistics due to its subjectively and the necessary reliance on a well-defined topic to which the stance is oriented.

In this work, we revisit and improve two well known argument mining tasks, **Argument Detection** and **Argument Stance Classification**. We remove a usual prerequisite of Argument Detection, the presence of a topic, and still demonstrate significantly better performance over baseline methods. Additionally, we introduce a novel computational argumentation task, **Claim Topic**

Extraction, which extracts the implicit or explicit topic being argued in a text, if an argument is present.

Task 1: Argument Detection

Let $\mathcal{S} = \{x_i, y_i\}$ be a piece of text, where x_i is a sequence of words, and $y_i \in \{0, 1\}$ is a binary label indicating whether x_i constitutes an argument or not. An argument is defined as a sequence of words that semantically contains at least one claim and at least one premise. We define the task of Argument Detection as follows: Given a sequence of words x_i, the goal is to determine whether x_i constitutes an argument or not, by checking whether it contains both a claim and a premise.

The task is learned through supervised fine-tuning of LLMs with the goal of minimizing the detection error on unseen argument instances. We outline in detail the novel formalization of this task and its translation to an Augmented Transition Network in our Sect. 3.

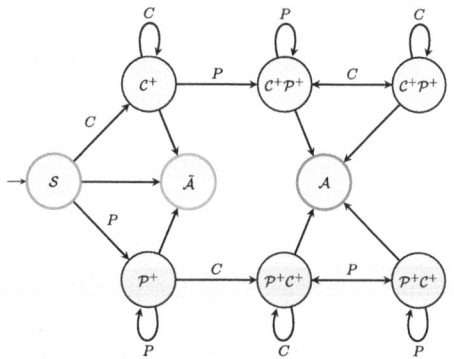

Fig. 2. Argument Detection Augmented Transition Network (ATN). $\mathcal{S}$ represents the start of the text classification, $\tilde{\mathcal{A}}$ represents 'Not an Argument', $\mathcal{C}$ represents Claim, $\mathcal{P}$ represents Premise, and $\mathcal{A}$ represents Argument. Looped arrows from $\mathcal{C}$ and $\mathcal{P}$ represent recursive calls to Claim and Premise Augmentation Networks. The symbol $^+$ indicate *at least one* I.e., $\mathcal{P}^+\mathcal{C}^+$ is the state where there is at least one premise with at least one claim.

Task 2: Claim Topic Extraction

Let $\mathcal{S} = \{x_i, t_i\}$ be a piece of text, where x_i is a sequence of words that may or may not constitute a claim made in an argument, and t_i is the topic being argued. The topic t_i may be explicitly mentioned in the sequence of words x_i, or it may be implicitly implied by the argument. We define the task of Claim Topic Extraction from arguments as follows:

Given a sequence of words x_i, the goal is to identify the topic t_i in x_i, whether the topic is explicitly mentioned or implicitly implied. Formally, we seek to learn a function $f : \mathcal{X} \rightarrow \mathcal{T}$ that maps a sequence of words $x \in \mathcal{X}$ to a topic $t \in \mathcal{T}$, where $\mathcal{X}$ is the set of all possible finite sequences of words, and $\mathcal{T}$ is the set of all possible topics. The function f identifies the topic t of the sequence of words x.

Claim Topic Extraction is a novel task within the scope of computational argument mining. In lay terms, Claim Topic Extraction is the "what" in the question, "*What* is being argued?" Given the formal definition of an argument described earlier, we focus here on the claim rather than on any premises, as the claim is the nucleus of the argument being made. The extraction of what is being argued is crucial in the field of argumentative stance detection. Stance detection not only depends on the presence of a target for attribution, but also is dependent on the contextual accuracy of the target, which in turn controls the accuracy of the model. Claim Topic Extraction distills the argument content and infers the context of the conclusion being made. The motivation for this task is twofold. First, it is often the case that in arguments expressed online, especially in informal exchanges on social media, the topic being argued is often implicit. For example:

> *"no one should be restricted to only one partner when multiple people may be equally right for them in different ways"*

In this example, there is no obvious explicit keyword or topic that is expressed. However, when a human looks at this sentence it is obvious that the topic being argued is monogamy/polygamy. Second, even in the cases where the topic being argued is explicitly mentioned in the claim, there may be more than one potential topics present. For example,

> *"The acquisition of nuclear weapons by dictators and despots makes revolution and social change impossible, creating more dire conditions"*

In this case, there are at least five different explicit topics that potentially could be detected by topic modeling methods: nuclear weapons, dictators, despots, revolution, and social change. Claim Topic Extraction identifies specifically that "nuclear weapons" is the topic of the claim, that is, the topic being argued. The other topics are relevant to the evidence or premises of the argument. This distinction is important when considering argument mining as a practical application, in which claim topics may be implicit. Contextual understanding and language generation are crucial for identifying argument topic, making LLMs the optimal architectural choice.

Task 3: Argument Stance Classification

Let $\mathcal{S} = \{x_i, t_i, s_i\}$ be a piece of text, where x_i is a sequence of words representing a text, t_i is the topic associated with the text, and $s_i \in \{0, 1, 2\}$ is the stance label, where $s_i = 0$ indicates that the text x_i does not contain an argument (No Argument); $s_i = 1$ indicates that the text x_i contains an argument in favor of the topic t_i (Argument in Favor); $s_i = 2$ indicates that the text x_i contains an argument against the topic t_i (Argument Against). We define the task of Argument Stance Classification as follows: Given a sequence of words x_i and a topic t_i, the goal is to classify the stance s_i of the text x_i towards the topic t_i as one of the three potential argument stances: No Argument, Argument in Favor, or Argument Against.

The task is learned through supervised fine-tuning of LLMs with the goal of minimizing the classification error on unseen argument instances. The model should be able to identify whether the given text constitutes an argument or not, and if it does, determine whether the argument is in favor of or against the provided topic.

Large Language Models

The following section outlines the basic details of the five Large Language Models we use to evaluate WIBA.

BART, first introduced in 2020, is a standard Transformer-based neural network [24] which, despite its simplicity, can be seen as generalizing BERT (due to the bidirectional encoder), GPT (with the left-to-right decoder), and other recent pretraining schemes [15].

Llama-2 7B/Llama-3 8B are pretrained LLMs released by Meta with 7 billion and 8 billion parameters, respectively [1,23]. Llama-2 is an updated version of Llama-1, with its pretraining data corpus being increased by 40%, doubled context length and grouped query attention. Llama-3 8B builds on this by incorporating an additional billion parameters and extra training data, achieving a 20-point improvement over Llama-2 7B on the MMLU Benchmark.

Mistral-7B is a 7-billion parameter LLM introduced by Mistral.AI, which leverages grouped-query attention and sliding window attention to increase both inference speed and also decoding memory requirements [13].

Yi-6B is an open-source transformer based LLM that also utilizes LLaMA architecture as the foundational framework. The model does not use LLaMA's weights however, and instead 01.ai created its own proprietary high-quality training datasets and training pipelines.

3   The WIBA Methodology

This section is split into three subsections, the first outlines the framework setup that is used in our Argument Detection and Argument Stance Classification tasks, the second introduces and outlines our proposed novel task, Claim Topic Extraction. And finally, we go over our methodology for Stance Classification. Each of the three proposed methods can be used independently, but can also be utilized sequentially to exploit the dependence: WIBA-Detect → WIBA-Extract → WIBA-Stance.

3.1   WIBA-Detect

Data Augmentation. Recent advances in LLM research propose the value of smaller, higher quality datasets over larger, noisier datasets for specific task fine-tuning [29]. Since modern LLMs have been pretrained on such a large knowledge base, there is no need for the scale and resources of large datasets for task alignment. We adopt this 'less is more' approach in our methodology, and train our Language Models on a dataset of only ~1000 instances, for 2 epochs, to prevent overfitting. Through hyperparameter optimization, we identify the LLaMA models alone experience performance improvements from an additional epoch of training. Therefore, Llama-2 & Llama-3 are fine-tuned for 3 epochs.

Argument Formalization. An advantage of LLMs is the ability to incorporate prompting instructions for guiding the agent into solving a specific task more effectively. We propose a novel formalization of Argument Detection for the advancement of LLM task prompt engineering. This formalization relies on the structure of a type of nondeterministic finite automaton, Augmented Transition Networks (ATN). This ATN is then translated into a pseudo-language for direct reasoning prompt engineering for both Argument Detection and Argument Stance Classification tasks. The foundation of the ATN is as follows:

S: The input text.

T: The set of possible tokens that can be derived from S. Each token represents a subset of words of arbitrary length, that can be one of four possible states, {Claim, Premise, Not Claim, Not Premise}. In other words, the set of tokens derived from S is $T(S) = \{t_1, t_2, t_3, ..., t_m\}$, where each token t_i is a concatenation of one or more sequential sub-texts.

Furthermore, a visualization of the ATN with the possible transition states and edge conditions is shown in Fig. 2 for greater clarity. Multiple $\mathcal{P}^+\mathcal{C}^+$ states are used to capture the functional distinctions and specific roles these states play in the transition processes involved in argument detection.

3.2   WIBA-Extract

For this task we fine-tune a Llama-3 8B model for 2 epochs, on our CTE dataset, so that it may learn to generate and identify the core topic of what is being argued. We optimize performance through the use of Chain-of-Thought prompting, which has been demonstrated to be useful for symbolic reasoning tasks [25]. Moreover, we use indirect reasoning prompting, through the use of contrapositives and contradictions in Chain-of-Thought prompting, which has been shown to enhance the overall accuracy of factual reasoning by 27.33% [28]. Due to lack of space, we showcase our complete prompt-engineering framework in our repository. For this task, we opt to re-merge our Low-Rank Adaptation (LoRA) [11] weights back into the base model for greater contextual understanding capabilities. The final product is a fine-tuned language model capable of extracting the topic being argued in a text, if one exists.

3.3 WIBA-Stance

We fine-tune each of our language models for the three class task of argument stance classification. Using LoRA Adapter, we only tune the linear layers, reducing the number of trainable parameters from ~7 billion to ~20 million. We opt to not re-merge the LoRA weights back into the base model, which has no influence on the performance of the model. This memory-efficient approach enables WIBA-Stance to be used in memory-constrained environments. The model is trained on an input that must contain a topic (provided by WIBA-Extract), and a text in order to anchor a corresponding argumentative stance towards the topic, if there is one. The models are trained similarly to WIBA-Detect, all for 2 epochs, except Llama-3 8B which improves until 4 epochs.

During our research, we observe a much higher order of different types of non-arguments, which initially our model struggles to identify, leading to many false positives. To overcome this issue and improve our F_1 performance, we append only to our validation data an additional 150 non-arguments from our IBM-ARG$_{HQ}$ validation set.

Argument Stance Formalization. After thorough experimentation of possible interpretations of a stance version of our Argument Transition Network, we find that the same ATN used for Detection performs the best for Stance as well. We therefore rely on the ATN logic in Fig. 2 for WIBA-Stance.

4 Evaluation

Our initial evaluation of WIBA against benchmark datasets returned many false-positives and false-negatives. But when examining the labels we found many instances where sentences in the benchmark data were coded as arguments that do not match our definition of what an argument is, as well as many argument sentences that were coded as not arguments. To resolve this issue, we conduct a blind human labeling of 500 random and even class distributed instances from the benchmark datasets. Each sample is blindly hand re-labeled according to our formal definition of an argument (we provide the Python code to do this blind classification in our replication materials). To validate benchmark methods, we evaluate UKP's BERT based tool on these new subsets of the data to get an accurate understanding of the improvement of the existing tools to WIBA. We next describe our datasets and summarize the results of our evaluation, shown in Table 1.

4.1 Datasets

For validation, we use three industry standard benchmark datasets and one we generate synthetically using Chat-GPT.

UKP$_{HQ}$: This corpus represents a subset of the full UKP Sentential Argument Mining Corpus by [21]. The original dataset contains 27,520 sentences spanning

Table 1. WIBA Evaluation Results

	Benchmark Corpus			
	UKP$_{HQ}$	GPT$_{HQ}$	DEBATE$_{HQ}$	IBM-ARG$_{HQ}$
Model	F_1	F_1	F_1	F_1
Task 1: Detection				
Existing Methods				
BERT$_{UKP}$ (+ topic)	58.7	41	59.7	—
WIBA-Detect (ATN)				
BART	23	44	34	—
Yi-6B	78.6	68	67.7	—
Mistral-7B	78.7	81.6	75	—
Llama-2 7B	**82.4**	83.3	78	—
LLama-3 8B	80.2	**85.8**	**79.6**	—
WIBA-Detect (No ATN)				
BART	23.2	27.3	35.5	—
Yi-6B	**77**	64	67.3	—
Mistral-7B	52.3	38.5	56.8	—
Llama-2 7B	73.6	**77.9**	54.1	—
Llama-3 8B	76.4	**81.9**	**69.4**	—
Task 3: Stance				
Existing Methods				
BERT$_{UKP}$ (+ topic)	53.2	57.3	—	58.4
WIBA-Stance (ATN)				
BART	13.6	18.2	—	20.3
Yi-6B	68.2	**78.3**	—	61.4
Mistral-7B	66.8	56.8	—	48.1
Llama-2 7B	68.3	77.1	—	**71.3**
Llama-3 8B	**71**	75.7	—	67.1

eight controversial topics. Each sentence is assigned a label of (No Argument, Argument In Favor, Argument Against) a topic. Upon manual investigation of this full dataset, we identified an inconsistency of the sentences with their labels, according to both [21] and our definition of what an argument is. We include a sample of over 200 detailed instances of these errors in our Github repository for greater transparency and explanation. We filter and select high quality examples for our train and validation data for a total of 333 instances across the three classes and an even distribution of the topics within those classes. For the test dataset we randomly selected 504 instances evenly selected across topics and labels, and in a blind-review hand-coded the labels carefully according to the WIBA formalization.

DEBATE$_{HQ}$: This IBM Debater dataset consists of 700 annotated argumentative sentences from recorded debates [20] over 20 different topics. Due to the dataset only containing 260 instances labeled as arguments, we select a subset of 250 non-arguments, and 250 arguments for a total of 500 instances to manually label in our blind-review. Since this dataset only contains a binary argument label, we do not evaluate argument stance classification against this dataset.

IBM-ARG$_{HQ}$: This dataset originally contained 5.3k arguments with an associate argument quality rank, a continuous value between [0,1] and a (pro, con) stance towards a topic [22]. We blind review a randomly selected sample of 500 instances with an evenly selected distribution of rank, and accessed each instance carefully to see if the text is an argument or not, towards the topic, and with the appropriate stance. Since this dataset is a multi-class stance dataset, we do not evaluate the argument detection methods against it.

GPT$_{HQ}$: This dataset consists of 800 short-form examples of synthetically generated arguments and non-arguments, across 400+ topics. The argumentative portion of the dataset contains examples from each of the five different types of reasoning used in argumentation: Inductive, Abductive, Deductive, Fallacious, and Analogical. Furthermore, for each reasoning type, there is a 50/50 split of informal and formal styles. We define the informal style of argumentation to contain a similar syntactical style to those found in online forums, such as Twitter and Reddit. Formal styles of argumentation are similar to those found in legal and political documents, such as congressional hearings or legal proceedings. For the non-arguments, we define a similar stylistic distinction of informal and formal. We prompt ChatGPT-3.5 with formal definitions for each type of argument [6], and ask it to generate complex and diverse *'argument type'* arguments across a variety of controversial issues. We specify lengths of 1–3 sentences, and either ask to emulate the style of political or legal texts such as congressional hearings, or those found on Reddit and Twitter. The full prompt design is provided in our repository.

CTE$_{Train}$: This dataset is designed for the task of claim topic extraction, created by combining the instances manually selected from the **IBM-ARG** dataset and the **UKP** dataset. Included in the train datasets are a total of 900 arguments with both implicit and explicit claim topics and 150 'No Topic' (No Argument). From the UKP dataset we collect an additional 445 'No Topic' instances. These are split into train, validation, and test splits.

4.2   Argument Detection (WIBA-Detect)

WIBA-Detect results appear in the top panel of Table 1.

The implementation of the WIBA ATN Formalization has a significant positive impact on the F_1 Score. By applying WIBAs ATN prompt engineering to assess each LLMs performance and contrasting it with their performance without such logic, a distinct and positive impact is evident, results in enhanced performance across all tested LLMs. BART shows the smallest improvement, only seeing an increase in F_1 in the GPT$_{HQ}$ dataset, likely due to its design limitations regarding system instructions. Mistral-7B experiences the most dramatic improvements in F_1 across all the datasets, ranging from 12%–31%. Llama-2 7B also see's a boost in performance ranging from 5%–18%, and Llama-3 8B from 4%–10%. Given these results across three diverse datasets, it is evident that WIBAs formalization significantly bolsters an LLMs capability to perform Argument Detection in a robust and effective manner.

WIBA Outperforms Existing Methods. Fine-tuning our LLMs on smaller, high-quality data, without the presence of a topic and only the sentence as the input, yielded an increase in performance compared to the BERT-based fine-tuned model released by the UKP Lab [18]. Unsurprisingly, BART is simply unable to adapt to the strict and small training environment of WIBA, as it is an infant language model. All the LLMs have roughly similar F_1 scores on the UKP$_{HQ}$ dataset, however the LLaMA based models excel on the GPT$_{HQ}$ and Debate$_{HQ}$ dataset indicating a superior ability to distinguish arguments based on their type and level of formality, and proving to be more adept at detecting formal, spoken arguments. WIBA-Detect consistently outperforms UKP methods using BERT, across all the datasets. For UKP$_{HQ}$ we see improvements of about 24%; for GPT$_{HQ}$ we see improvements ranging from 27% to 45%; for Debate$_{HQ}$ we see improvements from 8% to 20%.

WIBA Excels at Identifying Different Types of Arguments. Investigating the performance on the GPT$_{HQ}$ allows us to quantify WIBAs ability to identify different types of arguments with different formality levels. WIBA-Detect correctly identifies 100% of analogical arguments, 98% of fallacious argument, 92% of non-arguments, 87% of inductive arguments, 74% of deductive arguments, and 55% of abductive arguments. In total, evaluation upon this synthetic dataset reveals a strong ability to detect informal arguments, with an accuracy of 76% and formal arguments with an accuracy of 89%. Of the abductive and deductive arguments misclassified, 88% of the texts were very short, informal, and narrative-driven, making it difficult for formalization detection. While WIBAs stellar performance in detecting arguably more valuable argument types (analogical, fallacious) accentuates its robustness, evaluation on this synthetic dataset exposes potential gaps in the training data and methodology that can be addressed in future works.

4.3   Claim-Topic Extraction (WIBA-Extract)

We evaluate WIBA-Extract against popular Keyword Extraction and Topic Modeling methods, such as Flair's Named Entity Recognition (NER) methods

[2] and RAKE [19], by taking the average cosine-similarity score, where e represents the topic embedding, cos represents the cosine similarity calculation and N is the length of the test dataset.

$$\text{CTE Score} = \frac{1}{N} \sum_{i=1}^{N} \cos(e_{t_i^{gold}}, e_{t_i^{pred}})$$

We choose the Sentence Transformer mpnet-base-v2 as our embedding model due to the vast diversity of its training data. If a topic is generated when there is no argument, the cosine similarity for that specific instance will automatically be set to 0.

WIBA-Extract Excels in Short-form Text Topic Identification. Evaluating WIBA on GPT_{HQ}, which contains 402 unique and distinct topics and 800 argument/no-argument instances, WIBA-Extract achieves an impressive CTE Score of 64.8%. Investigating the performance further, of the instances WIBA correctly identified as arguments, it was able to extract the topic with a CTE Score of 74.2%. Comparing this to Flair's NER model, the model only generated 41/800 topics against the GPT_{HQ} dataset, displaying an inability to deal with short texts. Flair therefore only achieves a CTE Score of 2%; however for the topics it did generate, the similarity of those topics with the gold-label were 55.6%. RAKE generates 733/800 topics, and achieves an overall CTE Score of 40.4%, indicating a subpar ability to identify quality topics.

WIBA-Extract Outperforms Current Naïve Topic and Keyword Modeling Techniques. Evaluating NER against the CTE_{Test} dataset, we observe only 72 instances having a topic generated out of all 433 instances. Of the topics generated, Flair NER achieved a CTE Score of 20.1%. RAKE, although it generates a topic for 397/433 instances, only achieves a CTE Score of 29.3%. WIBA-Extract on the other hand generated 433 topics, one for every instance. Furthermore, our model demonstrates a remarkable overall CTE Score of 76.8%, over 40% higher than the current state-of-the-art techniques. For the correctly identified arguments, WIBA-Extract achieves a CTE score of 83.2%.

4.4 Argument Stance Classification (WIBA-Stance)

The WIBA-Stance results are in the bottom panel of Table 1.

The Implementation of the WIBA ATN Formalization has a Significant Positive Impact on the F_1 Score. Just like for the task of argument detection, the presence of a formalization has a significant impact on the LLM's ability to understand how to correctly classify stances for arguments. Mistral-7B experiences the largest impact in performance, with its F_1 score improving by 38% across all datasets. Llama-2 7B also saw significant improvements, ranging from 3% to 13%.

WIBA-Stance Outperforms Existing Methods. WIBA-Stance demonstrates improvements from UKPs Classification tool ranging from 13% to 37% across benchmark datasets. The most significant improvement was over the GPT_{HQ} dataset, indicating that WIBA-Stance was able to fill the gap in its ability to identify arguments in different forms.

Fig. 3. Our platform and API docs available at https://wiba.dev

5 Discussion

WIBA in Action. We envision WIBA as a method and tool that will enable the deeper understanding of opinions through argument mining. Users could include: a) researchers, b) politicians, c) public policy and marketing institutions. The goal of the user would be to understand in depth what its target group thinks and argues about an issue. We developed a user-friendly platform as well as an API endpoint, which allows anyone to either manually enter text, upload files, or easily make requests programmatically to perform WIBA tasks. Figure 3 shows our platform, with an example input and resulting output produced by WIBA-Extract as well as the location of our API Documentation.

What are the Limitations of WIBA? WIBA is limited by the challenges that any argument mining approach faces. These challenges include: (a) obscure, unconventional and vague language, (b) multiple intertwined topics and arguments, (c) detecting sophisticated verbal schemes, mostly sarcasm, and obscure niche pop-culture references. More specifically, our current method is limited to the English language, and does not innately enable researchers to control the granularity of topics extracted. However, we argue that our innovative three-stage decomposition of the argumentation will improve our ability to clearly identify arguments despite these hurdles compared to existing argument mining methods. Furthermore, our framework enables researchers to either create robust argument mining tools for different languages, or combine existing topic clustering tools like HDBSCAN for topic clustering.

Can We Increase Statistical Confidence? As with every classification study, more data and larger ground truth can increase the statistical confidence in our evaluation. We intend to expand the dataset and the ground truth in the future and continue to share that with the community.

Will LLMs Make WIBA Obsolete? Our answer to this question is twofold. Firstly, WIBA provides a transparent, structured, and theory-based approach to argument analysis, in contrast to the "black-box" solution that relying exclusively on an LLM brings. This clear, methodological framework ensures that WIBA retains its relevance and utility, despite the advancements in LLMs. Secondly, WIBA incorporates LLMs as a foundational component, meaning that enhancements in LLM capabilities directly contribute to the improvement of WIBA. For instance, our comparative analysis between Llama-3 8B and Llama-2 7B reveals that while LLMs continue to evolve, becoming larger and more sophisticated, the performance gains are modest. This suggests that while LLMs are crucial, they do not render WIBA obsolete, but rather enhance its effectiveness.

6   Related Work

While there is a great deal of work being conducted on individual components and task of computational argumentation, such as stance detection, claim detection, and counterargument generation, there is little work done on improving the foundational basis upon which all argument mining based tools are built from. Our contributions in this work involve going back to square one, and questioning the validity of the data and definitions a majority of argument mining research has been based on.

Argument Identification. There has been a robust amount of work done on creating models for identifying argumentative components, and for achieving various AM tasks, for example [3] explores the potential advantage of using zero-shot Large Language Models for various computational argumentation tasks. However, these works do not revisit the fundamentals of argument identification, nor investigate the training of these large language models for these tasks. [7] proposes an all-in-one framework for extracting argumentative components i.e., claims, evidence, evidence types, and stances. [14] proposes a task of context-dependent claim detection. [4] introduces a task for extracting arguments from two passages and identifying potential argument pairs. [17] introduced the concept of formalizing the argumentation process, but this technique was not generic and robust enough for the various argumentation environments available in the digital age.

Claim Topic Extraction. For our proposed novel task Claim Topic Extraction, there are various related concepts and works, such as [16] which proposes methods for extracting both the target and stance if neither are provided. This work does not require the text be an argument, which introduces a set of challenges for our task CTE. [26,27] demonstrate and define the use of generative aspects for aspect-based sentiment analysis and NER tasks.

7 Conclusion

The key contribution of our work is WIBA, a comprehensive and systematic approach for developing effective argument mining methods. Our solution consists of three state-of-the-art techniques that identify (a) whether a piece of text contains an argument, (b) the topic being argued, and (c) the stance of the argument towards that topic. Our method is grounded on argument theories, and we introduce systematic formalization of arguments and a computational framework which is represented in the form of an Augmented Transition Network. In addition, we introduce the task of Claim Topic Extraction, which to the best of our knowledge is novel. Another important innovation of our research design is that we evaluate WIBA on various types of arguments written in various styles to ensure that it is widely applicable even to the arguments that are short, of diverse types, and informal as frequently observed in online social media forums.

References

1. AI@Meta: Llama 3 model card (2024). https://github.com/meta-llama/llama3/blob/main/MODEL_CARD.md
2. Akbik, A., Bergmann, T., Blythe, D., Rasul, K., Schweter, S., Vollgraf, R.: FLAIR: an easy-to-use framework for state-of-the-art NLP. In: NAACL, 2019 Annual Conference of the North American Chapter of the Association for Computational Linguistics (Demonstrations), pp. 54–59 (2019)
3. Chen, G., Cheng, L., Tuan, L.A., Bing, L.: Exploring the potential of large language models in computational argumentation. arXiv preprint arXiv:2311.09022 (2023)
4. Cheng, L., Wu, T., Bing, L., Si, L.: Argument pair extraction via attention-guided multi-layer multi-cross encoding. In: Proceedings of the 59th Annual Meeting of the Association for Computational Linguistics and the 11th International Joint Conference on Natural Language Processing (Volume 1: Long Papers), pp. 6341–6353 (2021)
5. Cohen, J.: Deliberation and Democratic Legitimacy, pp. 17–34. Basil Blackwell (1989)
6. Dutilh Novaes, C.: Argument and argumentation. In: Zalta, E.N., Nodelman, U. (eds.) The Stanford Encyclopedia of Philosophy. Metaphysics Research Lab, Stanford University, fall 2022 edn. (2022)
7. Guo, J., Cheng, L., Zhang, W., Kok, S., Li, X., Bing, L.: Aqe: argument quadruplet extraction via a quad-tagging augmented generative approach. arXiv preprint arXiv:2305.19902 (2023)
8. Gutmann, A., Thompson, D.: Democracy and Disagreement: Why Moral Conflict cannot be Avoided in Politics, and What Should be Done About It. Princeton University Press (1996)
9. Habermas, J.: The Theory of Communicative Action, Volume One: Reason and the Rationalization of Society. Beacon Press, mccarthy edn. (1984)
10. Halpern, D., Gibbs, J.: Social media as a catalyst for online deliberation? Exploring the affordances of Facebook and Youtube for political expression. Computers in Human Behavior **29**, 1159–1168 (2013). https://doi.org/10.1016/j.chb.2012.10.008
11. Hu, E.J., Shen, Y., Wallis, P., Allen-Zhu, Z., Li, Y.: Lora: low-rank adaptation of large language models (2021)

12. Jefferson, T.: Statutes at Large in Virginia [1726], pp. 84–86 (1823)
13. Jiang, A.Q., et al.: Mistral 7b (2023)
14. Levy, R., Bilu, Y., Hershcovich, D., Aharoni, E., Slonim, N.: Context dependent claim detection. In: Proceedings of COLING 2014, the 25th International Conference on Computational Linguistics: Technical Papers, pp. 1489–1500 (2014)
15. Lewis, M., Liu, Y., Goyal, et al.: Bart: denoising sequence-to-sequence pre-training for natural language generation, translation, and comprehension. arXiv preprint arXiv:1910.13461 (2019)
16. Li, Y., Garg, K., Caragea, C.: A new direction in stance detection: target-stance extraction in the wild. In: Proceedings of the 61st Annual Meeting of the Association for Computational Linguistics, pp. 10071–10085 (2023)
17. Palau, R.M., Moens, M.F.: Argumentation mining: the detection, classification and structure of arguments in text. In: Proceedings of the 12th International Conference on Artificial Intelligence and Law, pp. 98–107 (2009)
18. Reimers, N., Schiller, B., Beck, T., Daxenberger, J., Stab, C., Gurevych, I.: Classification and clustering of arguments with contextualized word embeddings (2019).https://doi.org/10.48550/ARXIV.1906.09821
19. Rose, S., Engel, D., Cramer, N., Cowley, W.: Automatic keyword extraction from individual documents, pp. 1 – 20 (2010). https://doi.org/10.1002/9780470689646.ch1
20. Shnarch, E., Choshen, L., Moshkowich, G., Slonim, N., Aharonov, R.: Unsupervised expressive rules provide explainability and assist human experts grasping new domains. arXiv preprint arXiv:2010.09459 (2020)
21. Stab, C., Miller, T., Schiller, B., Rai, P., Gurevych, I.: Cross-topic argument mining from heterogeneous sources. In: Proceedings of the 2018 Conference on Empirical Methods in Natural Language Processing, Brussels, Belgium, pp. 3664–3674. Association for Computational Linguistics (2018). https://doi.org/10.18653/v1/D18-1402, https://aclanthology.org/D18-1402
22. Toledo, A., et al.: Automatic argument quality assessment–new datasets and methods. arXiv preprint arXiv:1909.01007 (2019)
23. Touvron, H., et. al.: Llama 2: open foundation and fine-tuned chat models (2023)
24. Vaswani, A., et al.: Attention is all you need (2023)
25. Wei, J., Wang, X., Schuurmans, D.: Chain-of-thought prompting elicits reasoning in large language models (2023)
26. Zhang, S., Shen, Y., Tan, Z., Wu, Y., Lu, W.: De-bias for generative extraction in unified NER task. In: Proceedings of the 60th Annual Meeting of the Association for Computational Linguistics (Volume 1: Long Papers), pp. 808–818 (2022)
27. Zhang, W., Li, X., Deng, Y., Bing, L., Lam, W.: Towards generative aspect-based sentiment analysis. In: Proceedings of the 59th Annual Meeting of the Association for Computational Linguistics and the 11th International Joint Conference on Natural Language Processing (Volume 2: Short Papers), pp. 504–510 (2021)
28. Zhang, Y., Sun, Y., Zhan, Y.: Large language models as an indirect reasoner: contrapositive and contradiction for automated reasoning (2024)
29. Zhou, C., Liu, P., et al.: Lima: Less is more for alignment (2023)

FReCS: A First Responder Classification System

Ademola Adesokan[1], Sanjay Madria[1(✉)], and Long Nguyen[2]

[1] Department of Computer Science, Missouri University of Science and Technology, Rolla, USA
{aaadfg,madrias}@mst.edu
[2] School of Applied Computational Sciences, Meharry Medical College, Nashville, USA
hlnguyen@mmc.edu

Abstract. In today's digital age, categorizing social media data, particularly from platforms like X, can be an effective strategy for identifying key first responders during emergencies, thereby improving overall emergency response efforts. In this study, we introduce a First Responder Classification System (FReCS), a framework that annotates and classifies disaster tweets from 26 crisis events. Our annotations cater for first reponders and their sub-layers. Furthermore, we proposed a classifier called RoBERTa-CAFÉ that integrates pre-trained RoBERTa with Cross-Attention and Focused-Entanglement components, improving the precision and reliability of classification tasks. The model is rigorously tested across publicly available disaster datasets. The RoBERTa-CAFÉ model outperformed state-of-the-art models in identifying relevant emergency communications, displaying its generalization, robustness, and adaptability. Our FReCS approach offers a pioneering technique for classifying first responders and enhances emergency management systems' operational capabilities, leading to more efficient and effective disaster responses. FReCS annotated dataset and code are available on GitHub (https://github.com/abdul0366/FReCS).

Keywords: Data Annotation · Social Media · Emergency Management · First Responder · Transformer

1 Introduction

First responders are crucial in disaster response, playing a vital role in safeguarding lives, properties, and communities as a whole [1]. Their prompt response to emergencies is significant, enabling swift and effective action in disaster situations [2]. This immediacy is vital for mitigating the impact of disasters, potentially addressing immediate needs in order to achieve the aforementioned roles [3]. The predominant focus of research in disaster management has been on the individuals and communities directly affected by calamitous events, with comparatively less emphasis on the experiences of professionals such as police officers and firefighters.

© The Author(s), under exclusive license to Springer Nature Switzerland AG 2025
L. M. Aiello et al. (Eds.): ASONAM 2024, LNCS 15211, pp. 355–372, 2025.
https://doi.org/10.1007/978-3-031-78541-2_22

Despite their importance, the effectiveness of first responders varies significantly across different types of disasters. For instance, the San Diego wildfires witnessed the effective deployment of first responders. Emergency managers and public health professionals played a crucial role in integrating their prevention and response efforts, effectively managing the significant disasters faced by the communities [4]. Similarly, the response to mental health calls by first responders following Hurricane Harvey provides insight into emergency service utilization during the disaster. This study examines the effects of Hurricane Harvey on mental health calls to Emergency Medical Services (EMS) and the Houston Police Department [5], demonstrating the critical role of first responders in managing complex emergencies.

However, the response to Hurricane Harvey also exposed some critical challenges, particularly in the context of Graduate Medical Education (GME) disaster planning at Corpus Christi Medical Center (CCMC). This situation underscored the need for more robust and effective disaster planning within GME programs, highlighting gaps in preparedness and response capabilities [6]. Another significant challenge encountered in disaster management, particularly highlighted during Hurricane Harvey, is the need for accurate data classification for first responders. This issue led to miscommunication between users and responders or volunteers, as evidenced in the event [7]. This gap in clear and accurate information exchange impeded the effective coordination of emergency response efforts, showcasing the need for improved data classification and communication strategies in disaster response. Furthermore, the response to Hurricane Maria brought to light significant challenges in managing disaster complexities and data management. This highlighted an urgent need for an automatic first responder classification system and communication strategy improvements during disaster response, emphasizing enhanced tools and methodologies for managing large-scale emergencies [8].

The absence of a structured classification system for responders in disaster management can lead to significant inefficiencies and heightened risks during emergency responses. As noted by [9], without a clear delineation and classification of roles, first responders may face challenges in coordination and communication, potentially leading to delayed response times, misallocation of resources, and increased risks to responders and affected populations. This lack of organization can worsen the impact of the disaster and impede recovery efforts [9].

The utilization of Social Media (SM) platforms, particularly X (formerly known as Twitter), in disaster management has been increasingly recognized. An online survey conducted among X users who sought help through tweets during Hurricane Harvey revealed a significant finding: 91% of these users reported that X was a valuable tool for facilitating the rescue of affected victims [10]. This statistic reinstated the growing importance of SM platforms in emergency response. SM data can effectively supplement traditional systems like dispatch calls, mostly used in emergency services [11].

Integrating SM data to classify first responders offer several transformative advantages, thereby enhancing disaster response's overall efficiency and effectiveness [1], such as: **1. facilitating the efficient allocation of resources** for informed and effective response strategies [12], **2. fostering greater public engagement** by responding effectively in times of crisis as a critical service [11], **3. Scalability of the traditional system using SM platforms** to handle large volumes of data, enabling the monitoring and response to multiple incidents simultaneously, which is too complicated and complex for manual systems [1], **4. Stabilization** plays a pivotal role in providing immediate assistance, ranging from medical aid and rescue operations to initial damage assessment [13]. **5. Psychological support,** in addition to physical assistance, first responders are instrumental in helping victims to calm, reassure, and assist individuals in shock or distress, which is essential in mitigating immediate psychological impacts [14].

The study aimed to provide answers to the following research questions:

- **Research Question 1:** How can we accurately classify the appropriate type of first responder for different emergency and crisis conditions?
- **Research Question 2:** Is it feasible to categorize first responders into subtypes that are customized to the specific needs and contextual demands of unique situations?

To address the challenges and questions mentioned earlier and explore the benefits of integrating X data for emergency response, we propose **FReCS**, a **F**irst **Re**sponder **C**lassification **S**ystem. The objective of FReCS is to re-annotate 26 crises for the purpose of first responder classification while also presenting a transformer-based model for classification tasks. Moreover, the model also includes a secondary classification to determine the specific sub-personnel required for crisis and emergency situations. Our FReCS system comprises four major classification tasks as shown in Fig. 1: (1.) Relevancy, (2.) Disaster Type, (3.) First Responder, and (4.) Secondary Classification.

To achieve this goal, we employ a blend of advanced deep-learning techniques, including a pre-trained transformer model and multiple custom attention mechanisms. Hence, our classifier, RoBERTa-CAFÉ, comprised of a **RoBERTa** with **C**ross, **A**daptive, and **F**ocused-**E**ntanglement Components which we used to classify different tasks and events based on textual X data for binary and multiclass classification. The following are our major contributions to this work:

- We annotated 27,933 disaster tweets for first responders using the CrisisLexT26 dataset [15]. This framework introduces specific categories for first responders such as Police, EMS, and Firefighters. These classes enable more accurate analysis and classification of crisis-related tweets, thereby improving model training for disaster management. The framework enhances the practical use of SM data in crisis scenarios and strengthens the efficiency of emergency response coordination via digital platforms.
- To achieve specificity and clarity in response, we introduced a secondary annotation to add sub-layers to the first responder category. This process specifies sub-personnel roles such as Mobile Medical Units, Crime Prevention Teams, and Urban Search and Rescue teams for various crises. This detailed annotation allows for more precise resource deployment. It improves the dataset's

utility for deeper analysis and modeling, enhancing the effectiveness of emergency management systems utilizing SM data.

- To ensure the reliability and accuracy of the dataset annotations, we used the Fleiss Kappa measure to assess the consistency of annotations among different annotators. Our inter-annotator agreement rating for first responder and secondary labels was 0.89 and 0.85, respectively.
- Our study introduces RoBERTa-CAFÉ, a modified pre-trained RoBERTa model enhanced with Cross-Attention and Focused-Entanglement Components to handle complex data better in crisis scenarios. This model incorporates Multi-Head Attention and an Adaptive Feed-Forward Network, which enhances its ability to filter and prioritize relevant SM information. The model demonstrates high effectiveness, with F1 and accuracy scores ranging from 86% to 100% across the four tasks. Our model significantly enhances the accuracy and reliability of automated disaster management systems in real-time applications.
- We validate the RoBERTa-CAFÉ model's effectiveness across diverse scenarios, showcasing its generalizability, consistency, robustness, and adaptability. The model effectively classifies data from various crises, tested on datasets such as CrisisLexT6 and CrisisBench, as well as specific events like the Nepal Earthquake and Queensland Floods. Its consistent high performance across different validation methods, including k-fold cross-validation, affirms its reliability as a tool for real-time crisis management.

2   Related Work

Recent evolutions in disaster management and first responder effectiveness have highlighted the role of integrating technology, policy development, and comprehensive training. These mutual efforts aim to improve response times, situational awareness, and overall outcomes during emergencies and disasters.

[16] identified a gap in real-time access to building system data for emergency responders, emphasizing the potential to significantly enhance situational awareness and reduce response times. They proposed a roadmap to overcome challenges in securely transmitting and processing building sensor data to first responders, emphasizing the need for a systemic approach to improve emergency response via informed decision-making.

Similarly, [1] introduced the ONSIDE, which leverages SM platforms to streamline disaster response coordination. By integrating Information-Centric Networking with a SM Engine, ONSIDE addresses the real-time analysis challenges of SM data, utilizing natural language processing to ensure rapid and relevant information delivery to first responders.

In aviation safety, [17] proposed an In-Time Aviation Safety Management System designed specifically for UAS and autonomous systems in emergency scenarios. This system emphasizes predictive modeling to proactively identify and mitigate risks, highlighting the need for scenario testing to identify new safety data requirements and operational hurdles.

Furthermore, emphasizing the role of education, [18] explored the impact of an Emergency First Response (EFR) training program at Tecnológico de Monterrey. Their findings underscored the importance of integrating emergency response training within higher education to enhance EFR skills across various disciplines, improving community and workplace safety.

Addressing the communication challenges in disaster scenarios, [19] presented ReDiCom, a resilient architecture designed to enhance first responder communication. By supporting network resilience and utilizing coded computation, ReDi-Com facilitates efficient information dissemination and resource management, underscoring the potential of technological advancements in improving disaster management.

On the policy front, [20] examined state-level policies addressing first responder mental health. Their study categorized policies into workers' compensation-related and non-workers' compensation-related, highlighting legislative efforts to support first responders facing adverse mental health outcomes due to occupational trauma and the need for systematic evaluations to establish evidence-based mental health care practices.

Lastly, [21] developed SOSFloodFinder, a system utilizing NLP and GPS technologies to classify urgency in emergency communications from flood victims. This innovation demonstrates how technology can enhance the precision and efficiency of first responder activities during floods, contributing to the broader goal of improving disaster management and response.

These studies demonstrate a comprehensive approach to enhancing the efficiency and effectiveness of emergency management and first responder activities. These efforts aim to improve safety, efficiency, and outcomes in disaster response and emergency situations by leveraging technology, policy development, and targeted training. However, FReCS stands apart from existing studies by incorporating sub-types into the first responder category. This allows for identifying specialized personnel (such as those in the police units responsible for criminal activities). This approach enables customized responses to unique emergency situations instead of a generalized approach that treats all situations with the same protocol.

3   Our Approach

This section outlines our methodology for accurately categorizing tweets for emergency response coordination. This process involves dataset annotation and a multi-level classification framework, as shown in Fig. 1. We further provide a detailed explanation of each step in the following subsection.

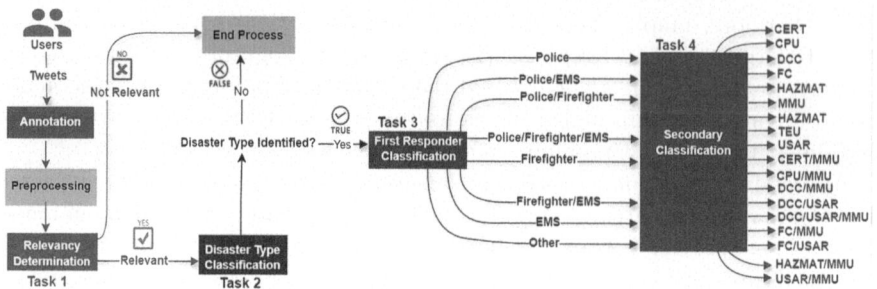

Fig. 1. FReCS Proposed System Framework.

3.1   Dataset and Annotation Process

In this study, we utilized the CrisisLexT26 dataset [15], comprising about 28,000 tweets across 26 crisis events from 2012 and 2013, initially annotated by crowd-sourced workers based on event types (e.g., Flood, Wildfire, Earthquake), informativeness, information types (e.g., caution and advice, infrastructure damage), and information sources (e.g., governments, NGOs). For detailed documentation on the crowdsourced annotations, see [15]. Notably, the dataset lacked annotations for first responders and secondary classifications. To fill this gap, we conducted a detailed annotation process over two months with a team of three students (two annotators and one experienced moderator). Our initial primary annotation encompassed four label classes for first responders: Police, EMS, Firefighter, and Other, aligning with FEMA standards. While recognizing that some regions classify additional agencies as first responders, we maintained these three primary categories for consistency across different jurisdictions. For secondary annotations, we initially introduced nine label categories: Mobile Medical Unit (MMU), Community Emergency Response Team (CERT), Crime Protection/Prevention Unit (CPU), Dispatch Call Center (DCC), Traffic Enforcement Unit (TEU), Hazardous Materials (HAZMAT), Fire Control (FC), Urban Search and Rescue (USAR), and Other. Table 1 shows the class-label distribution.

Table 1. Label Distribution for Task 3 and 4

First Responder Class Labels	Secondary Layer Class Labels
Police: 3953, **EMS:** 753, **Firefighter:** 1248, **Police/EMS:** 488, **Police/Firefighter:** 450, **Firefighter/EMS:** 181, **Police/Firefighter/EMS:** 290, **Other:** 20570	**FC/USAR:** 53, **MMU:** 481, **USAR/MMU:** 90, **USAR:** 236, **FC/MMU:** 65, **FC:** 634, **DCC/USAR/MMU:** 290, **DCC/MMU:** 385, **DCC/USAR:** 450, **DCC:** 2435, **CERT/MMU:** 158, **CERT:** 114, **CPU/MMU:** 103, **CPU:** 315, **HAZMAT:** 325, **HAZMAT/MMU:** 26, **TEU:** 80, **Other:** 20570

This classification allowed for a more nuanced assignment of resources, with DCC and TEU functioning as sub-layers of Police; CERT and MMU under EMS; and FC, HAZMAT, and USAR under Firefighter. This secondary classification aimed to enhance operational specificity and improve response efficiency by deploying the most suitable responder team to each unique emergency. During the annotation, it became evident that some tweets required the simultaneous deployment of multiple responder types. We addressed this complexity by assigning multiple labels where necessary, expanding the first responder and secondary classification labels from four to eight and nine to eighteen, respectively, as shown in Fig. 1. Following the initial annotation phase, we engaged in a rigorous review process. This collaborative approach involved annotators actively verifying each other's work, with the experienced moderator resolving any disagreements. We then assessed the consistency of these annotations through the inter-annotator agreement process. Each annotator rated their agreement with a score of 1 or disagreement with a score of 0. We employed the Fleiss Kappa statistical measure [22] to gauge the level of consensus among annotators. The results revealed a high consistency rate, with a Fleiss Kappa score of 0.89 and 0.85 for the first responder and secondary labels, respectively, indicating substantial agreement and affirming the reliability and accuracy of our annotations.

Our **preprocessing** steps include normalizing text [23], removing duplicates and links [24], special characters and stopwords [25] to ensure accurate and reliable model training.

3.2   RoBERTa-CAFÉ Classifier

Our classifier model, RoBERTa-CAFÉ (Cross-Attention and Focused-Entanglement) include classification of Task 1 (Relevancy), Task 2 (Disaster Type), Task 3 (First Responder) and Task 4 (Secondary layers), which integrates advanced attention mechanisms and RoBERTa's contextual embeddings to accurately classify disaster tweets for different tasks. The model has several core components, as shown in Fig. 2.

(1) RoBERTa contextual embeddings [26]: RoBERTa, an advanced version of BERT, forms the backbone of the RoBERTa-CAFÉ model, providing the ability to generate deep contextual embeddings. These embeddings enable the model to understand subtle language dynamics. The embeddings are expressed as follows:

$$\mathbf{E} = E_{\text{tokens}} + E_{\text{positions}} \tag{1}$$

We then process the input embeddings through multiple layers of transformer block, each block consisting of:

– Multi-Head Self-Attention, which allows the model to attend to different parts of the input sequence:

$$\text{MultiHead}(\mathbf{Q}, \mathbf{K}, \mathbf{V}) = \text{Concat}(head_1, \ldots, head_h)W^O \tag{2}$$

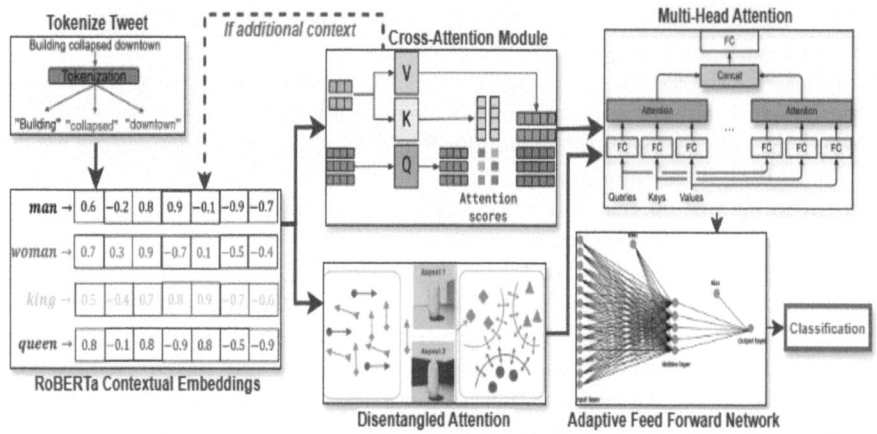

Fig. 2. RoBERTa-CAFÉ Classifier

- Additionally, a position-wise fully connected feed-forward network is applied to each position separately and identically in each layer:

$$\mathbf{FFN}(\mathbf{x}) = \max(0, xW_1 + b_1)W_2 + b_2 \tag{3}$$

- To aid in stabilizing the learning process, residual connections and layer normalization are used. The output is obtained by adding the output of the sublayer operation to the input, followed by layer normalization:

$$\mathbf{output} = \text{LayerNorm}(x + \text{Sublayer}(x)) \tag{4}$$

Where Sublayer(x) is the operation by the multi-head attention or the feed-forward network.

(2) Cross-Attention Module (CAM): integrates external contextual information with RoBERTa's embeddings [27]. This mechanism allows the model to focus on specific parts of the text by considering the additional context provided, thus enhancing its ability to adapt to various situations and datasets. The module is an extension of the self-attention mechanism and is applied between two different sets of inputs: the main input x and the context input context. Our CAM module comprises multiple steps:

- A linear transformation of the x (query) and context (key and value) inputs into query, key, and value spaces:

$$\mathbf{Q} = W_q x + b_q, \quad \mathbf{K} = W_k \text{context} + b_k, \quad \mathbf{V} = W_v \text{context} + b_v \tag{5}$$

Where W_q, W_k, W_v are weight matrices and b_q, b_k, b_v are biases for queries, keys, and values, respectively.

- Attention scores are computed by taking the dot product of the query with the key of each element in the context and dividing it by the square root of the dimension of keys to stabilize gradients during training:

$$\textbf{scores} = \frac{QK^T}{\sqrt{d_k}} \tag{6}$$

Where d_k is the dimensionality of the keys.
- We applied the softmax function to the scores to obtain the attention weights:

$$\textbf{attention weights} = \text{softmax(scores)} \tag{7}$$

- The output is computed as a weighted sum of the values with the weights given by the attention weights:

$$\textbf{output} = \text{attention weights} \cdot V \tag{8}$$

- Finally, the output is summed with the input to let the layer perform as a residual connection:

$$\textbf{attended output} = \text{output} + x \tag{9}$$

(3) Disentangled Attention (DeA): This component divides the attention mechanism into various paths (aspects) to aid the model in separately and simultaneously learning different kinds of features from the data [28], such as semantic and syntactic features. This division helps capture the diverse nature of language used in disaster-related communications more efficiently. In the DeA module, we processed two separate aspects of the input using sigmoid-activated linear transformations. It allows the layer to concentrate on different input aspects or features independently. The output is obtained by combining the two aspects and multiplying it with the input. The mechanism is represented as:

$$\textbf{aspect}_1(x) = \sigma(W_1 x + b_1) \ and \ \textbf{aspect}_2(x) = \sigma(W_2 x + b_2) \tag{10}$$

$$\textbf{output} = (\text{aspect}_1(x) + \text{aspect}_2(x)) \cdot x \tag{11}$$

where σ is the sigmoid function, W_1, W_2 are the weight matrices, b_1, b_2 are bias vectors, and x is the input.

(4) Multi-Head Attention (MHA): is a mechanism that allows models to attend to different representation subspaces at different positions, enabling them to capture a variety of dependencies in the input [29]. With multiple 'heads,' the model can capture a variety of dependencies in the input, such as those between different key terms in disaster data, which is crucial for accurate classification. Our MHA module divides the model's attention into multiple 'heads,' allowing it to attend to different parts of the input simultaneously. We represent the breakdown of the multi-head as follows:

$$\mathbf{Q} = W_Q X, \ \mathbf{K} = W_K X, \ \text{and} \ \mathbf{V} = W_V X \tag{12}$$

$$\mathbf{Attention}(Q, K, V) = \mathrm{softmax}\left(\frac{QK^T}{\sqrt{d_k}}\right) V \tag{13}$$

$$\mathbf{MultiHead}(\mathbf{Q}, \mathbf{K}, \mathbf{V}) = \mathrm{Concat}(head_1, \ldots, head_h)W^O \tag{14}$$

where heads are the individual attention outputs and W^O is another learned parameter matrix.

(5) **Adaptive Feed Forward Network (AFFN):** This is composed of feed-forward layers that utilize gating mechanisms, such as GLUs, to regulate the flow of information. This network adapts by enhancing or reducing feature representations, enabling it to focus on relevant features while discarding less important data [30].

$$\mathbf{x_{new}} = \mathrm{GLU}(W_i x + b_i) = (W_{i,1}x + b_{i,1}) \otimes \sigma(W_{i,2}x + b_{i,2}) \tag{15}$$

where $W_{i,1}, W_{i,2}$ and $b_{i,1}, b_{i,2}$ are the weights and biases of the linear transformations, σ is the sigmoid activation function, and $\otimes$ denotes element-wise multiplication.

(6) **Classification Layer:** The final layer of the model is a linear layer that maps the enriched text representations to the output classes, which correspond to different classifications under different tasks. This makes the model a valuable tool for automated disaster response systems.

Uniqueness: Our RoBERTa—CAFÉ is unique as it integrate enhanced attention mechanisms (CAM and DeA) that allows the model to focus on what is being said and how different aspects of the information related to external contexts and internal text structures. It also provides robust feature processing (AFFN and MHA) that ensures the model can efficiently process a wide array of textual features, enhancing the classifier's accuracy and flexibility across diverse disaster-related datasets.

4   Results

This section outlines the experimental validation of the methods introduced previously. The performance of our models was rigorously tested through train/test splits and k-fold cross-validation, employing metrics such as accuracy, precision, recall, and F1 score. These experiments were performed on a robust computational system featuring dual NVIDIA® Tesla V100 GPUs and an Intel® Xeon® Gold CPU, offering substantial computing power to meet the intensive processing requirements of our deep learning frameworks.

4.1   Model Training and Testing

Our RoBERTa-CAFÉ classifier employs advanced neural network architectures for accurate tweet classification. Recall in Sect. 3.2 that we use sophisticated attention mechanisms, including cross-attention, disentangled attention, multi-head attention, and an adaptive feed-forward network, to effectively handle complex textual dependencies.

Preprocessed tweets use RoBERTa's tokenizer to transform the text into token sequences. The tokens are managed by a custom PyTorch Dataset class for optimized batching and loading during training and testing. The RoBERTa-CAFÉ model is trained on a labeled dataset, split into different training and testing sets, using a DataLoader for efficient batch processing. We use *RAdam* and a learning rate scheduler to optimize training and achieve stable convergence in multiple epochs. The training process involves minimizing the loss for multi-class and binary classification tasks by adjusting model weights iteratively. Our hyperparameters are shown in Table 2.

The model's performance is assessed using the metrics described, including classification reports, after thorough training. This evaluation process examines the model's classification accuracy and its capacity to generalize to unseen data, ensuring that the RoBERTa-CAFÉ classifier effectively learns from the training data and remains robust when faced with new datasets. More information about our model training and hyperparameters can be found in our code on this link.

Table 2. Hyperparameters and their values for our Classification model

Hyperparameter	Value
Input Dimensions	768
Number of Heads in MHA	8
AFFN Dimension	2048
Depth of AFFN	2 layers
Dropout Rate	0.2
Number of Classes	Variable (as per task)
Batch Size	64
Optimizer	RAdam
Learning Rate	2×10^{-5}
Loss Function	CrossEntropyLoss/BCE
Learning Rate Scheduler	Step LR (gamma $= 0.1$, step size $= 10$)
Epochs	10
Max Length for Tokenization	128 characters

4.2   Train/Test Split Vs. K-Fold Cross Validation

For all four tasks, our study tested different train/test splits (90/10, 80/20, 70/30, 60/40) and cross-fold validations (5, 10, and 15). Our analysis indicates that the performance metrics results for different train/test splits ranging from 90/10 to 60/40 are highly consistent. This uniformity is illustrated in Table 3 by the matched F1 score and accuracy results across all tasks. Our analysis suggests that there are no significant differences in the performance metrics attributable to the proportion of the split, thus indicating the model's stability.

Moreover, the cross-fold validation results comprising 5, 10, and 15 folds, as demonstrated in Table 4, show comparable outcomes with negligible variances. The consistency of the results across various split ratios and cross-validation folds highlights the robustness of our model, which confirms its reliability irrespective of data segmentation methods.

Table 3. Result of our 4 tasks under 4 train/test splits using R–CAFÉ.

Tasks	Train (90)/Test (10)							Train (80)/Test (20)							Train (70)/Test (30)							Train (60)/Test (40)						
	Macro Avg			Wei. Avg				Macro Avg			Wei. Avg				Macro Avg			Wei. Avg				Macro Avg			Wei. Avg			
	Pr	Re	F1	Pr	Re	F1	Acc	Pr	Re	F1	Pr	Re	F1	Acc	Pr	Re	F1	Pr	Re	F1	Acc	Pr	Re	F1	Pr	Re	F1	Acc
Task 1	0.83	0.83	0.83	0.94	0.94	0.94	0.94	0.85	0.80	0.82	0.94	0.94	0.94	0.94	0.86	0.78	0.81	0.93	0.94	0.93	0.94	0.86	0.80	0.82	0.93	0.94	0.94	0.94
Task 2	1.00	1.00	1.00	1.00	1.00	1.00	1.00	0.99	1.00	1.00	1.00	1.00	1.00	1.00	0.99	0.99	0.99	0.99	0.99	0.99	0.99	1.00	1.00	1.00	1.00	1.00	1.00	1.00
Task 3	0.71	0.70	0.70	0.87	0.87	0.87	0.87	0.71	0.72	0.71	0.87	0.87	0.87	0.87	0.67	0.71	0.68	0.87	0.87	0.87	0.87	0.68	0.69	0.68	0.86	0.86	0.86	0.86
Task 4	0.58	0.66	0.60	0.87	0.86	0.86	0.86	0.65	0.54	0.57	0.86	0.86	0.85	0.86	0.61	0.59	0.59	0.86	0.85	0.85	0.85	0.59	0.57	0.56	0.86	0.84	0.85	0.84

Table 4. Evaluation of our 4 tasks under different folds using R–CAFÉ.

Tasks	5-Fold							10-Fold							15-Fold						
	Macro Avg			Wei. Avg				Macro Avg			Wei. Avg				Macro Avg			Wei. Avg			
	Pr	Re	F1	Pr	Re	F1	Acc	Pr	Re	F1	Pr	Re	F1	Acc	Pr	Re	F1	Pr	Re	F1	Acc
Task 1	0.85	0.81	0.82	0.93	0.94	0.93	0.94	0.85	0.81	0.83	0.93	0.94	0.93	0.94	0.85	0.81	0.83	0.93	0.94	0.93	0.94
Task 2	1.00	1.00	1.00	1.00	1.00	1.00	1.00	1.00	1.00	1.00	1.00	1.00	1.00	1.00	1.00	1.00	1.00	1.00	1.00	1.00	1.00
Task 3	0.68	0.72	0.70	0.87	0.86	0.86	0.86	0.69	0.72	0.70	0.88	0.87	0.87	0.87	0.67	0.71	0.69	0.87	0.86	0.87	0.86
Task 4	0.60	0.62	0.60	0.86	0.86	0.86	0.86	0.62	0.63	0.61	0.87	0.86	0.86	0.86	0.58	0.63	0.59	0.86	0.86	0.86	0.86

4.3   Task 1 (Relevancy) and Task 2 (Disaster Type)

Our RoBERTa-CAFÉ model outperforms [31] model in Task 1 as proven in Table 5, which focuses on relevancy. RoBERTa-CAFÉ achieved a recall improvement of 20% and an F1 score of 0.93, showing a 14% improvement over [31]'s F1 score. RoBERTa-CAFÉ's higher difference recall suggests that it is better at identifying relevant cases, while the increased precision suggests that its relevancy classification is more accurate. The higher F1 score confirms that RoBERTa-CAFÉ has a better balance of precision and recall, as shown in Table 5. Both the RoBERTa-CAFÉ and the [31] model achieved perfect scores in Task 2, which involves classifying disaster types. This indicates that both models are highly effective in accurately identifying all relevant cases without false positives or negatives. Furthermore, there is no significant difference in performance metrics between the two models for this task.

Table 5. Performance of Task 1 and 2

Models	Task 1			Task 2		
	P	R	F1	P	R	F1
Burel et al. [31]	0.87	0.74	0.79	1.00	1.00	1.00
R-CAFÉ	**0.93**	**0.94**	**0.93**	1.00	1.00	1.00

4.4 Task 3 (First Responder) and Task 4 (Secondary Classification)

Due to the uniqueness of our annotation labels, which creates a gap of not having related work to compare with, we evaluated the performance of the RoBERTa-CAFÉ model against four baseline classifiers: Decision Tree - DT, Naïve Bayes - NB, Support Vector Machine - SVM, and Logistic Regression - LR in Tasks 3 and 4, which involved first responder and secondary classification. The implementation of the four baselines is similar to the work of [32]. The results from Tables 6 and 7 showed that RoBERTa-CAFÉ provided a competitive approach with high and consistent scores across metrics, demonstrating its robustness in handling the tasks' unique requirements.

The model's performance was particularly notable for maintaining high accuracy and recall, which are crucial for reliable disaster response applications. RoBERTa-CAFÉ performed well in the recall, which is a critical factor in secondary classification tasks. It showed effectiveness in handling complex classification tasks, as demonstrated by its high F1 scores across all splits. The model's accuracies were also consistently high, indicating reliable performance across different data partitions.

Table 6. R–CAFÉ comparison with baseline models for Task 3 under 4 splits.

Models	Train (90)/Test (10)							Train (80)/Test (20)							Train (70)/Test (30)							Train (60)/Test (40)						
	Macro Avg			Wei. Avg				Macro Avg			Wei. Avg				Macro Avg			Wei. Avg				Macro Avg			Wei. Avg			
	Pr	Re	F1	Pr	Re	F1	Acc	Pr	Re	F1	Pr	Re	F1	Acc	Pr	Re	F1	Pr	Re	F1	Acc	Pr	Re	F1	Pr	Re	F1	Acc
DT	0.34	0.21	0.23	0.69	0.77	0.70	0.77	0.32	0.19	0.21	0.67	0.76	0.68	0.76	0.39	0.20	0.22	0.68	0.76	0.68	0.76	0.33	0.19	0.21	0.67	0.76	0.68	0.76
NB	0.64	0.38	0.45	0.80	0.81	0.80	0.81	0.59	0.36	0.41	0.79	0.80	0.78	0.80	0.65	0.37	0.42	0.80	0.81	0.79	0.81	0.64	0.34	0.39	0.79	0.80	0.78	0.80
SVM	0.68	0.53	0.59	0.82	0.84	0.83	0.84	0.68	0.53	0.58	0.82	0.84	0.82	0.84	0.70	0.53	0.59	0.82	0.84	0.82	0.84	0.69	0.52	0.58	0.82	0.83	0.82	0.83
LR	0.68	0.52	0.58	0.82	0.84	0.83	0.84	0.68	0.51	0.57	0.82	0.83	0.82	0.83	0.68	0.49	0.55	0.82	0.83	0.82	0.83	0.69	0.48	0.55	0.81	0.83	0.81	0.83
R–CAFÉ	**0.71**	**0.70**	**0.70**	**0.87**	**0.87**	**0.87**	**0.87**	**0.71**	**0.72**	**0.71**	**0.87**	**0.87**	**0.87**	**0.87**	**0.67**	**0.71**	**0.68**	**0.87**	**0.87**	**0.87**	**0.87**	**0.68**	**0.69**	**0.68**	**0.86**	**0.86**	**0.86**	**0.86**

4.5 Ablation Study

In our experiments, we analyzed the impact of custom attention layers on the performance of the RoBERTa-CAFÉ classifier. We compared the RoBERTa-CAFÉ with a version without custom attention layers (CAFÉ) across four tasks and train/test splits, using F1 scores as the benchmark. The results showed that incorporating custom attention layers improved the RoBERTa-CAFÉ's performance, as shown in Fig. 3.

Table 7. R–CAFÉ comparison with baseline models for Task 4 under 4 splits.

Models	Train (90)/Test (10)							Train (80)/Test (20)							Train (70)/Test (30)							Train (60)/Test (40)						
	Macro Avg			Wei. Avg				Macro Avg			Wei. Avg				Macro Avg			Wei. Avg				Macro Avg			Wei. Avg			
	Pr	Re	F1	Pr	Re	F1	Acc	Pr	Re	F1	Pr	Re	F1	Acc	Pr	Re	F1	Pr	Re	F1	Acc	Pr	Re	F1	Pr	Re	F1	Acc
DT	0.19	0.15	0.15	0.66	0.77	0.69	0.77	0.24	0.13	0.14	0.65	0.76	0.68	0.76	0.28	0.13	0.14	0.67	0.76	0.68	0.76	0.21	0.12	0.13	0.65	0.75	0.68	0.75
NB	0.50	0.22	0.27	0.79	0.81	0.78	0.81	0.45	0.22	0.26	0.76	0.80	0.77	0.80	0.40	0.21	0.25	0.76	0.80	0.77	0.80	0.42	0.20	0.24	0.76	0.80	0.76	0.80
SVM	0.60	0.44	0.48	0.82	0.84	0.82	0.84	0.61	0.40	0.46	0.81	0.83	0.81	0.83	0.60	0.38	0.45	0.81	0.83	0.81	0.83	0.61	0.38	0.44	0.80	0.83	0.81	0.83
LR	0.67	0.44	0.51	0.82	0.83	0.82	0.83	0.61	0.36	0.43	0.81	0.83	0.81	0.83	0.59	0.33	0.40	0.81	0.83	0.81	0.83	0.61	0.32	0.40	0.80	0.82	0.80	0.82
R–CAFÉ	0.58	**0.66**	**0.60**	**0.87**	**0.86**	**0.86**	**0.86**	**0.65**	0.54	0.57	**0.86**	**0.86**	0.85	**0.86**	0.61	0.59	0.59	**0.86**	0.85	0.85	0.85	0.59	0.57	0.56	**0.86**	0.84	0.85	0.84

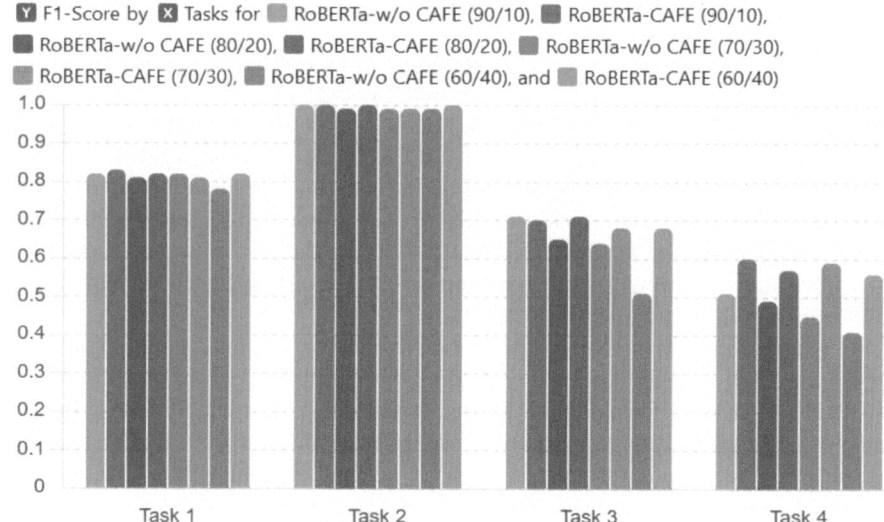

■ F1-Score by ✖ Tasks for ■ RoBERTa-w/o CAFE (90/10), ■ RoBERTa-CAFE (90/10), ■ RoBERTa-w/o CAFE (80/20), ■ RoBERTa-CAFE (80/20), ■ RoBERTa-w/o CAFE (70/30), ■ RoBERTa-CAFE (70/30), ■ RoBERTa-w/o CAFE (60/40), and ■ RoBERTa-CAFE (60/40)

Fig. 3. RoBERTa–CAFÉ Vs finetuned RoBERTa without CAFÉ

The RoBERTa-CAFÉ model consistently outperformed the RoBERTa w/o CAFÉ model in Task 1 for all train/test dataset splits. For Task 2, both models performed well, with the CAFÉ layers not showing a significant performance enhancement; this is attributed to the distinguishable characteristics inherent in this classification task. Task 3 demonstrated a noticeable improvement in performance for the RoBERTa-CAFÉ model compared to the RoBERTa w/o CAFÉ model, with the most significant increase in F1 score observed in the 90/10 data split. Task 4 showed a robust enhancement in the RoBERTa-CAFÉ model's performance, particularly in the 60/40 split, where the CAFÉ layers resulted in a 15% increase in F1 score over the RoBERTa w/o CAFÉ model.

4.6 Beyond FReCS: CrisisLexT6, CrisisBench, NEQ and QFL

In our study, we extended the RoBERTa-CAFÉ model to analyze its performance on various publicly available crisis-related datasets to evaluate its effectiveness and generalizability beyond the FReCS dataset. We considered four datasets: CrisisLexT6, CrisisBench, Nepal Earthquake (NEQ), and Queensland

Flood (QFL). Subsequently, we summarize our findings and highlight the comparison with existing models, thereby demonstrating the robust capabilities of RoBERTa-CAFÉ across diverse crisis communication scenarios.

NEQ dataset from Table 9, RoBERTa-CAFÉ, significantly outperformed [34] by increasing all metrics by 17 points, achieving 0.79 across these metrics. Similarly, in the QFL dataset, RoBERTa-CAFÉ's performance outperformed [34] by achieving an impressive score difference of 0.17 for precision, 0.16 for recall and F1 score, marking substantial improvements.

Table 8. Comparison with CrisisBench Dataset.

Models	Informativeness				Humanitarian			
	P	R	F1	Acc	P	R	F1	Acc
Alam et al. [33]	0.88	0.88	0.88	0.88	0.79	0.78	0.78	0.78
R-CAFÉ	0.88	0.88	0.88	0.88	0.79	0.78	0.78	0.78

Table 9. Comparison with NEQ and QFL Dataset.

Models	NEQ			QFL		
	P	R	F1	P	R	F1
Alam & Imran [34]	0.65	0.65	0.65	0.93	0.94	0.94
R-CAFÉ	**0.79**	**0.79**	**0.79**	**0.97**	**0.96**	**0.96**

Table 10. Comparison with CrisisLexT6 Dataset

Models	Acc	Models	M-F1
Jaoa [36]	0.94	Li et al. [37]	0.90
Li et al. [35]	**0.95**	Li et al. [35]	**0.95**
R-CAFÉ	**0.95**	R-CAFÉ	**0.95**

Regarding the CrisisLexT6 dataset from Table 10, RoBERTa-CAFÉ achieved an accuracy and macro-F1 score of 0.95, matching the performance of the best existing model by [35] and slightly outperforming [36] on accuracy. As for the CrisisBench dataset in Table 8, RoBERTa-CAFÉ mirrored the performance of [33] across all metrics.

These results affirm the versatility of our RoBERTa-CAFÉ model in handling a range of crisis-related communications with high precision and reliability. Its ability to adapt and maintain high performance across various datasets stresses its potential as a powerful crisis management and response tool.

5 Conclusion and Future Work

The effectiveness of FReCS, a First Responder Classification System that utilizes the advanced capabilities of the RoBERTa-CAFÉ model to scrutinize and classify SM data for emergency response purposes, has been demonstrated in this study. Our findings reveal that integrating refined custom attention mechanisms into the pre-trained RoBERTa model significantly enhances FReCS's precision and speed in identifying relevant emergency-related communications. The system's robust performance across various datasets highlights its potential to revolutionize the landscape of disaster management by providing timely and accurate information crucial for first responder deployment efficiency.

In future research, we plan to enhance the system's applicability and reliability across different geographical and cultural contexts; we intend to expand the dataset to include a more extensive range of languages. Additionally, integrating multimedia data, such as images and videos to enrich the system's contextual understanding and response accuracy.

Acknowledgement. This research project received support from the NSF - USA CNS-2219615, CNS-2219614, and the Kummer Institute for Student Success, Research, and Economic Development at the Missouri University of Science and Technology through the Kummer Innovation and Entrepreneurship Doctoral Fellowship.

References

1. Mittal, V., Jahanian, M., Ramakrishnan, K.K.: Online delivery of social media posts to appropriate first responders for disaster response. In: ACM International Conference Proceedings Series (2021). https://doi.org/10.1145/3427477.3429272
2. Khatoon, S., et al.: Development of social media analytics system for emergency event detection and crisis management. Comput. Mater. Continua **68**(3) (2021). https://doi.org/10.32604/cmc.2021.017371
3. Wu, K., Wu, J., Ye, M.: A review on the application of social media data in natural disaster emergency management. Prog. Geogr. **39**(8) (2020). https://doi.org/10.18306/dlkxjz.2020.08.014
4. Clements, B.W., Casani, J.A.P.: Disasters and Public Health: Planning and Response: Second Edition (2016). https://doi.org/10.1016/C2014-0-01322-6
5. Saunders, J., et al.: Emergency mental health calls to first responders following a natural disaster: examining the effects from Hurricane Harvey. Int. J. Acad. Med. **7**(1) (2021). https://doi.org/10.4103/IJAM.IJAM_71_20
6. Newman, B., Gallion, C.: Hurricane harvey: firsthand perspectives for disaster preparedness in graduate medical education. Acad. Med. **94**(9) (2019). https://doi.org/10.1097/ACM.0000000000002696
7. Zou, L., et al.: Social media for emergency rescue: an analysis of rescue requests on twitter during hurricane harvey. Int. J. Disaster Risk Reduction **85** (2023). https://doi.org/10.1016/j.ijdrr.2022.103513
8. Kress, M.M., Chambers, K.F., Hernandez Abrams, D.D., McKay, S.K.: Principles for data management, visualization, and communication to improve disaster response management: lessons from the Hurricane Maria response mission. J. Emerg. Manag. **19**(8) (2021). https://doi.org/10.5055/jem.0658

9. Haddow, G.D., Bullock, J.A., Coppola, D.P.: Introduction to Emergency Management (2020). https://doi.org/10.1016/B978-0-12-817139-4.01001-0

10. Mihunov, V.V., Lam, N.S.N., Zou, L., Wang, Z., Wang, K.: Use of twitter in disaster rescue: lessons learned from hurricane harvey. Int. J. Digit. Earth **13**(12) (2020). https://doi.org/10.1080/17538947.2020.1729879

11. Kirby, R.H., Reams, M., Lam, N.S.N.: The use of social media by emergency stakeholder groups: lessons learned from areas affected by hurricanes isaac and sandy. J. Homeland Secur. Emerg. Manag. **20**(2) (2023). https://doi.org/10.1515/jhsem-2021-0031

12. Koshy, R., Elango, S.: Utilizing social media for emergency response: a tweet classification system using attention-based BiLSTM and CNN for resource management. Multimedia Tools Appl. (2023). https://doi.org/10.1007/s11042-023-16766-z

13. Ein, N., et al.: Physical and psychological challenges faced by military, medical and public safety personnel relief workers supporting natural disaster operations: a systematic rev. Curr. Psych. **43**(2) (2024). https://doi.org/10.1007/s12144-023-04368-9

14. Dong, L., Bouey, J.: Public mental health crisis during COVID-19 pandemic, China. Emerg. Infec. Dis. **26**(7) (2020). https://doi.org/10.3201/eid2607.200407

15. Olteanu, A., Vieweg, S., Castillo, C.: What to expect when the unexpected happens: social media communications across crises. In: CSCW 2015 - Proceedings of the 2015 ACM International Conference on Computer-Supported Cooperative Work and Social Computing (2015). https://doi.org/10.1145/2675133.2675242

16. Holmberg, D.G., Raymond, M.A., Averill, J.: Delivering building intelligence to first responders. National Institute of Standards and Technology, Technical Note (2013)

17. Kirkman, D., et al.: Informing new concepts for UAS and autonomous system safety management using disaster management and first responder scenarios. In: AIAA/IEEE Digital Avionics Systems Conference - Proceedings, vol. 2021-October (2021). https://doi.org/10.1109/DASC52595.2021.9594356

18. Dieck-Assad, G., González Peña, O.I., Rodríguez-Delgado, J.M.: Evaluation of emergency first response's competency in undergraduate college students: enhancing sustainable medical education in the community for work occupational safety. Int. J. Environ. Res. Public Health **18**(15) (2021). https://doi.org/10.3390/ijerph18157814

19. Ramakrishnan, K.K., Yuksel, M., Seferoglu, H., Chen, J., Blalock, R.A.: Resilient communication for dynamic first responder teams in disaster management. IEEE Commun. Mag. (2022). https://doi.org/10.1109/MCOM.003.2200015

20. O'Dare, K., Mathis, A., Tawk, R., Atwell, L., Jackson, D.: State level policies on first responder mental health in the U.S.: a scoping review. Admin. Pol. Mental Health Mental Health Serv. Res. (2024). https://doi.org/10.1007/s10488-024-01352-8

21. Kamal, S.H., Aziz, A.A., Mustafa, W.A.: SOSFloodFinder: a text-based priority classification system for enhanced decision-making in optimizing emergency flood response. J. Aut. Intel. **7**(1) (2024). https://doi.org/10.32629/jai.v7i1.874

22. Fleiss, J.L.: Measuring nominal scale agreement among many raters. Psychol. Bull. **76**(5) (1971). https://doi.org/10.1037/h0031619

23. Adesokan, A., Madria, S., Nguyen, L.: HatEmoTweet: low-level emotion classifications and spatiotemporal trends of hate and offensive COVID-19 tweets. Soc. Netw. Anal. Mining **13**, 136 (2023). https://doi.org/10.1007/s13278-023-01132-6

24. Adesokan, A., Madria, S.: NeuEmot: mitigating neutral label and reclassifying false neutrals in the 2022 FIFA world cup via low-level emotion. In: Proceedings of the 2023 IEEE International Conference on Big Data, pp. 578–587 (2023)
25. Adesokan, A., Madria, S., Nguyen, L.: TweetACE: a fine-grained classification of disaster tweets using transformer model. In: Proceedings of the IEEE Applied Imagery Pattern Recognition Workshop (AIPR), pp. 1–9 (2023)
26. Semary, N.A., Ahmed, W., Amin, K., Pławiak, P., Hammad, M.: Improving sentiment classification using a RoBERTa-based hybrid model. Front. Hum. Neurosci. **17** (2023). https://doi.org/10.3389/fnhum.2023.1292010. ISSN: 16625161
27. Gheini, M., Ren, X., May, J.: Cross-attention is all you need: adapting pretrained transformers for machine translation. In: Proceedings of EMNLP 2021 - 2021 Conference on Empirical Methods in Natural Language Processing (2021). https://doi.org/10.18653/v1/2021.emnlp-main.132
28. He, P., Liu, X., Gao, J., Chen, W.: DEBERTA: decoding-enhanced bert with disentangled attention. In: Proceedings of ICLR 2021 - 9th International Conference on Learning Representations (2021)
29. Vaswani, A., et al.: Attention is all you need. In: Advances in Neural Information Processing Systems, vol. 30 (2017). ISSN: 10495258
30. Shazeer, N.: GLU Variants Improve Transformer, preprint arXiv:2002.05202 (2020)
31. Burel, G., Saif, H., Fernandez, M., Alani, H.: On semantics and deep learning for event detection in crisis situations. Presented at the Workshop on Semantic Deep Learning (SemDeep) (2017)
32. Das, S., Bhattacharyya, K., Sarkar, S.: Performance analysis of logistic regression, Naive Bayes, KNN, decision tree, random forest and SVM on hate speech detection from Twitter. Int. Res. J. Innov. Eng. Technol. **7**(3) (2023). https://doi.org/10.47001/irjiet/2023.703004
33. Alam, F., Sajjad, H., Imran, M., Ofli, F.: CrisisBench: benchmarking crisis-related social media datasets for humanitarian information processing. In: Proceedings of the International AAAI Conference on Web and Social Media, vol. 15 (2021). https://doi.org/10.1609/icwsm.v15i1.18115
34. Alam, F., Joty, S., Imran, M.: Domain adaptation with adversarial training and graph embeddings. In: ACL 2018 - 56th Annual Meeting of the Association for Computational Linguistics, Proceedings of the Conference (Long Papers), vol. 1 (2018). https://doi.org/10.18653/v1/p18-1099
35. Li, H., Caragea, D., Caragea, C.: Combining self-training with deep learning for disaster tweet classification. In: Proceedings of the International ISCRAM Conference, vol. 2021-May (2021). ISSN: 24113387
36. João, R.S.: On informative tweet identification for tracking mass events. In: Proceedings of ICAART 2021 - 13th International Conference on Agents and Artificial Intelligence, vol. 2 (2021). https://doi.org/10.5220/0010392712661273
37. Li, H., Li, X., Caragea, D., Caragea, C.: Comparison of word embeddings and sentence encodings as generalized representations for crisis tweet classification tasks. In: Proceedings of the ISCRAM Asian Pacific Conference (2018)

Exploring Behavioral Tendencies on Social Media: A Perspective Through Claim Check-Worthiness

Zeyu Zhang[✉], Zhengyuan Zhu, Haiqi Zhang, and Chengkai Li

University of Texas at Arlington, Arlington, USA
{zeyu.zhang,zhengyuan.zhu,haiqi.zhang}@mavs.uta.edu, cli@uta.edu

Abstract. This study examines how factual claims of different significance influence and reflect social media users' behavioral patterns. Leveraging "check-worthiness" as a measure of the factual significance of claims, we analyze the connection between factual claims and user behaviors on Twitter. Through a series of experiments using statistical methods such as correlation analysis and hypothesis testing, we provide insights into a few pivotal inquiries: (1) whether differences exist between users' tweeting tendencies toward check-worthiness, (2) the underlying reasons for such differences, (3) whether users tend to create, share, and endorse content with check-worthiness levels similar to their own tweets, and (4) whether users with similar tendencies toward check-worthiness exhibit heightened engagement. The experiments were conducted across three datasets, comprising over 48.5 million tweets and involving 15,000 users, spanning several domains and yielding statistically significant findings. Previous studies have primarily centered on examining the effectiveness and strategies of fact-checks rather than understanding people's behavioral tendencies toward factual claims. Our research pioneers understanding in this area, offering valuable insights for behavioral modeling and social sciences.

Keywords: check-worthiness · fact-checking · social media analytics

1 Introduction

In our present era, an unprecedented surge of falsehoods and partial truths has taken root within our society, posing a grave threat to national security, democratic principles, and public health. The digital worlds, notably prominent platforms such as Twitter (now called X), have become a breeding ground for fabricated news, a phenomenon that may have even cast its shadow over the presidential elections [2,3]. In response to the epidemic of misinformation, we have witnessed a significant proliferation of fact-checking endeavors globally. Numerous researchers and experts are currently engaged in diverse works concerning factual claims, which encompass statements based on verifiable information. These efforts span activities such as detecting [13], tracking [19], and

L. M. Aiello et al. (Eds.): ASONAM 2024, LNCS 15211, pp. 373–390, 2025.
https://doi.org/10.1007/978-3-031-78541-2_23

evaluating factual claims [23]. The practice of fact-checking has evolved into a pivotal interdisciplinary field, commanding attention across a spectrum of areas such as computer science, journalism, and communication.

While many researchers have delved into the realm of fact-checking and factual claims, a notable knowledge gap persists in our understanding of how factual claims, each possessing varying degrees of strength or importance, wield influence over people's interactions on social media. Observations and explanations of people's behaviors pertaining to factual claims are needed in order to fill this gap. This would entail answering many crucial questions, including whether individuals exhibit behavioral tendencies toward specific factual claims and the underlying factors that drive and differentiate these tendencies. Moreover, can we apply the age-old adage "Birds of a feather flock together" to denizens of social media, particularly concerning their responsiveness to factual claims? These unknowns offer us a new perspective to study social media. The answers to these unknowns may facilitate studies such as behavioral modeling and recommendation systems by providing noteworthy new features. Moreover, it may foster research in fields such as psychology and sociology by introducing new human behavioral patterns on social media.

To study this subject, a suitable instrument is necessary to measure the strength or importance of a claim. In the field of automated fact-checking, restricting claims to those that are objectively fact-checkable makes the task more amenable to automation while reducing the volume of content needing manual fact-checking. Researchers have forged invaluable tools to detect check-worthy factual claims [14,28]. These efforts provide a yardstick for evaluating the importance of a claim to be fact-checked—denoted as "check-worthiness." As stated in [12], the initial work defined the concept of check-worthiness, check-worthy claims are those of which the general public would be interested in knowing the veracity—whether the claims are true or false. For instance, as displayed in Fig. 1, (a) depicts a claim of significant check-worthiness, as it is highly probable that the public is interested in its veracity. In contrast, (b) conveys relatively low check-worthiness, as the public's interest in verifying the claim is limited. Finally, (c) exhibits the lowest check-worthiness, as there is no factual claim in the statement. By harnessing the concept of check-worthiness, a pathway emerges for investigations into how factual claims of varying importance influence and reflect people's behaviors on social media such as Twitter.

Individuals' behaviors on social media mainly consist of posting, commenting, sharing, liking, and following. They make up the communication and information diffusion on social media. Hence, many studies explored factors that influence these behaviors. For example, Comarela et al. [8] found several factors influencing user response or retweet probability, including previous responses to the same user, the user's posting rate, age, and tweet content. Firdaus et al. [9] discovered that a user's emotion towards a topic is a useful feature in modeling their retweet decision. Hopcroft et al. [15] observed that geographic distance, common friends, social status overlap, and interactions between two users (e.g., retweeting and replying) are correlated with two-way following relationships. Although a lot

(a) Tweet with High Check-worthiness

(b) Tweet with Relatively Low Check-worthiness

(c) Tweet with Low Check-worthiness

Fig. 1. Tweets with Different Check-worthiness Levels

of studies explored the factors correlating with posting, sharing, and following behaviors on social media, none of them has looked into the impact of check-worthiness. A few slightly related studies mainly focused on the strategies and effectiveness of fact-checks [6,17,21] rather than people's behavioral tendencies related to check-worthiness.

Considering our limited understanding of the impact of check-worthiness on social media behaviors, it is meaningful to give an investigation into it. Therefore, in this study, we conduct a range of experiments aimed at uncovering the underlying connections between check-worthiness and user behaviors on social media, particularly focusing on how check-worthiness influences/reflects individuals' tweeting, liking, and following actions. To achieve this goal, we collected 3 datasets from Twitter, comprising approximately 48.5 million tweets and 15,000 users. These datasets encompass a range of general topic domains including literature, arts, religion, politics, etc.

Our experiments on these datasets identified several pronounced results— (1) People do express different behavioral tendencies toward factual claims; (2) People in domains such as media, politics, and technology tend to have more factual claims, while people related to literature, arts, and religion generate fewer factual claims; (3) Individuals tend to share and like posts with similar check-worthiness levels as their own posts; (4) In instances where individuals followed each other, there is a higher likelihood of similar preferences towards factual claims when compared to one-way following relationships. In summary, our contributions can be delineated as follows:

- Pioneering investigation into the interplay between check-worthiness and user behaviors within social media.
- Provision of an expansive dataset comprising around 48.5 million tweets and 15,000 users, encompassing different types of tweets and domains.
- The outcomes of our experiments yield several noteworthy revelations that have the potential to stimulate diverse studies such as behavioral modeling and recommendation systems, particularly those pertaining to population behavior patterns in the context of social media platforms.

2   Related Work

Claim detection, which involves identifying claims worth fact-checking, is a task in the workflow of fact-checking in which check-worthiness is conceptualized. It can be viewed as a binary classification task and further as a ranking task on the claims. Numerous works have been dedicated to various modeling methodologies and their evaluations for claim detection [11,14,18,27,28]. Furthermore, researchers have used it for various application contexts, including automated check-worthy claims collection [19], factual claims visualization [22], fake news detection [26], and so on.

Research on identifying factors that influence posting, sharing, liking, and following behaviors in social media is directly related to our study. There are many existing works on this topic. Comarela et al. [8] conducted an extensive characterization of a large Twitter dataset which includes all users, social relations, and messages posted from the beginning of the platform up to August 2009. They identified several factors that influence user responses or retweet probability, such as previous interactions with the same tweeter, the tweeter's posting frequency, and entities in tweets such as hashtags and mentions. Firdaus et al. [5] uncovered that a user's emotional state can influence their retweeting behavior. They demonstrated this by constructing a retweet prediction framework based on an emotion detection model and conducting experiments using the Stanford Twitter Sentiment dataset [16] and the Obama–McCain Debate dataset [24]. Hopcroft et al. [15] identified that geographic distance, common friends, social status overlap, and the interactions between two users are correlated with two-way following relationships.

While there exists an abundance of research that separately addresses check-worthiness and human behaviors on social media, to the best of our knowledge, no existing work has linked them together. There are works with a focus on analyzing the relationship between fact checks and audience behaviors. For example, Kim et al. [17] analyzed 914 news articles with fact checks in South Korea. They found that news articles triggered more audience comments when they mentioned the importance of fact-checking the claim under scrutiny and conveyed negative content. Clayton et al. [6] evaluated the effectiveness of strategies for designing fact-checks by conducting experiments among 2,994 participants recruited from Amazon Mechanical Turk. Their experiments found the "Rated false" tag is more effective than the "Disputed" tag and the effect of a general warning is small compared to these two tags. Park et al. [21] discovered the unexpected and diminished effect of fact-checking due to cognitive biases. They found that claims labeled "Lack of Evidence" were often treated as false, revealing an uncertainty-aversion bias, and users who initially disapproved of a claim were less likely to change their views when presented with opposite fact-checking labels, indicating a disapproval bias. All these studies concentrated on scrutinizing the connections between fact-checks and their audiences' behaviors, primarily with the goal of understanding people's responses to truthfulness of claims. In contrast, our study is centered on discerning the correlation between check-worthiness of claims and people's behaviors.

3   Research Questions

Our research focuses on investigating how check-worthiness of claims impacts or reflects social media users' behaviors. There are many different angles and means to tackle this topic such as retweeting analysis and content analysis. Regardless of the approaches, the essence of this topic lies in determining whether check-worthiness can serve as an indicator to capture people's behavioral patterns in common and to differentiate among groups of individuals. In this paper, we study it by observing the tweeting, following, and liking behaviors among different groups of Twitter users since those behaviors make up most of their activities on Twitter.

To investigate the impact of check-worthiness, our initial inquiry revolves around determining whether individuals exhibit varying behavioral tendencies when confronted with factual claims of differing check-worthiness levels (i.e., showing higher engagement with claims of low/high check-worthiness). If such disparity arises, we then dig into the underlying rationales and figure out whether this disparity has an influence on individuals' behaviors or reflects unique characteristics of the respective populations. To unravel these unknowns, we introduce the following specific research questions for our study.

Q1. Do people exhibit different behavioral tendencies toward check-worthiness?

Q2. What factors and commonalities within the population might account for such behavioral tendencies?

Q3. Do people maintain similar check-worthiness levels between the tweets they post and those they favor?

Q4. Do people tend to follow others who exhibit similar behavioral tendencies toward check-worthiness?

4   Datasets

The smallest research unit in this study is a tweet, which is a social media post published by a Twitter user. For the convenience of wording, we categorize tweets in our datasets into 3 types: *original-tweet*—a tweet that is initially created by a Twitter user; *retweet*—a tweet that is reposted by a Twitter user from an original-tweet; and *liked-tweet*—a tweet that is liked by a Twitter user. As shown in Fig. 1, (a) is an example of an original-tweet, (b) is an example of a retweet, and (c) is an example of a liked-tweet. We collected 3 datasets to support our study, which are accessible at *Zenodo*.[1] Due to Twitter's content redistribution policy, the datasets only include tweet IDs and user IDs instead of tweet content and user profile details. Below are detailed descriptions of the datasets.

Randomly Sampled Users Dataset (RSU). This dataset consists of $11,173$ users collected through Twitter's APIs. We collected $10,000$ random English tweets in February 2023 using Twitter's Volume Stream API. The tweets were posted by around $3,000$ users. For each user, we collected up to 100 of its most recent followees using Twitter's Following API. Through the Timeline and Liking APIs, for each user, we collected their most recent tweets (up to $3,200$ tweets due to Twitter's limit) and liked-tweets (up to $3,200$ too). We then filtered out users that have insufficient tweets (less than 100 original-tweets or less than 80 retweets/liked-tweets) to ensure that the sample sizes are statistically significant in our analyses. Finally, we have $11,173$ users along with $40,405,150$ tweets. To prevent potential sampling biases in the collected data, we randomly selected 50 users and examined their profiles and their most recent 30 tweets to detect any biases (e.g., concentration in their backgrounds and topics). The results showed no significant concentration in the users' backgrounds or topics.

Politics Dataset (POL). This dataset contains all tweets from selected U.S. news media and U.S. politicians including *Senators, House Members, US Governors, US Secretaries of State, US Cabinet*, and *US Election Officials* at collection time. We used Twitter's Timeline API to collect the most recent tweets (up to $3,200$) of the target accounts. The dataset was collected in May 2023, with $8,153,745$ tweets and $3,784$ Twitter accounts.

Humanities Dataset (HUM). This dataset contains $341,285$ tweets and 498 Twitter accounts from selected Twitter lists including *Book Author, Christianity, Artists, Buddhism, Musician*, and *Philosophers*. We use Twitter's List and Timeline APIs to collect the accounts and their most recent tweets (up to $1,000$). The dataset was collected in January 2024.

5   Methodology

Each tweet in our datasets is associated with a corresponding check-worthiness score to indicate its check-worthy level. We employed the ClaimBuster [14] API[2]

---

[1] https://zenodo.org/records/11081026.
[2] https://idir.uta.edu/claimbuster/api.

to obtain check-worthiness scores. Given a tweet, the API returns a score ranging from 0 to 1, corresponding to how likely the tweet contains a check-worthy factual claim.

ClaimBuster has been used by researchers and fact-checkers in various contexts. For instance, the Duke Reporters' Lab[3] used ClaimBuster to create daily email alerts to professional fact-checkers with the most check-worthy claims from TV program transcripts and social media. These alerts have led to at least 33 claims featured in 30 different articles by fact-checking outlets, including one from The Washington Post that was discussed in a news report [10]. It has been applied in real-time for the live coverage of all primary election and general election debates of the U.S. presidential elections since 2016. Post-hoc analysis of the claims checked by professional fact-checkers at *CNN*, *PolitiFact.com*, and *FactCheck.org* reveals a highly positive correlation between ClaimBuster and fact-checkers in deciding which claims to check [13].

Although ClaimBuster has been widely applied in presidential debates, political speeches, and interviews, it is worth assessing its effectiveness on tweets, which are less formal and noisier. We did not use public datasets such as CLEF CheckThat![4] for evaluation because their data includes multimodal features (e.g., images) and does not align perfectly with our evaluation criteria. Our labels differ from theirs by 20% in a random sample of 100 tweets from their dataset. To this end, we conducted a human evaluation on a random sample of 200 tweets selected from our datasets. Each tweet was annotated by 3 annotators who labeled them as either check-worthy or non-check-worthy. All of the 3 annotators possess the concept and experience of check-worthiness evaluation as they all have contributed to factual claims detection tasks. The final label of each tweet was decided by majority vote. We used a check-worthiness score threshold of 0.5 to classify the tweets: If a tweet received a ClaimBuster score above 0.5, it would be classified as check-worthy; otherwise, non-check-worthy. This simple classifier has an accuracy of 0.84, indicating that ClaimBuster is effective in identifying check-worthy tweets and thus it can be used as a reliable tool for analyses in our study.

Our study frequently utilizes correlation analysis and hypothesis testing in the experiments as they are simple and useful tools for identifying and verifying underlying connections between variables. In this study, we use a scatter plot to visualize the relationship between two variables and use the Pearson correlation coefficient [7] to measure the direction and strength of a linear relationship between two variables.

Hypothesis testing is widely used in verifying statistical conjectures by examining data samples. We use it to validate our presumption about individuals' behavioral tendencies toward check-worthiness. We refer to H_0 as the null hypothesis, for which we test whether to accept it. If we reject it, we will accept the alternative hypothesis H_a. In this study, we primarily use hypothesis testing to assess the equality of check-worthiness distributions across thousands of

---

[3] http://reporterslab.org/tech-and-check.
[4] https://checkthat.gitlab.io/clef2024/task1.

user sets, aiming to ascertain the similarities or differences between individuals' tweeting behaviors. When conducting the same hypothesis test many times using different data, one may observe some statistically significant results just by chance, even if there is no true effect. As we are performing some hypothesis tests thousands of times in the experiments in Sect. 6, the false discovery rate (FDR) using the Benjamini-Hochberg procedure [1] was applied to control false significant results by adjusting the p-values.

Generally speaking, when determining if two samples originate from the same distribution, our preference would be *Z-test* or *T-test* in instances where we have equally sized samples and can make the assumption that the underlying populations adhere to normal distributions with known variances. Nonetheless, the check-worthiness of a Twitter user's posts hardly conforms to a normal distribution. We substantiated this claim by performing Shapiro–Wilk tests [25] on the randomly sampled users dataset RSU (Sect. 4). Shapiro–Wilk test is one of the most popular hypothesis tests for examining how close the sample data fit to a normal distribution by ordering and standardizing the sample. Given each user in the RSU dataset, we performed Shapiro–Wilk test with significance level $\alpha = 0.05$ on the check-worthiness scores of the user's original-tweets, retweets, and liked-tweets, respectively. The result, as displayed in Table 1, shows that only a few of the null hypotheses were accepted across all users and tweet types. This suggests a very low probability that the check-worthiness scores of a user's original-tweets, retweets, or liked-tweets follow a normal distribution.

Table 1. Normality Test on Check-worthiness Distributions

H_0 ($\alpha = 0.05$)	Accept	Reject
The check-worthiness scores of a user's original-tweets are normally distributed	37	11136
The check-worthiness scores of a user's retweets are normally distributed	259	10914
The check-worthiness scores of a user's liked-tweets are normally distributed	58	11115

Given that it is highly unlikely the check-worthiness scores follow normal distributions, *Z-test* and *T-test* become less applicable. Hence, we select two non-parametric tests that are applicable under less rigorous conditions—Brunner Munzel test [4] and Kolmogorov-Smirnov test [20]. Both tests possess the capability to assess the stochastic equality of two random variables—whether one is "larger" than another—without rigorous assumptions such as identical distribution type and equal variances. The Kolmogorov-Smirnov test is more strict since it tests whether two samples are from the same distribution, while the Brunner Munzel test only examines the stochastic equality of two samples. We articulate the formal null hypotheses of these two tests as follows.

- **H_0 of Brunner Munzel (BM) test**: For randomly selected values X and Y from two populations, the probability of X being greater than Y is equal to the probability of Y being greater than X.
- **H_0 of Kolmogorov-Smirnov (KS) test**: Two sets of samples are drawn from the same (but unknown) probability distribution.

6 Experiments

6.1 Q1: Individuals' Behavioral Tendencies Toward Check-Worthiness

The very first question we want to answer is whether people exhibit different behavioral tendencies toward check-worthiness. The RSU dataset contains a large number of random users and their corresponding tweets, making it a suitable dataset for investigating this query.

The most straightforward way of checking an individual's behavioral tendency toward check-worthiness is the overall check-worthiness of their posts. Hence, for each user in the RSU dataset, we computed the median check-worthiness score of their tweets, denoted as *individual check-worthiness*. We chose the median because check-worthiness scores are typically not normally distributed and tend to be skewed. We present in a histogram (Fig. 2) the distribution of individual check-worthiness of all users in the RSU dataset. It shows that, although individual check-worthiness mostly concentrates between 0.3 and 0.4, there are people who exhibit a particular tendency toward higher or lower check-worthiness. That motivates us to explore more about the underlying rationales and behavioral consequences of those preferences.

6.2 Q2: Causes of Different Behavioral Tendencies Toward Check-Worthiness

Knowing the difference between people's behavioral tendencies toward check-worthiness prompts us to speculate whether there are some common attributes correlated with those preferences. To investigate this question, we conducted correlation analyses on various features of users in the RSU dataset.

First of all, we analyzed the numeric features—posts count, favorites count, followers count, followees count, listed count (number of lists containing the user), and media count (number of posts containing images or videos). These features primarily reflect a user's popularity and activity level. We performed a univariate correlation analysis by calculating the Pearson correlation coefficients between individual check-worthiness and log-transformed feature values. The results, as Fig. 3 shows, are all weak correlations along with most p-values less than 0.005. In addition, a multivariate regression analysis on these features yielded both tiny coefficients and an R^2 value of 0.25, indicating a weak correlation between individual check-worthiness and these features. Based on the results, we cannot conclude that features related to popularity and activity level are indicators of individuals' behavioral tendencies toward check-worthiness.

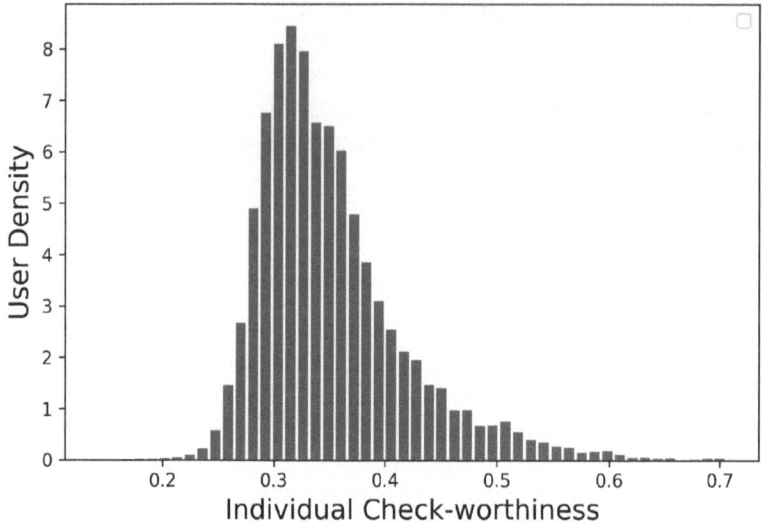

Fig. 2. Individual Check-worthiness Distribution

Besides those numerical features representing popularity and activity levels, there are other more complex features that could affect/indicate the tendencies. Such features may include occupation, political spectrum, and educational background. Since these features are either hidden or challenging to identify using automatic methods, we decided to conduct a content analysis to get some insights.

Firstly, among the 11,173 users in the RSU dataset, we selected all users with individual check-worthiness scores less than 0.25 or greater than 0.55 as they encompass two tails of the individual check-worthiness distribution, and thereby represent two small groups with weak and strong behavioral tendencies toward check-worthiness. The group denoted as U_0 comprises 146 users with low individual check-worthiness, while the group denoted as U_1 consists of 169 users with high individual check-worthiness. For both groups, we gathered all the user profile descriptions and tweets from the users, and conducted word frequency analysis. More specifically, for all the user profile descriptions and tweets respectively, we tokenized, removed stopwords, lemmatized, and counted word frequencies. The result is interesting, as Table 2 shows. The top frequent words in tweets from U_0 are general and irrelevant to specific people/events/affairs (e.g., love, life, like, god, good), while the top frequent words in tweets from U_1 are more concrete and highly related to trending topics/events (e.g., russian, ukraine, cannabis). The analysis of the user profile descriptions further enhances this observation. The top frequent words in user profile descriptions from U_0 are more related to literature/art, life/entertainment, and religion, while the top frequent words in user profile descriptions from U_1 are more related to journalism, politics, and technology.

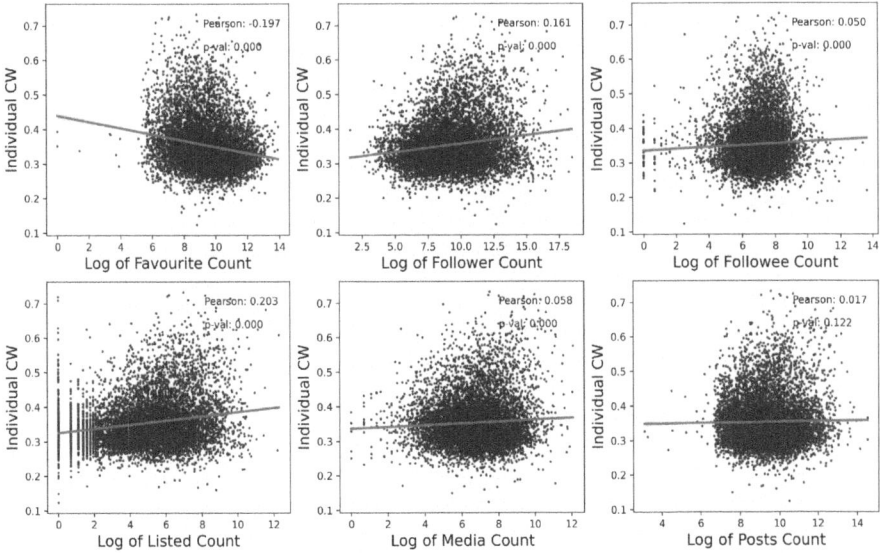

Fig. 3. Correlation between Individual CW and Popularity/Activity Features

The results from the word frequency analyses appear to suggest that individuals' professions, backgrounds, and interests are potentially related to their behavioral tendencies toward check-worthiness. To confirm this conjecture, we randomly selected 100 users from U_0 and U_1 respectively, and then we annotated each user account based on their backgrounds and interests. The results, as shown in Table 3, reveal that a majority of users in U_0 lack explicit backgrounds, though a considerable portion comprises writers and influencers. Their primary focus lies in sharing their ideologies and daily lives. On the other hand, users in U_1 are prominently associated with the media, with over half of the selected 100 users actively doing media-related jobs. Additionally, users with backgrounds in research and politics are also notably present. The dominant interests within U_1 encompass politics, general news, public interests, and technology/science, validating our initial conjecture.

To further solidify this conclusion, we also compared the individual check-worthiness distributions of three specific groups of Twitter users using the aforementioned datasets. The first group contains all users from the HUM dataset, which encompasses individuals related to humanities such as literature, arts and religion. The second group consists of all users from the POL dataset, which represents individuals related to politics and journalism. The third group consists of all users from the RSU dataset which are randomly sampled user accounts. Figure 4 depicts the individual check-worthiness distributions of these three groups of users. The figure shows a left-skewed distribution for the HUM dataset, a right-skewed distribution for the POL dataset, with the RSU in the middle. This finding suggests that users in the HUM dataset generally possess

Table 2. Top Words in Tweets/Profiles from U_0 and U_1

U_0's Tweets	U_1's Tweets	U_0's Profiles	U_1's Profiles
people (13112)	new (9591)	author (15)	news (23)
love (12462)	russian (6841)	podcast (7)	reporter (12)
life (12419)	ukraine (5135)	com (7)	newsletter (9)
one (10759)	people (4152)	book (7)	com (9)
like (9956)	energy (3906)	producer (6)	world (8)
god (9138)	report (3667)	life (5)	public (7)
us (8654)	said (3576)	people (5)	tech (7)
time (8249)	russia (3485)	views (5)	government (7)
get (7506)	cannabis (3215)	buddhist (5)	research (6)
good (7368)	us (3135)	writer (5)	policy (6)

Table 3. Top-ranked Backgrounds/Interests in U_0 and U_1

U_0's BGs	U_1's BGs	U_0's Interests	U_1's Interests
unknown (36)	media (33)	ideology (40)	politics (25)
writer (19)	reporter (13)	daily life (39)	general news (14)
influencer (12)	research (9)	religion (7)	public good (12)
pastor (3)	politician (6)	entertainment (3)	tech&science (12)
speaker (3)	analyst (5)	photography (2	climate (9
singer (2)	journalist (4)	writing (2)	energy (8)
photographer (2)	unknown (4)	general (2)	security (7)
consultant (2)	writer (4)		business (6)
student (2)	advocate (4)		war (3)
teacher (2)	editor (3)		economics (2)

lower individual check-worthiness, whereas those in POL tend to exhibit higher levels of check-worthiness.

6.3   Q3: Impact of Check-Worthiness on Tweeting Behaviors

With the conclusion that people express different behavioral tendencies toward check-worthiness, one may ask how well it aligns with people's tweeting behaviors. More specifically, it would be useful to know whether people tend to share and like posts with similar check-worthiness as their own posts. To answer this question, we conducted experiments on the RSU dataset.

We define O, R, and L as random variables of the check-worthiness of a randomly picked original-tweet, retweet, and liked-tweet from a given user. Moreover, given the dataset, we define X and P as random variables of the check-worthiness of a random tweet and a random popular tweet (liked or retweeted by

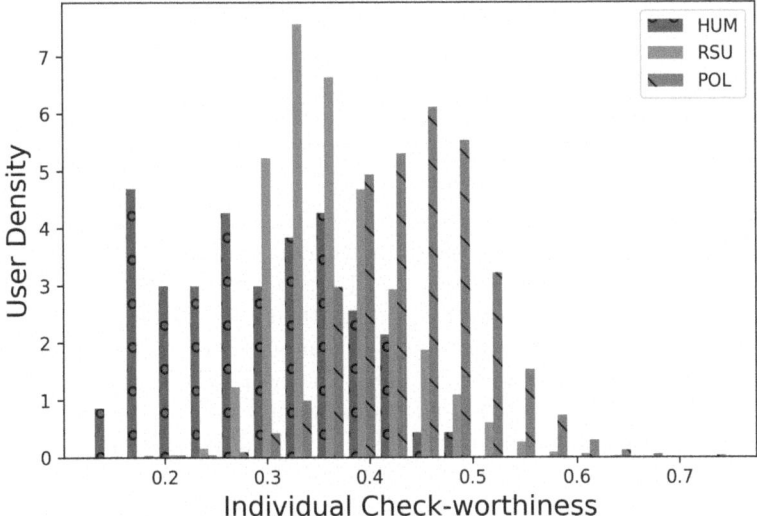

Fig. 4. Individual CW Distributions of HUM, RSU, and POL

anyone) from that dataset. Hence, with the RSU dataset, we have 4 hypotheses defined as follows to test the stochastic equality between O and other random variables:

- **Hyp1** $\begin{cases} H_0: & P(O > R) = P(O < R) \\ H_a: & P(O > R) \neq P(O < R) \end{cases}$

- **Hyp2** $\begin{cases} H_0: & P(O > L) = P(O < L) \\ H_a: & P(O > L) \neq P(O < L) \end{cases}$

- **Hyp3** $\begin{cases} H_0: & P(O > P) = P(O < P) \\ H_a: & P(O > P) \neq P(O < P) \end{cases}$

- **Hyp4** $\begin{cases} H_0: & P(O > X) = P(O < X) \\ H_a: & P(O > X) \neq P(O < X) \end{cases}$

For each user in the RSU dataset, we performed Brunner Munzel (BM) test and Kolmogorov-Smirnov (KS) test on Hyp1-4, as explained in Sect. 5. Table 4 shows the results of the acceptance for those hypotheses with alpha (significance level) equal to 0.05. We can see that the acceptance rates of Hyp1-2 are greater than that of Hyp3-4 for all the tests, which means the check-worthiness distribution of a user's original-tweets is more likely to have the same shape as the check-worthiness distributions of same user's retweets and liked-tweets, in comparison to random and popular tweets from arbitrary users.

In addition, we also performed a correlation analysis on the median check-worthiness of original-tweets, retweets, and liked-tweets for all the users in the RSU dataset. As Fig. 5 shows, where each point represents a user account, there

Table 4. Acceptances of Hyp1-4

Test	Hyp1	Hyp2	Hyp3	Hyp4
BM Test	1797/16.1%	2896/25.9%	939/8.4%	1013/9.1%
KS Test	1126/10.1%	1831/16.4%	248/2.2%	284/2.5%

exist strong correlations between the median check-worthiness scores of the users' original-tweets, retweets, and liked-tweets. To reduce the effects of content redundancy (such as when a user retweets or likes their own original-tweet, or when a retweet is also liked by the retweeter), we computed the overlap ratios among these three types of tweets for each user. The result shows that, on average, merely 1.3% of original-tweets are additionally retweeted by their authors, and less than 1% of original-tweets are also liked by the authors. Additionally, about 16% of retweets are also liked by the retweeters. Following the removal of overlapping tweets, we conducted the correlation analysis again, and the results as shown in Fig. 5 remain largely unchanged. Therefore, we are able to conclude that people overall have the behavioral tendency to post, share, and favor tweets with similar check-worthiness levels.

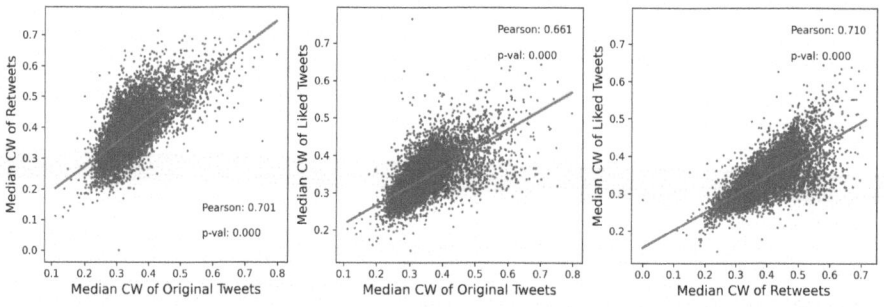

Fig. 5. Correlation in Median Check-Worthiness Among Three Types of Tweets

6.4   Q4: Impact of Check-Worthiness on Following Behaviors

Besides tweeting activities, another important activity on social media is following, which influences a large portion of the information a user receives. Therefore, it is natural and crucial to find out whether people tend to follow others with similar tendencies toward check-worthiness. More specifically, we want to examine whether the check-worthiness distribution of a user's tweets is more similar to that of its followers than other users.

We define U, V, F, and X as random variables of the check-worthiness of a randomly picked tweet from a given user, one of its followers, one of its friends (being both follower and followee), and a random user respectively. Here we have the hypotheses defined as follows to test the stochastic equality between U and other random variables:

- **Hyp5** $\begin{cases} H_0: & P(U > V) = P(U < V) \\ H_a: & P(U > V) \neq P(U < V) \end{cases}$

- **Hyp6** $\begin{cases} H_0: & P(U > F) = P(U < F) \\ H_a: & P(U > F) \neq P(U < F) \end{cases}$

- **Hyp7** $\begin{cases} H_0: & P(U > X) = P(U < X) \\ H_a: & P(U > X) \neq P(U < X) \end{cases}$

In the RSU dataset, we have $10, 402$ (follower, followee) pairs and 351 friend pairs, with a total of $9, 124$ distinct accounts involved. Table 5 shows the results of the acceptance for hypotheses Hyp5-7 with alpha (significance level) equal to 0.05. We can see that the acceptance rates of Hyp5 are greater than Hyp7, meaning the check-worthiness distribution of a user's tweets is more likely to have the same shape as the check-worthiness distribution of its followers' tweets compared with a random user's tweets. However, this likelihood is not strong since the acceptance rates of Hyp5 do not exceed Hyp7 by a lot. A more substantial result comes from the acceptance rates of Hyp6, which are much higher. This indicates a higher likelihood of check-worthiness similarity between tweets from a pair of users in a two-way following relationship than in a one-way following relationship.

Table 5. Acceptances of Hyp5-7

Test	Hyp5	Hyp6	Hyp7
BM Test	1043/10%	59/16.9%	696/7.6%
KS Test	335/3.2%	50/14.3%	130/1.4%

To further verify this conclusion, we again performed a correlation analysis on the individual check-worthiness of users and their friends. As Fig. 6 shows, there exists a weak correlation between the individual check-worthiness of followers and followees. However, the correlation becomes stronger when we compare the individual check-worthiness of users with a two-way following relationship. Similar to what we discussed in Sect. 6.3, our calculation shows that the average ratio of tweet overlap between each pair is less than 1%. This means the possibility of the result being influenced by overlapping tweets among friends is low. Therefore, the result is solid and aligns with our conjecture.

7   Limitation

While we have examined certain social media behaviors and identified several patterns, a few questions about the observations remain unanswered. For exam-

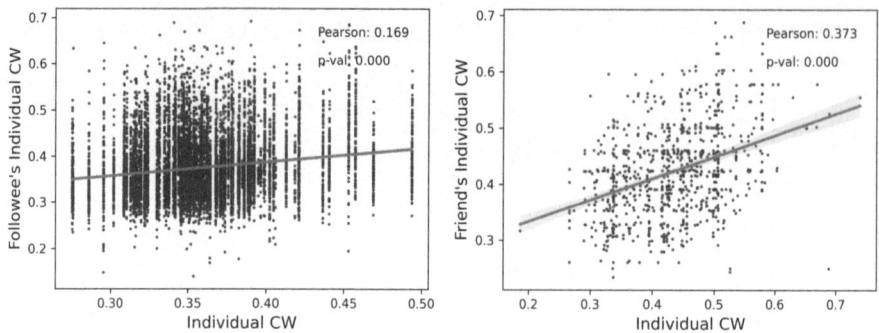

Fig. 6. Correlation between Following Parties' Individual Check-worthiness

ple, since most users may not consciously choose high or low check-worthy content when posting, sharing and liking tweets, the observations cannot be interpreted as definitive indicators of users' behavioral patterns. However, our statistics reveal that a significant portion of users demonstrate consistent behavioral patterns associated with check-worthiness, suggesting a possible subconscious adherence to such patterns. Undoubtedly, a more nuanced investigation is warranted, particularly concerning where the observed patterns come from. The inherent value of our study lies not only in the answers it provides but also in the thought-provoking questions it raises. This study possesses the potential to not only address current gaps in knowledge but also to act as a catalyst, possibly inaugurating a new line of research endeavors.

8   Conclusions

This work identified the existence of the difference between individuals' behavioral tendencies toward factual claims. In particular, the population from domains such as politics, general news, and technology is more likely to engage with check-worthy content compared to those associated with arts, literature, and religions. Through a set of experiments, the research has established a strong correlation between these tendencies and users' posting, sharing, and liking behaviors, indicating a conspicuous pattern to engage with content of similar check-worthiness. Furthermore, the findings emphasize the heightened efficacy of two-way following relationships in reflecting shared preferences towards factual claims.

The concept of check-worthiness emerges as a potent tool for understanding human behaviors within the realm of social media. Our results not only provide valuable insights into the impact and adaptability of check-worthiness but also lay the groundwork for future investigations to delve deeper into the various dimensions of its influence on social media behaviors and its potential applications across diverse domains.

Acknowledgement. This work is partially supported by the National Science Foundation award #2346261. The authors acknowledge the Texas Advanced Computing Center (TACC) at The University of Texas at Austin for providing HPC resources that have contributed to the research results reported within this paper.

References

1. Benjamini, Y., Hochberg, Y.: Controlling the false discovery rate: a practical and powerful approach to multiple testing. J. Roy. Stat. Soc.: Ser. B (Methodol.) **57**(1), 289–300 (1995)
2. Bovet, A., Makse, H.A.: Influence of fake news in twitter during the 2016 us presidential election. Nat. Commun. **10**(1), 7 (2019)
3. Brummette, J., DiStaso, M., Vafeiadis, M., Messner, M.: Read all about it: the politicization of "fake news" on twitter. Journ. Mass Commun. Q. **95**(2), 497–517 (2018)
4. Brunner, E., Munzel, U.: The nonparametric behrens-fisher problem: asymptotic theory and a small-sample approximation. Biometrical J. J. Math. Methods Biosci. **42**(1), 17–25 (2000)
5. Chen, J., Liu, Y., Zou, M.: User emotion for modeling retweeting behaviors. Neural Netw. **96**, 11–21 (2017)
6. Clayton, K., et al.: Real solutions for fake news? Measuring the effectiveness of general warnings and fact-check tags in reducing belief in false stories on social media. Polit. Behav. **42**, 1073–1095 (2020)
7. Cohen, I., et al.: Pearson correlation coefficient. In: Noise Reduction in Speech Processing, pp. 1–4 (2009)
8. Comarela, G., Crovella, M., Almeida, V., Benevenuto, F.: Understanding factors that affect response rates in twitter. In: Proceedings of the 23rd ACM Conference on Hypertext and Social Media, pp. 123–132 (2012)
9. Firdaus, S.N., Ding, C., Sadeghian, A.: Topic specific emotion detection for retweet prediction. Int. J. Mach. Learn. Cybern. **10**, 2071–2083 (2019)
10. Funke, D.: This Washington post fact check was chosen by a bot (2018)
11. Hansen, C., Hansen, C., Alstrup, S., Grue Simonsen, J., Lioma, C.: Neural checkworthiness ranking with weak supervision: Finding sentences for fact-checking. In: Companion Proceedings of the 2019 World Wide Web Conference, pp. 994–1000 (2019)
12. Hassan, N., Li, C., Tremayne, M.: Detecting check-worthy factual claims in presidential debates. In: Proceedings of the 24th ACM International on Conference on Information and Knowledge Management, pp. 1835–1838 (2015)
13. Hassan, N., Tremayne, M., Arslan, F., Li, C.: Comparing automated factual claim detection against judgments of journalism organizations. In: Computation+ Journalism Symposium, pp. 1–5 (2016)
14. Hassan, N., et al.: Claimbuster: the first-ever end-to-end fact-checking system. Proc. VLDB Endow. **10**(12), 1945–1948 (2017)
15. Hopcroft, J., Lou, T., Tang, J.: Who will follow you back? Reciprocal relationship prediction. In: Proceedings of the 20th ACM International Conference on Information and Knowledge Management, pp. 1137–1146 (2011)
16. Hu, X., Tang, L., Tang, J., Liu, H.: Exploiting social relations for sentiment analysis in microblogging. In: Proceedings of the sixth ACM International Conference on Web Search and Data Mining, pp. 537–546 (2013)

17. Kim, H.S., et al.: Fact-checking and audience engagement: a study of content analysis and audience behavioral data of fact-checking coverage from news media. Digit. Journ. **10**(5), 781–800 (2022)
18. Lespagnol, C., Mothe, J., Ullah, M.Z.: Information nutritional label and word embedding to estimate information check-worthiness. In: Proceedings of the 42nd International ACM SIGIR Conference on Research and Development in Information Retrieval, pp. 941–944 (2019)
19. Majithia, S., et al.: Claimportal: integrated monitoring, searching, checking, and analytics of factual claims on twitter. In: Proceedings of the 57th Annual Meeting of the Association for Computational Linguistics: System Demonstrations, pp. 153–158 (2019)
20. Massey, F.J., Jr.: The kolmogorov-smirnov test for goodness of fit. J. Am. Stat. Assoc. **46**(253), 68–78 (1951)
21. Park, S., Park, J.Y., Chin, H., Kang, J.h., Cha, M.: An experimental study to understand user experience and perception bias occurred by fact-checking messages. In: Proceedings of the Web Conference 2021, pp. 2769–2780 (2021)
22. Rony, M.M.U., Hoque, E., Hassan, N.: Claimviz: visual analytics for identifying and verifying factual claims. In: 2020 IEEE Visualization Conference (VIS), pp. 246–250. IEEE (2020)
23. Samadi, M., Talukdar, P., Veloso, M., Blum, M.: Claimeval: integrated and flexible framework for claim evaluation using credibility of sources. In: Proceedings of the AAAI Conference on Artificial Intelligence, vol. 30 (2016)
24. Shamma, D.A., Kennedy, L., Churchill, E.F.: Tweet the debates: understanding community annotation of uncollected sources. In: Proceedings of the First SIGMM Workshop on Social Media, pp. 3–10 (2009)
25. Shapiro, S.S., Wilk, M.B.: An analysis of variance test for normality (complete samples). Biometrika **52**(3/4), 591–611 (1965)
26. Shu, K., Cui, L., Wang, S., Lee, D., Liu, H.: defend: explainable fake news detection. In: Proceedings of the 25th ACM SIGKDD International Conference on Knowledge Discovery & Data Mining, pp. 395–405 (2019)
27. Vasileva, S., Atanasova, P., Màrquez, L., Barrón-Cedeño, A., Nakov, P.: It takes nine to smell a rat: neural multi-task learning for check-worthiness prediction. arXiv preprint arXiv:1908.07912 (2019)
28. Wright, D., Augenstein, I.: Claim check-worthiness detection as positive unlabelled learning. arXiv preprint arXiv:2003.02736 (2020)

You Must Be a Trump Supporter: Political Identity Projections on the Social Web

Shubh Mittal[1] , Tisha Chawla[1] , and Ashiqur R. KhudaBukhsh[2]([⊠])

[1] Vellore Institute of Technology, Vellore, Tamil Nadu 632014, India
{shubh.mittal2020,tisha.chawla2020}@vitstudent.ac.in
[2] Rochester Institute of Technology, Rochester, NY 14623, USA
axkvse@rit.edu

Abstract. ⚠This paper contains offensive content. This paper assesses the extent of political polarization in the United States by demonstrating the phenomenon of political identity projection, where individuals attribute political affiliations to others based on political discourse. This aspect of political behavior, often found in interactions between authors on various social media platforms, remains relatively unexplored. To address this gap, our research utilizes a comprehensive dataset of comments on YouTube news videos from three prominent US cable news networks (Fox News, CNN, and MSNBC) to interpret expressions of political polarization. First, we assess the accuracy of LLMs in identifying political identity projections, exploring the potential biases these models may incorporate. Second, we conduct a user engagement analysis that highlights interaction patterns and their implications for understanding political identity projections across different news outlets.

Keywords: Political Identity Projections · Large Language Models · Social Web

1 Introduction

Polarization [10,34,42,50] has emerged as a pressing concern in the United States, with evident adverse effects [59] on multiple fronts, including the educational system [3], job market [2], healthcare facilities [58], religious institutions [8], and the political arena [22]. Scholarship disparities [25], alongside challenges in healthcare access and divisions within religious communities, underscores the breadth of this issue. Moreover, political polarization significantly shapes public discourse [38,39] and policy-making processes [23,60]. Particularly within social media platforms, discussions often reflect this divide, leading to detrimental outcomes such as misinformation [32,36] and social fragmentation [44]. Studies highlight the pivotal role of platforms like Facebook [24], YouTube [20,46], and Twitter [49] in shaping online engagement and influencing public opinion and policy debates [17].

© The Author(s), under exclusive license to Springer Nature Switzerland AG 2025
L. M. Aiello et al. (Eds.): ASONAM 2024, LNCS 15211, pp. 391–404, 2025.
https://doi.org/10.1007/978-3-031-78541-2_24

Table 1. Illustrative examples highlighting different identity projections in the YouTube comment sections of MSNBC (left column), CNN (middle column), and Fox News (right column).

MSNBC	CNN	FOX
Calling someone you don't know a freeloader? That seems a little presumptuous, don't you think? Not to be so myself, but I'm thinking you must be a Trump supporter. Is it his policy or his cult of personality that you are drawn to?	And how did you come to that brilliant conclusion? Several paid agitators (many likely from out of state) organized to interrupt a policy speech at set intervals with their anti-white communist rhetoric, and suddenly the whole state hates him? You and anyone who believes you must be anencephalic.	This message is for Isa, I think you hate President Trump because he beat Hillary plus you must be a little snow flake, you should move to Disneyland to live a fake life and you can dance and sing all they long.
Ohh, so Fox News is owned by Democratic elites? The channel that claims to be the number 1 watched channel? Are you kidding me? You must be a troll because your logic is totally twisted around. Who are the elites who have purchased most of the radio waves in the US? It's name is Sinclair Corp and it is all right wing programming. I think the media has it's problems for sure. But to label it all left is crazy. I also think it's crazy to get your news from one news source (Fox)! Spread your wings, Cary!	It's a volcanic land mass. "Fact". Facts don't have anything to do with emotions or opinion. Funny how your care so much for my opinion on why I won't visit or live in a particular area. You must be Republican. Twisting things into a narrative so that you may assume some kind of none existant moral high ground. I don't do stupid shit, you also don't know a damn thing about me.	Really?? You must be a Hilary supporter and a CNN listener also. Trump talks about what he does to all the crowds at his rallies, at least he hasn't murdered anybody or lied to his supporters, unlike Hilary. I'm dillusional?? I think you are if you support Hilary and her crime sprees which makes you unpatriotic. Please don't use rhetoric which you don't understand just to make a comment.
you must be blind if you think hillary is involved in this russia nonsense. but thats what you want. keep saying her name as trump goes to jail	They are not killing all life...just black ones....That mean black lives matters too! That is all that means...and its people like you the problem....oh you must be a cop!	haha you must be thinking of downtown Chicago. Our economy is much more stable than Europe, the DOW is reaching record highs, and the Trump administration is about to create national infrastructure for a minimum of 4 years. Lack of commodities? We have the biggest amount of consumerism on the planet.

Additionally, analyses of major US cable news networks' [13, 27, 41] content and audience engagement on platforms like YouTube reveal a distinct partisan [9, 34, 35, 55] and ideological split [63]. This polarization extends beyond online spaces, further exacerbating societal divisions [31] and challenging efforts to foster constructive dialogue and governance [28]. Despite the apparent ubiquity of these discussions, there remains a lack of comprehensive analysis focusing on political identity projection where user $\mathcal{A}$ assumes a political identity for user $\mathcal{B}$. Consider the comment *if you find this movie offensive, you must be a sensitive snowflake ....* In this comment, the comment author is projecting a political identity (*sensitive snowflake*) to her target. Table 1 lists a few illustrative examples.

How often do we notice such political identity projections happen in the social web political discourse? What is the broad nature of such identity projections? This research investigates political identity projection via a massive dataset with documented political dissonance [34, 35] of more than 80 million YouTube comments on news videos hosted by the official YouTube channels of CNN, MSNBC, and FOX. Our analysis seeks to understand the pervasive nature of political polarization and how it shapes user interactions and discussions in different settings. Specifically, our contributions are as follows:

- First, we identify a set of political identity projections frequently used on the social web. We examine how well large language models (LLMs) can identify these political identity projections.
- Second, we analyze user engagement patterns in the presence of political identity projections.

Our analyses reveal that

1. while humans can discern political leanings in identity projections consistently, not all large language models are equally astute;
2. beyond mere name-calling, social web posts containing political identity projections also contain politically polarizing texts;
3. political identity projections elicit more user endorsement (in the form of likes) than non-political identity projections; and
4. political identity projections often trigger an escalating behavior in the discourse.

2   Related Work

Political polarization in the US has reached a level where hyper-partisanship significantly impacts both governance and societal unity [19,29,40]. Research indicates that these ideological divides intensify during election cycles and are amplified by the dynamics of social media and traditional news consumption [24, 60]. This escalation in polarization is not just a matter of differing opinions but influences legislative gridlock [4,6] and reduces the effectiveness of governance [43,54]. Moreover attempts to expose individuals to opposing political views on social media can unintentionally intensify their original partisan biases [3], known as the backfire effect, complicating efforts to mitigate polarization through digital platforms. Content moderation efforts to keep the online environment safe can also become highly subjective to political leanings [52,61].

Stereotypes in US politics [7], particularly those surrounding Democrats and Republicans, play a crucial role in shaping public perceptions [14] and voter behaviour [12]. Media portrayals often exacerbate these stereotypes, leading to polarized public opinion which, in turn, influences electoral outcomes [48] and political engagement [5,45]. These stereotypes can drastically simplify complex political narratives, leading to a binary and often confrontational political landscape. Our research builds on these findings by investigating political identity projection, where individuals attribute political affiliations based on observed behaviors and discussions online [37,56]. We investigate how digital interactions reflect and reinforce perceived political identities, exploring the impact of media narratives on these perceptions.

3 Dataset

The dataset comprises a collection of user comments on videos hosted by the official YouTube channels of three major US cable news networks: CNN, Fox News, and MSNBC. Recognized for their significant viewership [26] and considered among the most viewed cable news outlets in the United States, these networks collectively cater to a diverse audience spectrum [15]. Our corpus consists of a vast volume of data, including 46,990,892 comments and 35,177,044 replies, across 191,205 videos which span from 2015 to 2020 detailed in Table 2 and Table 1. This dataset is a reliable snapshot of US political discourse and has been previously used to study political polarization [34], election misinformation [35], and health misinformation [62].

In this research, we deal with comments and replies having the projection *'you must be'*. We acknowledge that identity projections can happen in several other ways (e.g., *it seems I am talking to*). That said, we believe this phrase is one of the most common ways to project political identities and the sheer scale of our data ensures a meaningful analysis. Prior literature also indicates similar text template-based searches to limit the focus to a small portion of relevant data (see, e.g., [21,33]).

Table 2. Distribution of our dataset.

News channel	Videos	Comments	Replies
FOX News	64,696	20,156,002	12,650,882
CNN News	95,230	19,290,236	15,183,800
MSNBC News	31,279	10,241,259	7,342,362

4 Political Identity Projection Identification

We first narrow our search down to frequently used identity projections (e.g., *you must be a Trump supporter* or *you must be a Soros hack*). We next manually annotate these examples (one male and one female annotator) with three labels: *liberal*, *conservative*, and *unrelated*. This process mostly yielded consensus labels. The inter-annotator agreement was measured using Cohen's Kappa, resulting in a score of 0.98, indicating almost perfect agreement. Very few remaining disagreements were resolved by an expert social scientist[1].

RQ 1: *Can large language models detect political identity projections?*

Our human annotation exercise reveals humans are highly consistent in categorizing political identity projections. But how good are large language models (LLMs) in this task? We sample the most frequent 1,000 instances of political

---

[1] Publicly available at Github link: https://bit.ly/46oQZHu.

identity projections evenly distributed across the three categories (*liberal, conservative,* and *unrelated*). We assess the accuracy of LLMs in identifying political identity projections (shown in Table 3). We consider the following LLMs: GPT-4 [1], Mistral [30], and Gemini [57]. Table 4 details the LLM performance. The results demonstrate a significant difference in the performance of LLMs for political identity projection identification. We observe that GPT-4, with an accuracy of 95%, is much better suited for political identity projection identification. In contrast, Mistral and Gemini considerably underperform with an accuracy of 43% and 21%, respectively. Gemini's poor performance perhaps suggests vulnerability to social biases [47]. The performance disparity highlights that while humans can easily interpret political identity projections, not all LLMs are equally adept at this.

Table 3. Example annotations associated with different political identity projections, showcasing high-confidence predictions from GPT-4 trained model to identify the political labels.

Political Identity Projections	Annotations
you must be a trump supporter	conservative
you must be a Foxnews idiot	conservative
you must be a republican conman	conservative
you must be a hillary bot	liberal
you must be a democrat fucktard	liberal
you must be a soros hack	liberal
you must be a fucking genius	unrelated
you must be a god	unrelated
you must be a nice person	unrelated

Table 4. Accuracy of LLMs for identification of political identity projections.

LLM	Accuracy
GPT-4	95%
Mistral	43%
Gemini	21%

RQ 2: *Beyond the projected identity, do comments containing conservative or liberal identity projections exhibit polarization?*

We were curious to investigate if the political identity projections are mere name-calling, or if they also contain additional polarizing linguistic signals. We investigate this through a recent framework that leverages classification accuracy

Table 5. Example replies to feed into the content classifier with the target label '*you must be*' removed and categorized into two classes: liberal (left column) and conservative (right column).

Liberal	Conservative
~~You must be a demorat!~~? If this was a N.Korea you would not be here today! Are you a Clinton supporter? Cinton is dangerous for everybody think before judging your president.	This is the procedure of the rules passed by the Republicans during the Bengazi inquiry , moronic fool.you remember that nothing burger ran by Gowdy and Pompeo that went nowhere.so get your facts straight before you make yourself look like a fool.ah, it's too late.~~you must be Republican~~.
where the hell do you get your disinformation ? ~~You must be Obama loser.~~ Trump is way more of a leader than Obama ever was. Our economy is the strongest it has been in thirty years. Under Obama unemployment was in double digits and you couldn't find a job and welfare was the highest it's ever been.	How is the weather in Siberia? If not, then ~~you must be a Trumptard~~ wallowing in deflection and whataboutism as always. Attempting to relay facts and reason with your kind is like trying to administer medicine to the dead, it's just futile. However, you keep attempting to evoke "ghetto" in a derogatory as if your people don't reside their too-meth infested trailer parks are the worst! Lastly, your orange Nazi-in-chief are the worst thing ever for this country, so you can't throw policy stones when you live in a cheap glass house. Jeez!
Who brought up "the left" or "nazis"?How are either related to my post about Israeli exceptionalism and influence in US politics and foreign policy?What an interesting knee jerk reaction to criticism of Israel- you must be a leftist or a nazi! Lmao is this the power of the Israeli online defense force? You criticize Israel? ~~You're a leftist nazi!~~The war in Iraq cost the US 2 trillion dollars and is projected to cost a total 6 trillion, and for what? Nothing. ISIS was formed, millions died in the conflicts after the war, and only Israel and Saudi Arabia benefited. The US paid the bill while our "greatest ally" didn't send a single soldier to help.	did i ever state prisons are full of innocent people lol ~~you must be republican~~ huh ive noticed a trend where repubs connect there own dots and put words in peoples mouths. What i said is that harsh punishments for minor crimes arent helping anyone i also stated selling a bit of weed not getting caught with a little weed intent to distribute carries a harsher punishment i would expect an expert on everyone and everything to understand that lol

as a proxy for polarization [13]. Consider $\mathcal{D}_1$ and $\mathcal{D}_2$ are two datasets. If $\mathcal{D}_1$ and $\mathcal{D}_2$ are distributionally highly similar, a model that takes an instance of any of these two corpora as input and tries to predict which of the two corpora the instance is coming from, will have a very hard time predicting. On a balanced test set with an equal number of samples from $\mathcal{D}_1$ and $\mathcal{D}_2$, the accuracy of the model will be close to chance. In contrast, if $\mathcal{D}_1$ and $\mathcal{D}_2$ are distributionally highly dissimilar, the classification accuracy will be considerably better than chance.

We consider each comment to have two components: political identity projection and targeted content. For example, in our example *if you find this movie offensive, you must be a sensitive snowflake*, the projected identity is *sensitive snowflake* and the content stripped of the identity projection is *if you find this movie offensive*. If we create two corpora, one with the comments with liberal identity projections ($\mathcal{D}_{liberal}$) and the other with conservative identity projections ($\mathcal{D}_{conservative}$), we can run a similar experiment to estimate how linguistically dissimilar these two corpora are. However, if we do not remove the identity projections, the classification will be trivial due to shortcut learning [18]. If we strip both $\mathcal{D}_{liberal}$ and $\mathcal{D}_{conservative}$ off the projected identity and run a similar experiment, an accuracy considerably higher than chance possibly indicates the presence of polarizing content that extends beyond political identity projections.

To this end, we develop a content classifier using BERT [11]. We remove the identity projections from $\mathcal{D}_{liberal}$ and $\mathcal{D}_{conservative}$ and retain all comments containing more than 20 words to ensure that comments have sufficiently long context. This method evaluated how well the classifier could identify political projections after removing our target phrase as shown in Table 5. We achieved an accuracy of 70%, considerably more than chance. This result indicates that comments with political identity projections also contain polarizing texts beyond the juvenile name-calling. Few illustrative examples are listed in Table 5.

5   User Engagment

5.1   Counter-Projecting

We identified 6,944 comments across the three news channels where political identity projections occurred more than once within the replies section (Table 6 provides a detailed breakdown by channel). This highlights recurring patterns of targeted projections within the dataset. Further investigation into the reply thread revealed that, on average, the second instance of a target or insulting reply occurs after 17 subsequent replies to the initial comment. This finding suggests a probable escalation of negative interactions, indicating that the initial instance of target projections may predispose the conversation towards subsequent negative replies. The concept of counter-projections emerges in the analysis, referring to the reactionary patterns observed within the replies. These patterns indicate that subsequent replies may either oppose or affirm the political identity projections introduced in earlier conversations, contributing to a complex political interaction within online discussions.

Table 6. Total political identity projections associated with each news channel.

News Channel	Total political identity projections
CNN	3,337
FOX	2,473
MSNBC	1,134

5.2   Annual Disaggregation

The analysis in Fig. 1 shows an increase in negative comments during U.S. presidential election years, particularly in 2016 and 2020.

– *Election vs. Non-Election Years:* Across the three news networks, substantial variations in targeted comments have been observed during the election cycles. While FOX experienced a significant peak in political identity projecting comments during the 2020 election year, CNN displayed its highest level of such comments in 2019, contrary to an expected rise in an election year like 2020. This suggests that CNN's audience engagement and the nature of discourse may have been influenced by events leading up to the election year. MSNBC, on the other hand, showed a continuous increase in political identity projections, culminating in 2020, indicating heightened viewer engagement during the election period.

– *Comparative Channel Analysis:* Comparing the channels, FOX (30,889 identity projections) and CNN (38,455 identity projections) had higher instances of politically projecting comments than MSNBC (15,684 identity projections). This could be reflective of the different editorial content and audience engagement strategies employed by these networks. The higher numbers in FOX and CNN show that these channels may foster more polarized environments, which can increase during critical political periods.

5.3   Participation Disparities in Contributors

Over 44 users heavily contribute to the comment section, surpassing 50 comments, indicating concentrated activity among top contributors. Additionally, there are 531 middle-tier users with comment counts ranging from 10 to 50, contributing significantly, though less than the top tier but more than the less engaged users. Moreover, 17,670 users, have made fewer than 10 comments,

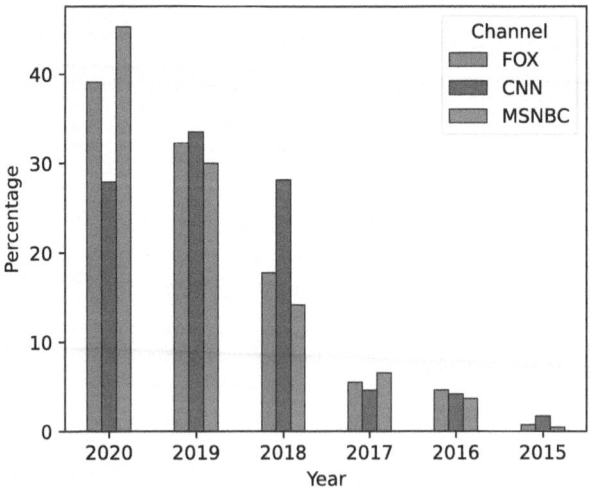

Fig. 1. Graphical representation of the yearly political identity projection count for the news channels.

illustrating a long tail of less involved users. This distribution follows a Pareto pattern, where a few users are highly active while the majority are minimally engaged.

5.4 User Reaction Distribution

We observe distinct patterns in the mean like counts of political identity projection comments across the three news channels (shown in Table 7), reflecting the political inclinations of their respective audiences. FOX, known for its Republican-leaning viewership [53], exhibits higher mean like counts for comments projecting liberals (2.0384) compared to those on CNN (1.0124) and MSNBC (0.8195), which are channels with predominantly Democratic-leaning audiences [16]. Conversely, comments about conservatives receive more likes on CNN (1.7175) and MSNBC (2.9395), considerably higher than that on FOX (1.3656).

This trend highlights that viewers are more likely to engage with and endorse comments that align with identity projections of the opposing political party, reinforcing the echo chamber effect [51] where ideological biases are perpetuated. The analysis also shows that comments categorized as 'Unrelated,' which contain identity projections not specific to any political group, receive relatively uniform mean like counts across all channels (CNN: 1.4989, FOX: 1.5738, MSNBC: 1.7981). This indicates that non-partisan identity projections are roughly equally likely to be engaged with, regardless of the audience's political leanings.

Table 7. Mean like count of political identity projections across our data.

Class	μ(CNN)	μ(FOX)	μ(MSNBC)
Liberal	1.0124	2.0384	0.8195
Conservative	1.7175	1.3656	2.9385
Unrelated	1.4989	1.5738	1.7981

6 Conclusions and Discussion

This paper analyzes political identity projections in comments/replies on YouTube videos from prominent news channels such as CNN, FOX, and MSNBC. We utilize well-known LLMs such as GPT-4, Mistral, and Gemini to label the political identity projections in user comments with GPT-4 demonstrating a superior performance as compared to other LLMs. Our results show a significant increase in political identity projections during U.S. election years, illustrating how political events intensify user engagement and alter online interactions. Our analysis also highlights that political leanings may influence user engagement with political identity projections suggesting political polarization.

Our research raises the following points:

- **Effectiveness of LLMs in Target Projection Identification:** We compared the performance of GPT-4, Mistral, and Gemini in labelling political projections and concluded that GPT-4 provides the results that best align with human labels. On the other hand, Mistral and Gemini prediction show lesser alignment with human labels. This result indicates that before using large language models for tasks requiring understanding of political subtleties, we need to carefully evaluate the LLM's capability to understand political nuances.
- **Bias in Political Discourse:** Our analysis indicates that channels like CNN and FOX display a higher incidence of targeted comments against specific political groups, which may be influenced by the channel's editorial biases or the perceived political leanings of their audiences. This finding calls for further examination of how news media can influence or reflect political biases in public discourse.
- **Influence of Election Cycles on User Engagement:** The observed spike in user activity and polarized projections during election years suggests that these periods not only increase user engagement but also heighten political identity projections. Future research could explore how moderation strategies and community management can lessen polarization during election cycles, helping to mitigate divisiveness on social media platforms.

Our study has several limitations. We primarily focused on the phrase '*you must be*' to identify political identity projections within YouTube comments, potentially missing other similar expressions like '*you should be*' or '*you are an*'. Also, the vast scale of the dataset poses challenges in manually reviewing and identifying every potential expression of identity projections. Our reliance on a template-based method, while common in linguistic analysis, may not fully capture the distinct ways in which individuals project political identities. Further, while our dataset is a reasonable snapshot of US political discourse, it does not include other platforms like Reddit or Twitter, which could provide additional contexts for understanding online political discourse.

7   Ethical Statement

This research analyzes publicly available data from YouTube procured using publicly available YouTube API. We initially annotate this data by humans to identify instances of political identity projections, with subsequent labelling supported by LLMs. We consider both male and female annotators and compare multiple LLMs. We report results in aggregate form and reveal no individual user data. The objective of this research is not to defame any political group, news channel, or community. Instead, our goal is to shed light on the extent of political polarization evident in public discourse on major news channels' YouTube platforms, aiming to contribute to a broader understanding of political polarization through the lens of political identity projections.

References

1. Achiam, J., et al.: Gpt-4 technical report. arXiv preprint arXiv:2303.08774 (2023)
2. Autor, D., et al.: The polarization of job opportunities in the us labor market: implications for employment and earnings. Center Am. Progress Hamilton Project **6**, 11–19 (2010)
3. Bail, C.A., et al.: Exposure to opposing views on social media can increase political polarization. Proc. Natl. Acad. Sci. **115**(37), 9216–9221 (2018)
4. Barber, M., McCarty, N., Mansbridge, J., Martin, C.J.: Causes and consequences of polarization. Pol. Negotiat. Handbook **37**, 39–43 (2015)
5. Barcelos, R.H.: To engage or not engage? the features of video content on YouTube affecting digital consumer engagement. J. Consum. Behav. **20**(1), 15–26 (2021)
6. Binder, S.A.: Stalemate: Causes and Consequences of Legislative Gridlock. Rowman & Littlefield (2004)
7. Bordalo, P., Tabellini, M., Yang, D.Y.: Stereotypes and politics (2020)
8. Castle, J.J., Stepp, K.K.: Partisanship, religion, and issue polarization in the united states: a reassessment. Pol. Beh. 1–25 (2021)
9. Chen, K., et al.: Partisan us news media representations of Syrian refugees. In: Proceedings of the International AAAI Conference on Web and Social Media, vol. 17, pp. 103–113 (2023)
10. Demszky, D., et al.: Analyzing polarization in social media: method and application to tweets on 21 mass shootings (2019)
11. Devlin, J., Chang, M.W., Lee, K., Toutanova, K.: Bert: pre-training of deep bidirectional transformers for language understanding. arXiv preprint arXiv:1810.04805 (2018)
12. Ditonto, T.: Direct and indirect effects of prejudice: Sexism, information, and voting behavior in political campaigns. Pol. Groups Identities **7**(3), 590–609 (2019)
13. Dutta, S., Li, B., Nagin, D.S., KhudaBukhsh, A.R.: A murder and protests, the capitol riot, and the Chauvin trial: estimating disparate news media stance. In: Raedt, L.D. (ed.) Proceedings of the Thirty-First International Joint Conference on Artificial Intelligence, IJCAI 2022, pp. 5059–5065. ijcai.org (2022). https://doi.org/10.24963/IJCAI.2022/702
14. Eagly, A.H., Nater, C., Miller, D.I., Kaufmann, M., Sczesny, S.: Gender stereotypes have changed: a cross-temporal meta-analysis of us public opinion polls from 1946 to 2018. Am. Psychol. **75**(3), 301 (2020)
15. Emelu, M.N.: The us cable televisions' framing of mass shooting: a grounded discovery of competing narratives. Front. Commun. **8**, e1174946–e1174946 (2023)
16. Ensor, K.: The Partisan Delivery of News: A Content Analysis of CNN and Fox, vol. 16, no. 11. Johnson & Wales University, ScholarsArchive@ JWU (2018)
17. Garimella, V.R.K., Weber, I.: A long-term analysis of polarization on twitter. In: Proceedings of the International AAAI Conference on Web and Social Media, vol. 11, pp. 528–531 (2017)
18. Geirhos, R., et al.: Shortcut learning in deep neural networks. Nature Mach. Intell. **2**(11), 665–673 (2020)
19. Grossmann, M., Hopkins, D.A.: Ideological republicans and group interest democrats: the asymmetry of Am. party politics. Perspect. Polit. **13**(1), 119–139 (2015)
20. Halpern, D., Gibbs, J.: Social media as a catalyst for online deliberation? exploring the affordances of Facebook and YouTube for political expression. Comput. Hum. Behav. **29**(3), 1159–1168 (2013)

21. Halterman, A., Keith, K.A., Sarwar, S.M., O'Connor, B.: Corpus-level evaluation for event QA: the indiapoliceevents corpus covering the 2002 gujarat violence. In: Findings of the Association for Computational Linguistics: ACL/IJCNLP 2021, Online Event, August 1-6, 2021. Findings of ACL, vol. ACL/IJCNLP 2021, pp. 4240–4253. Association for Computational Linguistics (2021)

22. Hare, C., Poole, K.T.: The polarization of contemporary American politics. Polity **46**(3), 411–429 (2014)

23. Hefeker, C., Neugart, M.: Policy rules and political polarization (2024)

24. Heiss, R., Schmuck, D., Matthes, J.: What drives interaction in political actors' Facebook posts? profile and content predictors of user engagement and political actors' reactions. Inf. Commun. Soc. **22**(10), 1497–1513 (2019)

25. Hetherington, M.: Review article: putting polarization in perspective. Br. J. Pol. Sci. **39**, 413 – 448 (2009). https://doi.org/10.1017/S0007123408000501

26. Hong, J., et al.: Analysis of faces in a decade of us cable tv news. In: KDD'21: Proceedings of the 27th ACM SIGKDD Conference on Knowledge Discovery & Data Mining (2021)

27. Hyun, K.D., Moon, S.J.: Agenda setting in the partisan TV news context: attribute agenda setting and polarized evaluation of presidential candidates among viewers of NBC, CNN, and fox news. J. Mass Commun. Quart. **93**(3), 509–529 (2016)

28. Iandoli, L., Primario, S., Zollo, G.: The impact of group polarization on the quality of online debate in social media: a systematic literature review. Technol. Forecast. Soc. Chang. **170**, 120924 (2021)

29. Iyengar, S., Sood, G., Lelkes, Y.: Affect, not ideology: a social identity perspective on polarization. Public Opin. Q. **76**(3), 405–431 (2012)

30. Jiang, A.Q., et al.: Mistral 7b. arXiv preprint arXiv:2310.06825 (2023)

31. Jost, J.T., Baldassarri, D.S., Druckman, J.N.: Cognitive-motivational mechanisms of political polarization in social-communicative contexts. Nat. Rev. Psychol. **1**(10), 560–576 (2022)

32. Juneja, P., Bhuiyan, M.M., Mitra, T.: Assessing enactment of content regulation policies: a post hoc crowd-sourced audit of election misinformation on YouTube. In: Proceedings of the 2023 CHI Conference on Human Factors in Computing Systems, pp. 1–22 (2023)

33. KhudaBukhsh, A.R., Bennett, P.N., White, R.W.: Building effective query classifiers: a case study in self-harm intent detection. In: Proceedings of the 24th ACM International Conference on Information and Knowledge Management, CIKM 2015, Melbourne, VIC, Australia, October 19 - 23, 2015, pp. 1735–1738. ACM (2015)

34. KhudaBukhsh, A.R., Sarkar, R., Kamlet, M.S., Mitchell, T.: We don't speak the same language: interpreting polarization through machine translation. In: Proceedings of the AAAI Conference on Artificial Intelligence, vol. 35, pp. 14893–14901 (2021)

35. KhudaBukhsh, A.R., Sarkar, R., Kamlet, M.S., Mitchell, T.M.: Fringe news networks: dynamics of us news viewership following the 2020 presidential election. In: 14th ACM Web Science Conference 2022, pp. 269–278 (2022)

36. Kouzy, R., et al.: Coronavirus goes viral: quantifying the covid-19 misinformation epidemic on twitter. Cureus **12**(3) (2020)

37. Kubin, E., Von Sikorski, C.: The role of (social) media in political polarization: a systematic review. Ann. Int. Commun. Assoc. **45**(2), 188–206 (2021)

38. Kubin, E., Von Sikorski, C.: The role of (social) media in political polarization: a systematic review. Ann. Int. Commun. Assoc. **45**(3), 188–206 (2021)

39. Lee, C., Shin, J., Hong, A.: Does social media use really make people politically polarized? direct and indirect effects of social media use on political polarization in south korea. Telematics Inform. **35**(1), 245–254 (2018)
40. Lee, F.E.: How party polarization affects governance. Annu. Rev. Polit. Sci. **18**, 261–282 (2015)
41. Marozzo, F., Bessi, A.: Analyzing polarization of social media users and news sites during political campaigns. Soc. Netw. Anal. Min. **8**, 1–13 (2018)
42. McCarty, N., Poole, K.T., Rosenthal, H.: Polarized America: The Dance of Ideology and Unequal Riches. MIT Press (2016)
43. McCoy, J., Rahman, T., Somer, M.: Polarization and the global crisis of democracy: common patterns, dynamics, and pernicious consequences for democratic polities. Am. Behav. Sci. **62**(1), 16–42 (2018)
44. Minh Pham, T., Kondor, I., Hanel, R., Thurner, S.: The effect of social balance on social fragmentation. J. R. Soc. Interface **17**(172), 20200752 (2020)
45. Munaro, A.C., Hübner Barcelos, R.: YouTube as a source of patient information for coronavirus disease (COVID-19): a content-quality and audience engagement analysis. Rev. Med. Virol. **30**(3), 2105 (2021)
46. Munaro, A.C., Hübner Barcelos, R., Francisco Maffezzolli, E.C., Santos Rodrigues, J.P., Cabrera Paraiso, E.: To engage or not engage? the features of video content on YouTube affecting digital consumer engagement. J. Consum. Behav. **20**(5), 1336–1352 (2021)
47. NPR: Google races to find a solution after AI generator Gemini misses the mark (2024). https://www.npr.org/2024/03/18/1239107313/google-races-to-find-a-solution-after-ai-generator-gemini-misses-the-mark
48. Papakyriakopoulos, O., Zuckerman, E.: The media during the rise of trump: Identity politics, immigration, "Mexican" demonization and hate-crime. In: Proceedings of the International AAAI Conference on Web and Social Media, vol. 15, pp. 467–478 (2021)
49. Park, C.S.: Does twitter motivate involvement in politics? Tweeting, opinion leadership, and political engagement. Comput. Hum. Behav. **29**(4), 1641–1648 (2013)
50. Poole, K.T., Rosenthal, H.: The polarization of American politics. J. Pol. **46**(4), 1061–1079 (1984)
51. Ross Arguedas, A., Robertson, C., Fletcher, R., Nielsen, R.: Echo chambers, filter bubbles, and polarisation: a literature review (2022)
52. Sap, M., Swayamdipta, S., Vianna, L., Zhou, X., Choi, Y., Smith, N.A.: Annotators with attitudes: how annotator beliefs and identities bias toxic language detection. In: Proceedings of the 2022 Conference of the North American Chapter of the Association for Computational Linguistics: Human Language Technologies, NAACL 2022, pp. 5884–5906. Association for Computational Linguistics (2022)
53. Schroeder, E., Stone, D.F.: Fox news and political knowledge. J. Public Econ. **126**, 52–63 (2015)
54. Somer, M., McCoy, J.L., Luke, R.E.: Pernicious polarization, autocratization and opposition strategies. Democratization **28**(5), 929–948 (2021)
55. Stanley, A.: How msnbc became fox's liberal evil twin. The New York Times (2012). https://www.nytimes.com/2012/08/31/us/politics/msnbc-as-foxs-liberal-evil-twin.html. Accessed 01 Sept 2020
56. Szmuda, T., Syed, M.T., Singh, A., Ali, S.: YouTube as a source of patient information for coronavirus disease (COVID-19): a content-quality and audience engagement analysis. Rev. Med. Virol. **30**(3), 1–9 (2020)
57. Team, G., et al.: Gemini: a family of highly capable multimodal models. arXiv preprint arXiv:2312.11805 (2023)

58. Tesler, M.: The spillover of racialization into health care: how president Obama polarized public opinion by racial attitudes and race. Am. J. Pol. Sci. **56**(3), 690–704 (2012)
59. Tucker, J.A., et al.: Social media, political polarization, and political disinformation: a review of the scientific literature. In: Political Polarization, and Political Disinformation: A Review of the Scientific Literature, 19 March 2018 (2018)
60. Weber, T., et al.: Political polarization: challenges, opportunities, and hope for consumer welfare, marketers, and public policy. J. Public Policy Mark. **40**(2), 184–205 (2021)
61. Weerasooriya, T.C., Dutta, S., Ranasinghe, T., Zamperi, M., Homan, C.M., KhudaBukhsh, A.R.: Vicarious offense and noise audit of offensive speech classifiers: unifying human and machine disagreement on what is offensive. In: Proceedings of the 2020 Conference on Empirical Methods in Natural Language Processing (EMNLP), pp. 11648–11668 (2023)
62. Yoo, C.H., KhudaBukhsh, A.R.: Auditing and robustifying COVID-19 misinformation datasets via anticontent sampling. In: Thirty-Seventh AAAI Conference on Artificial Intelligence, AAAI 2023, pp. 15260–15268. AAAI Press (2023)
63. Gil de Zúñiga, H., Correa, T., Valenzuela, S.: Selective exposure to cable news and immigration in the us: the relationship between fox news, CNN, and attitudes toward Mexican immigrants. J. Broadcast. Electron. Media **56**(4), 597–615 (2012)

Node Generation for Node Classification in Sparsely-Labeled Graphs

Hang Cui[(✉)] and Tarek Abdelzaher

University of Illinois, Urbana Champaign, Champaign, USA
{hangcui2,zaher}@illinois.edu

Abstract. In the broader machine learning literature, data-generation methods demonstrate promising results by generating additional informative training examples via augmenting sparse labels. Such methods are less studied in graphs due to the intricate dependencies among nodes in complex topology structures. This paper presents a novel node generation method that infuses a small set of high-quality synthesized nodes into the graph as additional labeled nodes to optimally expand the propagation of labeled information. By simply infusing additional nodes, the framework is orthogonal to the graph learning and downstream classification techniques, and thus is compatible with most popular graph pretraining (self-supervised learning), semi-supervised learning, and meta-learning methods. The contribution lies in designing the generated node set by solving a novel optimization problem. The optimization places the generated nodes in a manner that: (1) minimizes the classification loss to guarantee training accuracy and (2) maximizes label propagation to low-confidence nodes in the downstream task to ensure high-quality propagation. Theoretically, we show that the above dual optimization maximizes the global confidence of node classification. Our Experiments demonstrate statistically significant performance improvements over 14 baselines on 10 publicly available datasets.

1 Introduction

Node classification in sparsely-labeled graphs is a challenging problem in social network applications. In *graph literature*, the recent state-of-the-art can be summarized into propagation-based methods and infusion-based methods. The propagation-based methods enhance the propagation of graph models via long-range propagation networks [1,2] and the discovery of extra propagation patterns [3,4]. Infusion-based methods transfer knowledge from external tasks/domains (often known as base classes) to the new sparse classes (novel classes), such as knowledge graphs [5] and meta classes of abundant labels [1,6–8], via meta-learning and semi-supervised methods. Both methods aim to improve the propagation of labeled information to unlabeled nodes. Previous work suffered from two major drawbacks: (1) Reliance on external knowledge or dataset characteristics: Infusion-based methods require abundant labels from

L. M. Aiello et al. (Eds.): ASONAM 2024, LNCS 15211, pp. 405–421, 2025.
https://doi.org/10.1007/978-3-031-78541-2_25

external sources, whereas propagation-based methods rely on dataset characteristics for long-range propagation patterns. (2) Poor adaptation from self-supervised signals. Self-supervised models demonstrate remarkable success on pre-training graph representations prior to downstream tasks. However, previous approaches could not effectively utilize self-supervised signals and are shown to underperform when the labels are sparse [9].

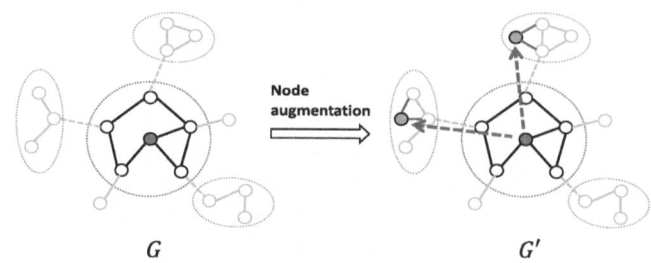

G G'

Fig. 1. Node generation framework: bridging GNN propagation and node augmentation. G represents the original graph where the red circle denotes the sparsely labeled node's (red) local neighborhood, and the green circle denotes low-score regions in the graph. G' is the graph with two generated nodes (orange), which is obtained from augmenting the labeled node (red). (Color figure online)

In the broader machine learning literature, generative methods have shown promising results in numerous machine-learning tasks, including [10–13]. The generative methods generate additional informative training examples by augmenting the sparse labels. By operating merely on input data (namely, augmenting the data set), they have the advantage of being orthogonal to and combinable with a variety of other learning methods, including self-supervised, graph meta-learning, and semi-supervised methods. Despite the recent success of generative methods in end-to-end training tasks, there are key challenges in adapting them to graph neural networks (GNN): (1) topology dependence: message-passing models depend on their local neighborhood in the GNN training/inference process; and (2) the over-smoothing dilemma: shallow GNNs have limited propagation ranges, especially when the labels are sparse, while deep GNNs generally suffer from over-smoothing. As a result, the generated nodes must be carefully placed to propagate high-quality information while minimizing over-smoothing. Simple adaptations fail to generate high-quality augmented data in the graph settings without considering the underlying network topology. Therefore, the goal of this paper is to study the **optimal set of generated nodes that maximizes the propagation of (sparse) labeled information.**

Our method bridges two popular research areas on graph learning: GNN propagation and node augmentation. Multiple prior efforts [1,3,4] have attempted to improve the propagation of labels in a graph, using methods such as edge/substructure inference [3,4] and graph diffusion [1]. However, the availability of such high-quality hyper-relations is often limited. Augmentations [14–16] is frequently used in graph self-supervised learning to improve the robustness of self-supervised representations. This paper bridges the above two pieces of literature by actively exploring high-quality hyper-relations via augmentations, aiming to expand GNN propagation.

Our novel task-aware node generation method is plug-and-play and is compatible with any graph learning method. Given the learned (pre-trained) node representations and downstream node classification task, the goal is to actively generate a small set of high-quality augmented nodes to significantly expand the propagation range of labeled information. We show that for any node-level metrics (i.e., derived from self-supervised signals) for node classification, such as prediction confidence, the set of generated nodes can be optimally determined as a closed-form solution via dual optimizations: (1) Minimize the classification loss of the augmented node; (2) Maximize information propagation to low-score nodes of the node-level metric (i.e., if we use prediction confidence, then the propagation is maximized to low-confidence nodes). The minimization part ensures the classification accuracy of augmented nodes, aiming to reduce cross-class oversmoothing and improve classification accuracy. The maximization part ensures the quality of generated nodes capable of propagating information to key graph regions.

The joint optimization problem can be understood as an optimal 'cover' problem on the set of low-score nodes. Although the cover problem is known to be NP-hard, we propose an efficient approximate algorithm that actively searches for the set in a greedy fashion. In addition, we provide several optimizations that further improve our framework in terms of both performance and time complexity. The contribution of this paper includes:

– The first node generation framework to improve node classification in sparsely-labeled graphs, by optimally exploring the set of generated nodes that maximize propagation to low-score areas in the graph.
– The method is orthogonal to (and thus combinable with) other popular graph learning techniques, including self-supervised learning and semi-supervised learning, and supplements support and testing sets for graph meta-learning.

2   Related Works

2.1   Graph Representation Learning and Node Classification

Node classification is a fundamental problem in graph learning. Graph neural networks (GNN) [17] become the backbone of modern graph learning models. Examples include graph convolutional networks [17] and graph attention networks [18]. Recently, graph self-supervised learning has emerged as a popular approach, where the graph/node representations are learned unsupervised before knowing the downstream tasks. Popular solutions include graphautoencoders [19,20], and graph contrastive learning [21,22]. Although self-supervised learning provides a promising path towards graph unsupervised learning, it still requires sufficient labeled information to train the classifiers or fine-tuning.

In **sparsely labeled settings**, previous works attempt to improve the propagation of labeled information to unlabeled nodes, which can be summarized into *propagation-based methods* and *infusion-based methods*. Propagation-based methods explicitly expand propagation by long-range propagation models [1,2], connectivity-based edge sampling [4], and topological relational inference [3]. Infusion-based methods transfer knowledge from external sources, such as knowledge graphs and base classes of abundant labels (often known as few-shot meta-learning) [9,23–25].

2.2 Data Generation for Sparse Labeled Tasks

Gold-labeled samples are usually limited in real-world tasks due to the high cost of labeling. **Data-generation-based** methods generate additional training examples by augmenting labeled data. *Feature-oriented methods* [12] generate diverse but label-invariant synthetic training samples via feature optimization. *Decoupling methods* [11]: decouples task-dependent components and task-invariant components, and then generates training samples via reconstruction. *Interpolation methods* [13] generates training examples via interpolation of labeled data. *Input perturbation* [26] performs perturbations on the labeled data, such as masking, rotation, and augmentation.

3 Motivation and Preliminary

3.1 Problem Formulation

Given an input attribute graph $G(V, E)$, where V, E are the vertex and edge sets, denote the adjacency matrix as A and node feature matrix as $X \in \mathcal{R}^{|V| \times d}$; the goal is to classify unlabeled nodes into c classes, where c is a known parameter. We study the sparse labeled setting, where only a small (e.g., 1%) subset of nodes are labeled. Note that, we do not assume any base class with abundant labels. Throughout this paper, we use X as the input features, Y as the labels, and H as the learned embedding. We use $\sim$ to represent node generation variables, for example, $\tilde{V}$ as the set of generated nodes and $\tilde{H}$ as the embedding matrix after node generation.

3.2 Motivation

In machine learning literature, data-generation methods reported state-of-the-art performance in many sparsely labeled machine learning tasks, including named entity recognition [12], rationalization [11], and meta-task generation [13]. The core principle is to generate additional informative training examples from the few labeled samples. The key challenges to applying data generation in graphs lie in the intricate topology dependence of message-passing models and the over-smoothing dilemma of GNNs. Thus the problem is to **generate an optimal set of synthesized nodes that significantly increases the propagation of labeled nodes to unlabeled nodes.** Our method bridges the two research topics in the following:

GNN Propagation: Previous attempts propose propagating sparse labels information following the principle of homophily: two nodes tend to be similar (share the same label) if connected via a prediction metric $\hat{A} = pred(v_i, v_j)$ that agrees with the label matrix. The prediction metric is often (pre-) trained alongside the GNNs to discover additional propagation patterns. Examples include:

- Propagation network [1] utilizes diffusion to build the label propagator:

$$\hat{A} = \sum \alpha^{(k)} A^{(k)}, A^{(k)} = TA^{(k-1)} \tag{1}$$

where $\alpha^{(k)}$ and T are the diffusion parameters.
- Edge sampling [4] utilizes link prediction models to generate additional connections:

$$\hat{A} = Connect(V, E) \tag{2}$$

where $Connect$ is a pre-trained link prediction module such as MLP.
- Subgraph inference [3] injects new edges among two nodes if the subgraph characteristics of their k-hop neighborhoods are sufficiently close.

$$\hat{A}_{uv} = \frac{exp(d_W(G_u, G_v))}{\sum_{v \in N_k(u)} exp(d_W(G_u, G_v))} \tag{3}$$

The key challenge of the above methods is the limited availability of high-quality connectivity samples in $\hat{A}$.

Feature-Based Node Augmentation: The most commonly used augmentation objective is the classification loss:

$$\mathcal{L}(f(\tilde{v}), y_v) \tag{4}$$

where $f()$ is the downstream classifier, y_v is the golden label of node v, $\mathcal{L}()$ is a loss function such as cross entropy. The objective measures the classification deviation from augmenting v to $\tilde{v}$. The popular feature-based node augmentation methods [26] are:

- perturbation-based: $X_{\tilde{v}} = X_v + \delta$
- interpolation-based: $X_{\tilde{v}} = (\alpha_1 X_{v_1} + \alpha_2 X_{v_2})/(\alpha_1 + \alpha_2)$
- mixup-based: mixup features of labeled nodes

However, previous augmentation methods in graphs do not explicitly improve the limited propagation range of sparse labels.

To counter the above challenges, we propose to bridge the above two lines of research: **using node augmentations to actively explore high-quality**

connectivity samples. The labeled nodes are augmented to satisfy both conditions: (1) align with the downstream classifier in Eq. (4); (2) generate high-quality connectivity samples in $\hat{A}$. We show that the augmentations significantly increase the availability of high-quality connectivity samples to expand GNN propagation while also regulated by the augmentation objective to align with the downstream tasks. Our framework is demonstrated in Fig. 1.

4   Our Method

Given the embedding matrix H from any graph learning methods, denote the downstream classification loss as $\mathcal{L}(f(H), Y)$, where $\mathcal{L}$ is the loss function, $f()$ is the classifier, and Y is the label (ground truth) matrix. Following the discussion in Sect. 3.2, the labeled nodes are actively augmented to (1) align with the downstream task classifier and (2) generate high-quality connectivity samples:

$$\tilde{V} = argmin_{\tilde{V}} \sum_{\tilde{v} \in \tilde{V}} \mathcal{L}(f(X_{\tilde{v}}), y_v) - \alpha \times Evaluate(\hat{A}|\tilde{V}) \tag{5}$$

where the first term is the augmentation objective that minimizes the prediction deviation of augmented nodes $\tilde{v}$ from the original node's label y_v; the second term evaluates the propagation quality of $\hat{A}$ generated by the set of augmented nodes $\tilde{V}$. The exact formulation of $Evaluate()$ is derived in the following subsections.

Laplacian Smoothness. Recent works [27,28] demonstrated that the message passing of GNNs can be unified into solving the Laplacian smoothness:

$$H = argmin_H ||H - X||_F^2 + \lambda \cdot tr(H^T L H) \tag{6}$$

where H is the embedding matrix, X is the input feature matrix, L is the normalized Laplacian matrix of the graph, and λ is a hyperparameter that controls the smoothness of the graph signal. The above can be rewritten from a node-centric view:

$$H = argmin_H \sum_{i \in \mathcal{V}} ||H_i - X_i||_2^2 + \lambda \sum_{i \in \mathcal{V}} \sum_{j \in N_i} ||H_i - H_j||_2^2 \tag{7}$$

where N_i is the set of neighboring nodes of i.

The above formulate provides a closed-form derivation of message passing after node generation:

$$\tilde{H} = argmin_{\tilde{H}} \sum_{i \in \mathcal{V}} ||\tilde{H}_i - X_i||_2^2 + \lambda \sum_{i \in \mathcal{V}} \sum_{j \in N_i} ||\tilde{H}_i - \tilde{H}_j||_2^2$$

$$+ \tilde{\lambda} \sum_{\tilde{i} \in \tilde{V}} \sum_{j \in N_{\tilde{i}}} ||X_{\tilde{i}} - \tilde{H}_j||_2^2 \tag{8}$$

where $\tilde{V}$ denotes the set of generated nodes, $X_{\tilde{v}}$ denotes the features of generated nodes, $\tilde{H}$ represents the feature matrix after node generation ($\tilde{H}$ does not include the generated nodes since they are not part of the classification task).

Evaluate Propagation of Generated Nodes. The expected classification performance on unlabeled nodes is commonly estimated as the confidence score $Conf()$ of the classifier [29]. We adopt the increment of confidence score $\Delta Conf()$ (by default reciprocal of variance) as the metric for $Evaluate()$:

$$\tilde{V} = argmin_{\tilde{V}} \sum_{\tilde{v} \in \tilde{V}} \mathcal{L}(f(X_{\tilde{v}}), y_v)$$
$$- \alpha \times [Conf \circ f(\tilde{H}) - Conf \circ f(H)] \tag{9}$$

where $\circ$ denotes function composition.

Proposition 1. *Given a pre-trained link prediction model $Connect(v_i, v_j)$ and embedding matrix H, assume the number of generated nodes is small, s.t. $|V_s| << |V|$, and the edges of generated nodes are created following Erdos–Renyi (ER) model proportional to the $Connect(\tilde{v}, v_i)$, then the expected $\tilde{H}$ after generating node $\tilde{v}$ with feature $X_{\tilde{v}}$ is:*

$$\mathbb{E}[\tilde{H}_i] = \underbrace{(1 - \frac{\tilde{\lambda}}{1 + d_i\lambda + \tilde{\lambda}} Connect(\tilde{v}, v_i))H_i}_{original\ embedding}$$
$$+ \underbrace{\frac{\tilde{\lambda}}{1 + d_i\lambda + \tilde{\lambda}} Connect(\tilde{v}, v_i)X_{\tilde{v}} \tilde{v}}_{propagation\ of} \tag{10}$$

where d_i is the degree of node i and $X_{\tilde{v}}$ is the feature of the generated node.

Proposition 1 gives a closed-form solution of $\tilde{H}$, which can be understood as an interpolation between the original embedding H_i and the feature of the generated node $X_{\tilde{v}}$.

A greedy solution

Proposition 2. *If the classifier $f()$ and the confidence score metric $conf()$ are convex, and for $\forall \tilde{v} \in \tilde{V}$, $\forall v_i \in V_s$, $Conf \circ f(\tilde{v}) > Conf \circ f(v_i)$, then:*

$$\sum_{v_i \in V_s} Conf \circ f(\mathbb{E}[\tilde{H}_i]) - Conf \circ f(H_i) \tag{11}$$

is submodular and monotone in terms of $\tilde{V}$.

Proposition 2 grants a greedy approximation of $1 - 1/\epsilon$, such that the nodes are generated sequentially one at a time:

$$\tilde{v} = argmin_{\tilde{v}} \mathcal{L}_g = \mathcal{L}(f(X_{\tilde{v}}), y_v)$$
$$-\alpha \times \sum_{v_i \in V_s} [Conf \circ f(\mathbb{E}[\tilde{H}_i]) - Conf \circ f(H)_i] \qquad (12)$$

where V_s is the set of low-confident nodes.

Algorithm 1. A greedy approximation

Require: Graph $G = (V, E)$, where denotes V_l the set of few-shot labeled nodes, V_s the set of low-confident nodes from downstream classification tasks, $\tilde{V}$ the set of generated synthesized nodes, $\mathcal{L}_g(\cdot)$ the loss function in Eq. (12), $f(\cdot)$ the classifier, $Connect(\cdot, \cdot)$ the pre-trained link prediction model, $Conf(\cdot)$ confident score from downstream classifier, $g(\cdot)$ node augmentation function, hyperparameter for objective loss α, learning rate τ, stopping threshold δ, number of generated synthesized node m.

1: Initialize $\tilde{V} = \emptyset$
2: **for** $t = 1 \cdots m$ **do**
3:   **for** $v \in V_l$ **do**
4:     $\tilde{v} \leftarrow g_\theta(v)$
5:     $grad_{\tilde{v}}(\mathcal{L}_g) \leftarrow \mathcal{L}_g.backward()$
6:     **while** $grad_{\tilde{v}}(\mathcal{L}_g) > \delta$ **do**
7:       $\tilde{v} \leftarrow \tilde{v} - \tau \cdot grad_{\tilde{v}}(\mathcal{L}_g)$
8:       $grad_{\tilde{v}}(\mathcal{L}_g) \leftarrow \mathcal{L}_g.backward()$
9:     **end while**
10:    $\tilde{V} \leftarrow \tilde{V} + \tilde{v}$
11:    **for** $v_i \in V_s$ **do**
12:      $Conf(v_i) \leftarrow Conf \circ f(\mathbb{E}[\tilde{H}_i])$
13:    **end for**
14: **end for**
15: **end for**

The detailed pseudocode is shown in Algorithm 1. Line 3–9 generates an augmented node via gradient descent for a randomly sampled labeled node. The greedy search is initialized with an (optional) pre-defined augmentation function $g_\theta(\cdot)$ as discussed in Sect. 3.2. The optional function enables ad-hoc adaptation from previous data-generation methods for desired characteristics of downstream tasks. After obtaining the set of generated nodes, we generate edges using ER model (proportional to the link prediction probability).

Time Complexity we denote d_{in}, d_h, and d_{out} as the input, hidden, and output feature size. Assuming $Connect(,)$ is a 2-layer MLP, then the time complexity of our method is $O(|V_s|d_{out}^4 d_{MLP}^2 + epo \times |E|d_{in}d_{out})$, where epo is the number of epoch of fine-tuning. For larger graphs, the complexity becomes $O(|E|)$.

4.1 An Simpler Approximation

Algorithm 1 is tesseracted on d_{out}, which can be expensive in large-scale practical applications.

By Jensen's inequality,

$$Conf \circ f(\mathbb{E}[\tilde{H}_i]) - Conf \circ f(H_i) \tag{13}$$

$$\leq (1 - \frac{\tilde{\lambda}}{1 + d_i\lambda + \tilde{\lambda}} Connect(\tilde{v}, v_i)) Conf \circ f(H_i) \tag{14}$$

$$+ \frac{\tilde{\lambda}}{1 + d_i\lambda + \tilde{\lambda}} Connect(\tilde{v}, v_i) Conf \circ f(\tilde{X}_{\tilde{v}}) \tag{15}$$

The second term can be further relaxed to:

$$\frac{\tilde{\lambda}}{1 + d_i\lambda + \tilde{\lambda}} Connect(\tilde{v}, v_i) \tag{16}$$

given that the confidence scores of generated nodes are high. Therefore, Eq. (12) is relaxed to

$$\mathcal{L}_{apx} = \underbrace{\mathcal{L}(f(X_{\tilde{v}}), y_v)}_{\text{augment loss}} + \underbrace{\frac{\tilde{\lambda}}{1 + d_i\lambda + \tilde{\lambda}} Connect(\tilde{v}, v_i)]}_{\text{edge probability}} \tag{17}$$

$$- \alpha \times \sum_{v_i \in V_s} \underbrace{[(1 - \frac{\tilde{\lambda}}{1 + d_i\lambda + \tilde{\lambda}} Connect(\tilde{v}, v_i)) Conf \circ f(H_i)}_{\text{propagation to low confident nodes}} \tag{18}$$

$\mathcal{L}_{apx}$ has complexity $O(|V_s| d_{out}^2 d_{MLP}^2)$.

Understand. $\mathcal{L}_{apx}$ has three terms, which govern: (1) augmentation loss of preserving the original class; (2) propagation to low-confident nodes; (3) edge probability between the generated node and the low-confident nodes.

4.2 Fine-Tuning

After obtaining the generated node-set and the corresponding edges, the graph representation is fine-tuned by predicting the generated node set to its labels. We follow the general data-generation training strategy [11–13]. Algorithm 1 uses multiple reusable components, $Conf(\cdot), g_\theta(\cdot), Connect(\cdot)$, which enables efficient multi-view augmentations. Therefore, we generate multiple sessions of node sets with randomized V_a and initialization, then use a fusion function (such as mean-pooling) to summarize them into the final representation.

4.3 Optimization and Heuristics

This section discusses several optimization choices and heuristics:

Protypical Networks. The back-propagation of $\mathcal{L}_g$ can be expensive when the set V_s is large. We utilize the prototypical network [30], which computes a set of prototypes to summarize V_s through an embedding function, such as k-means. We then only use the prototype set as $\hat{V}_s$ to derive the generated nodes.

Dynamic Pruning. In most scenarios, we prefer a few highly confident connection samples over many less-confident connections. To achieve this goal, we use a dynamic pruning module that prunes the tail k_t nodes in the link prediction results, every few epochs.

Hyperparameter. The hyper-parameter α is vital to balance the trade-off between receptive range and accuracy, which can be tricky to set up. We introduce a dynamical function to update α online. In details, let $W_1 = \mathcal{L}(f(X_{\tilde{v}}), y_v)$ and $W_2 = \sum_{v_j \in V_s} m(\tilde{V}, v_j)(1 - Conf(v_j))$. We use a controller $Con()$ that stabilizes the ratio of both objectives W_1/W_2 to a pre-defined value η.

5 Experiments

In this section, we demonstrate our method as a task-aware node generation framework compatible with the existing graph learning methods. For each method, we first obtain their (pre-trained) graph representations, then use our method to generate a set of augmented nodes, and finally fine-tune the graph representation to obtain the classification results.

5.1 Setup and Pipelines for Self and Semi-supervised Methods

Datasets. We use 3 small-scale benchmarks [31] for node classification: *CiteSeer*, *Cora*, and *PubMed*; and 3 large-scale graphs [32]: *ogbn-Arxiv*, *ogbn-Products*, and *ogbn-Product*. For the 3 benchmarks, we use the few-shot setup of 0.5%, 1%, and 2%. For the large-scale graphs, we use 1%, 2%, and 5%. Note that, we do not assume any base classes of abundant labels. In other words, every class in the dataset is considered sparsely labeled. The training and testing split is randomly selected in 10 separate trails. The detailed statistics is shown in Table 1.

Table 1. Dataset statistics

Datasets	#Nodes	#Edges	#Features
Cora	2,485	5,069	1,433
Citeseer	2,110	3,668	3,703
PubMed	19,717	44,324	300
ogbn-Arxiv	169,343	1,166,243	128
ogbn-Products	2,449,029	61,859,140	100
ogbn-Papers100M	111,059,956	1,615,685,872	128

Baselines. For the three small-scale benchmarks, we compare our methods with self-supervised learning methods: GraphMAE2 [33], GRACE [34], BGRL [35], MVGRL [16], and CCA-SSG [36]; semi-supervised learning methods: GCN [37], GAT [18], Meta-PN [1], DPT [2], edge sampling [4], and subgraph inference [3]. Unless specified, we run the above baselines with the same experimental setups described in their original papers.

For the three large-scale datasets, we compare our methods with baselines, which perform well in small-scale benchmarks and include a released implementation scalable to large datasets. The baselines are GraphMAE2, GRACE, BGRL, and CCA-SSG. We also include Simplified Graph Convolution (SGC) [38], a simplified version of GCN for scalability. If available, we use the same source code provided by the above methods; if not, we use GraphSAINT [39] sampling strategy with GNN initialization. Due to high computation cost, We only apply our node generation to the results of SGC to demonstrate our performance.

Pipelines. Since our method does not specify any backbone representation learning method, we use the above baselines to produce the initial representation and then use our method for node generation. To produce a fair comparison, we also fine-tune the self-supervised methods with labeled data on the pre-trained representations. The fine-tuning process uses semi-supervised objectives with the same linear classifiers on default or the objective stated in the baselines if applicable. For semi-supervised models, we use the standard semi-supervised pipelines as default [11–13], or the pipelines described in their original papers.

After the initial classification, we use the reciprocal of prediction variance as the confidence score and generate the set of low-confident nodes V_s using a pre-defined threshold. We use a two-layer MLP as the pre-trained link prediction model. We then output the set of generated nodes (by default 5 per labeled nodes) following Eq. (12) and perform fine-tuning on the representation in 10 separate trials. Note that, the fine-tuning is significantly cheaper as the set of synthesized nodes is small. Finally, we output the classification results using the fine-tuned representation.

Results. The node classification results on the three benchmarks are shown in Table 2. Our method (represents with '+') improves the node classification performance of self-supervised and semi-supervised learning baselines by a large margin (6.7%-20.5%). In addition, our method also reduces the standard deviation of the results over 10 random trials. We observe a larger performance gain on extremely sparsely-labeled scenarios (0.5%), demonstrating the significance of generating high-quality nodes in a task-aware manner. The fine-tuning results close the gap between the baseline performances, showing our method's independence from any particular pre-training methods. Another notable observation is: self-supervised methods with our fine-tuning techniques can match and outperform semi-supervised methods designed for sparse tasks (Meta-PN and DPT).

Table 3 shows the node classification results on the three large-scale datasets. Again, we observe large performance gains from our fine-tuning methods, especially under the few-shot scenario(1%). The overall performances of all the baselines are higher because the datasets are denser. In *Arxiv* and *Papers100M*, we

Table 2. Classification Accuracy% of Benchmarks. + presents the fine-tuning results using **our** proposed method

Backbone	Cora			CiteSeer			PubMed		
	0.5%	1%	2%	0.5%	1%	2%	0.5%	1%	2%
Self-supervised approaches									
GraphMAE2	45.2 ± 2.1	55.8 ± 1.4	67.6 ± 1.4	43.7 ± 1.1	56.7 ± 0.9	59.4 ± 0.8	56.0 ± 1.9	62.7 ± 0.8	71.2 ± 2.6
+	69.3 ± 1.2	71.8 ± 0.8	78.8 ± 0.6	64.4 ± 0.9	65.5 ± 0.6	66.4 ± 0.6	62.2 ± 1.1	77.1 ± 0.4	78.7 ± 1.0
GRACE	45.8 ± 2.4	62.0 ± 3.6	69.9 ± 1.6	42.6 ± 4.5	52.6 ± 3.6	59.7 ± 3.2	60.9 ± 2.4	67.9 ± 1.4	76.9 ± 3.0
+	65.9 ± 1.4	72.7 ± 1.9	79.8 ± 1.1	59.3 ± 1.0	62.3 ± 0.5	67.0 ± 0.4	68.8 ± 1.2	78.8 ± 0.6	83.8 ± 1.2
BGRL	46.6 ± 3.2	60.5 ± 1.4	66.2 ± 2.6	47.9 ± 3.3	52.4 ± 2.7	56.4 ± 2.4	60.4 ± 1.9	66.4 ± 1.1	71.5 ± 2.4
+	66.2 ± 1.8	70.6 ± 0.8	74.8 ± 0.4	59.9 ± 1.4	66.0 ± 0.6	71.5 ± 0.4	69.5 ± 1.2	77.6 ± 0.5	81.4 ± 0.4
MVGRL	46.4 ± 2.4	60.9 ± 0.9	66.9 ± 2.1	43.2 ± 1.0	53.5 ± 0.8	60.1 ± 1.0	60.2 ± 2.3	68.4 ± 1.5	77.4 ± 2.4
+	65.9 ± 1.6	71.4 ± 0.9	75.2 ± 0.8	60.2 ± 0.9	66.5 ± 0.8	71.5 ± 0.5	70.9 ± 1.4	76.9 ± 0.6	80.5 ± 0.4
CCA-SSG	48.5 ± 2.0	62.6 ± 1.4	67.9 ± 2.4	47.2 ± 1.4	54.1 ± 1.0	62.2 ± 1.2	61.3 ± 2.0	68.2 ± 0.9	77.5 ± 1.9
+	63.2 ± 1.1	70.5 ± 0.6	77.6 ± 0.9	60.3 ± 0.8	61.3 ± 0.6	66.0 ± 0.5	69.9 ± 1.2	77.6 ± 0.5	81.1 ± 0.4
Semi-supervised approaches									
GCN	44.2 ± 4.4	48.6 ± 1.8	67.3 ± 2.4	40.2 ± 5.4	53.6 ± 4.0	62.6 ± 2.6	58.5 ± 3.6	66.5 ± 1.5	70.3 ± 4.0
+	60.2 ± 2.4	68.1 ± 1.2	73.1 ± 1.9	58.5 ± 2.7	62.8 ± 2.1	69.8 ± 1.4	65.9 ± 2.4	74.2 ± 1.0	79.8 ± 2.6
GAT	46.1 ± 4.0	50.4 ± 1.8	69.2 ± 1.4	43.1 ± 5.0	54.2 ± 3.6	63.5 ± 2.4	59.1 ± 3.4	67.0 ± 1.4	70.9 ± 3.4
+	61.4 ± 2.0	69.4 ± 1.2	74.9 ± 0.6	59.9 ± 2.2	63.0 ± 1.9	70.0 ± 1.1	66.2 ± 2.0	74.2 ± 0.9	79.6 ± 1.9
Meta-PN	50.1 ± 4.0	70.5 ± 1.8	77.6 ± 1.4	50.1 ± 5.0	68.9 ± 3.6	74.2 ± 2.4	71.2 ± 3.4	77.8 ± 1.4	80.5 ± 3.4
+	65.4 ± 2.0	74.9 ± 1.2	79.1 ± 0.6	66.2 ± 2.2	70.0 ± 1.9	76.3 ± 1.1	74.1 ± 2.0	79.6 ± 0.9	82.4 ± 1.9
DPT	55.2 ± 4.0	70.4 ± 1.8	75.9 ± 1.4	58.1 ± 5.0	68.4 ± 3.6	72.1 ± 2.4	71.4 ± 3.4	77.8 ± 1.4	80.4 ± 3.4
+	64.4 ± 1.6	73.1 ± 1.1	78.2 ± 0.6	67.4 ± 1.6	69.8 ± 1.2	75.9 ± 1.1	74.7 ± 1.5	79.1 ± 1.1	81.4 ± 1.6
Edge	51.2 ± 5.2	70.5 ± 2.9	74.6 ± 2.6	48.1 ± 5.5	66.4 ± 4.2	73.8 ± 3.8	68.2 ± 4.2	76.2 ± 2.6	80.9 ± 3.2
Subgraph	53.6 ± 4.5	71.9 ± 2.6	75.1 ± 2.0	49.1 ± 4.8	67.0 ± 3.8	74.1 ± 3.0	68.3 ± 3.5	76.6 ± 2.5	81.5 ± 2.6

observe a performance gain of (4%–5.1%) and (4.5%–6.8%), which demonstrates the effective of our method on large datasets. On *Product* however, the performance gain is relatively small (2%–3.6%), because the performance is closing to the upper bound of the dataset according to the ogb leaderboard.

5.2   Computation Time

This section demonstrates the computation time of pre-training and our node generation fine-tuning. We use baselines: GCN, GraphMAE2 (feature-masking method), and GRACE (graph-augmentation method). The fine-tuning is conducted with 20 epochs for small benchmarks and 50 for large-scale datasets, compared to 200 and 1000 epochs for the initial self-supervised learning. Note that, each epoch of self-supervised learning is also dymatically more expensive than GCN. Table 4 shows the computation time(s) of GraphMAE2, Grace, and GCN with and without our method on RTX 4090. For larger datasets (ogbn), our approach only adds 2% overhead for self-supervised methods and 5% for GCN.

Table 3. Classification Accuracy% of large-scale datasets.

Baseline	Arxiv			Products			Papers100M		
	1%	2%	5%	1%	2%	5%	1%	2%	5%
GraphMAE2	66.9 ± 0.5	68.3 ± 0.5	70.2 ± 0.3	77.0 ± 0.4	79.1 ± 0.4	80.5 ± 0.2	58.7 ± 0.4	60.2 ± 0.4	62.9 ± 0.6
GRACE	64.3 ± 0.5	66.8 ± 0.4	69.2 ± 0.5	77.1 ± 0.6	78.0 ± 0.6	79.7 ± 0.5	55.5 ± 0.2	57.9 ± 0.2	59.4 ± 0.2
BGRL	65.1 ± 1.2	66.6 ± 1.0	69.0 ± 0.3	76.3 ± 0.5	78.0 ± 0.5	79.5 ± 0.4	55.1 ± 0.2	57.9 ± 0.2	60.4 ± 0.5
CCA-SSG	64.1 ± 0.2	66.9 ± 0.3	68.3 ± 0.3	75.9 ± 0.4	77.2 ± 0.4	78.5 ± 0.4	55.7 ± 0.2	57.2 ± 0.2	59.8 ± 0.1
SGC	63.9 ± 0.5	65.2 ± 0.5	67.0 ± 0.3	73.0 ± 0.4	74.6 ± 0.4	75.6 ± 0.4	59.2 ± 0.2	61.2 ± 0.4	63.9 ± 0.3
Our	**72.1 ± 0.3**	**73.5 ± 0.3**	**74.2 ± 0.3**	**80.6 ± 0.3**	**81.5 ± 0.2**	**82.5 ± 0.2**	**65.5 ± 0.5**	**66.1 ± 0.4**	**67.4 ± 0.4**

Table 4. Running time (s) of training + our fine-tuning. OOM denotes out of memory.

Baselines	GraphMAE2	GRACE	GCN
Cora	12.91+0.56(4%)	2.76+0.33(12%)	0.82+0.11(13%)
Citeseer	17.89+0.67(4%)	4.55+0.42(9%)	2.01+0.24(12%)
PubMed	25.11+1.10(4%)	41.3+0.94(2%)	4.78+0.49(10%)
Arxiv	196.8+6.2(3%)	612.2+6.1(1%)	70.9+0.42(6%)
Products	1615+20(1%)	OOM	311+17(5%)
Papers	7296+120(2%)	OOM	2414+117(5%)

5.3   Setup and Pipelines for Meta-learning Methods

A key distinction exists between our problem formulation and meta-learning-based methods: we do not assume any base classes with abundant labels, such that the number of labels in every class is few. However, we identify that our node generation approach can be adapted to graph meta-learning by generating additional nodes in training (when labels in base classes are not so abundant) and testing phase. We conduct experiments on four meta-learning baselines: TEG [23], TLP [9], Meta-GNN [24], and Tent [25].

Datasests. We use popular meta-learning datasets [23]: Corafull, Amazon-clothing, Amazon-electronics, and dblp instead of self-supervised learning datasets because meta-learning methods favor datasets with more classes.

Pipelines. As discussed at the beginning of this subsection, meta-learning-based methods require base class with abundant labels while our problem formulation contradicts the above requirement. In particular, if the base class is also k-shot, generating the training episodes is impossible. As a result, we relax our formulation by giving the base classes additional labeled nodes. We conducted three experiments by giving the base classes 3x and 5x labeled nodes. The training episodes are then sampled from the labeled nodes in the base classes. Each method is tested using the released codes and hyperparameters. We use our node generation method to generate additional nodes for base and novel classes (by default 5 per nodes). In detail, we use Grace as the default self-supervised method and then train a logistic classifier on labeled nodes. We then generate additional nodes following Algorithm 1. We conduct 10 independent trials for each method.

Table 5. Classification Accuracy% on 5-way k-shot meta-learning. + presents the results after node generation using our method

Backbone	Corafull		Amazon-clothing		Amazon-electronics		dblp	
	3-shot	5-shot	3-shot	5-shot	3-shot	5-shot	3-shot	5-shot
3x labeled nodes in base classes								
TEG	68.2 ± 2.5	72.4 ± 2.3	82.2 ± 2.5	86.4 ± 2.5	81.6 ± 1.2	86.0 ± 2.3	80.4 ± 1.0	83.9 ± 1.2
+	78.5 ± 2.6	80.2 ± 2.7	90.3 ± 2.2	91.2 ± 1.8	87.8 ± 1.8	88.5 ± 1.7	86.9 ± 1.0	86.8 ± 1.2
TLP	64.9 ± 2.5	71.9 ± 3.5	83.1 ± 3.5	87.0 ± 3.4	77.6 ± 3.0	83.2 ± 3.2	79.1 ± 3.1	84.8 ± 3.0
+	76.0 ± 3.0	79.0 ± 3.0	87.8 ± 2.8	91.5 ± 2.9	88.0 ± 2.6	88.7 ± 2.5	84.6 ± 2.2	86.4 ± 2.2
Meta-GNN	55.3 ± 2.4	60.6 ± 3.4	73.7 ± 2.5	76.5 ± 2.1	62.2 ± 1.9	67.8 ± 1.8	68.0 ± 1.9	72.8 ± 2.0
+	67.4 ± 2.2	70.4 ± 1.6	83.7 ± 2.2	83.2 ± 1.9	69.7 ± 1.6	70.8 ± 1.6	75.6 ± 1.8	76.9 ± 1.4
Tent	62.2 ± 3.1	66.4 ± 2.4	77.1 ± 2.6	82.2 ± 2.6	77.1 ± 1.9	81.1 ± 1.8	79.2 ± 1.9	83.0 ± 1.8
+	71.4 ± 2.2	75.5 ± 2.5	86.2 ± 2.6	87.9 ± 1.9	81.0 ± 2.0	84.0 ± 1.9	85.5 ± 1.2	85.8 ± 1.4
5x labeled nodes in base classes								
TEG	69.4 ± 2.9	75.5 ± 2.7	87.2 ± 2.2	88.4 ± 2.5	82.2 ± 1.1	86.3 ± 2.3	82.0 ± 1.1	85.9 ± 2.1
+	78.3 ± 2.5	80.2 ± 2.4	89.8 ± 1.6	92.7 ± 1.6	90.0 ± 2.0	89.2 ± 1.3	86.2 ± 2.3	87.4 ± 2.6
TLP	65.4 ± 3.5	72.4 ± 3.1	83.7 ± 3.4	87.7 ± 3.1	78.1 ± 2.9	85.2 ± 2.8	79.1 ± 2.0	84.4 ± 2.4
+	76.4 ± 2.8	79.2 ± 2.8	87.8 ± 2.6	92.7 ± 2.4	88.5 ± 2.6	89.3 ± 2.6	84.5 ± 2.1	86.9 ± 2.4
Meta-GNN	57.4 ± 2.6	65.1 ± 3.4	79.7 ± 2.6	80.8 ± 2.2	66.8 ± 2.1	69.1 ± 1.8	71.2 ± 1.9	76.8 ± 1.8
+	66.5 ± 2.4	74.4 ± 1.8	88.2 ± 2.2	86.8 ± 1.9	73.8 ± 1.9	72.2 ± 1.8	78.1 ± 1.8	80.8 ± 1.4
Tent	63.7 ± 2.1	70.6 ± 2.5	80.2 ± 2.4	84.4 ± 2.5	77.9 ± 1.8	81.5 ± 1.8	81.3 ± 1.9	85.3 ± 2.0
+	73.1 ± 2.1	75.4 ± 2.2	86.3 ± 2.2	90.0 ± 1.9	83.4 ± 2.0	84.8 ± 1.6	85.0 ± 1.5	86.4 ± 2.4

Results. Table 5 shows the classification accuracy of meta-learning methods with 3x, 5x, 10x labeled nodes in base classes. Overall, our method improves performance by $\sim 8\%$, $\sim 7\%$, and $\sim 5\%$, respectively. The standard deviation does not reduce as significantly as in the previous section due to the random sampling of training episodes. Our experiment stops at 10x labeled nodes because the baseline performance is already similar to abundant labels in base classes.

6   Conclusion

This work explores node-generation frameworks for few-shot node classification on graphs. The applications include fine-tuning self-supervised graph representation learning, few-shot semi-supervised learning, and supplementing base and novel classes in graph meta-learning. Our work bridges two lines of research, GNN propagation and graph augmentation, to derive an optimal set of generated nodes that can propagate few-shot information to key areas on the graph.

Our framework augments the few labeled nodes to actively explore high-quality connectivity samples to selected low-confident areas on the graph. We propose an efficient greedy solution to approximately find the set and several heuristics to improve the scalability and adaptability of our method. We demonstrate our performance on 10 datasets on 10 baselines.

References

1. Ding, K., Wang, J., Caverlee, J., Liu, H.: Meta propagation networks for graph few-shot semi-supervised learning. In: Proceedings of the AAAI Conference on Artificial Intelligence, vol. 36, no. 6, pp. 6524–6531 (2022)
2. Liu, Y., Ding, K., Wang, J., Lee, V., Liu, H., Pan, S.: Learning strong graph neural networks with weak information. In: Proceedings of the 29th ACM SIGKDD Conference on Knowledge Discovery and Data Mining, pp. 1559–1571 (2023)
3. Chen, Y., Coskunuzer, B., Gel, Y.: Topological relational learning on graphs. In: Advances in Neural Information Processing Systems, vol. 34, pp. 27 029–27 042 (2021)
4. Tan, Z., Ding, K., Guo, R., Liu, H.: Supervised graph contrastive learning for few-shot node classification. In: Joint European Conference on Machine Learning and Knowledge Discovery in Databases, pp. 394–411. Springer (2022)
5. Yao, H., et al.: Graph few-shot learning via knowledge transfer. In: Proceedings of the AAAI Conference on Artificial Intelligence, vol. 34, no. 04, pp. 6656–6663 (2020)
6. Giannone, G., Winther, O.: Scha-VAE: hierarchical context aggregation for few-shot generation. In: International Conference on Machine Learning, pp. 7550–7569. PMLR (2022)
7. Liu, Y., Li, M., Li, X., Giunchiglia, F., Feng, X., Guan, R.: Few-shot node classification on attributed networks with graph meta-learning. In: *Proceedings of the 45th International ACM SIGIR Conference on Research and Development in Information Retrieval*, pp. 471–481 (2022)
8. Wang, S., Tan, Z., Liu, H., Li, J.: Contrastive meta-learning for few-shot node classification. In: Proceedings of the 29th ACM SIGKDD Conference on Knowledge Discovery and Data Mining, pp. 2386–2397 (2023)
9. Tan, Z., Wang, S., Ding, K., Li, J., Liu, H.: Transductive linear probing: a novel framework for few-shot node classification. In: Learning on Graphs Conference, pp. 4–1. PMLR (2022)
10. Duan, S., Li, W., Cai, J., He, Y., Wu, Y.: Query-variant advertisement text generation with association knowledge. In: Proceedings of the 30th ACM International Conference on Information & Knowledge Management, pp. 412–421 (2021)
11. Liu, G., Zhao, T., Xu, J., Luo, T., Jiang, M.: Graph rationalization with environment-based augmentations. In: Proceedings of the 28th ACM SIGKDD Conference on Knowledge Discovery and Data Mining, pp. 1069–1078 (2022)
12. Sohn, H., Park, B.: Robust and informative text augmentation (RITA) via constrained worst-case transformations for low-resource named entity recognition. In: Proceedings of the 28th ACM SIGKDD Conference on Knowledge Discovery and Data Mining, pp. 1616–1624 (2022)
13. Yao, H., Zhang, L., Finn, C.: Meta-learning with fewer tasks through task interpolation. arXiv preprint arXiv:2106.02695 (2021)

14. Zhu, Y., Xu, Y., Yu, Liu, Q., Wu, S., Wang, L.: Deep graph contrastive representation learning. arXiv preprint arXiv:2006.04131 (2020)

15. Velickovic, P., Fedus, W., Hamilton, W.L., Liò, P., Bengio, Y., Hjelm, R.D.: Deep graph infomax. ICLR (Poster) **2**(3), 4 (2019)

16. Hassani, K., Khasahmadi, A.H.: Contrastive multi-view representation learning on graphs. In: International Conference on Machine Learning, pp. 4116–4126. PMLR (2020)

17. Zhang, S., Tong, H., Xu, J., Maciejewski, R.: Graph convolutional networks: a comprehensive review. Comput. Soc. Networks **6**(1), 1–23 (2019). https://doi.org/10.1186/s40649-019-0069-y

18. Veličković, P., Cucurull, G., Casanova, A., Romero, A., Lio, P., Bengio, Y.: Graph attention networks. arXiv preprint arXiv:1710.10903 (2017)

19. Kipf, T.N., Welling, M.: Variational graph auto-encoders. arXiv preprint arXiv:1611.07308 (2016)

20. Hou, Z., et al.: Graphmae: self-supervised masked graph autoencoders. In: Proceedings of the 28th ACM SIGKDD Conference on Knowledge Discovery and Data Mining, pp. 594–604 (2022)

21. Zhu, Y., Xu, Y., Yu, F., Liu, Q., Wu, S., Wang, L.: Graph contrastive learning with adaptive augmentation. In: Proceedings of the Web Conference 2021, pp. 2069–2080 (2021)

22. You, Y., Chen, T., Shen, Y., Wang, Z.: Graph contrastive learning automated. In: International Conference on Machine Learning, pp. 12 121–12 132. PMLR (2021)

23. Kim, S., Lee, J., Lee, N., Kim, W., Choi, S., Park, C.: Task-equivariant graph few-shot learning. arXiv preprint arXiv:2305.18758 (2023)

24. Zhou, F., Cao, C., Zhang, K., Trajcevski, G., Zhong, T., Geng, J.: Meta-GNN: on few-shot node classification in graph meta-learning. In: Proceedings of the 28th ACM International Conference on Information and Knowledge Management, pp. 2357–2360 (2019)

25. Wang, S., Ding, K., Zhang, C., Chen, C., Li, J.: Task-adaptive few-shot node classification. In: Proceedings of the 28th ACM SIGKDD Conference on Knowledge Discovery and Data Mining, pp. 1910–1919 (2022)

26. Ding, K., Xu, Z., Tong, H., Liu, H.: Data augmentation for deep graph learning: a survey. ACM SIGKDD Explorations Newsl **24**(2), 61–77 (2022)

27. Ma, Y., Liu, X., Zhao, T., Liu, Y., Tang, J., Shah, N.: A unified view on graph neural networks as graph signal denoising. In: Proceedings of the 30th ACM International Conference on Information & Knowledge Management, pp. 1202–1211 (2021)

28. Liu, H., Han, H., Jin, W., Liu, X., Liu, H.: Enhancing graph representations learning with decorrelated propagation. In: Proceedings of the 29th ACM SIGKDD Conference on Knowledge Discovery and Data Mining, pp. 1466–1476 (2023)

29. Liu, G., Zhao, T., Inae, E., Luo, T., Jiang, M.: Semi-supervised graph imbalanced regression. arXiv preprint arXiv:2305.12087 (2023)

30. Li, J., Zhou, P., Xiong, C., Hoi, S.C.: Prototypical contrastive learning of unsupervised representations. arXiv preprint arXiv:2005.04966 (2020)

31. Yang, Z., Cohen, W., Salakhudinov, R.: Revisiting semi-supervised learning with graph embeddings. In: International Conference on Machine Learning, pp. 40–48. PMLR (2016)

32. Hu, W., et al.: Open graph benchmark: Datasets for machine learning on graphs. In: Advances in Neural Information Processing Systems, vol. 33, pp. 22 118–22 133 (2020)

33. Hou, Z., et al.: Graphmae2: a decoding-enhanced masked self-supervised graph learner. In: Proceedings of the ACM Web Conference 2023 (WWW'23) (2023)
34. Zhu, Y., Xu, Y., Yu, F., Liu, Q., Wu, S., Wang, L.: Deep graph contrastive representation learning. In: ICML Workshop on Graph Representation Learning and Beyond 2020. http://arxiv.org/abs/2006.04131
35. Thakoor, S., et al.: Large-scale representation learning on graphs via bootstrapping. International Conference on Learning Representations (ICLR) (2022)
36. Zhang, H., Wu, Q., Yan, J., Wipf, D., Yu, P.S.: From canonical correlation analysis to self-supervised graph neural networks. Adv. Neural. Inf. Process. Syst. **34**, 76–89 (2021)
37. Kipf, T.N., Welling, M.: Semi-supervised classification with graph convolutional networks. arXiv preprint arXiv:1609.02907 (2016)
38. Wu, F., Souza, A., Zhang, T., Fifty, C., Yu, T., Weinberger, K.: Simplifying graph convolutional networks. In: International Conference on Machine Learning, pp. 6861–6871. PMLR (2019)
39. Zeng, H., Zhou, H., Srivastava, A., Kannan, R., Prasanna, V.: Graphsaint: graph sampling based inductive learning method. arXiv preprint arXiv:1907.04931 (2019)

The Dawn of Decentralized Social Media: An Exploration of Bluesky's Public Opening

Erfan Samieyan Sahneh[1] , Gianluca Nogara[1] , Matthew R. DeVerna[2] ,
Nick Liu[2], Luca Luceri[3] , Filippo Menczer[2] , Francesco Pierri[4] ,
and Silvia Giordano[1(✉)]

[1] ISIN - DTI, SUPSI, Lugano, Switzerland
{erfan.samieyan,silvia.giordano}@supsi.ch
[2] Observatory on Social Media, Indiana University, Bloomington, USA
[3] Information Sciences Institute, USC, Los Angeles, CA, USA
[4] Dipartimento Elettronica, Informazione e Bioingegneria, Politecnico di Milano, Milan, Italy

Abstract. Bluesky is a Twitter-like decentralized social media platform that has recently grown in popularity. After an invite-only period, it opened to the public worldwide on February 6th, 2024. In this paper, we provide a longitudinal analysis of user activity in the two months around the opening, studying changes in the general characteristics of the platform due to the rapid growth of the user base. We observe a broad distribution of activity similar to more established platforms, but a higher volume of original than reshared content, and very low toxicity. After opening to the public, Bluesky experienced a large surge in new users and activity, especially posting English and Japanese content. In particular, several accounts entered the discussion with suspicious behavior, like following many accounts and sharing content from low-credibility news outlets. Some of these have already been classified as spam or suspended, suggesting effective moderation.

Keywords: Bluesky · decentralization · online social media · misinformation

1 Introduction

Bluesky Social[1] is a novel decentralized social media platform for microblogging based in the United States. Originally based on invite-only subscription, Bluesky officially opened to the public on February 6th, 2024 [29]. With only a few thousand users active in January 2024, the platform registered more than one million new users on the first day of its opening [6].

Bluesky is built upon the Authenticated Transfer (AT) protocol [17], which aims to enable modern social media and online conversations to function similarly to the early web, where individuals could easily create blogs or use

---

[1] https://bsky.social/about.

L. M. Aiello et al. (Eds.): ASONAM 2024, LNCS 15211, pp. 422–437, 2025.
https://doi.org/10.1007/978-3-031-78541-2_26

Really Simple Syndication (RSS) feeds to follow multiple blogs. This approach is intended to foster a more open and decentralized online community [20] compared to "walled-garden" platforms like Facebook and Twitter/X. The AT protocol is an open framework for creating social applications, offering users insight into their construction and development. It establishes a standard format for user identity, following mechanisms, and data across social applications, facilitating interoperability between apps and empowering users to transfer their accounts effortlessly. Thus, Bluesky's model aims to promote competition among developers, who are free to build various interfaces, filters, and additional services. According to Bluesky, this competitive environment should decrease the necessity for censorship as the best solutions naturally gain prominence [5].

In this paper, we provide the first exploration of how opening the Bluesky platform to the public affected key metrics of interest. We investigate patterns of user activity, their political leaning, and the quality of information circulating on the platform. Crucially, we explore these characteristics both before and after the platform was opened to the public. For a general analysis of the activity on the platform during the first year, we refer the reader to contemporary work [13,27].

2   Related Work

Distributed social networks aim for decentralization, allowing users to have more control and privacy. Early efforts like LifeSocial.KOM [14] and PeerSON [8] were based on the peer-to-peer model but faced challenges in performance and reliability. This led to a shift toward server-based federated models like Mastodon [28,32]. This approach balances flexibility and ease of use while maintaining some decentralization [7].

Mastodon, created in 2016, is a free and open-source social media platform that allows users to create their own servers ("instances") and connect with others across the globe. It is a decentralized social network, meaning that it is not owned by a single entity, but rather a network of independent servers that are connected together. A number of studies have identified striking features that make up Mastodon's distinct "fingerprint," distinguishing it from better-known online social networks [7,10]. Mastodon has, however, suffered from some natural pressures towards centralization, which can lead to potential points of failure [28].

To overcome this weakness, Bluesky developed its own AT protocol in order to provide decentralization, with several features that distinguish it from other authentication protocols. Scalability, security, and ease of use make it an attractive option for building open and decentralized social media applications that prioritize user privacy and data security [17]. Using standard web technologies and re-using existing data models from the Web 3.0 protocol family also contribute to its efficiency and reliability [12]. Additionally, its federated networking model bolsters security by dispersing data across numerous servers, mitigating the risk of a single point of failure [28]. As mentioned above, two contemporary work on Bluesky have been published: [13,27]. Unlike this work, they provide a general analysis of Bluesky's first-year activities.

Table 1. Basic statistics of the Bluesky dataset.

	original	reply	repost	total	%
messages	30,235,716	20,842,322	20,530,919	71,608,957	
with links	3,215,963	339,921	2,683,874	6,239,758	8.57%
active users				2,734,569	
sharing messages				1,752,083	
with links				389,077	14.22%
follow actions				39,214,164	
block actions				2,799,597	

3   Methods

The present analysis is based on data collected in a two-month period around the time that the platform opened to the public.

3.1   Data Collection

We collected data by accessing Bluesky's public and free "Firehose" endpoint, which provides developers with real-time access to atomic actions performed on Bluesky such as user posts, follows, likes, etc. [2,4]. If a real-time connection with Bluesky is interrupted, the endpoint enables data collectors to resume collection from the moment the connection was lost, retrieving data from up to 72 h prior. This ensures comprehensive data coverage throughout the observed timeframe. We utilized the `dart` library from the AT protocol [3] to fetch data by invoking the `com.atproto.sync.subscribeRepos` endpoint, also known as the Firehose endpoint [16]. This process, facilitated by an existing open-source project [9], is straightforward and flexible due to the absence of user authentication requirements in the AT protocol.

Bluesky enables tracking of various user activities on the platform. Our analyses focus on several key actions: posts, replies, reposts, follows, and blocks. These actions represent the main forms of user interaction and communication. Users follow others to stay updated, create posts to share content or repost content created by others, and reply to engage in discussions. While the Bluesky terms of service do not impose privacy restrictions on the data collection, we collect only public information about users, posts, and any attached metadata in accordance with Bluesky's Privacy Policy.[2] We do not make the data publicly available and only provide anonymized information in this paper, except for a few prominent accounts in §4.8.

Table 1 provides basic statistics of our dataset, which spans 56 d, from Jan 9th to March 4th 2024, and comprises 114 million activities (message actions plus other actions like follow and block).

---

[2] https://bsky.social/about/support/privacy-policy.

3.2   News Source Labeling

To assess the reliability of news outlets shared on Bluesky, we label web domains using NewsGuard[3] ratings, following a consolidated approach in the literature [25,30]. NewsGuard is a reputable and unbiased organization that employs experts to evaluate news sources based on factors such as transparency, accountability, adherence to journalistic standards, and error correction to determine credibility. NewsGuard ratings range from 0 (highly unreliable) to 100 (highly reliable). Approximately 6.2 million posts (8.7% of all posts) contained a URL. Of these, we were able to label around 1 million (16% of posts with URLs, and 1.4% of all posts) with a NewsGuard rating. The average rating of the posts with links in the social network is high, close to 94. We further define low-credibility websites as news outlets with a NewsGuard rating of 30 or lower, following previous literature [21,24,26]. We also leveraged political bias ratings from Media Bias/Fact Check,[4] an independent organization that rates news media sources, to label the political leaning of news websites shared on Bluesky. Information sources are categorized across a seven-point political spectrum: Extreme Left (-3), Left (-2), Left-Center (-1), Least Biased (0), Right-Center (1), Right (2), and Extreme Right (3).

4   Results

4.1   Online Activity

We observe in Table 1 that original posts are the most common user activity on the platform, indicating that users tend to create more original content rather than reshare or interact with existing posts. This contrasts with centralized social media platforms, such as Twitter/X, where resharing through retweets is more prevalent [1,19], and is likely due to the sudden increase in the user-base.

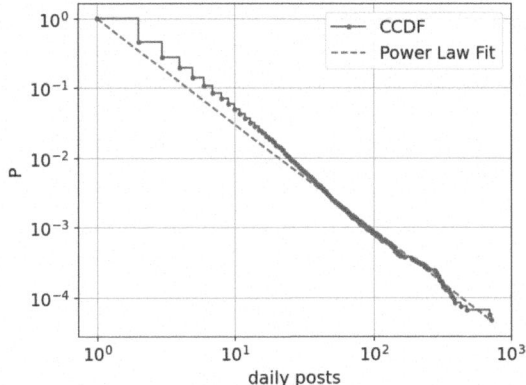

Fig. 1. Complementary cumulative distribution of user activity (number of original posts) on Bluesky during one day (March 4, 2024). The dashed line is a power-law fit of the distribution, which yields a slope of 1.53 ($x_{min} = 1, x_{max} = 1,000$).

The daily average number of posts per active user is 3.16 (95% CI: [3.13, 3.20]). However, the user activity is quite heterogeneous. Figure 1 illustrates that the number of daily posts x follows a broad, power-law distribution $P(x) \sim x^{-\alpha}$ with $\alpha \approx 2.53$. This exponent value is consistent across different days, ranging between 2.50 and 2.71. This extreme level of heterogeneity indicates that the average is not a good statistical descriptor of the activity

---

[3] https://www.newsguardtech.com/.

[4] https://mediabiasfactcheck.com/.

distribution, as the fluctuations are very high: most users post infrequently but a non-negligible fraction posts several hundred messages per day.

4.2   Temporal Patterns

To investigate the impact of the Bluesky opening on its user base, Fig. 2 plots the daily number of active users and following actions. The platform's opening on February 6th (dashed line) resulted in spikes of activity, up to 1 million active users and over 7 million follow actions on the following day. These spikes in behavior represent an almost six-fold increase in active users and an almost 35-fold increase in following actions, relative to the day before the opening. Both trends decreased rapidly in the following days, stabilizing at levels slightly higher than those seen before the opening, likely as the initial excitement around the new platform subsided.

During the entire observation period, 287,539 users blocked a total of 758,681 accounts, yielding a ratio of 2.7 block actions per user. The average number of block actions per user is higher (9.5) because this blocking activity is heterogeneous, with some users blocking many accounts. Figure 2 shows that an increase in users and following activities is associated with a rise in blocking activities, with a 3.6-fold increase from 33,269 to 120,054 instances of blocking. After a few days, the frequency of blocking activities stabilized but remained slightly higher than the levels observed before February 6.

Figure 3 illustrates the volume of shared posts over time. We distinguish user-sharing activities into original posts, replies, and reposts, as described in §3.1. Table 1 previously highlighted that original posts are the most prevalent type of shared activity, a trend that becomes particularly prominent following the platform's public launch.

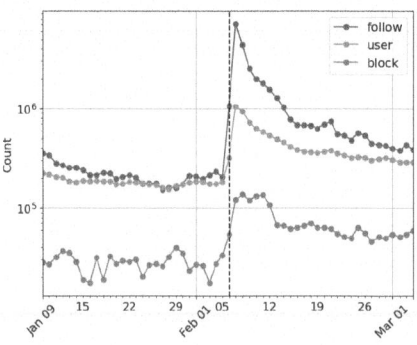

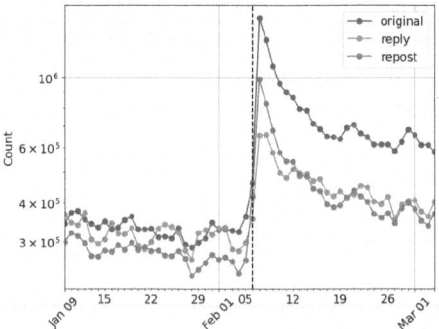

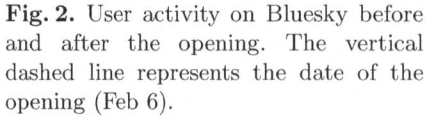

Fig. 2. User activity on Bluesky before and after the opening. The vertical dashed line represents the date of the opening (Feb 6).

Fig. 3. Sharing activity on Bluesky before and after the opening. The vertical dashed line represents the date of the opening.

As expected, we observe a significant increase in the volume of shared content coinciding with the platform's opening on February 6th. This surge is reflected in all types of sharing activities, likely due to the influx of new users joining during this period, as shown in Fig. 2. Specifically, the volume of original posts increased approximately 4.3 times, from 362k on February 5th to 1.5M the following day. Reposts and replies also rose from 262,201 and 297,717 to 990,835 and 656,063, respectively. Despite these substantial initial increases, the volume of each activity type diminished in the subsequent days.

4.3   Languages

To examine the language distribution in user-shared content, we removed irrelevant text (e.g., URLs, emojis, etc.) and applied a language classifier using the `langdetect`[5] NLP library.

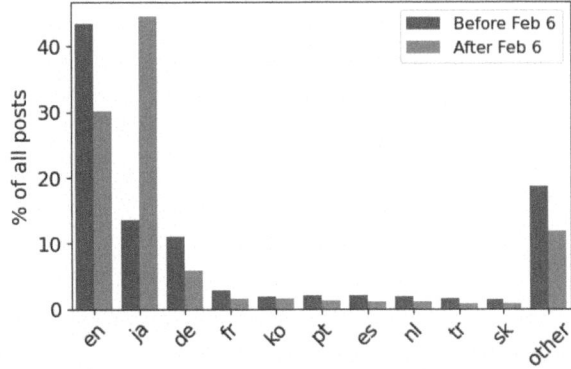

Figure 4 displays the prevalence of the top 10 languages in user posts, highlighting a dominance of English and Japanese, which together comprise

Fig. 4. Top 10 languages in Bluesky. Bars of the same color sum to one.

more than two-thirds of all content. Prevalence of English content decreased from 43% to 30% after the platform's opening, while the share of Japanese content increased from 14% before the opening to 44% afterward.

Figure 5 illustrates the trends of the top five languages used during the observation period. We note a significant increase in Japanese posts just after the opening, making it the most used language. Meanwhile, content in other languages, including English, remains relatively similar to how it was before the opening, experiencing only small fluctuations following the opening.

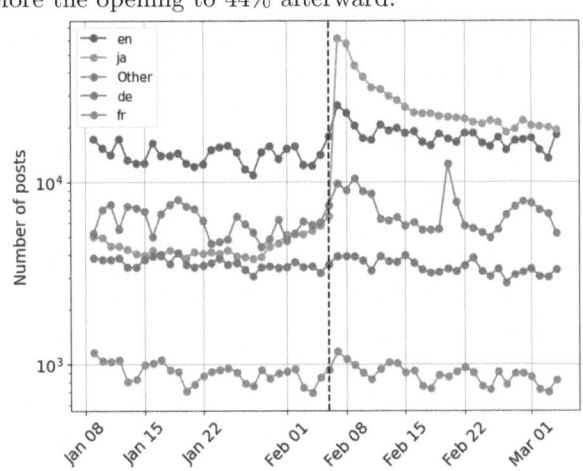

Fig. 5. Trend of 5 top languages on Bluesky during the observation period.

---

[5] https://pypi.org/project/langdetect/.

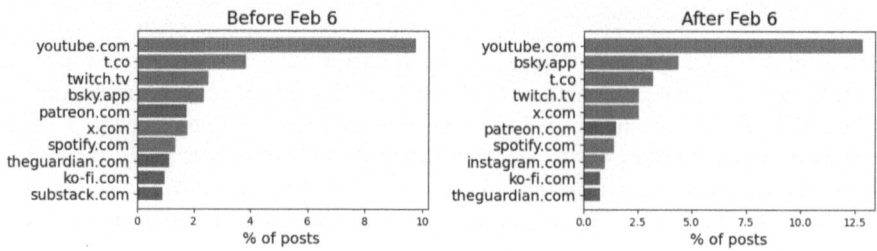

Fig. 6. Websites shared on Bluesky: in blue, we highlight the websites that are neither social media nor shortened links. (Color figure online)

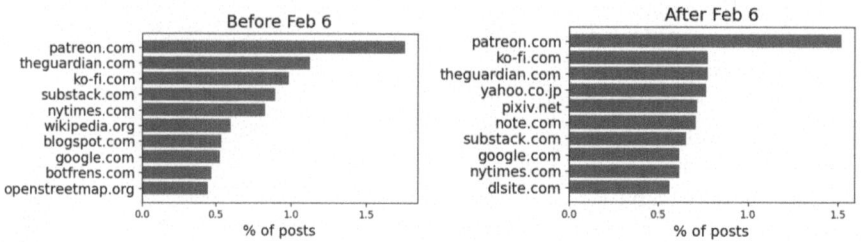

Fig. 7. Websites shared on Bluesky excluding social media and shortened links.

4.4   Information Sources

Figure 6 illustrates the ten domains most shared before and after February 6. We observe that the most common links are either to other social media platforms, Bluesky itself, or link-shortening services like `bit.ly` and `t.co`, Twitter's built-in service. This appears to suggest that a great deal of content on Bluesky is in reference to other, more mainstream platforms.

Figure 7 presents the same information after removing links to social media platforms (spotify.com, twitter.com, x.com, twitch.tv, youtu.be, youtube.com, bsky.app, instagram.com) and link shorteners (t.co, bit.ly, t.me). This procedure allows us to observe that content from news and information outlets like *The Guardian*, *The New York Times*, and *Wikipedia* constituted a larger share of the content before the platform's opening to the public. Consistent with the language analysis (§4.3), we observe a surge in Japan-related activity, with a high percentage of links to Japanese websites following the opening (yahoo.co.jp, pixiv.net, note.com, and dlsite.com). Patreon and Ko-fi, two platforms for crowdfunding and content creation, rank as the most shared web domains beyond social media and news platforms. While such sites have been used to launch fundraising campaigns in times of distress [31], manual inspection reveals that Patreon is primarily used to promote adult content on Bluesky.

4.5 Political Leaning

Figure 8 shows the distribution of the estimated political alignment of active Bluesky users, defined as those engaging in at least five posts with links to rated websites. The political alignment of each user is obtained by averaging the political alignment of the websites they share during the observation period (see §3.2). We used a Mann-Whitney test and did not find a statistically significant difference between the

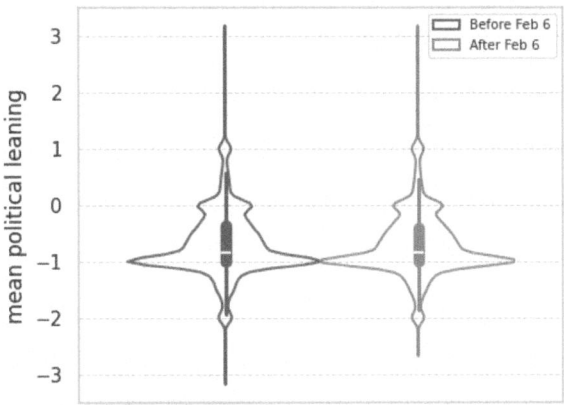

Fig. 8. Political leaning of Bluesky users (with at least 5 rated posts) before and after February 6. Leanings are calculated by averaging the political alignment of websites shared by each user. The violin plots also display the interquartile range (box) and median (horizontal line inside the box).

distributions of the political leanings before and after the opening ($p = 0.05$).

Similarly, we analyzed the distribution of the estimated political alignment of posts (Fig. 9). While no significant difference was found at the user level before and after the opening, some variation is observed at the post level, particularly among right-center (Mann-Whitney: $p = 0.03$), right ($p = 0.046$), and extreme-right posts ($p = 0.008$).

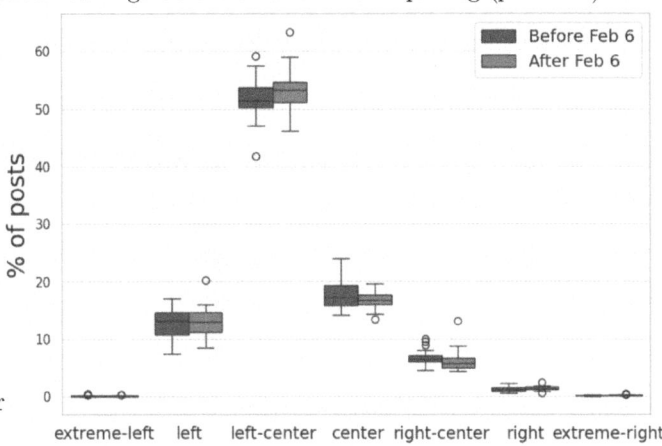

Fig. 9. Political leaning of Bluesky posts before and after Feb. 6.

4.6 Credibility

The number of users sharing any link with a credibility rating is 36,713 before and 47,612 after the opening. A small fraction of these (0.12%) shared low-credibility content.

Figure 10 illustrates the production of low-credibility content before and after the opening. While the volume of this content on Bluesky remained low (around 0.4% of all posts on an average day), we find that it significantly increased after the opening (Mann-Whitney: $p < 0.001$).

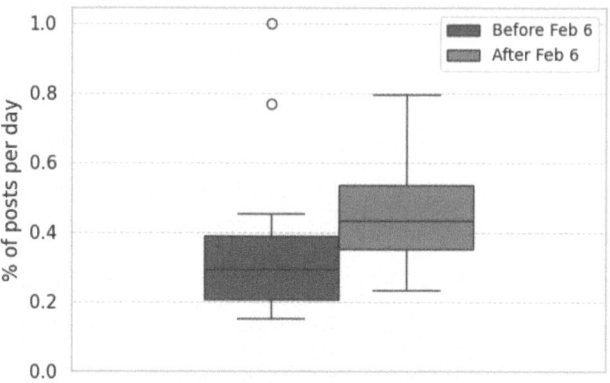

Fig. 10. Distributions of the percentage of posts linking to low-credibility sources, out of all posts linking to rated sites, for each day in the periods before and after the platform opened.

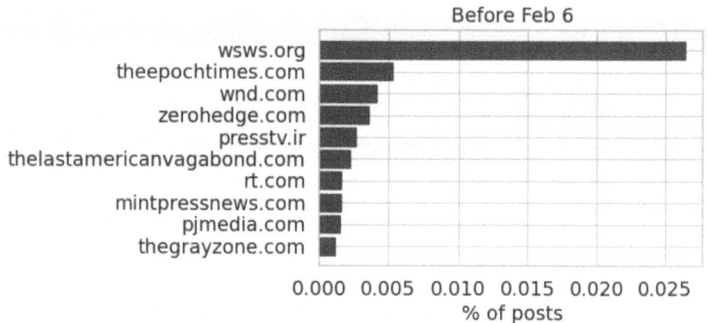

Fig. 11. Most shared low-credibility websites before Feb. 6.

Figure 11 reports on the most shared low-credibility websites before the opening. The list remained similar after the opening, with the notable exception of the top website, `wsws.org`, whose shares decreased to half of the percentage recorded prior to the opening. This is not due to a decrease in the share of this domain, but rather to the increase in the share of all domains.

Similar to what has been reported for Twitter [11,15,23,26], we identified a small subset of users responsible for most of the unreliable content shared in

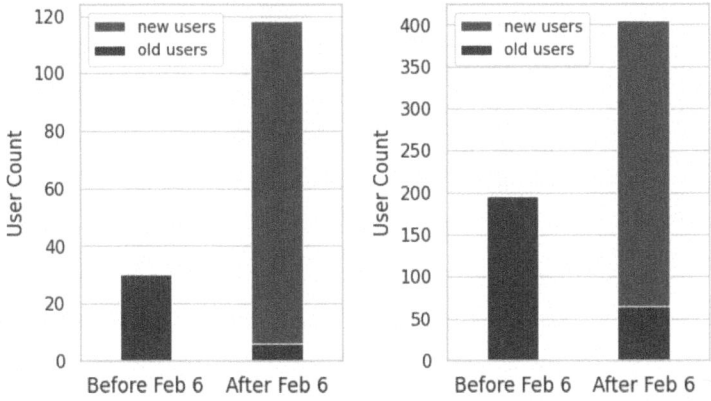

Fig. 12. Number of users sharing links with credibility ratings before and after the opening for different subsets of users. Left: Users who only shared links to low-credibility sources. Right: Users who shared at least one low-credibility link.

Bluesky's brief existence. In particular, we observe that ten accounts (1.8% of all users who spread low-credibility content) are responsible for spreading 62% of links to low-credibility sources. Manual inspection of each account confirms that all were created before the platform opened to the public and remain active at the time of writing. Furthermore, they exhibit suspicious behavior, posting a high volume of content almost exclusively from low-credibility news outlets, possibly in automated fashion.

Figure 12 shows the number of users sharing links to low-credibility sources before and after the opening, distinguishing between new and existing users. We observe that the number of users who only shared low-credibility links quadrupled after the opening. The number of those who shared at least one low-credibility link approximately doubled.

4.7   Toxicity

We used the Perspective API [18] to analyze the toxicity of shared content in both English and Japanese.

As illustrated in Fig. 13 (top), English content tends to have higher toxicity scores. Figure 13 (bottom) shows that while, on average, the toxicity of English content is three times higher than that of Japanese content, the values remain stable over time.

4.8   Follower Network

In Fig. 2, we observed a significant increase in the *following* activities. This increase is reflected in various follower network statistics before and after the opening, as detailed in Table 2. While the density of the follower network slightly decreased after the opening, the size of the strongly connected component more

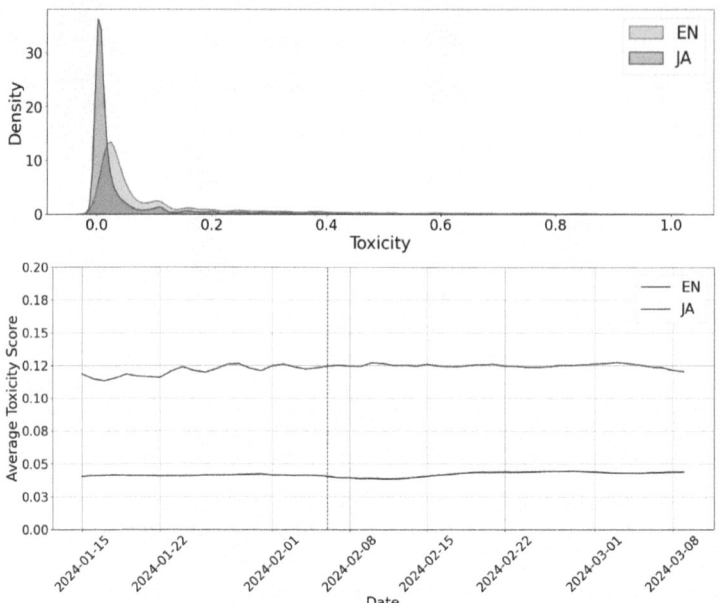

Fig. 13. Toxicity of Bluesky posts in English (EN) and Japanese (JA). Top: Distribution of toxicity for the entire observation period. Bottom: Weekly running average of daily toxicity scores. The dashed red line indicates the platform opening. (Color figure online)

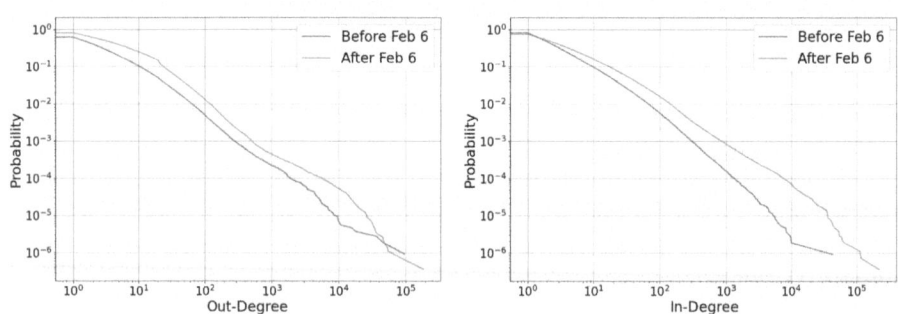

Fig. 14. Complementary cumulative out-degree distributions of node in the follower network.

Fig. 15. Complementary cumulative in-degree distributions of node in the follower network.

than tripled, and the average degree more than doubled, indicating that Bluesky users tend to follow more accounts after the opening. The out-degree distributions in Fig. 14 confirm this trend.

Table 2. Follower network statistics. LSCC stands for the largest strongly connected component.

	Before Feb. 6	After Feb. 6	Difference
Number of nodes	1,088,539	2,751,272	1,662,733
Number of edges	5,230,054	28,838,739	−23,608,685
Density	4.4×10^{-6}	3.8×10^{-6}	$-.6 \times 10^{-6}$
Avg. in-degree	9.6	20.3	10.7
Avg. out-degree	7.4	14.9	7.5
LSCC size	~200k	~650k	~450k

Table 3 presents the ten accounts that gained the most new followers alongside the ten accounts that followed the most other users before and after the platform's opening. For comparison, the average number of followers before and after February 6th is 9.6 and 20.3, respectively, while the median is 3.0 and 4.0, respectively.

The average number of accounts followed by Bluesky users before and after February 6th is 7.4 and 14.9, respectively, while the median is 2.0 and 4.0, respectively. This analysis points to a potentially spammy behavior: some accounts followed a suspiciously large number of users right after joining the Bluesky, as illustrated in Fig. 14.

The crossover in the tail (just before degree 10^5) in the out-degree CCDF curves is due to the uneven distribution of follow activities and to the discrepancy in the number of users before and after the opening. In fact, both before and after the opening, we have only one outlier user who performed a high number of follow actions (respectively, 50,570 and 167,220 actions, which is about 2–3 times the number of actions of other users in Table 3). Furthermore, since the users who performed follow actions before the opening are about 1/3 compared to the users who performed follow actions after (see Table 2), the probability in the CCDF tail for the first is higher than the one for the latter. When we remove these two outlier users, this crossover between the two CCDF curves disappears.

The CCDF curves for the in-degree distributions before and after the opening are illustrated in Fig. 15. Table 3 report the ten top users for both periods.

Among the users that gained the most followers, several are consistent across both periods, with the majority being news outlets, such as *Washington Post*, *New York Times*, and *Bloomberg*. This prevalence of news outlets among the accounts with the most new followers suggests that Bluesky may be evolving into another platform for news dissemination, potentially replacing Twitter. However, during the observed period, three new Japanese accounts emerged in this category, with one reaching the second position, behind the official Bluesky account.

Table 3. Annotated list of the ten users with the most new followers and the ten users who performed most follow activities before and after the platform's opening. Only the real names of prominent accounts are included for privacy reasons.

Accounts with most new followers				Accounts who followed most accounts			
Before Feb. 6		After Feb. 6		Before Feb. 6		After Feb. 6	
Account	Num.	Account	Num.	Account	Num.	Account	Num.
Bluesky ↔	43,801	Bluesky ↔	120,709	user	50,570	user 🗂 ✹	167,220
user ↔	10,256	user 🗂 ⊡	88,438	user	24,588	user 🗂 ⊡	85,075
user	9,886	user ↔	66,535	user ♣ˣ	23,482	user 🗂 ⊡	55,073
Wash. Post ↔	8,508	NY Times ↔	62,087	user	21,907	user 🗂 ⊡	45,848
NY Times ↔	8,393	Wash. Post ↔	60,904	user	13,414	user 🗂 ♣-	44,588
user	6,809	user ↔	55,081	user	12,310	user ♣ˣ	44,460
Bluesky CEO	6,314	user 🗂 ⊡	54,944	user	10,689	user 🗂 ♣-	43,786
user ↔ ⊞	5,744	Bloomberg ↔	54,800	user	9,950	user 🗂 ✹	40,926
user ↔ ⊞	5,621	user ↔ ⊞	52,455	user	8,211	user 🗂 ⊡	37,209
Bloomberg ↔	5,332	user 🗂 ⊡	49,697	user	7,588	user 🗂 ⊡	36,189

Account type: ✹spam; ♣deleted; ♣ˣsuspended; ⊡ Japanese; ⊞Journalist.
Time key: 🗂created after Feb. 6; ↔present before and after Feb. 6.

Before the platform's opening, the users who followed the most new accounts were primarily English-language accounts and appeared to engage in normal activity, except for one user suspended by Bluesky. However, after the opening, the composition of these users changed significantly: half were Japanese, two accounts were deleted, one was suspended, and two were classified as spam by Bluesky. Unlike the overall network behavior, these users were more engaged in reposting activities than in creating original posts.

5   Discussion

We provided the first large-scale analysis of how opening the Bluesky platform to the public affected user activity and network structure. We observe a broad distribution of activity similar to that of more established platforms; however, there is a higher volume of original content compared to reshared content. This, coupled with the very low toxicity we observed, contrasts with other (centralized) social media platforms. After opening to the public, Bluesky experienced a large surge in new users, especially posting English and Japanese content, and activities. The significant increase in Japanese users deserves further investigation; it might signify a conversational system that is more appealing to this population.

We observed no significant changes in the toxicity level and in the political leaning of users, but some variation is observed at the post level. Further, we discovered some users exhibiting suspicious behavior, such as connecting with many other users and sharing content from low-credibility news outlets. We also identified a small subset of users responsible for the majority of unreliable content shared, echoing similar findings on platforms like Twitter/X. Additionally, after the platform's opening, a subset of users that followed the most new accounts during this time were flagged as spam or were even deleted or suspended. These findings suggest some attempts to misuse Bluesky. However, the fact that some of the suspicious and misbehaving actors have already been banned or tagged as spam indicates that content moderation is taking place on Bluesky.

This work has some limitations. First, it is based on a relatively short period (56 d), during which a significant change occurred—allowing anyone to join. Second, it was not possible to trace 8% of the reposts back to their original posts as they predated our data collection period. Third, the websites for which we have credibility and political bias scores make up a small fraction of all links shared. Finally, our analysis of toxicity is limited only to English and Japanese content.

Future work could incorporate a more extended multi-language analysis through tools like Perspective API, which can evaluate toxicity in various languages [22], as well as longitudinal analyses of user behavior and network structure.

Acknowledgments. This work was partially supported by the Swiss National Science Foundation (grant number CRSII5_209250) and the Italian Ministry of Education (PRIN PNRR grant CODE prot. P2022AKRZ9 and PRIN grant DEMON prot. 2022BAXSPY).

References

1. Alshaabi, T., et al.: The growing amplification of social media: measuring temporal and social contagion dynamics for over 150 languages on Twitter for 2009–2020. EPJ Data Sci. **10**(1), 15 (2021)
2. AT Protocol: event Stream. https://atproto.com/specs/event-stream
3. atprotodart.com: Bluesky | Dart package. Software library. https://pub.dev/packages/bluesky. Accessed July 2024
4. Bluesky: Firehose. https://www.docs.bsky.app/docs/advanced-guides/firehose. Accessed29 July 2024
5. Bluesky: FAQ: What is Bluesky? (2024). https://bsky.social/about/faq. Accessed 17 April 2024
6. Bluesky: One million new users since we opened Bluesky yesterday! (2024). https://bsky.app/profile/bsky.app/post/3kkv3clm3su2h, post on Bluesky. Accessed 16 April 2024
7. Bono, C.A., La Cava, L., Luceri, L., Pierri, F.: An exploration of decentralized moderation on mastodon. In: Proceedings of the 16th ACM Web Science Conference, pp. 53–58 (2024)
8. Buchegger, S., Schiöberg, D., Vu, L.H., Datta, A.: PeerSoN: P2P social networking: early experiences and insights. In: Workshop on Social Network Systems (2009). https://api.semanticscholar.org/CorpusID:15841016
9. Burghardt, K.: Data collection with Bluesky firehose endpoint. https://github.com/KeithBurghardt/bluesky_firehose/tree/main. Accessed July 2024
10. Cava, L.L., Greco, S., Tagarelli, A.: Understanding the growth of the Fediverse through the lens of mastodon. Appl. Netw. Sci. **6** (2021). https://api.semanticscholar.org/CorpusID:235669708
11. DeVerna, M.R., Aiyappa, R., Pacheco, D., Bryden, J., Menczer, F.: Identifying and characterizing superspreaders of low-credibility content on Twitter. PLOS ONE **19**(5), e0302201 (2024). https://doi.org/10.1371/journal.pone.0302201

12. Edwards, C.: Social media and the distributed self. Eng. Technol. **18**(4), 40–46 (2023)
13. Failla, A., Rossetti, G.: "I'm in the Bluesky Tonight": insights from a year worth of social data. arXiv preprint arXiv:2404.18984 (2024)
14. Graffi, K., Gross, C., Stingl, D., Hartung, D., Kovacevic, A., Steinmetz, R.: Life-Social.KOM: a secure and P2P-based solution for online social networks. In: 2011 IEEE Consumer Communications and Networking Conference (CCNC), pp. 554–558 (2011). https://api.semanticscholar.org/CorpusID:10789841
15. Grinberg, N., Joseph, K., Friedland, L., Swire-Thompson, B., Lazer, D.: Fake news on Twitter during the 2016 U.S. presidential election. Science **363**(6425), 374–378 (2019). https://doi.org/10.1126/science.aau2706
16. Kato, S.: Bluesky firehose endpoint. https://atprotodart.com/docs/lexicons/com/atproto/sync/subscriberepos/. Accessed July 2024
17. Kleppmann, M., et al.: Bluesky and the AT protocol: usable decentralized social media. arXiv preprint arXiv:2402.03239 (2024). https://arxiv.org/pdf/2402.03239.pdf
18. Lees, A., et al.: A new generation of Perspective API: efficient multilingual character-level transformers. In: The 28th ACM SIGKDD Conference on Knowledge Discovery and Data Mining (KDD 2022), pp. 3197–3207 (2022)
19. Luceri, L., Cardoso, F., Giordano, S.: Down the bot hole: actionable insights from a one-year analysis of bot activity on Twitter. First Monday (2021)
20. Masnick, M.: Protocols, not platforms: a technological approach to free speech. Technical Report. Knight First Amendment Institute at Columbia University (2019). https://knightcolumbia.org/content/protocols-not-platforms-atechnological-approach-to-free-speech. Accessed 16 April 2024
21. Nogara, G., Pierri, F., Cresci, S., Luceri, L., Giordano, S.: Misinformation and Polarization around COVID-19 vaccines in France, Germany, and Italy. In: Proceedings 16th ACM Web Science Conference 2024 (2024)
22. Nogara, G., Pierri, F., Cresci, S., Luceri, L., Törnberg, P., Giordano, S.: Toxic Bias: perspective API misreads German as more toxic. arXiv preprint arXiv:2312.12651 (2023)
23. Nogara, G., Vishnuprasad, P.S., Cardoso, F., Ayoub, O., Giordano, S., Luceri, L.: The disinformation dozen: an exploratory analysis of Covid-19 disinformation proliferation on Twitter. In: Proceedings 14th ACM Web Science Conference 2022 (2022). https://api.semanticscholar.org/CorpusID:249993446
24. Pierri, F.: The diffusion of mainstream and disinformation news on Twitter: the case of Italy and France. In: Companion Proceedings of The Web Conference (WWW), pp. 617–622 (2020)
25. Pierri, F., DeVerna, M.R., Yang, K.C., Axelrod, D., Bryden, J., Menczer, F.: One year of COVID-19 vaccine misinformation on Twitter: longitudinal study. J. Med. Internet Res. **25** (2023)
26. Pierri, F., et al.: VaccinItaly: monitoring Italian conversations around vaccines on Twitter and Facebook. arXiv preprint arXiv:2101.03757 (2021)
27. Quelle, D., Bovet, A.: Bluesky: network topology, polarisation, and algorithmic curation. arXiv preprint arXiv:2405.17571 (2024)
28. Raman, A., Joglekar, S., Cristofaro, E.D., Sastry, N.R., Tyson, G.: Challenges in the decentralised web: the mastodon case. In: Proceedings of the Internet Measurement Conference (2019). https://api.semanticscholar.org/CorpusID:202565923
29. Silberling, A.: Bluesky is now open for anyone to join (2024). https://techcrunch.com/2024/02/06/bluesky-is-now-open-for-anyone-to-join/. Accessed 16 April 2024

30. Yang, K.C., et al.: The COVID-19 Infodemic: Twitter versus Facebook. Big Data Soc. **8**(1) (2021)
31. Ye, J., Jindal, N., Pierri, F., Luceri, L.: Online networks of support in distressed environments: solidarity and mobilization during the Russian invasion of Ukraine. In: Companion Proceedings of the International AAAI Conference on Web and Social Media (ICWSM) (2023)
32. Zignani, M., Gaito, S., Rossi, G.P.: Follow the "Mastodon": structure and evolution of a decentralized online social network. In: International Conference on Web and Social Media (2018). https://api.semanticscholar.org/CorpusID:49418536

Weaponizing the Wall: The Role of Sponsored News in Spreading Propaganda on Facebook

Daman Deep Singh[1]([✉]), Gaurav Chauhan[1], Minh-Kha Nguyen[2], Oana Goga[3], and Abhijnan Chakraborty[4]

[1] Indian Institute of Technology Delhi, New Delhi, India
damandeepddsb@gmail.com
[2] Université Grenoble Alpes, Grenoble, France
[3] CNRS, Inria, Institut Polytechnique de Paris, Paris, France
[4] Indian Institute of Technology Kharagpur, Kharagpur, India

Abstract. A large fraction of people today consume most of their news online, and social media platforms like Facebook play a significant role in directing traffic to news articles. While news organizations often use Facebook advertising to drive traffic to their websites, this practice can inadvertently lead to biases in what articles users get exposed to, or worse, could be used as a mechanism for manipulation. In this work, we examine the impact of sponsored news on Facebook on the dissemination of propaganda. Propaganda is a method of persuasion that is frequently employed to advance some sort of goal, such as a personal, political, or business objective. By analyzing 17 Million+ Facebook posts and 6 Million sponsored advertisements gathered over 182 days, we observe that advertisers of all kinds, including politicians, media houses, and commercial corporations, publish thousands of ads/boosted posts every day on Facebook. However, Facebook does not include advertisements from news organizations in their public ad archive, even when these ads address political and social issues. This exemption places news organizations in a unique position where they can publish paid political opinions without any transparency requirements. The risk is that news organizations or other third-party interest groups can selectively promote news articles that support their agenda, giving these ads an appearance of legitimacy because they link to established news sites. In this paper, we explore how such sponsored news on Facebook can be a powerful tool for spreading propaganda. Through this work, we hope to raise awareness among users about the potential biases in sponsored news and the need to critically evaluate the information they see on Facebook.

Keywords: Facebook Ads · Propaganda Detection · Sponsored News

D. D. Singh and G. Chauhan—Equal contribution.

L. M. Aiello et al. (Eds.): ASONAM 2024, LNCS 15211, pp. 438–454, 2025.
https://doi.org/10.1007/978-3-031-78541-2_27

1 Introduction

In our increasingly interconnected world, the way we consume news has undergone a profound transformation. A significant portion of today's population turns to the internet as their primary source of information, and while online news outlets are experiencing a surge in web traffic, a substantial share of article views stems from social media referrals[1]. Platforms like Facebook and X (Twitter) have become the primary drivers for the dissemination of news, reshaping the way information flows through our digital society.

Traditionally, the propagation of news articles on social media was largely organic, driven by user engagement through likes, shares, and retweets. However, in recent times, news organizations have evolved their strategies to include the paid promotion of specific articles, harnessing the power of advertising to reach wider audiences [23]. The concept of promoting news articles on platforms like Facebook isn't inherently problematic, but it carries the potential for unintended consequences. For instance, it can inadvertently introduce biases into the information users encounter [7], or worse, become a means of manipulation [37]. The careful selection of news articles for promotion, combined with the micro-targeting capabilities offered by platforms like Facebook [46], can be exploited to advance various agendas, shaping public opinion and eliciting emotional responses from the audience. This approach often falls under the realm of propaganda, a method of persuasion frequently employed to further personal, political, or business agenda.

While advertisers from diverse backgrounds publish countless ads and boosted posts on Facebook daily, there is a key distinction in the content of the ads posted by news organizations. Most of news ads don't merely encourage users to subscribe or visit their homepage but explicitly promote individual news articles. The challenge lies in the fact that the titles (and snippets) of these ads carry message/information to the users encountering them. Unlike when users visit a news website, where they can exercise control over the articles they read, they have no such control over the sponsored news that populates their Facebook timeline. This lack of control leaves them vulnerable to the influence of the messaging, whether consciously or subconsciously.

Furthermore, a complex issue arises from Facebook's decision to exclude ads from news organizations from their 'social issues, elections, or politics' ad archive[2], even when these ads touch on political, electoral, or social issues. This omission negates the transparency requirements[3] imposed on other advertisers, placing news organizations in a unique position to publish paid political opinions without accountability. Consequently, there is a concerning potential for news organizations or third-party interest groups to meticulously select news articles

---

[1] https://www.zdnet.com/article/social-media-is-key-driver-for-news-consumption.

[2] https://www.fb.com/business/help/issuesandpolitics
https://www.fb.com/business/help/313752069181919?id=288762101909005.

[3] Including mandatory disclosure of amount spent to publish the ad, name of the entity/person responsible, etc. [28].

that align with their agenda, all the while appearing legitimate because these ads direct users to well-known news outlets. When combined with the micro-targeting capabilities of sponsored post platforms [37], this becomes a potent tool for influencing users and spreading propaganda.

In this work, by gathering extensive longitudinal data from Facebook, we try to analyze the role of sponsored posts in spreading propaganda. Leveraging a reliable, state-of-the-art propaganda detection model, we demonstrate that sponsored news posts indeed contain more propaganda compared to non-sponsored news posts on Facebook, worryingly bringing in higher engagement from the audience. We also check how this trend varies across the political spectrum. Additionally, we present a comprehensive analysis of propaganda spread by Facebook news pages around the time of the US Capitol attack (in January 2021), illustrating an escalation of propaganda during this period across both left-leaning and right-leaning media outlets.

With 2024 being touted as the biggest election year in history with more than half the world's population participating in polls[4], and the recent concerns regarding the potential misuse of large language models (LLMs) in spreading propaganda at scale [9,18], we believe that our work will help raise awareness about the potential issues with sponsored news and the need to critically evaluate all the information users get on social media sites like Facebook.

2   Related Works

Propaganda on Social Media. The pervasiveness of social media platforms has created a fertile ground for proliferation of propaganda. Subsequently, several studies have aimed at analyzing and evaluating the impact of these platforms in influencing and reinforcing mass opinion across a variety of socially significant topics, including elections [15,38], controversies [20], political-affiliations [3], and even ethnic violence [31]. Prior works have focused on propaganda dissemination on different social media platforms, such as Reddit [3], Twitter [19], and Facebook [30,33,40].

Detecting Propaganda. Propaganda is a subtle yet impactful way of influencing opinions that is hard to detect. As a consequence, considerable effort have been put into precisely defining propaganda and elaborating on nuanced propaganda techniques [12,29]. The research community has proposed useful benchmark datasets [2,11,22,24,44], as well as developed effective propaganda detection methods, including BERT-based models [1,3,12], Large Language Models [42], among others. Additionally, various methods have been devised to tackle specific challenges in propagandistic content, including addressing code-switched social media text [39], employing multimodal approaches [32], and adapting strategies for multi-lingual propaganda [34,41]. In this work, we utilize this line

---

[4] https://www.economist.com/interactive/the-world-ahead/2023/11/13/2024-is-the-biggest-election-year-in-history.

Table 1. Eighteen fine-grained propaganda techniques proposed by [12].

Propaganda Type	Definition
Doubt	Questioning the credibility of someone or something
Appeal-to-Fear/Prejudice	Attempt to increase opposition to a position by spreading fear/terror among the populace
Exaggeration/ Minimization	Attempting to make something seem either less or more significant than it truly is by employing exaggerations to diminish or amplify its importance
Causal Oversimplification	Escaping the complexities of a situation by scapegoating a specific person or group, offering a superficial explanation that absolves others of responsibility
Flag Waving	Emphasizing a profound sense of duty to intense national or group sentiment, such as race, gender, or political preference
Loaded Language	Influencing a group by using emotionally charged (positive or negative) words and phrases
Name Calling or Labeling	Assigning derogatory labels or names to individuals or groups as a means of discrediting them or their ideas
Slogans	Succinct and impactful phrases with an emotional appeal, utilizing labels and stereotypes
Thought-terminating Cliché	Usage of phrases or words to stifle meaningful conversation and critical thought on a subject, offering simplistic answers or diverting attention from crucial ideas
Repetition	Repeating a message with the expectation that the audience will eventually accept it
Bandwagon	Showcasing a majority's support for a certain belief to persuade others
Black and White Fallacy	Presenting two apparent solutions as the sole options, despite the existence of the only ones available when, in fact, there are more options
Whataboutism	Instead of presenting solid evidence to challenge an opponent's argument, this approach seeks to weaken their perspective by alleging hypocrisy
Obfuscation, Intentional Vagueness & Confusion	Employing vague generalizations to prompt the audience to draw their own conclusions
Appeal to Authority	Depending on expert opinion without concrete evidence about the incident or event
Red Herring	Introducing an irrelevant topic into the discussion shift individuals' focus
Reductio ad Hitlerum	Proposing a conclusion solely based on the origin of something or someone, rather than considering its current meaning or context
Straw Man	Substituting a comparable proposition for an opponent's, typically an extreme version, and refuting it instead of the original statement

of work to select the best performing algorithm for detecting propaganda in news posts.

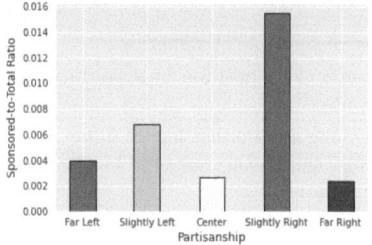

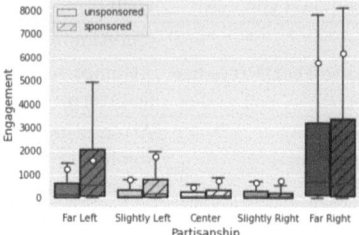

Fig. 1. Ratio of sponsored posts to all posts in each partisanship category.

Fig. 2. Box plot of engagement. Black lines represent medians, and white dots represent means.

Interdisciplinary Efforts for Combating Propaganda. Concerns over online propaganda's reach have sparked broad research across disciplines. Psychologists and linguists have studied persuasion tactics such as emotional triggers and cognitive biases [4,6,45,48,49]. Sociologists have delved into the cultural and societal conditions that facilitate the influence of propaganda [8,26,48]. Legal scholars have advocated for regulating the use of technology towards responsible online spaces [10,21]. Besides these efforts, various studies have focused on counteracting propaganda's impact through enhancing media literacy [17]. Our work complements these efforts by pointing out a novel source of propaganda and offers policy suggestions to tackle their spread.

3   Dataset Gathered

Following the data collection framework by [16], we gathered extensive longitudinal data comprising news posts and sponsored ads posted by media organizations on Facebook over a period of 182 days, starting from 1st August 2020 till 30th January 2021. First, we utilized the Meta Ad Library[5] to collect active ads, i.e., the ads that were posted/run on Facebook during the data collection period, resulting in a total of 6,741,422 advertisements in the given time frame. Concurrently, we employed Meta's CrowdTangle API [43] to gather news posts from the Facebook pages of various news media organizations.

To identify these organizations, we took a two-pronged approach. First, we relied on an independent media watchdog 'Media Bias/Fact Check' which surveys news outlets and provides qualitative information about them, such as their political leaning and news quality [13]. Following this, we identified 2,863 Facebook pages corresponding to the media organizations covered by Media

---

[5] https://transparency.fb.com/en-gb/researchtools/ad-library-tools/.

Bias/Fact Check. Subsequently, recognizing the rise of 'social media only' news channels that might elude traditional media watchdog groups [5, 36], we considered all Facebook pages which claimed to be 'News Media' in their 'About' section and posted at least one advertisement. In total, we compiled 10, 492 Facebook news channels, including details like their page_id, name, city, country, website, followers_count, creation time, misinformation status, and partisanship indications.

Table 2. Normalized engagement across sponsored and unsponsored posts for different partisanship groups.

Partisanship	Sponsored (S)	Un-Sponsored (US)	Ratio (S/US)
Far Left	0.001925	0.00068	2.848
Slightly Left	0.002767	0.00062	4.498
Center	0.001090	0.00092	1.181
Slightly Right	0.001557	0.00088	1.768
Right	0.011814	0.00668	1.767

Overall, we collected 17,815,182 Facebook posts (on average, 97,885 posts per day) and 6,741,422 ads (37,040 ads per day). Even after we filter out non-English posts[6] and Facebook pages with less than 10 posts, we end up with a collection of 12,506,833 posts which provides us with a rich and comprehensive dataset for the analysis.

Notably, the time period under examination holds particular significance for the study of social media news, as it spans the 2020 US Presidential Election, encompassing both pre-election and post-election posts shared by news outlets on Facebook. Particularly noteworthy inclusion is the time of the US Capitol attack (January 6, 2021) following the defeat of the former President Donald Trump. The dataset serves as a valuable resource for examining the combined impact of social media news, sponsorship, and the potential dissemination of propaganda within this context.

4   Sponsored News Posts on Facebook

The dataset includes different types of ads (or sponsored posts) posted by various entities, including businesses, celebrities, political parties, and news media outlets. We first identified the sponsored news posts within this dataset by comparing the destination URL of each advertisement with the destination URL

---

[6] We keep only English posts as our propaganda classifier (explained in later section) was trained on English data alone.

of the collected news posts. A match between the two URLs indicates a sponsored post. While this methodology may not capture all sponsored posts, it ensures the accuracy of identified sponsored content. Overall, we found 72,113 sponsored posts (out of the total 12,506,833 posts) using this approach. For a comprehensive analysis, we categorize the identified sponsored posts into various partisanship groups. To this end, we employ the partisanship classifications from Media Bias/Fact Check [13], focusing on 2,863 Facebook pages they cover, categorized as far *left, slightly left, center, slightly right*, and *far right*. Next, we examine how various media outlets leverage Facebook ads to expand their reach.

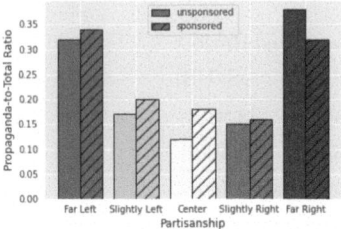

Fig. 3. Ratio of propaganda posts to total posts by sponsorship and partisanship.

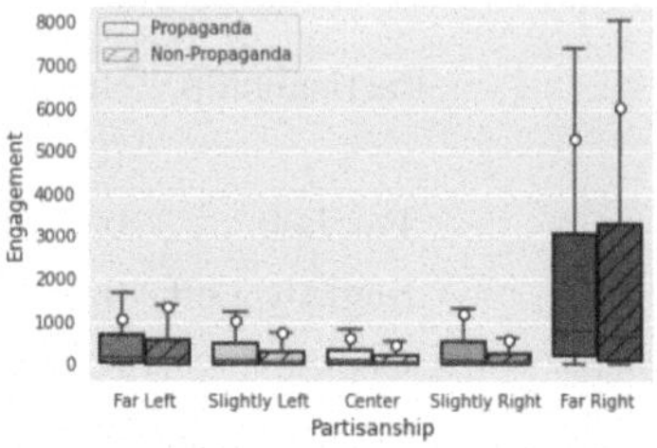

Fig. 4. Engagement of propagandistic and non-propagandistic across various partisanships.

Sponsored-to-Total-Post Ratio Across Political Spectrum. Figure 1 reveals the relative prevalence of sponsored content across partisanship categories, expressed as the ratio of sponsored posts to total posts. Our findings indicate a notable trend: news channels affiliated with either the far left, slightly left, or slightly right political leanings tend to exhibit higher sponsored-to-total post ratios compared to the center. This suggests a potential association between political orientation and susceptibility to sponsorship, with some leaning groups attracting more commercial partnerships than others. However, it is important to note that this trend does not appear to extend to the far-right channels, which deviate from the pattern by displaying a lower sponsorship ratio. We also observed that newly established Facebook pages with a political inclination tend to exhibit a higher ratio of sponsored to total posts, indicating a potential strategy to quickly and effectively disseminate information and increase reach.

Analyzing Engagement with Sponsored Posts. We analyze the engagement of sponsored and unsponsored posts, by measuring engagement as the sum of the corresponding likes, reactions, and comments. As depicted by Fig. 2, across all partisanship categories (with a negligible exception favoring unsponsored posts in *slightly-right* leaning pages), sponsored posts demonstrate higher reach compared to unsponsored posts of the same partisanship. The engagement of sponsored posts in left and right-leaning categories surpasses that of center-oriented pages.

Table 2 presents the engagement levels for each partisanship category, normalized by the follower count of the corresponding Facebook page, for both sponsored and non-sponsored posts. While there is only a slight increase in engagement for the sponsored posts of center-leaning pages, there is a notable increase in the corresponding engagement for right- and left-leaning pages, which suggests a deliberate approach in the design of sponsored posts to encourage higher user engagement. Hence, based on the above analysis, it is clear that sponsored posts on Facebook is an effective strategy for gaining higher user engagement as compared to unsponsored posts.

Table 3. Comparison of classifiers on the QCRI test for sentence-level classification. Best results highlighted in bold.

Model	Recall	Precision	F1-score
Random	0.501	0.245	0.329
BERT	0.556	0.557	0.556
ALBERT	0.560	0.527	0.543
XLNet	**0.601**	0.518	0.556
T5	0.4535	**0.607**	0.519
MGN-ReLU	0.593	0.554	**0.577**

5   Propaganda Detection

Next, we focus on identifying propaganda posts on Facebook. Propaganda detection has typically been posed as a multi-class classification problem [3,12,42], where the classification labels differ based on the dataset under consideration. For our analysis, we utilize the multi-granularity propaganda detection model (MGN-ReLU) proposed by [12]. This model has been trained on a large, high-quality benchmark dataset that features both sentence-level and span-level annotations, i.e., each sentence is manually annotated at the sentence level as either *propaganda* or *non-propaganda*, and each text span within the sentence belonging to any of the 18 propaganda classes (described in Table 1) is annotated with the corresponding propaganda technique. We refer to this dataset as the QCRI dataset[7]. As shown by [12], due to the extra supervision provided by the span-level annotations, fine-grained classification enhances the model's performance at the sentence level. Consequently, the model not only provides binary classification on input sentences but also offers insights into the involved propaganda techniques. This also improves the model's explainability, thereby enhancing its overall reliability. The reliability of this model is corroborated by subsequent studies such as [3], which show that the model is robust to topical biases in the

---

[7] Since it was contributed by researchers from the Qatar Computing Research Institute (QCRI) [12].

annotated dataset and learns linguistic patterns in the dataset rather than being influenced by topical confounds, which is desirable.

For establishing the suitability of the MGN-ReLU model for our analysis, we perform a comparison against several prominent classification models. We fine-tune a suite of well-known transformer-based classifiers such as BERT [14], ALBERT [25], XLNet [47], and T5 [35] on the sentence-level QCRI dataset using the hyperparameters described in [12]. As an additional baseline, we also report the performance of a Random classifier that classifies each sentence uniformly at random. Note that we do not fine-tune the MGN-ReLU model; rather we utilize the MGN-ReLU model provided by [12], which is already trained on the QCRI dataset. Table 3 shows the performance of the models for the binary, sentence-level classification on the QCRI dataset. As expected, all models perform significantly better compared to Random classification. Notably, the MGN-ReLU model emerges as the best classifier, achieving the best F1-score, and reasonably good Precision and Recall scores. Recently, large language models (LLMs) have also been employed for propaganda detection. However, they were found to perform worse than BERT-based models on the QCRI dataset [42]. Hence, we proceed with the MGN-ReLU classifier in this work.

To further assess the reliability of the MGN-ReLU classifier on our dataset, we randomly selected 100 posts and manually annotated them. We then evaluated the classifier's performance on this annotated subset and found that the classifier exhibited strong performance, achieving precision of 0.76, recall of 0.59, and F1-score of 0.67. Notably, this performance surpassed that observed on the QCRI test set, providing compelling evidence in support of confidently using this classifier for our analysis.

Table 4. Fractions of Top-5 propaganda types spread across sponsorship and partisanship. (S) and (US) stand for sponsored posts and unsponsored posts respectively.

Propaganda Type	Overall (S)	Overall (US)	Far Left (S)	Far Left (US)	Slight Left (S)	Slight Left (US)	Center (S)	Center (US)	Slight Right (S)	Slight Right (US)	Far Right (S)	Far Right (US)
Loaded Language	0.14	0.12	0.24	0.24	0.15	0.13	0.13	0.08	0.12	0.11	0.25	0.18
Name Calling/ Labelling	0.07	0.06	0.14	0.14	0.06	0.06	0.06	0.04	0.06	0.05	0.16	0.1
Flag Waving	0.03	0.02	0.06	0.04	0.03	0.02	0.03	0.01	0.02	0.02	0.08	0.04
Exaggeration/Minimisation	0.02	0.02	0.05	0.03	0.03	0.02	0.02	0.01	0.02	0.02	0.04	0.02
Doubt	0.02	0.02	0.06	0.05	0.03	0.02	0.02	0.01	0.02	0.02	0.07	0.18

6   Propaganda in Facebook Posts

In this section, we conduct a thorough analysis of the posts in our dataset, considering the classification labels (*propaganda* or *non-propaganda*) assigned through the process described in the previous section. Our investigation includes a comprehensive study of *sponsorship*, *partisanship* (Far Left, Slightly Left, Center, Slightly Right, and Far Right), and *engagement* (reactions, comments, etc.) across both propaganda and non-propaganda labels. Note that the fine-grained classification of posts in our dataset follows the same approach as that used for the news articles in the QCRI dataset [12]. However, an entire Facebook post, typically comprising a few sentences, is classified as a *propaganda* post only if it contains at least one sentence predicted to be propagandistic by the classifier.

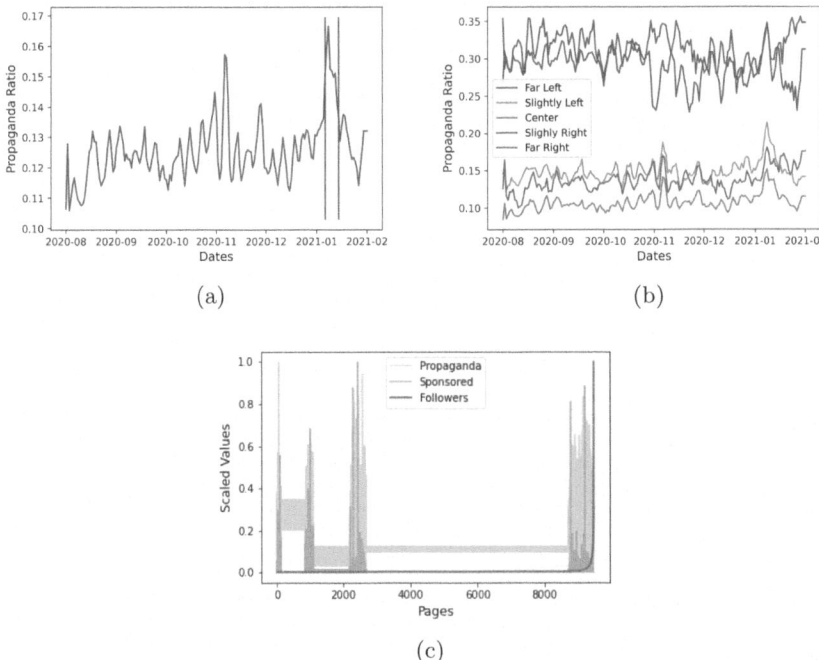

Fig. 5. (a) Ratio of propaganda posts to total posts over time by all Facebook channels. (b) Plot of ratio of propaganda posts to total posts over time by US-based Facebook channels alone. (c) Line plot showing the distribution of the fraction of propaganda and sponsored posts for Facebook pages sorted by their followers' count.

RQ1. Do sponsored posts spread more propaganda? We find that approximately 19% of sponsored posts are classified as propaganda, compared to around 16% of unsponsored posts. This suggests that sponsored posts are more likely to contain propaganda. To statistically validate this observation, while considering the substantial difference in the total number of unsponsored (around 12.4 million) and sponsored (near 72K) posts, we performed a Chi-square test. The test reveals a significantly higher proportion of propagandistic posts among sponsored posts compared to unsponsored posts, with a χ^2 value of 264.29 and $p \ll 0.05$. Next, when we examine the distribution of propaganda posts across various partisan affiliations in both sponsored and unsponsored posts, we observe in Fig. 3 that the overall partisanship distribution remains quite similar. However, there's a noticeable bias towards Far Left and Far Right leanings. Notably, almost 70% of the propaganda posts originate from news channels with Far Left and Far Right political leanings, regardless of the type of sponsorship. This correlation strongly suggests that the political bias of these channels is positively associated with their use of propaganda.

We also analyzed various propaganda techniques present in the posts. Table 4 displays the top 5 propaganda techniques found in our dataset. This table provides insights into the distribution of posts associated with each technique across different partisan affiliations and sponsorship categories. A noteworthy observation is the general trend of sponsored posts exhibiting a relatively higher prevalence of propaganda compared to unsponsored posts across all partisanships. The most significant disparity in terms of sponsorship is particularly evident in the case of Far Right-leaning propaganda posts, with sponsored posts being up to twice as propagandistic in terms of numbers.

RQ2. Do propagandistic posts generate more engagement? Recall that, in the previous sections, we've already established that sponsored posts lead to more engagement and that sponsored content is more likely to contain propaganda. Here, we examine the engagement levels of propaganda and non-propaganda posts (across various partisanship categories). To gauge engagement, we utilize the total sum of reactions and comments on a post, normalized by the number of followers of the corresponding channel. Figure 4 illustrates the engagement of propaganda posts categorized by their partisanship. Regardless of the partisanship, propaganda posts consistently exhibit higher engagement compared to non-propaganda posts (with a minor exception in the 'Far Right' partisanship group).

RQ3. Do influential facebook pages tend to post more propaganda? Now we study whether highly influential Facebook pages, indicated by a substantial follower count[8] tend to share more propagandistic content. Figure 5c shows the distribution of the fraction of propaganda as well as sponsored posts posted by Facebook pages within our dataset, arranged in a non-decreasing order of their followers' count[9]. Although a clear linear or monotonic correlation is not evident between a page's follower count[10] and its frequency of posting sponsored or propagandistic content, it can be seen that propaganda is more prevalent on pages with either very low or very high follower counts (as depicted by the co-occurring spikes of propagandistic and sponsored contents for such pages). Specifically, the bottom 25% and top 25% pages by follower count post significantly higher ($p < 0.05$ in Mann-Whitney U Test [27]) propaganda content compared to the overall distribution. Delving deeper, we also find that any page with >20% sponsored posts tends to spread about 2.31 times more propaganda compared to pages posting <20% sponsored content.

---

[8] We observed a high correlation between the 'engagement' and 'followers count' of a page, hence, using 'engagement' as a measure of a page's influence gives qualitatively similar results.

[9] For consistency, all values shown in the figure have been scaled between 0 and 1 using min-max scaling.

[10] Low Pearson (0.096) and Spearman (0.383) correlation coefficients.

7   Case Study: 2020 US Presidential Election

Table 5. Facebook posts before the US Elections from the Left and Right classified news channels, these posts clearly indicate the type of propaganda being spread out through the news channels on Facebook.

Channel	Partisanship	Post Content	Propaganda Type
The New Civil Rights Movement (Followers: 377679)	Left	**"Do all of Putin's operatives spread disinfo that can so easily be fact checked?"** Richard Grenell, President Donald Trump's former Acting Director of National Intelligence, is under fire after posting a 2019 photo of Joe Biden and attacking him as **phony** for not wearing a mask Since the photo was clearly taken months before the coronavirus was even discovered, Grenell is ...'Disinformation **Grifter**': Ex-Trump Intel Chief's Anti-Biden Stunt Backfires, Earns Him a 'Manipulated Media' Label	Doubt*, Loaded Language**, Name-Calling***
The Stranger (Followers: 128486)		—Election week 2020 begins tonight —Biden sweeps Dixville Notch —When will we know if we're still living in a democracy?After four long years that felt like a thousand **endless nightmares***: Election week 2020 is finally here. East coast polls (plus Georgia and Indiana) close at 4 p.m. PST, and we'll start seeing some results about a half-hour or an hour afterwards. Slog goes live shortly thereafter. Stay tuned for upda... Slog AM: Alright **You Nervous Little Freaks**, It's Time to Boot the **Bad President** and End **This American Carnage**	Exaggeration*, Name calling**, Flag waving***
Pamela Geller (Followers: 1291098)	Right	It's **one bombshell after*** another now. BOMBSHELL AUDIO! Hunter Biden Confesses Partnership With **"The F**king Spy Chief of China"** ...Joe Biden Named In Criminal Case Witness - Geller Report News	Loaded Language*, Name calling**
PJ Media (Followers: 405847)		"This will get ugly." In the Battle for Florida, It Looks Like Democrats are Heading for a **Bloodbath***	Loaded Language*

While so far we have reported the analyses of propaganda at a longer timescale, we now delve into a particular event. Our data collection period encompasses significant events like the US Presidential Election on November 3rd, 2020, and the subsequent US Capitol attack in Washington, D.C. on January 6th, 2021. Our investigation entails a comprehensive scrutiny of propaganda disseminated through news pages on Facebook, spanning the periods preceding the election, the election itself, and the aftermath. This analysis seeks to offer insights into the role and importance of propaganda in the political landscape during this pivotal period.

7.1   Pre-election and Election Period

The influence of social media on political opinions is substantial, often contributing to polarization by recommending content aligning with users' existing views.

Table 6. Facebook posts after the US elections and before the Capitol Attack.

Channel	Post Content	Propaganda Type
100% Fed up	**This isn't about Democrats or Republicans; this is about America***. American Patriots are 100% Fed Up with the corruption! Now arrives the hour of action! We've got to do this now! 100% Fed Up – The Biden Campaign's primary defense is don't hear the evidence. That is why the pu... The Gateway Pundit: **WOW! Stop The Steal!**** Arrives... The Hour Of Action	Flag waving*, Slogan**
The Globe and Mail (Followers: 786k)	Vice President Pence Can Stop the Steal and Keep the Peace "All that is necessary for the triumph of evil is that good men do nothing." (Edmund Burke) I suspect Vice President Mike Pence has quoted that many times. January 6 might be his opportunity to live out his day to **be the good man who stopped evil***. About **half of our nation**** understands that Trump's ...Vice President Pence Can Stop The Steal And Keep The Peace	Name calling*, Flag waving**

Particularly on Facebook, where news channel pages boast large followings, there is a tendency to disseminate propagandistic content that can sway the opinions of their followers. Table 5 presents detailed examples of propaganda dissemination by both left-wing and right-wing news channels during pre-election period.

Examining Fig. 5a, which tracks the propaganda trends from news pages over six months corresponding to our dataset's timeline, reveals a noteworthy pattern. The graph illustrates the proportion of propaganda posts relative to total posts. A noticeable peak emerges during the first week of November, falling between the two most statistically significant changepoints[11], aligning with the election date (November 3, 2020). This surge in propaganda posts seems to be initiated around mid-September, reaches its zenith in early November, and undergoes subsequent decline in December. This observation strongly suggests that the news pages on social media exhibit a preference for employing propaganda as a subtle tool to influence their audience, particularly during pivotal events like elections.

7.2   Post Elections: Capitol Attack

The aftermath of the 2020 US presidential elections witnessed the alarming events of the Capitol attack, where around 2000 individuals, seemingly fueled by the then president's speech, marched to the Capitol to protest alleged voting fraud[12] We analyze the influence of social media on this incident to understand and emphasize the extent of propaganda spread and its possible ramifications. In

---

[11] We performed the changepoint analysis using the `Dynp` algorithm (https://centre-borelli.github.io/ruptures-docs/user-guide/detection/dynp/)

[12] https://www.usatoday.com/in-depth/news/2021/02/01/civil-war-during-trumps-pre-riot-speech-parler-talk-grew-darker/4297165001/.

Table 6, we present manually labeled examples of propaganda dissemination by Facebook channels during the post-election and pre-attack period. During this phase, we observed a surge in propaganda employing techniques like "slogan", "name-calling", and "flag-waving". These tactics aimed at provoking individuals to take action against the election results without presenting substantiated arguments. Concurrently, various Facebook campaigns emerged, encouraging people to protest by promoting "slogan" and "flag-waving" propaganda. A notable example is the relentless promotion of the slogan "Stop the Steal!" to such an extent that Facebook intervened to mitigate posts containing this slogan[13]

Revisiting Fig. 5a, the correlation between the Facebook propaganda dissemination and the timing of the Capitol attack is quite evident. The highest peak in propaganda around the second week of January precisely aligns with the date of the Capitol attack (7th January 2020). Figure 5b further elucidates the dynamic response to the surge in propaganda. Particularly we see an increase in right-wing propaganda following the period of the Capitol attack in order to pacify the protest, while the left-wing propaganda declines. Interestingly, we also witness a peak in propaganda across all partisan groups during the period surrounding the attack, specifically in the first and second weeks of January, 2020. This suggests a link between the surge in propaganda on social media and real-world events like the attack, emphasizing the influential role that online propaganda on social media can play in shaping offline actions and consequences.

8   Conclusion

In this work, we emphasized the influence of sponsored news posts on propaganda dissemination on Facebook. Through extensive analysis and experiments on a curated dataset of Facebook posts, using advanced propaganda detection methods, we found that sponsored news posts are more likely to be propagandistic and elicit increased user engagement. Additionally, we conducted a detailed analysis of Facebook posts surrounding the 2021 US Capitol Attack, revealing patterns in propaganda dissemination by various news channels during that time.

Limitations. We acknowledge that although we have made extensive efforts to justify our hypotheses, our analysis is constrained by the labels assigned through an imperfect propaganda classifier. Additionally, despite the comprehensive scope of the dataset, which spans a significant timeframe, the dynamic nature of the socio-political landscape introduces a potential limitation in its temporal generalizability.

Reproducibility. Our codebase is available at https://github.com/ddsb01/sponsored-propaganda.

---

[13] https://www.nytimes.com/2021/01/11/us/facebook-stop-the-steal.html.

Acknowledgments. This research was supported in part by the French National Research Agency (ANR) ANR-21-CE23-0031-02, ANR-22-CE38-0017, ANR-21-CE39-0019 and ANR-22-PECY-0002 grants, and by the EU 101041223 and 101021377 grants.

References

1. Abdullah, M., Altiti, O., Obiedat, R.: Detecting propaganda techniques in English news articles using pre-trained transformers. In: ICICS 2022 (2022). https://doi.org/10.1109/ICICS55353.2022.9811117
2. Baisa, V., Herman, O., Horak, A.: Benchmark dataset for propaganda detection in Czech newspaper texts. In: RANLP 2019. INCOMA Ltd. (2019). https://aclanthology.org/R19-1010
3. Balalau, O., Horincar, R.: From the stage to the audience: propaganda on Reddit. In: EACL. ACL (2021)
4. Barfar, A.: A linguistic/game-theoretic approach to detection/explanation of propaganda. Expert Syst. Appl. **189**, 116069 (2022)
5. Bengani, P.: Hundreds of 'pink slime' local news outlets are distributing algorithmic stories and conservative talking points. Columbia Journalism Rev. (2019)
6. Cai, W.: Analysis of the promotion of consumer psychological analysis and marketing to sports propaganda. Revista de Psicología del Deporte (J. Sport Psychol.) (2023)
7. Chakraborty, A., Ghosh, S., Ganguly, N., Gummadi, K.P.: Dissemination biases of social media channels: on the topical coverage of socially shared news. In: ICSWM (2016)
8. Chaudhari, D.D., Pawar, A.V.: Propaganda analysis in social media: a bibliometric review. Information Discovery Delivery **49**(1), 57–70 (2021)
9. Chen, C., Shu, K.: Combating misinformation in the age of LLMs: opportunities and challenges (2023)
10. Corbin, C.M.: The unconstitutionality of government propaganda. Ohio St. LJ (2020)
11. Da San Martino, G., Barrón-Cedeño, A., Wachsmuth, H., Petrov, R., Nakov, P.: SemEval-2020 task 11: detection of propaganda techniques in news articles. In: SemEval-2020. ICCL (2020)
12. Da San Martino, G., Yu, S., Barrón-Cedeño, A., Petrov, R., Nakov, P.: Fine-grained analysis of propaganda in news articles. In: EMNLP-IJCNLP (2019)
13. Van Zandt, D.: Media biasfact checkk - search and learn the bias of news media (2024). https://mediabiasfactcheck.com
14. Devlin, J., Chang, M.W., Lee, K., Toutanova, K.: BERT: pre-training of deep bidirectional transformers for language understanding. arXiv preprint arXiv:1810.04805 (2018)
15. Dmitry Chernobrov, E.B.: Competing propagandas: how the United States and Russia represent mutual propaganda activities. Politics (2022)
16. Edelson, L., Nguyen, M.K., Goldstein, I., Goga, O., McCoy, D., Lauinger, T.: Understanding engagement with U.S. (mis)information news sources on Facebook. In: IMC 2021. ACM, New York, NY, USA (2021). https://doi.org/10.1145/3487552.3487859
17. The Council of Europe: Dealing with propaganda, misinformation and fake news (2023). https://www.coe.int/en/web/campaign-free-to-speak-safe-to-learn/dealing-with-propaganda-misinformation-and-fake-news

18. Goldstein, J.A., Sastry, G., Musser, M., DiResta, R., Gentzel, M., Sedova, K.: Generative language models and automated influence operations: emerging threats and potential mitigations (2023)
19. Guarino, S., Trino, N., Celestini, A., Chessa, A., Riotta, G.: Characterizing networks of propaganda on twitter: a case study. Appl. Netw. Sci. (2020)
20. Guimaraes, A., Balalau, O., Terolli, E., Weikum, G.: Analyzing the traits and anomalies of political discussions on reddit. In: ICWSM (2019)
21. Swieboda, H., Kuczabski, M., Szpyra, R., Zawadzki, T., Walecki, T., Stobiecki, P.: Social control in the face of digital propaganda. Eur. Res. Stud. J. (2021)
22. Huang, K.H., McKeown, K., Nakov, P., Choi, Y., Ji, H.: Faking fake news for real fake news detection: propaganda-loaded training data generation. In: ACL (2023)
23. Johnson, K.A., St. John III, B.: News stories on the Facebook platform: millennials' perceived credibility of online news sponsored by news and non-news companies. Journalism Pract. (2020)
24. Kmetty, Z., Vincze, V., Demszky, D., Ring, O., Nagy, B., Szabo, M.K.: Partelet: a Hungarian corpus of propaganda texts from the Hungarian socialist era. In: LREC 2020. European Language Resources Association, Marseille, France (2020). https://aclanthology.org/2020.lrec-1.290
25. Lan, Z., Chen, M., Goodman, S., Gimpel, K., Sharma, P., Soricut, R.: ALBERT: a lite BERT for self-supervised learning of language representations. arXiv preprint arXiv:1909.11942 (2019)
26. Malhan, M., Dewani, P.P.: Propaganda as communication strategy: historic and contemporary perspective. Acad. Mark. Stud. J. (2020)
27. McKnight, P.E., Najab, J.: Mann-Whitney U Test. Wiley, New York (2010). https://doi.org/10.1002/9780470479216.corpsy0524
28. Meta: Increasing our ads transparency, February 2023. https://about.fb.com/news/2023/02/increasing-our-ads-transparency/
29. Miller, C.R.: The techniques of propaganda. From "how to detect and analyze propaganda" (1939)
30. Moravec, P.L., Collis, A., Wolczynski, N.: Countering state-controlled media propaganda through labeling: evidence from Facebook. Inf. Syst. Res. (2023)
31. Mozur, P.: A genocide incited on Facebook, with posts from Myanmar's military (2018). https://www.nytimes.com/2018/10/15/technology/myanmar-facebook-genocide.html
32. Ng, V., Li, S.: Multimodal propaganda processing. In: AAAI (2023)
33. Pierri, F., Luceri, L., Jindal, N., Ferrara, E.: Propaganda and misinformation on Facebook and Twitter during the Russian invasion of Ukraine. In: WebSci 2023. ACM (2023)
34. Purificato, A., Navigli, R.: APatt at SemEval-2023 task 3: the Sapienza NLP system for ensemble-based multilingual propaganda detection. In: Ojha, A.K., Doğruöz, A.S., Da San Martino, G., Tayyar Madabushi, H., Kumar, R., Sartori, E. (eds.) SemEval-2023. ACL (2023)
35. Raffel, C., et al.: Exploring the limits of transfer learning with a unified text-to-text transformer (2023)
36. Ribeiro, F., et al.: Media bias monitor: quantifying biases of social media news outlets at large-scale. In: ICWSM (2018)
37. Ribeiro, F.N., et al.: On microtargeting socially divisive ads: a case study of Russia-linked ad campaigns on Facebook. In: FAccT (2019)
38. Rizoiu, M.A., Graham, T., Zhang, R., Zhang, Y., Ackland, R., Xie, L.: DebateNight: the role and influence of socialbots on Twitter during the 1st 2016 U.S. presidential debate (2018)

39. Salman, M.U., Hanif, A., Shehata, S., Nakov, P.: Detecting propaganda techniques in code-switched social media text (2023)
40. Seo, H., Ebrahim, H.: Visual propaganda on Facebook: a comparative analysis of Syrian conflicts. Media War Conflict **9**(3), 227–251 (2016). https://doi.org/10.1177/1750635216661648
41. Solopova, V., Popescu, O.I., Benzmüller, C., Landgraf, T.: Automated multilingual detection of Pro-Kremlin propaganda in newspapers and telegram posts (2023)
42. Sprenkamp, K., Jones, D.G., Zavolokina, L.: Large language models for propaganda detection (2023)
43. Tess: Crowdtangle api (2024). https://help.crowdtangle.com/en/articles/1189612-crowdtangle-api
44. Tian, J., Gui, M., Li, C., Yan, M., Xiao, W.: MinD at SemEval-2021 task 6: propaganda detection using transfer learning and multimodal fusion. In: SemEval-2021. ACL (2021). https://aclanthology.org/2021.semeval-1.150
45. Torok, R.: Symbiotic radicalisation strategies: propaganda tools and neuro linguistic programming (2015)
46. Venkatadri, G., et al.: Privacy risks with Facebook's PII-based targeting: auditing a data broker's advertising interface. In: 2018 IEEE Symposium on Security and Privacy (SP). IEEE (2018)
47. Yang, Z., Dai, Z., Yang, Y., Carbonell, J., Salakhutdinov, R.R., Le, Q.V.: XLNet: generalized autoregressive pretraining for language understanding. In: Wallach, H., Larochelle, H., Beygelzimer, A., d'Alché-Buc, F., Fox, E., Garnett, R. (eds.) NeurIPS (2020)
48. Young, K., Lindstrom, F.B., Hardert, R.A.: Kimball young on social psychology, rural sociology, and anthropology at Wisconsin, 1926–1940. Sociological Perspectives (1989). http://www.jstor.org/stable/1389124
49. Zienkowski, J.: Propaganda and/or ideology in critical discourse studies. DiscourseNet Collaborative Working Paper Series (2021)

Exploring Relationships Between Cryptocurrency News Outlets and Influencers' Twitter Activity and Market Prices

Meysam Alizadeh[1(✉)], Yasaman Asgari[2], Zeynab Samei[3], Sara Yari[1],
Shirin Dehghani[4], Mael Kubli[1], Darya Zare[5], Juan Diego Bermeo[1],
Veronika Batzdorfer[6], and Fabrizio Gilardi[1]

[1] Digital Democracy Lab, University of Zurich, Zurich, Switzerland
{alizadeh,syarim,kubli,jbermeo,gilardi}@ipz.uzh.ch
[2] Digital Society Initiative and Department of Mathematical Modeling and Machine Learning, University of Zurich, Zurich, Switzerland
yasaman.asgari@math.uzh.ch
[3] Department of Computer Science, IPM, Tehran, Iran
zsamei@ipm.edu
[4] Department of Computer Engineering, Allameh Tabataba'i University, Tehran, Iran
sh.dehghani@atu.ac.ir
[5] Department of Informatics, University of Zurich, Zurich, Switzerland
daryazm@uzh.ch
[6] Institute for Sociology and Computational Social Science, Karlsruhe Institute of Technology, Karlsruhe, Germany
vbatzdorfer@kit.de

Abstract. Academics increasingly acknowledge the predictive power of social media for a wide variety of events and, more specifically, for financial markets. Anecdotal and empirical findings show that cryptocurrencies are among the financial assets that have been affected by news and influencers' activities on Twitter. However, the extent to which Twitter crypto influencer's posts about trading signals and their effect on market prices is mostly unexplored. In this paper, we use LLMs to uncover buy and not-buy signals from influencers and news outlets' Twitter posts and use a VAR analysis with Granger Causality tests and cross-correlation analysis to understand how these trading signals are temporally correlated with the top nine major cryptocurrencies' prices. Overall, the results show a mixed pattern across cryptocurrencies and temporal periods. However, we found that for the top three cryptocurrencies with the highest presence within news and influencer posts, their aggregated LLM-detected trading signal over the preceding 24 h granger-causes fluctuations in their market prices, exhibiting a lag of at least 6 h. In addition, the results reveal fundamental differences in how influencers and news outlets cover cryptocurrencies.

Keywords: Social Prediction · NLP · LLM · Prompt Engineering · Time Series Analysis

© The Author(s), under exclusive license to Springer Nature Switzerland AG 2025
L. M. Aiello et al. (Eds.): ASONAM 2024, LNCS 15211, pp. 455–471, 2025.
https://doi.org/10.1007/978-3-031-78541-2_28

1   Introduction

Cryptocurrencies are a form of digital money that utilizes blockchain technology, a groundbreaking, decentralized, and encryption-based system that allows for the digital establishment of trust [26]. Since the introduction of Bitcoin in 2009, the disruptive capabilities of cryptocurrencies have sparked significant growth and interest [10]. This surge was mainly driven by media coverage highlighting the extraordinary returns on cryptocurrency investments, which led to a modern-day gold rush [16]. Moreover, the current international regulatory landscape for cryptocurrencies remains sparse, as they are mostly not yet recognized as a fully developed asset class [26]. This lack of regulation, combined with their immense popularity and absence of institutional backing, contributes to the cryptocurrency market's extreme volatility and unpredictability, earning it the nickname "Wild West" [16].

The instability of the cryptocurrency market is significantly driven by news and social media posts [2]. This dynamic is exacerbated by investors' challenges in verifying the accuracy of such information [16]. Given the relatively recent emergence of the cryptocurrency market, traditional media often lag in reporting developments, making social media a key information source for investors. Twitter (now re-branded as X), in particular, serves as a crucial platform for obtaining real-time updates and sentiment analysis on cryptocurrencies, allowing users to express their opinions and feelings [12].

According to behavioral economics, emotions and sentiment can greatly influence individual actions and decision-making processes [11]. Previous research explored the effect of public Twitter sentiment on cryptocurrency prices [16], Elon Musk's Twitter activity on the cryptocurrency market [2], Twitter effect on stock market decisions during pandemics [27], and the predictive power of Twitter for Bitcoin price [25] (see [5] for a recent review).

Given the abundance of Twitter data that captures the sentiments of cryptocurrency participants, the extent to which Twitter posts can help traders make informed decisions is mostly unexplored. Previous research on the association of Twitter posts with the cryptocurrency market suffers from at least two significant drawbacks. First, they mostly fail to account for fake accounts, such as bots and marketing campaigns. Indeed, previous research has shown that about 14% of the active accounts in the cryptocurrency space are bots [16]. Second, the efficiency of the traditional sentiment analysis methods in capturing the true opinion of users is questionable, and more explicitly defined machine learning models that generate buy and sell signals [17] may produce better results.

In this paper, our goal is to determine whether LLM-detected trading signals from news outlets and crypto influencers' Twitter posts reflect cryptocurrency market dynamics. Our study utilizes text classification, network science methods, and times series analysis to analyze the retweet and co-mention networks. This structure consists of undirected links between cryptocurrencies that appear together in a tweet. In summary, we aim to answer two key research questions:

1. How do the cryptocurrencies that are frequently co-mentioned together correlate with each other in their market prices? Do the links in the co-mention

network reveal shared characteristics among cryptocurrencies, such as their technology or use cases?

2. Do the LLM-detected trading signals (buy or not-buy) of top cryptocurrencies obtained from news outlets and influencers' Twitter posts demonstrate a (lagged) correlation with their market prices?

2 Background

2.1 Effect of Sentiment on Financial Markets

Kaplanski and Levy [13] describe sentiment as any misconception that may lead to a deviation from the true fundamental value of an asset. Baker and Wurgler [3] note that the key factor distinguishing highly speculative assets is the difficulty and subjectivity involved in valuing them accurately. Fama et al. [6] argued that stock prices are unpredictable due to the erratic nature of news. According to Peterson [24], financial markets are considerably influenced by news, which in turn impacts sentiment. Traditionally, sentiment has been measured through investor surveys like those by the American Association of Individual Investors or Investor Intelligence. However, these methods are somewhat constrained by their dependence on achieving representative sample sizes. Recent advancements have seen a rise in the use of Natural Language Processing (NLP) to assess sentiment from news sources [20].

2.2 Twitter Sentiment Analysis

The effectiveness of using Twitter sentiment analysis to forecast financial markets is prominently illustrated in the study by Bollen et al. [4], and further supported by Li et al. [19]. In their research, Li et al. [19] employ a Naive Bayes sentiment classifier and regression models to demonstrate that Tweets about stocks can predict daily stock returns. The findings also indicate that increases in Twitter activity follow higher market volatility from the previous day, positioning Twitter sentiment as both a predictor and a result of market movements. Furthermore, Mao et al. [22] found that, unlike traditional investor sentiment, Twitter sentiment significantly predicts stock returns over the following 1–2 days. In general, the predictive power of Twitter sentiment for financial markets is generally observed to be the strongest between 1–4 days [4,22,28].

2.3 Twitter and Cryptocurrency Market

Social media has become the main source of information for cryptocurrencies. Several studies have analyzed user discussions on forums like Reddit and Bitcointalk.org (e.g. [14,15]). Many researchers recognize the short-term (1–7 days) and long-term (30–90 days) forecasting abilities of social media and news sentiment on Bitcoin's price and trading volumes. The quantity of posts [21] and Elon Musk's activity on Twitter [2] have been linked to the trading volume of

Bitcoin [21]. Similarly, Zou et al. [29] proposed a multimodal model for Bitcoin extreme price movement prediction using Twitter.

Twitter sentiment analysis has been used to predict changes in Bitcoin price. Georgoula et al. [8] utilized a SVM to predict these fluctuations. Garcia and Schweitzer [7] applied a lexicon-based method along with a Vector Autoregressive (VAR) model and Granger-causality tests, discovering that rises in Twitter sentiment polarity often precede shifts in Bitcoin prices. Mai et al. [21] performed intraday analysis and found Twitter posts effective in forecasting Bitcoin's hourly returns. Overall, the most effective number of lags is observed from 1–5 lags for interday analysis and 2–4 lags for intraday analysis [14,21].

2.4   Limitations of Current Literature

The extant literature suffers from at least three major drawbacks: (1) much of the current literature is dedicated to Bitcoin, and other major coins have been neglected; (2) the vital role of the Twitter influencers has not been explored. Nearly all previous work has not distinguished between the original content produced by crypto experts and the content produced by lay users or even bots and marketing campaigns. (3) the meaning of "Twitter sentiment" is rather vague as it may refer to emotion, stance, or buy or sell signals.

3   Methods and Data

3.1   Twitter Data

We collect four different Twitter datasets. First, we used Twitter Academic API between 07/01/2022 and 14/01/2022 and between 01/03/2022 and 07/03/2022 and queried for 'crypto' and a list of the top 200 cryptocurrencies to collect all tweets containing them (11.83M tweets posted by 2.1M unique users). Second, we manually searched for all authentic news outlets that have finance or cryptocurrency sections and have active Twitter accounts (74 accounts). Next, we collected all tweets posted by these 74 accounts between 01/01/2023 and 16/09/2023 (122,837 tweets). These two datasets serve for our influencer identification (see Section 3.4) and model training purposes (see Sects. 3.3 and 3.5). Our third and fourth datasets are created by continuously crawling Twitter data of the crypto-related influencers and news outlets between 10/06/2023 and 02/28/2024.

3.2   Cryptocurrency Data

We consider nine major cryptocurrencies according to their market capitalization. This includes Bitcoin (BTC), Ethereum (ETH), Ripple (XRP), Solana (SOL), Dogecoin (DOGE), Binance coin (BNB), Ada (ADA), Polkadot (DOT), and Shiba Inu (SHIB). The financial information for these nine studied cryptocurrencies was obtained from CoinMarketCap (https://coinmarketcap.com)

from October 2023 to March 2024. CoinMarketCap is commonly used as a reference for cryptocurrency prices as it aggregates prices from numerous exchanges, providing a composite and more stable price estimate that avoids the biases of individual exchange prices.

3.3 Twitter Data Pre-processing

To prune out irrelevant tweets from our corpus, we trained and compared two classifiers and five LLMs on 500 annotated tweets. Tweets were classified as "relevant" if they were directly related to cryptocurrencies or related to economic or policy topics related to cryptocurrency market (e.g., US interest rate and cryptocurrency regulations). Out of the six classification models (*CyptoBERT, BERTweet, GPT-3.5* trained with two distinct prompts, *Random Forest with CryptoBERT Embedding*, and *Random Forest and CryptoBERT-kk08 Embedding*), we achieved the maximum accuracy of 0.94 on a 20% unseen out-of-sample test data for GPT-3.5 with optimized prompt. Therefore, for all four datasets we used GPT-3.5 with an optimized prompt to exclude irrelevant tweets (Table 1).

Table 1. Comparing the accuracy of various classifiers for pruning irrelevant tweets.

model	balancing method	accuracy	f1-score
CoinBERT	unbalance	92%	91%
BERTweet	gpt3.5 augmentation	92%	91%
CoinGPT	–	**96%**	**94%**
GPT-4	–	93%	91%
Random forest+cryptobert embedding	smote	88%	82%
Random forest+cryptobert-kk08 embedding	smote	86%	81%

3.4 Influencer Identification

We use four different methodologies to identify crypto-related influencers on Twitter. First, we used our nearly 12 million tweets dataset to construct a retweet network. Following Kwak et al. [18], we calculate three centrality metrics, including *PageRank, Betweenness*, and *Closeness* scores for all users within the retweet network. From each centrality measure, we select the top 1000 accounts. After combining these lists, we obtained 1,283 users.

Second, we examined the marketing campaigns of 10 coins that were launched in 2023. Through the retweet network analysis, we identified the influential accounts. This led to identifying 87 new influencers based on their *PageRank* scores. Third, we crafted a list of 50 non-major cryptocurrency coins and their related keywords and collected all tweets posted between 01/01/2024 to 01/02/2024. This process yielded 11,408 users who were potential influencers.

Finally, we used LunarCrush (https://lunarcrush.com/) website and scraped the list of top 1000 influencers for each of the leading 100 coins. After removing mutual influencers, we obtained 36,311 unique users from LunarCrush.

After combining these four lists and excluding duplicates, we collected user profile information (bio descriptions and follower counts) and the 10 most recent tweets for each account. Users with irrelevant descriptions, fewer than 5000 followers, no tweets in the last 3 months, and users with an average engagement of fewer than 200 in their last 10 tweets were excluded. Overall, we obtained 2,687 cryptocurrency influencers on Twitter.

3.5 Trading Signal Detection

We use CryptoBERT [17] to detect buy, sell, and neutral signals from the text of the influencers' tweets. CryptoBERT is trained on a corpus of over 3.2 million crypto-related social media posts. We compare its performance with *Random Forests, Logistic Regression, SVM,* and *XGBoost.* Our feature engineering includes content features such as top unigrams, top bigrams, top hashtags, top mentions, number of mentions, number of emojis, and number of hashtags. Considering the previous findings on the ability of ChatGPT [9] and open-source LLMs [1] in text annotation tasks, we included GPT-4, GPT-3.5, LlaMa-2 (7b), LlaMa-2 (70b), FLAN-T5 (L), and FLAN-T5 (XL) in our comparison. We randomly selected 600 tweets from influencers and 200 tweets from news outlets' datasets, and labeled them for buy, sell, and neutral signals (Table 2). As can be seen in Table 2, Since the frequency of 'sell (bearish)' signal in the influencer's data is low (5.5% of total labeled data), we combined the 'sell' and 'neutral' classes into one class and called it the 'not-buy' class.

Table 2. The distribution of signals across annotated datasets.

Dataset	Class names	% samples
Twitter Influencers	Buy (Bullish)	56.8
	Sell (Bearish)	5.50
	Neutral	37.7
News Outlets	Buy (Bullish)	30.5
	Sell (Bearish)	25.5
	Neutral	44.0

In Fig. 1, we see the CryptoBert model outperforms all other models for the influencers dataset. However, for the news outlets' data, GPT-4 and GPT-3.5 are the two best-performing models and slightly outperform CryptoBERT. Since CryptoBERT's performance is strong across both datasets, and it is open-source (and thus more ethical for academic purposes) and free to use, we selected it as our final choice.

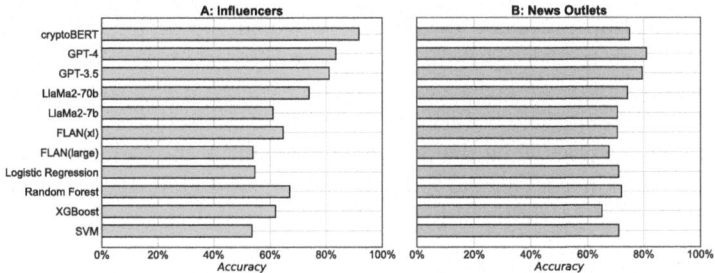

Fig. 1. Choosing the best AI model for the trading signal detection task.

3.6 Constructing Co-mention Network

In a co-mention network, nodes represent individual cryptocurrency coins, and edges symbolize the two coins that were mentioned together in a tweet. The network is undirected and weighted, characterized by symmetric adjacency matrix A with entries $w_{ij} \in \mathbb{N}$ corresponding to the frequency of co-mentions between coins i and j.

3.7 Time Series Processing

Time series data consist of a sequence of values or observations recorded over time. Mathematically, a time series can be defined as a set of ordered observations $X = \{x_1, x_2, \ldots, x_n\}$, where each x_i corresponds to the value of the series at the ith time point.

The Granger causality test assesses whether one time series can predict another, a method commonly used to investigate economic time series interactions. The null hypothesis states that X does not cause Y in the Granger sense. The procedure involves formulating a vector autoregression (VAR) model for the time series under study. For two variables X and Y, the model can be specified as:

$$Y_t = \alpha + \sum_{i=1}^{k} \beta_i Y_{t-i} + \sum_{i=1}^{k} \gamma_i X_{t-i} + \epsilon_t \tag{1}$$

$$X_t = \delta + \sum_{i=1}^{k} \phi_i X_{t-i} + \sum_{i=1}^{k} \psi_i Y_{t-i} + \eta_t \tag{2}$$

Here, α and δ represent constant terms, β_i, γ_i, ϕ_i, and ψ_i are the coefficients, and ϵ_t and η_t denote the error terms of the model. The parameter k signifies the number of lagged observations in the model. An F-statistic is calculated to evaluate the null hypothesis. Granger causality testing identifies patterns of lagged correlation but does not confirm causality. This concept is akin to how observing cloud cover, which often precedes rain, can help predict rain but does not cause it.

Cross-correlation evaluates the degree to which two series are correlated at different time lags k. If we denote another time series as $Y = \{y_1, y_2, \ldots, y_n\}$, the cross-correlation γ at lag k can be computed using (3). The numerator of the formula represents the covariance between the x series shifted by k units in time and the y series. The denominator normalizes this value, ensuring that the cross-correlation coefficient γ lies between -1 and 1, inclusive. The sign and magnitude of γ at different values of k can reveal the leading and lagging relationships between the two-time series.

$$\gamma = \frac{\sum_i (x_{i+k} - \bar{X})(y_i - \bar{Y})}{\sqrt{\sum_i (x_{i+k} - \bar{X})^2}\sqrt{\sum_i (y_i - \bar{Y})^2}} \tag{3}$$

4   Results

4.1   Description of Influencers and News Outlets Twitter Data

The final Twitter dataset used in this experiment contains 470,658 English tweets from influencers and news outlet's Twitter accounts, between 10/06/2023 and 02/28/2024. Tweets written by influencers dominate the dataset with 413,071 tweets (87.8%), while news outlets account for 57,587 tweets (12.2%). This underscores the pivotal role of influencers in shaping discussions and influencing trading decisions within the cryptocurrency market (Fig. 2-A). Moreover, Bitcoin appears in 340,193 tweets (34.8%), followed by Ethereum and Solana with 39,742 (9.9%) and 27,688 (6.9%) tweets, respectively (Fig. 2-B). Figure 2-C illustrates the coin mention share distribution, highlighting structural variations between influencers and news outlets. An interesting observation here is that the share of Bitcoin mentions in news outlets' tweets (61.9%) is higher than influencers' tweets (48.2%).

Figure 2-D shows the dataset predominantly contains 'not buy' signal, representing 54.39% (256,016 tweets), while 'buy' signals account for 45.60% (214,629 tweets). Figure 2-E details the signal distribution across nine major coins. BTC, XRP, and DOT predominantly feature 'not buy' signals, whereas ETH, SOL, ADA, DOGE, SHIB, and BNB mainly show 'buy' signals. Notably, DOGE has 2.53 times more 'buy' than 'not buy' signals, the highest ratio amongst all examined coins.

4.2   Analyzing the Cryptocurrencies Co-mention Network

Exploratory Analysis of the Co-mention Network. This study focuses on the co-mention network of the top 100 coins according to their market capitalization. The dataset includes tweets mentioning between zero (generally discussing the cryptocurrency financial market) to over 40 coins, with the most common scenario involving single-coin mentions (44.4% of tweets). We also examine the differences in how influencers and news outlets reference cryptocurrencies. According to Fig. 3-A, influencers tend to mention more coins per tweet (0.9)

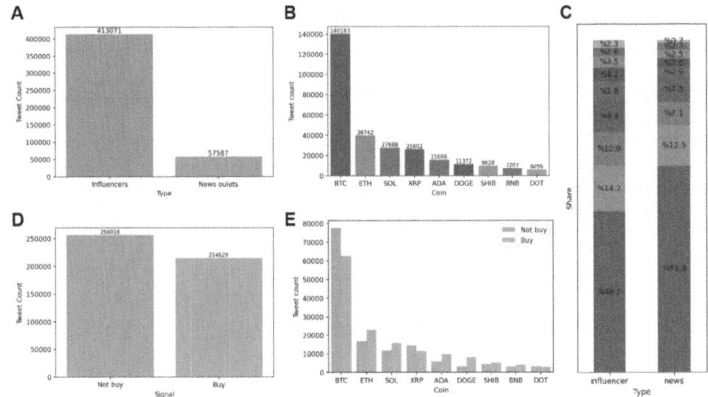

Fig. 2. Exploratory Analysis: (A) User Type Distribution (B) Frequency of Tweets by Major Coin, with Bitcoin as the Most Mentioned (C) Comparison of 9-major coins' mentions in News Outlets versus Influencer Tweets (D) Signal Distribution (E) 'Not Buy' signals predominant for BTC, XRP, and DOT, while 'Buy' Signals are more common for ETH, SOL, ADA, DOGE, SHIB, and BNB.

compared to news outlets (0.56). This suggests that influencers may discuss a broader range of coins, likely to engage a larger audience or to demonstrate their knowledge across various market segments.

Furthermore, Fig. 3-B reveals distinct characteristics in tweets signaling 'buy' versus 'not buy' intentions. Tweets encouraging buying typically mention at least one specific coin, suggesting a direct investment recommendation. In contrast, 'no buy' tweets often do not specify any coins and focus more on general market conditions or cautions. The average number of coins mentioned in 'buy' tweets is higher (0.96) compared to 'not buy' tweets (0.77), showing a more targeted approach to investment suggestions.

We reveal key patterns in the co-mention network using the k-core method to filter the network, where nodes must have connections involving at least 1% of the total degree sum. Figure 3-C shows this filtered k-core network of 24 nodes. The edge thickness represents the edge weight, while the node size indicates the node's degree of centrality. The data highlights several key connections: BTC to ETH with 14,062 connections (5.16%), BTC to USDT (3.54%), BTC to BCH (2.37%), ETH to SOL (2.14%), BTC to XRP (1.81%), BTC to SOL (1.79%), BTC to ADA (1.25%), and XRP to ETH (1.18%). Bitcoin (BTC) and Ethereum (ETH) are frequently featured, with a notable connection between Ethereum Classic (ETC) and Non-Fungible Tokens (NFT).

Interplay Between the Co-mention Network and Cryptocurrency Attributes. Next, We investigate whether the connections in the co-mention network indicate common features among cryptocurrencies, such as their technology or applications. Following Mungo et al. [23], we utilize tags from Coinmarket-

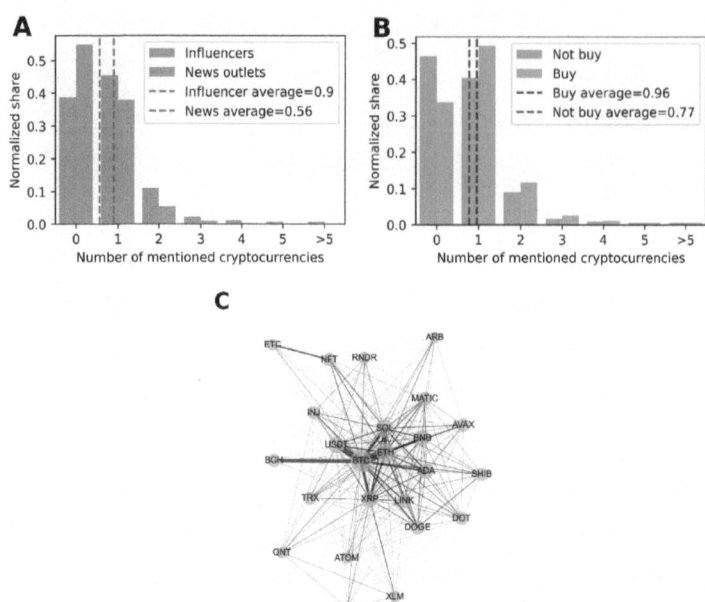

Fig. 3. Structure of the cryptocurrencies co-mention network: (A) Influencers mention more coins in their tweets than news outlets do. (B) Tweets signaling a buy mention more coins than tweets not signaling a buy. (C) The co-mention network of top 24 coins. There exists a notable connection between NFT and ETC.

cap that describe various cryptocurrencies' primary attributes. These tags detail each cryptocurrency's blockchain technology and information about its ecosystem, such as whether the cryptocurrency is built on an independent or existing blockchain and is involved in decentralized finance (DeFi) projects. Additionally, the tags describe the specific functions and utilities of the cryptocurrencies, for instance, their use in distributed storage, as fan tokens, or as digital stores of value like digital gold.

Following the methodology suggested by Mungo et al. [23], we assign a vector x_i to each cryptocurrency, where each tag j is represented as $x_{i,j} = 1$ if it applies to the cryptocurrency and $x_{i,j} = 0$ otherwise. We then calculate the cosine similarity in the characteristics space between each pair of cryptocurrencies to create a similarity matrix. Next, we compute a Pearson correlation between the weighted adjacency and similarity matrices. The correlation coefficient of 0.1735 with a p-value of < 0.001 suggests that the patterns observed in the co-mention network are only slightly correlated with the coin characteristics, and there are other mechanisms run by the influencers and news outlets on the mentioning patterns within the social media.

Interplay Between the Co-mention Network Structure and Cryptocurrency Price Returns. We analyze pairs of coins in the co-mention network that constitute at least 1% of the total connections. From the initial set of 24 cryptocurrencies considered, we exclude two: NFT and USDT. NFTs, or non-fungible tokens, differ from cryptocurrencies as they do not have standard prices listed on platforms like CoinMarketCap due to the unique nature of each NFT, which contrasts with fungible assets like Bitcoin or Ethereum where each unit is interchangeable. Additionally, USDT is excluded because it is a stablecoin.

We construct a 22×22 correlation matrix, denoted as $\mathbf{C}$, in which each cell $\mathbf{C}_{ij}$, represents the cross-correlation at zero lag between coins i and j. This correlation is calculated differently depending on the time scale: for pairs where $i > j$, we use hourly log returns, while for $i < j$, we use log returns for mean weekly prices. This dual-time approach allows us to capture both short-term and long-term interdependencies between cryptocurrencies. In Fig. 4, the lower triangle displays the correlations computed from hourly data, and the upper triangle shows the correlations from weekly data. Prior to conducting the correlation analysis, an Augmented Dickey-Fuller test confirmed that all price log return series are stationary, each with a p-value of < 0.001. All the correlation values are statistically significant with p-value< 0.001.

The correlation values for short-term dynamics consistently exceed 0.3 across all pairs except for SHIB. However, for the long-term test, several pairs show zero or negative correlations. The highest correlation is 0.89 for the long term between XLM and XRP and 0.78 for the short term. Following closely behind for short-term dynamics are DOT-ADA (0.73) and BTC-ETH (0.72). Conversely, the strongest correlations for long-term dynamics are noted for SHIB-DOT (0.87) and DOT-ATOM (0.84). The lowest short-term correlation values are predominantly between SHIB and all other coins, with the lowest long-term correlation being between ARB and RNDR (-0.35).

We hypothesize that a higher weight between pairs in the co-mention network leads to higher similarity in the logarithmic returns of their hourly and mean weekly prices. To test this hypothesis, we perform a Pearson correlation analysis between the adjacency matrix of the co-mention network and the correlation matrix $\mathbf{C}$ for the log returns of hourly and weekly prices. The resulting correlation coefficient for hourly prices is 0.154, with a p-value of 0.019, and for weekly prices is 0.16, with a p-value of 0.015. This indicates that while there is a statistically significant relationship, the effect size is relatively small. This suggests that factors other than social media co-mentions likely play a more significant role in influencing financial market movements. Examples of these influential factors can be shared with Investors because investors often move in herds; thus, if a significant investor adjusts their portfolio across several assets, it could lead to simultaneous price movements in those assets. Similar Technological Foundations or common applications could also be another factor.

4.3   Correlation of LLM-Detected Signals and Cryptocurrency Prices

In this section, we examine the existence of Granger causality and lagged cross-correlation between the market prices of nine major cryptocurrencies and their

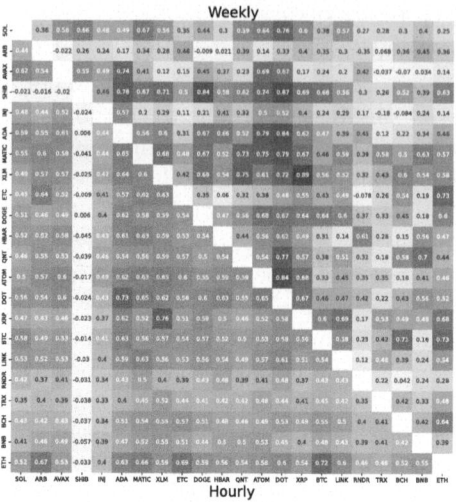

Fig. 4. Correlation matrix for log returns of prices for pairs of coins in the network holding at least 1% degree share. Upper triangle: Long-term weekly cross-correlation analysis ($k = 0$). Lower triangle: Short-term hourly cross-correlation analysis ($k = 0$).

associated social signals. Our goal is to identify the presence of these relationships and the specific lags where they become significant. This analysis also investigates if the optimal lag differs among cryptocurrencies, indicating that the interplay between market dynamics and social signals may be coin-specific.

The 24-hour Twitter social signal for each cryptocurrency i at time t is quantified as (4). An increase in SS_t^i is hypothesized to correlate with a rise in the cryptocurrency's price.

$$SS^i(t) = \left(\frac{1 + N_{\text{Buy}}^i(t)}{1 + N_{\text{Not buy}}^i(t)} \right) \tag{4}$$

We can also include information about the cryptocurrency market (Tweets related to crypto but without mentioning any coin) in our analysis because the market status influences prices simultaneously.

$$SS_{\text{Crypto}}^i(t) = \left(\frac{1 + N_{Buy}^i(t) + N_{\text{Buy}}^{\text{Crypto}}(t)}{1 + N_{\text{Not buy}}^i(t) + N_{\text{Not buy}}^{\text{Crypto}}(t)} \right) \tag{5}$$

The log return of the social signal for cryptocurrency i from time $t - 1$ to t is denoted by $r_{\text{SS}}^i(t)$. Similarly, cryptocurrency prices exhibit non-stationary behavior, and the log-returns of the prices are denoted by $r_{\text{CP}}^i(t)$.

Table 3 illustrates a Granger causality analysis between Twitter social signals, $r_{\text{SS}_{\text{crypto}}}^i(t)$, and price changes of nine major cryptocurrencies, $r_{\text{CP}}^i(t)$, highlighting how these signals may predict price movements. BTC and ETH demonstrate significant predictive correlations at multiple hourly lags of less than 10 h, whereas

DOGE exhibits signs of correlation after 10 h, and SOL for the entire 24 h, suggesting a robust effect of influencers and news outlets on coins prices. In contrast, ADA, XRP, and DOT show significant Granger causal links predominantly within the first three hours, indicating a rapid market response to new social data. Conversely, BNB exhibits week predictive lags, implying that Twitter data does not hugely influence price changes. SHIB reveals predictive power specifically at 21–22 hour lags, showcasing delayed market reactions to social signals. DOGE and SOL also present varied predictive lags, reflecting their community-driven market dynamics. This analysis underscores the diverse ways different cryptocurrencies respond to social media, driven by unique investor behaviors and market conditions.

Table 3. Granger causality analysis between Twitter social signal $r^i_{SS_{crypto}}(t)$ and prices changes of 9 major cryptocurrencies $r^i_{CP}(t)$. Blue color shows p−value < 0.01, orange shows p−value < 0.05 and green, p−value < 0.1.

	ADA	BNB	BTC	DOGE	DOT	ETH	SHIB	SOL	XRP
1H	0.040	0.111	0.001	0.017	0.043	0.002	0.812	0.001	0.021
2H	0.077	0.080	0.003	0.025	0.085	0.003	0.898	0.001	0.053
3H	0.047	0.090	0.001	0.006	0.063	0.003	0.978	0.001	0.047
4H	0.063	0.083	0.002	0.013	0.103	0.008	0.961	0.003	0.070
5H	0.100	0.118	0.001	0.013	0.185	0.010	0.972	0.004	0.101
6H	0.091	0.180	0.001	0.020	0.215	0.006	0.887	0.003	0.058
7H	0.094	0.230	0.002	0.031	0.297	0.012	0.236	0.006	0.072
8H	0.104	0.300	0.004	0.015	0.353	0.015	0.324	0.005	0.065
9H	0.144	0.383	0.006	0.021	0.435	0.024	0.320	0.008	0.091
10H	0.187	0.466	0.008	0.001	0.266	0.023	0.412	0.011	0.141
11H	0.209	0.560	0.012	0.003	0.334	0.025	0.490	0.014	0.105
12H	0.230	0.640	0.014	0.004	0.414	0.019	0.523	0.009	0.127
13H	0.282	0.647	0.023	0.006	0.429	0.026	0.438	0.011	0.158
14H	0.339	0.655	0.034	0.009	0.509	0.041	0.275	0.005	0.193
15H	0.296	0.687	0.046	0.008	0.545	0.051	0.330	0.007	0.214
16H	0.318	0.735	0.057	0.009	0.469	0.042	0.330	0.003	0.253
17H	0.375	0.793	0.078	0.008	0.493	0.062	0.241	0.004	0.292
18H	0.435	0.843	0.111	0.013	0.521	0.065	0.215	0.002	0.263
19H	0.366	0.794	0.099	0.009	0.580	0.078	0.277	0.002	0.261
20H	0.192	0.795	0.116	0.006	0.502	0.059	0.272	0.001	0.105
21H	0.217	0.808	0.137	0.006	0.543	0.079	0.046	0.001	0.126
22H	0.263	0.823	0.158	0.009	0.591	0.090	0.042	0.001	0.161
23H	0.291	0.861	0.166	0.013	0.633	0.110	0.051	0.002	0.150
24H	0.398	0.859	0.355	0.007	0.386	0.060	0.169	0.001	0.112

Table 4. Highest correlation coefficient with associated lags for 9 major coins with different formulas

Price	Social signal	ADA	BNB	BTC	DOGE	DOT	ETH	SHIB	SOL	XRP
$CP^i(t)$	$SS^i(t)$	$0.09^{(7D)}$	$-0.15^{(5D)}$	$-0.15^{(6D)}$	$-0.25^{(2H)}$	$-0.35^{(3D)}$	$0.23^{(1H)}$	$-0.05^{(6D)}$	$0.31^{(6D)}$	$0.06^{(2H)}$
$CP^i(t)$	$SS^i_{\text{crypto}}(t)$	$0.25(0H)$	$0.06^{(0H)}$	$-0.03^{(6D)}$	$0.22^{(1H)}$	$0.19^{(0H)}$	$0.23^{(0H)}$	$0.25^{(0H)}$	$0.33^{(10H)}$	$0.44^{(2H)}$
$r^i_{\text{CP}}(t)$	$r^i_{\text{SS}}(t)$	$0.03^{(6D)}$	$0.03^{(18H)}$	$0.08^{(0H)}$	$0.04^{(0H)}$	$0.05^{(0H)}$	$0.07^{(1H)}$	$0.03^{(20H)}$	$0.04^{(1H)}$	$0.04^{(8H)}$
$r^i_{\text{CP}}(t)$	$r^i_{\text{SS, crypto}}(t)$	$0.04^{(0H)}$	$0.05^{(0H)}$	$0.08^{(0H)}$	$0.05^{(0H)}$	$0.04^{(0H)}$	$0.06^{(1H)}$	$0.04^{(7H)}$	$0.06^{(1H)}$	$0.04^{(20H)}$

Table 4 illustrates the highest cross-correlation values between each cryptocurrency's price and its LLM-detected trading signal aggregated on influencers and news outlets using hourly (24 h) and daily lags (up to 7 days) to capture short-term and long-term market behaviors. Each row represents a different computation formula for the price and social trading signals, and each cell below the cryptocurrency names columns shows the highest correlation value for a cryptocurrency with the associated lag. In general, we can see that the coefficients for price are much larger than the coefficients for the log return when market social signals are included. For most of the cryptocurrencies, the cross-correlation values are highest when considering no lag, meaning that the 24-hour social signal instantly affects the price. Adding the information about the crypto-market increases the correlation coefficient. Additionally, the correlation coefficient and the associated lags differ from one coin to another, spanning from hour delays to daily ones. This suggests both short-term and long-term impacts of social signals from news outlets and influencers, warranting further investigation in future studies.

5   Conclusion

In this paper, we investigated an ecosystem of 2,687 cryptocurrency influencers and 74 financial news outlets on Twitter. We have collected all tweets posted by influencers and news outlets between 10/06/2023 and 02/28/2024 and used ChatGPT and CryptoBERT to exclude irrelevant content (94% accuracy) and detect 'buy' and 'not-buy' signals (84% accuracy) from their tweets text respectively. We have also gathered time series data of the top 9 major cryptocurrency prices and explored whether (1) aggregated LLM-detected 'buy' and 'not-buy' signals correlate with price and (2) co-mentioned cryptocurrencies exhibit correlation in their prices.

Our exploratory results showed that (1) influencers post about 9 times as many tweets as news outlets about cryptocurrencies; (2) of the 470,658 tweets posted by influencers and news outlets between 10/06/2023 and 02/28/2024, 54% were classified as implying a 'buy' signal and 66% implying a 'not-buy' signal; (3) 44.4% of all tweets posted by influencers and news outlets only contain a single coin name; and (4) tweets containing a 'buy' signal typically mention at least one specific coin, whereas, 'not-buy' tweets often do not specify any coin name and usually focus on general market conditions.

Our time series analysis results showed that while there is a statistically significant correlation between the attributes of the co-mentioned cryptocurrencies, the effect size is small (0.17). This suggests that influencers and news outlets run other mechanisms to influence the mentioning patterns within their Twitter posts. With respect to the correlation between the co-mentioned cryptocurrencies and their market prices, our analyses revealed that while the correlation values for short-term dynamics consistently exceed 0.3 except for SHIB across all pairs, when examining long-term dynamics, several pairs show correlations close to zero or even negative values. Finally, our analysis of cryptocurrency prices and social signals revealed mixed results, with some cryptocurrencies showing a positive correlation with trading signals and others showing zero or negative correlations.

Furthermore, this research limited its scope by testing only a few formulas for obtaining social signals. However, there are additional possibilities to explore. For instance, one could assign more weight to users with larger followings or those who are more central within the follower network according to centrality measures. Similarly, greater weight could be given to tweets with numerous likes and retweets. Another avenue for improvement is developing methods to identify fake accounts, bots, misinformation spreaders, and campaigns attempting to manipulate the market. Future research should study other social media platforms where cryptocurrency supporters and influencers are actively posting, such as Reddit, Telegram, and YouTube, as well as exploring mechanisms that could explain the sort of relationships we found in this paper.

Acknowledgments. We thank Atena Jafari, Daria Stetsenko, Mahdis Abbasi, Dina Della Casa, Zahra Baghshahi, Fabio Melliger, Sarvenaz Ebrahimi, and Maria Korobeynokiva for their outstanding research assistance. Yasaman Asgari thanks the University of Zurich and the Digital Society Initiative for (partially) financing this project.

Disclosure of Interests. The authors declare that they have no competing interests.

References

1. Alizadeh, M., et al.: Open-source large language models outperform crowd workers and approach ChatGPT in text-annotation tasks. arXiv preprint arXiv:2307.02179 (2023)
2. Ante, L.: How Elon Musk's Twitter activity moves cryptocurrency markets. Technol. Forecast. Soc. Chang. **186**, 122112 (2023)
3. Baker, M., Wurgler, J.: Investor sentiment in the stock market. J. Econ. Perspect. **21**(2), 129–151 (2007)
4. Bollen, J., Mao, H., Zeng, X.: Twitter mood predicts the stock market. J. Comput. Sci. **2**(1), 1–8 (2011)
5. Cano-Marin, E., Mora-Cantallops, M., Sánchez-Alonso, S.: Twitter as a predictive system: a systematic literature review. J. Bus. Res. **157**, 113561 (2023)
6. Fama, E.F., Fisher, L., Jensen, M.C., Roll, R.: The adjustment of stock prices to new information. Int. Econ. Rev. **10**(1), 1–21 (1969)

7. Garcia, D., Schweitzer, F.: Social signals and algorithmic trading of bitcoin. R. Soc. Open Sci. **2**(9), 150288 (2015)
8. Georgoula, I., Pournarakis, D., Bilanakos, C., Sotiropoulos, D., Giaglis, G.M.: Using time-series and sentiment analysis to detect the determinants of bitcoin prices. Available at SSRN 2607167 (2015)
9. Gilardi, F., Alizadeh, M., Kubli, M.: ChatGPT outperforms crowd workers for text-annotation tasks. Proc. Natl. Acad. Sci. **120**(30), e2305016120 (2023)
10. Guidi, B.: When blockchain meets online social networks. Pervasive Mob. Comput. **62**, 101131 (2020)
11. Kahneman, D.: Maps of bounded rationality: psychology for behavioral economics. Am. Econ. Rev. **93**(5), 1449–1475 (2003)
12. Kaminski, J.: Nowcasting the bitcoin market with twitter signals. arXiv preprint arXiv:1406.7577 (2014)
13. Kaplanski, G., Levy, H.: Sentiment and stock prices: the case of aviation disasters. J. Financ. Econ. **95**(2), 174–201 (2010)
14. Karalevicius, V., Degrande, N., De Weerdt, J.: Using sentiment analysis to predict interday bitcoin price movements. J. Risk Finance **19**(1), 56–75 (2018)
15. Kim, Y.B., Lee, J., Park, N., Choo, J., Kim, J.-H., Kim, C.H.: When bitcoin encounters information in an online forum: using text mining to analyse user opinions and predict value fluctuation. PloS One **12**(5), e0177630 (2017)
16. Kraaijeveld, O., De Smedt, J.: The predictive power of public twitter sentiment for forecasting cryptocurrency prices. J. Int. Finan. Markets. Inst. Money **65**, 101188 (2020)
17. Kulakowski, M., Frasincar, F.: Sentiment classification of cryptocurrency-related social media posts. IEEE Intell. Syst. **38**(4), 5–9 (2023)
18. Kwak, H., Lee, C., Park, H., Moon, S.: What is Twitter, a social network or a news media? In: Proceedings of the 19th International Conference on World Wide Web, pp. 591–600 (2010)
19. Li, T., van Dalen, J., van Rees, P.J.: More than just noise? Examining the information content of stock microblogs on financial markets. J. Inf. Technol. **33**(1), 50–69 (2018)
20. Li, X., Xie, H., Chen, L., Wang, J., Deng, X.: News impact on stock price return via sentiment analysis. Knowl.-Based Syst. **69**, 14–23 (2014)
21. Mai, F., Bai, Q., Shan, J., Wang, X.S., Chiang, R.H.L.: The impacts of social media on bitcoin performance (2015)
22. Mao, H., Counts, S., Bollen, J.: Predicting financial markets: comparing survey, news, Twitter and search engine data. arXiv preprint arXiv:1112.1051 (2011)
23. Mungo, L., Bartolucci, S., Alessandretti, L.: Cryptocurrency co-investment network: token returns reflect investment patterns. EPJ Data Sci. **13**(1), 11 (2024)
24. Peterson, R.L.: Trading on Sentiment: The Power of Minds Over Markets. Wiley, New York (2016)
25. Shen, D., Urquhart, A., Wang, P.: Does Twitter predict bitcoin? Econ. Lett. **174**, 118–122 (2019)
26. Sockin, M., Xiong, W.: Decentralization through tokenization. J. Financ. **78**(1), 247–299 (2023)
27. Valle-Cruz, D., Fernandez-Cortez, V., López-Chau, A., Sandoval-Almazán, R.: Does Twitter affect stock market decisions? Financial sentiment analysis during pandemics: a comparative study of the H1N1 and the Covid-19 periods. Cogn. Comput. **14**, 372–387 (2022)

28. Zhang, X., Fuehres, H., Gloor, P.A.: Predicting stock market indicators through Twitter "i hope it is not as bad as i fear". Procedia-Soc. Behav. Sci. **26**, 55–62 (2011)
29. Zou, Y., Herremans, D.: PreBit–a multimodal model with Twitter FinBERT embeddings for extreme price movement prediction of Bitcoin. Expert Syst. Appl. **233**, 120838 (2023)

AmGNN: A Framework for Adaptive Processing of Inter-layer Information in Multi-layer Graph

Huaisheng Zhu, Zongyu Wu, Tianxiang Zhao, and Suhang Wang[✉]

The Pennsylvania State University, University Park, PA 16802, USA
{hvz5312,zongyuwu,tkz5084,szw494}@psu.edu

Abstract. Graphs play a vital role in various applications. Graph Neural Networks (GNNs) excel at capturing topology information by using a message-passing mechanism to enrich node representations with local neighborhood information. Despite their success in modeling single-layer graphs, real-world scenarios often involve multi-layer graphs where nodes can have multiple edges or relationships represented as different layers. Existing methods of multi-layer graph learning struggle to efficiently process inter-layer information, as they mainly focus on preserving similar layers or shared invariant information, which may not be suitable for all situations. We propose a novel framework called Adaptive Multi-layer Graph Neural Networks (AmGNN) to address this challenge. AmGNN learns shared invariant information for nodes that need it and selectively preserves relevant layers' information for nodes not requiring shared invariance. We introduce multi-layer graph contrastive learning to efficiently capture invariant information and learn weights for adaptive processing. Our experiments on real-world multi-layer graphs validate the effectiveness of AmGNN in node classification tasks.

Keywords: Multi-layer Graph · Node Classification · Graph Neural Networks

1 Introduction

Graphs are pervasive in the real world, such as social networks [24], recommendation systems [2], and knowledge graphs [21]. To capture the topology information in graphs, Graph Neural Networks (GNNs) typically follow the message-passing mechanism, which aggregates the neighborhood representation of a node to enrich the node's representation. This process enriches node representations and preserves both node feature characteristics and topological structures, benefiting various tasks like node classification [14], link prediction [36] and clustering [29].

Despite the great success of GNNs in modeling graphs, the majority of existing works concentrate on single-layer graphs, where only one edge/relationship exists between any pair of nodes. However, in the real world, a pair of nodes

L. M. Aiello et al. (Eds.): ASONAM 2024, LNCS 15211, pp. 472–488, 2025.
https://doi.org/10.1007/978-3-031-78541-2_29

can have multiple edges or relationships, forming what is known as a *multi-layer graph*. For example, Fig. 1 shows a 3-layer university's social network with three different types of relations: club interactions (Layer 1), pre-university friendships (Layer 2), and same-major student relations (Layer 3). Simply adopting GNNs designed for single-layer graphs can't achieve good performance on multilayer graphs. This is because of limited structural information within a single layer and empirical evidence indicating the inadequate performance of single-layer GNNs [18]. Hence, several efforts have been made to model the relations across different layers in multi-layer graphs by aggregating both intra-layer and inter-layer information [18,32]. Generally, they adopt existing GNN models for single-layer graphs to model intra-layer information. For inter-layer aggregation, they aim to preserve shared invariant information (certain common patterns existing in different layers) across layers by relying on correlations between different layers. They typically retain the most similar layers' information to learn representations.

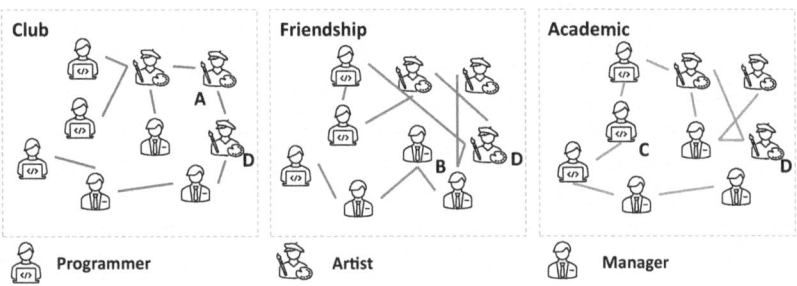

Fig. 1. A motivation example of adaptively processing inter-layer information for multilayer graph learning. Some nodes require shared invariant information, while others do not.

In real-world applications, shared invariant information based on similarity across layers might not always be meaningful for label information, as some nodes' labels could be relevant to specific layers' information. Note that shared invariant information could encompass several layers with higher correlations or might involve most of the similar local neighbors across different layers. Consider the scenario depicted in Fig. 1, nodes correspond to students within the university's social network, while edges symbolize interactions occurring in distinct social environments. The labels pertain to the occupations of these students. It shows that individuals' occupations after graduation might be influenced by specific relations or shared patterns across diverse relations. For example, student A's engagement in the Art club could lead to connections with students aspiring to become artists. If we use club relations for predictions, their label might be accurately predicted based on their friends' information. However, the preservation of the most similar or invariant information across layers, like the neighbor patterns of student A in both friendship and academic relations (where the neighbors are labeled as managers), could lead to inaccurate predictions. Similarly, student B is more inclined to prioritize the preservation of friendship relations and student C is more inclined to prioritize the preservation of academic relations. In contrast, certain individuals, like student D, have occupations that are influenced by shared invariant information from three relations, exhibiting similarities in their local neighborhoods and having most of their friends across different relations belonging to the artist occupation. Assigning higher weights to just one layer could introduce noise, as half of D's neighbors belong to different classes compared to D. This underscores the importance of identifying nodes that need to preserve similar layers' information. Hence, a crucial step is to determine which nodes require shared invariant information across layers and which nodes specifically need information from certain layers. Once this differentiation is made, an adaptive selection process is employed to incorporate important information from different layers for nodes that may benefit from specific layers' information for accurate label prediction. Though promising, the work on adaptively learning layers' information for each node to predict labels of nodes in multilayer graphs is limited.

Therefore, in this paper, we explore a novel problem of node classification in multi-layer graphs, focusing on recognizing nodes that need to preserve shared invariant information while selectively leveraging information from other layers. In essence, there are two main challenges: (i) How to determine the nodes requiring invariant information and preserve shared invariant information for node classification?; (ii) How to assist nodes, not in need of invariant information, in selecting relevant layer information for node classification? To resolve these challenges, we introduce a novel framework called Adaptive Multi-layer Graph Neural Networks (AmGNN). Firstly, to learn the weights determining whether to preserve invariant information, AmGNN assigns higher weights to nodes with consistent predicted results from pretrained GNN models on single layer relations. By doing so, preserving shared information for these nodes is more likely to yield correct prediction results. Moreover, to efficiently capture invariant information, we deviate from previous approaches that rely on layers' similarity. Instead, we introduce a multi-layer graph contrastive learning method that allows us to assign different weights to the contrastive learning loss for each node, providing control over whether nodes preserve invariant information. Additionally, for nodes without consistent prediction results, we encourage them to preserve the information from specific layers where the pre-trained GNN models yield high-confidence predictions. Our main **contributions** are: (i) We explore a novel problem of adaptive processing for inter-layer information for multi-layer graphs; (ii) We introduce AmGNN, which learns shared invariant information exclusively for nodes that require it and adaptively preserves the relevant layers' information for nodes that may not need shared invariant information; and (iii) Experiments on real-world multi-layer graphs demonstrate the effectiveness of the proposed framework AmGNN.

2   Related Work

Graph Neural Networks. GNNs are popular for node representation learning on graph-structured data and can be categorized into two types: spectral-based [8,9,14,28,33] and spatial-based [5,6,30,34]. Spectral-based GNNs use graph signal processing and apply convolutional operations to graph data in the spectral domain. GCN [14], which uses a first-order approximation, is an example of a spectral-based GNN. Spatial-based GNNs aggregate information from neighboring nodes to update the representation of a given node. For example, Graph Attention Network (GAT) [30] utilizes attention mechanisms to update node representations using different weights.

Graph Contrastive Learning. Contrastive methods have been widely used in graph self-supervised learning [6,7,15]. Graph AutoEncoders [15], for instance, learns node embeddings unsupervisedly by reconstructing the adjacency matrix, while GMI [23] maximizes the mutual information of both features and edges between inputs and outputs. Inspired by the success of contrastive learning, recent work has explored creating positive pairs using reliable graph information instead of data augmentations. For example, SUGRL [20] utilizes two encoder models, one based on GCN and the other an MLP, to generate two sets of node embeddings from different sources. The positive pair for each node is then constructed using its GCN output and MLP output. Similarly, AFGRL [16] and AF-GCL [31] consider nodes in the target node's multi-hop neighborhood as candidate positive examples and employ well-designed similarity measures to select the most similar nodes as positive examples.

GNNs for Multilayer Graphs. Multi-layer graph, also known as a multiplex, multi-view, or multi-dimensional graph, takes into account multiple relationships among nodes [1,17,18,25]. Various approaches have been proposed to handle the complexity of multi-layer graphs. Some methods, such as MVE [25] and HAN [32], leverage attention mechanisms to effectively combine embeddings from different views. mGCN [18] addresses interactions within and across views to improve node classification. Node embedding techniques have been explored for node clustering and classification tasks in multi-layer graphs [3,26]. For instance, VANE [3] employs adversarial training to enhance node representation learning comprehensiveness and robustness. Contrastive learning has also found application in learning expressive representations for multi-layer graphs [11,12]. For instance, HDMI [12] learns network embeddings for multi-layer networks by incorporating high-order mutual information and a fusion module based on an attention mechanism. Also, SSDCM [19] maximizes mutual information between local and contextualized global graph summaries using an InfoMax learning strategy, facilitating effective joint modeling of nodes and clusters while utilizing cross-layer links to regularize embeddings across different layers.

Our work is inherently different from existing works: (i) Existing works on multi-layer GNNs concentrate on preserving layers' invariant information across layers; while our study focuses on a novel problem of efficiently processing inter-layer information for multi-layer graphs; (ii) We propose a novel framework that identifies which nodes require shared invariant information and which nodes need specific information from individual layers. Additionally, our framework guides the inter-layer aggregation process adaptively, allowing for selective and effective information propagation across different layers.

3   Problem Definition and Notations

We use $\mathcal{G} = \{\mathcal{V}, \mathcal{E}_1, \ldots, \mathcal{E}_L\}$ to denote an L-layer attributed graph, where $\mathcal{V} = \{v_1, \ldots, v_N\}$ is the set of N nodes, $\mathcal{E}_l \subseteq \mathcal{V} \times \mathcal{V}$ is the set of edges in the l-th layer, and $\mathbf{X}$ is the node attribute matrix with $\mathbb{X}[j,:] \in \mathbf{R}^{1 \times d}$ being node attribute vector for node v_j. $\mathbf{A}^l$ is the adjacency matrix for the l-th layer. $A^l_{i,j} = 1$ if nodes v_i and v_j are connected in layer l, otherwise $A^l_{i,j} = 0$. Specifically, in semi-supervised node classification, only a subset of nodes are labeled. We denote the labeled set as $\mathcal{V}_L \in \mathcal{V}$ with $\mathcal{Y}_L$ being the corresponding label set of the labeled nodes. The remaining nodes $\mathcal{V}_U = \mathcal{V} \backslash \mathcal{V}_L$ are the unlabeled set. The problem is formally defined as: *Given a multi-layer graph* $\mathcal{G} = \{\mathcal{V}, \mathcal{E}_1, \ldots, \mathcal{E}_L\}$ *and the partial labels* $\mathcal{Y}_L$, *we aim to learn a node classifier* $Q_\theta(\mathcal{V}, \mathbf{A}^1, \ldots, \mathbf{A}^L, \mathbf{X}) \rightarrow \mathcal{Y}$.

4   Methodology

In this section, the proposed framework is designed for adaptively learning different layers' information of multi-layer graphs. An illustration of the proposed framework is shown in Fig. 2. Capturing inter-layer information is crucial for multi-layer graph neural networks to effectively handle node classification tasks. To achieve this, we propose a multi-layer Graph Neural Network framework that incorporates intra-layer aggregation, inter-layer aggregation, and multi-layer graph contrastive learning to model and preserve shared invariant information. However, as depicted in Fig. 1, not all nodes necessarily require invariant information; some nodes may benefit from it, while others may require specific layers' information. To address this issue, we introduce node-specific weights for multi-layer graph contrastive learning, enabling us to adaptively control the importance of invariant information for each node. Moreover, AmGNN guides the inter-layer aggregation process, allowing nodes that need specific information to learn relevant layers' information more effectively. Next, we introduce the details.

4.1   Multi-layer Graph Neural Network

To enhance representation learning for multi-layer graphs by aggregating both inter-layer and intra-layer information, we introduce the Multi-layer GNN. Each layer in the graph may possess distinct structural characteristics that can be beneficial for downstream tasks. The design of this module is inspired by mGCN [18].

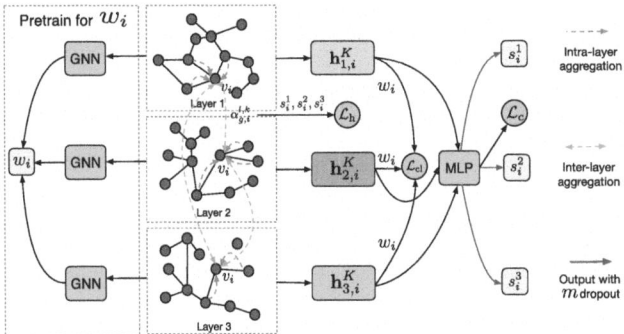

Fig. 2. An overview of the proposed AmGNN. We use a multi-layer graph with $L = 3$ layers as an example.

The first step of Multi-layer GNNs is to model intra-layer information, conventional message passing for single-layer graphs is used to model the intra-layer relations as:

$$\mathbf{T}_l^k = \mathbf{H}^{k-1}\mathbf{W}_l^k, \quad \mathbf{F}_l^k = \sigma\left(Conv(\mathbf{T}_l^k)\right), \quad \mathbf{E}_l^k = Conv_{lin}(\mathbf{T}_l^k), \tag{1}$$

where l is the layer of graphs and k is the layer of the model. $Conv$ means the convolution operation in GNN and $Conv_{lin}$ denotes the linear operation in GNN to make sure $\mathbf{F}_l^k$ and $\mathbf{E}_l^k$ have the same dimension. $\mathbf{W}_l^k$ is the learnable weight matrix utilized to transform the representation of the fused representation to the layer l and σ means activation function. $\mathbf{H}^{k-1}$ is the learned representation matrix from the last layer of the model and $\mathbf{H}^0 = \mathbf{X}$. However, the structure information within a single layer is limited. Thus, independently learning and fusing information from different layers may not adequately capture the diverse layers' information for each node. To tackle this challenge, we propose to enhance the message-passing process by aggregating inter-layer information. Specifically, we aggregate information from other layers for each node using different weights, enabling a more effective modeling of diverse information across the layers. Subsequently, inspired by [18], we combine the aggregated intra-layer and inter-layer information to create a comprehensive representation for each node:

$$\mathbf{H}_l^k = \text{Concat}(\mathbf{F}_l^k, \sum_{g=1, g \neq l}^{L} \mathbf{G}_g^{l,k}\mathbf{E}_g^k), \quad \boldsymbol{\alpha}^{l,k} = \text{softmax}([\boldsymbol{z}_1^k, ...\boldsymbol{z}_{l-1}^k, \boldsymbol{z}_{l+1}^k, ..., \boldsymbol{z}_L^k]),$$
$$\tag{2}$$

where $\boldsymbol{z}_g^k = \text{MLP}(\mathbf{E}_g^k, \mathbf{E}_l^k)$ and softmax represents the row-wise softmax operation. $\boldsymbol{\alpha}^{l,k}$ represents weights used to aggregate information from other layers to layer l. $\boldsymbol{G}_g^{l,k}$ is a diagonal matrix $diag(\boldsymbol{\alpha}_g^{l,k})$ where $\boldsymbol{\alpha}_g^{l,k}$ denotes the weights to aggregate information from layer g to l. Concat represents the concatenation operation. Subsequently, in order to capture information from all layers, following [18], we aggregate the representations from different layers to obtain the representation for the next model layer:

$$\mathbf{H}^k = \sigma \left(\mathbf{W}^k \cdot \text{Concat} \, {}_{l=1}^{L} \mathbf{H}_l^k \right), \tag{3}$$

where $(\cdot)$ means matrix multiplication and $\mathbf{W}^k$ is the learnable parameter. Finally, we obtain the predicted label matrix by the final model layer:

$$\hat{\mathbf{Y}} = \text{MLP}(\mathbf{H}^K), \tag{4}$$

where $\hat{\mathbf{Y}} \in \mathbb{R}^{N \times C}$ is the predicted probability matrix and C is the number of classes. We adopt cross-entropy to train the model for the node classification as:

$$\mathcal{L}_c = - \sum_{v_i \in \mathcal{V}_L} \sum_{c=1}^{C} Y_i^c \log \hat{Y}_i^c, \tag{5}$$

where Y_i^c is the c-th element of the one-hot encoding of v_i's ground-truth label.

4.2 Multi-layer Graph Contrastive Learning

To effectively aggregate information from different layers in a multi-layer Graph Neural Network using learnable weights, it is vital to develop supervised signals that guide the learning process across these layers. The shared invariant information among different layers holds essential characteristics of both the graph's structure and node features [18]. One common way to preserve shared invariant information across layers is to employ contrastive learning methods, i.e., we treat node representations from various layers as positive samples, promoting the preservation of shared and similar information within their representations. Moreover, we incorporate negative samples to further reinforce the preservation of shared invariant information. To achieve this goal, we introduce a multi-layer graph contrastive learning method across different layers for v_i of the model's final K-th layer as:

$$\mathcal{R}_{\mathcal{N}}^{i,K} = \sum_{l=1}^{L} \sum_{g \neq l}^{L} \log \frac{e^{\cos\left(\mathbf{h}_{l,i}^K, \mathbf{h}_{g,i}^K\right)}}{e^{\cos\left(\mathbf{h}_{l,i}^K, \mathbf{h}_{g,i}^K\right)} + \sum_{v_j \in \mathcal{N}_n^{g,i}} e^{\cos\left(\mathbf{h}_{l,i}^K, \mathbf{h}_{g,j}^{K-}\right)}},$$

where $\mathcal{N}_n^{g,i}$ is set of negative on the layer g for v_i, $\mathbf{h}_{g,j}^K$ is v_j's representation on g-th graph and K-the model layer, and $\cos(\cdot, \cdot)$ denotes the cosine similarity function between two vectors. While many graph contrastive learning methods typically employ data augmentation techniques to generate negative samples, such approaches can be computationally and memory-intensive. To mitigate these problems, we opt for a simpler approach by randomly selecting some negative samples for node v_i that do not have links based on the original graph structure. During the training process, these negative samples form a negative sampling set $\mathcal{N}_n^{g,i}$ for the layer g of the node v_i.

4.3 Adaptive Inter-layer Aggregation

Contrastive learning for multi-layer graphs is effective in preserving invariant information across layers. However, not all nodes necessarily require this shared invariant information. To address this, we propose to assign different weights to the multi-layer contrastive learning loss of different nodes, allowing for selective preservation of shared invariant information based on each node's specific needs. To calculate these weights, as shown in Fig. 2, we first train L GNN models on L graphs respectively to obtain predicted class probabilities for nodes in different layers. Note that this only need to be done once. Let $\mathbf{p}_i^l \in \mathbb{R}^C$ denote the predicted class probability vector of node v_i in layer l, where C denotes the number of classes for node classification. Nodes with the same labels across different layers can be denoted as $\mathbf{p}_i^l = \mathbf{p}_i^{l'}$ for l' ranging from 1 to L and $l \neq l'$. For nodes with consistent predictions, the predicted label can be treated as invariant information across layers. These consistent prediction results are highly confident and have a high probability of accurately representing the ground truth label for node v_i, as models trained separately exhibit similar prediction outcomes. To represent the level of consistent prediction results for nodes in different layers, we first calculate the average prediction results $\mathbf{p}_i$ from pretrained GNN models on different layers:

$$\mathbf{p}_i = \frac{1}{L}\sum\nolimits_{l=1}^{L}\mathbf{p}_i^l, \tag{6}$$

To quantify the level of consistent prediction results, we utilize entropy as a measure, where high entropy indicates low-level consistency and low entropy indicates high-level consistency. Accordingly, we use normalized entropy as the weight to represent the level of nodes with consistent prediction results:

$$w_i = 1 - b_i, \quad b_i = -\frac{\sum_{c=1}^{C}\mathbf{p}_{v_i,c}\log\mathbf{p}_{v_i,c}}{\log C}, \tag{7}$$

where b_i is the normalized entropy of $\mathbf{p}_i$. To ensure that lower or higher weights correspond to higher or lower predicted consistency, and since $b_i \in [0,1]$, we use $w_i = 1 - b_i$ to denote the weight assigned to nodes for deciding the level of preserving shared invariant information. To help nodes preserve invariant information, we use the multi-layer graph contrastive learning with different weights w_i for different nodes on the model's final Kth layer:

$$\mathcal{L}_{\mathrm{cl}} = -\frac{w_i}{N}\mathcal{R}_N^{i,K}, \tag{8}$$

Training the multi-layer graph contrastive learning may still be time-consuming so we update our model's parameters using this loss every 30 epochs.

Another challenge is about how to preserve useful information for nodes with low-level consistent prediction results because invariant shared information for these nodes may not be reliable as shown in Fig. 1. To address this challenge, we propose to propagate trustworthy information across layers and mitigate the influence of uncertain information. Firstly, we denote the set of nodes which don't have consistent prediction results from separately trained GNN models on different layers as $\mathcal{N}_h$. And we call these nodes inconsistent nodes. We show the empirical results in Fig. 3. Specifically, we visualize the predicted probability distribution of nodes' ground-truth classes in relation to the ratio of correctly

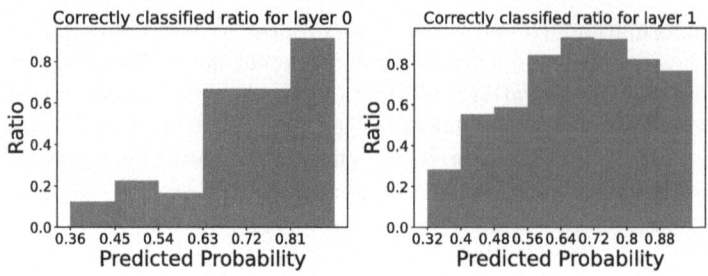

Fig. 3. Predicted probability distribution (correctly classified ratio) for inconsistent nodes on the dataset ACM.

classified nodes. This visualization is obtained from GNN models trained independently on individual layers. The predicted probability of node v_i' ground-truth classes can be denoted as p_{i,c_g}^l, where c_g is the ground-truth label of the node v_i. Figure 3 shows how the ratio of correctly classified nodes varies across different intervals of predicted probability. From the figure, we observe that predicted results from pretrained GNN models on layers with high-confidence probabilities (higher predicted probabilities for the ground-truth label) hold a high correctly classified ratio. Based on these observations, we will encourage each layer's representation to aggregate layers with highly confident prediction results for nodes $v_i \in \mathcal{N}_h$.

To accurately measure the confidence level of predicted results, relying solely on the value of predicted probability may not provide a comprehensive evaluation. One commonly used approach to assess confidence is by employing dropout techniques [4,13]. Dropout introduces randomness by deactivating neurons during training, creating diverse subnetworks with unique active neurons that lead to prediction variations. When a model is uncertain, distinct subnetworks yield significantly different outputs for the same input, while confident models produce more consistent outputs across dropout instances. Specifically, we start by obtaining different prediction results by applying dropout several times to the model for the layer l. This yields a set of representation vectors $\{\mathbf{H}_{1,l}^K, \ldots, \mathbf{H}_{m,l}^K\}$ from the model. Subsequently, we input these representation vectors into the prediction layer (MLP layer) in Eq. (4) and obtain the results $\{\mathbf{y}_1^l, \ldots, \mathbf{y}_m^l\}$.

To quantify the confidence level of nodes, we calculate the level of consistency among the different prediction results. This consistency measure helps us assess the stability and reliability of the model's predictions, providing valuable insights into the uncertainty estimation of the model. To represent the consistency level of prediction results, we first obtain the average prediction results by generating m results using dropout:

$$\overline{\mathbf{y}}^l = \frac{1}{m} \sum_{a=1}^{m} \mathbf{y}_a^l, \tag{9}$$

Furthermore, the entropy measure, as discussed in Eq. (7), can also be utilized to assess the uncertainty level of predicted results obtained from different dropout ratios. This allows us to determine the level of trustable information across layers. The confidence score for the hard classified node v_i is:

$$s_i^l = 1 - \frac{\sum_{c=1}^{C} \overline{y}_{i,c}^l \log \overline{y}_{i,c}^l}{\log C}, \tag{10}$$

where the value of s_i^l can serve as an indicator of which layers' information is considered trustable. To reduce computational overhead, we update s_i^l every 10 epochs.

With s_i^l, we can now encourage the inter-layer aggregation to effectively aggregate trustable information. This is achieved with the following loss function

$$\mathcal{L}_h = -\frac{1}{L |\mathcal{N}_h|} \sum_{v_i \in \mathcal{N}_h} \sum_{k=1}^{K} \sum_{l=1}^{L} \sum_{g=1, g \neq l}^{L} s_i^g \log \alpha_{g,i}^{l,k}, \tag{11}$$

Table 1. Statistics of the Datasets

Datasets	Num of Nodes	Edge Types	Num of Edges per Layer	Feature Dimension	Num of Classes
ACM	3025	PSP	2210761	1830	3
		PAP	29281		
IMDB	3550	MAM	66428	1007	3
		MDM	13788		
DBLP	7907	PAP	144783	2000	4
		PPP	90145		
		PATAP	57137515		
Amazon	7621	IVI	266237	2000	4
		IBI	1104257		
		IOI	16305		

where $\alpha_{g,i}^{l,k}$ is the weight to control the inter-layer aggregation for the node v_i. Note that the above loss function works for $L \geq 3$ but does not work for $L = 2$. This is because $\alpha_{g,i}^{l,k}$ obtained from Eq.(2) is 1 for a two-layer graph, making $\mathcal{L}_h$ being 0.

4.4 Final Objective Function of AmGNN

Putting everything together, the final objective function of AmGNN is given as:

$$\min_{\Theta} \mathcal{L} = \mathcal{L}_c + \lambda \mathcal{L}_{cl} + \gamma \mathcal{L}_h, \tag{12}$$

where θ represents the learnable parameters for our model, λ and γ are hyper-parameters to control the loss for multi-layer graph contrastive learning and adaptive inter-layer aggregation.

5 Experiments

In this section, we conduct experiments on real-world multi-layer graphs to demonstrate the effectiveness of AmGNN. In particular, we aim to answer the following research questions: (i) **RQ1** Can AmGNN provide accurate node classification for multi-layer graphs? (ii) **RQ2** What are the contributions of each component for AmGNN? (iii) **RQ3** How reliable is confidence score?

5.1 Datasets

Following previous work [12,22], we conduct experiments on four publicly available real-world multi-layer graphs. The statistics of these datasets are summarized in Table 1 and their details are given below: (i) **ACM** [32]: There are two kinds of edges: paper-author-paper (PAP) and paper-subject-paper (PSP). Each paper has a node feature extracted from its abstract. papers have one of the following labels: Database, Wireless Communication, and Data Mining; (ii) **IMDB**[1]: There are two edge types: movie-actor-movie (MAM) and movie-director-movie (MDM). The movie feature is the bag-of-words feature of movie plots. Each movie belongs to one of the following classes: Action, Comedy, Drama; (iii) **DBLP** [27]: It contains three types of edges: paper-paper-paper (PPP), paper-author-paper (PAP), and paper-author-term-author-paper (PATAP). The bag-of-words feature of paper's abstract represents each paper. Each node has one of the following labels: Data Mining, Artificial Intelligence, Computer Vision, and Natural Language Processing; (iv) **Amazon** [10]: It is sourced from Amazon.com and represents a three-layer graph based on three relations: (1) item-viewed-item (IVI): two items are viewed by the same customer; (2)item-bought-item (IBI): two items are bought by the same customer; (3) item-together-item (IOI): two items are co-bought by a customer. Each node is classified into one of the following classes: Beauty, Automotive, Patio Lawn and Garden, and Baby. Both DBLP and Amazon are three-layer graphs while ACM and IMDB are two-layer graphs. Note that though our $\mathcal{L}_h$ only works for graphs of three or more layers, we also include two-layer graphs to show that our method can also achieve good results on two-layer graphs.

5.2 Experimental Setup

Baselines. We compare AmGNN with representative and state-of-the-art methods for node classification, which include:

---

[1] https://www.imdb.com/.

- **GCN** [14]: GCN is one of the most popular spectral GNN models based on graph Laplacian, which has shown great performance for node classification.
- **GAT** [30]: GAT adopts an attention mechanism for enhanced neighborhood aggregation during message-passing.
- **GIN** [35]: Graph Isomorphism Networks (GIN) employ multi-layer perceptrons to process the aggregated information from neighbors at each layer, enabling the model to learn more potent and expressive node representations.
- **GCN/GAT/GIN-C**: As GCN/GAT/GIN can only deal with single layer graph, for GCN/GAT/GIN-C, we combine edge information from all layers to construct one adjacency matrix. Then, we adopt them on the combined adjacency matrix for node classification.
- **mGCN** [18]: mGCN utilizes GCN to extract node embedding for each layer of the multi-layer graphs and then combine them via the attention mechanism.
- **HAN** [32]: The model learns node embeddings specific to different metapaths from different relations. It then employs the attention mechanism to combine these embeddings into a single vector representation for each node.
- **SSDCM** [19]: It uses an InfoMax to maximize mutual information between local and contextualized global graph summaries. It also leverages cross-layer links to regularize the embeddings across different layers of the graph.
- **HDMI** [12]: HDMI adopts a new contrastive learning loss by capturing high-order information across different layers.
- **X-GOAL** [11]: It includes a GOAL framework, which learns node embeddings for each graph layer, and an alignment regularization to model and propagate information across different layers. After obtaining node embeddings for both HDMI and X-GOAL, we employ MLPs to perform node classification.

Table 2. Node Classification Performance. The best and second-best performances are marked with boldface and underlined.

Dataset	ACM		IMDB		DBLP		Amazon	
Metric	Accuracy	F1-score	Accuracy	F1-score	Accuracy	F1-score	Accuracy	F1-score
GCN	91.08±0.39	91.14±0.38	66.95±0.28	66.97±0.32	82.82±0.63	83.64±0.68	75.82±0.15	75.22±0.04
GAT	90.53±0.17	90.51±0.16	67.30±0.26	67.20±0.24	81.95±0.26	82.74±0.29	75.65±0.78	75.10±0.85
GIN	89.54±0.04	89.54±0.04	62.44±0.85	62.32±0.92	81.92±0.80	82.75±0.79	74.89±0.23	74.26±0.39
GCN-C	67.31±0.44	63.98±0.98	62.42± ±0.68	62.62±0.70	77.13±0.51	77.20±0.47	72.09±0.24	67.24±0.54
GAT-C	68.71±0.29	62.79±0.75	62.59±0.47	61.38±0.93	77.42±0.63	77.60±0.54	71.49±0.31	67.82±0.65
GIN-C	69.52±0.31	63.42±0.49	62.19±0.59	62.77±0.93	76.98±0.43	77.11±0.38	66.39±0.48	62.60±0.26
HAN	92.87±0.26	92.94±0.35	69.61±0.32	69.05±0.54	84.07±0.43	84.12±0.31	77.57±0.46	77.28±0.27
mGCN	92.63±0.58	91.70±0.34	67.84±0.47	67.93±0.48	83.91±0.39	83.27±0.46	85.98±0.87	85.59±0.97
SSDCM	92.99±0.41	92.32±0.20	68.65±0.22	67.55±0.27	83.75±0.28	83.76±0.37	84.18±0.26	86.11±0.29
HDMI	92.72±0.10	92.78±0.10	67.44±0.56	67.71±0.54	**84.08±0.11**	**84.83±0.10**	86.49±0.56	86.78±0.46
X-GOAL	93.56±0.07	93.38±0.06	69.48±0.32	69.71±0.33	83.54±0.13	84.57±0.14	89.04±0.12	89.94±0.13
AmGNN	93.58±0.06	93.61±0.05	69.93±0.99	69.98±1.05	83.02±0.20	83.70±0.25	89.33±0.67	89.18±0.68
Ours+x-GOAL	**94.87±0.45**	**94.91±0.43**	**70.43±0.31**	**70.51±0.35**	83.33±0.14	84.08±0.20	**90.36±0.43**	**90.19±0.39**

Configurations. All experiments are conducted on a machine with Nvidia GPU (NVIDIA RTX A6000, 48 GB memory) and a machine with Nvidia GPU

(NVIDIA RTX A100, 80 GB memory). The learning rate is initialized to 0.001. Besides, all models are trained until convergence, with the maximum training epoch being 1000. The implementations of all baselines are based on Pytorch Geometric or their source code. Train/eval/test splits are set to 3/1/6. The hyperparameters of all methods are tuned on the validation set.

Evaluation Metrics. Following existing works on node classification [11,14], we adopt Accuracy and Macro F1-score as the evaluation metrics.

5.3 Node Classification Performance

In this subsection, we compare the performance of the proposed method with baselines for node classification on the multi-layer graphs, which aims to answer **RQ1**. Each experiment is conducted 5 times. Means and standard deviations are reported in Table 2. From the table, we observe: (**i**) Compared with GCN/GAT/GIN, the performance of GCN/GAT/GIN-C becomes worse on most datasets. It demonstrates that simply combining information from other layers can have a negative effect on the node classification task. Our proposed method can further outperform both GCN/GAT/GIN-C and GCN/GAT/GIN, which verifies the effectiveness of our method to greatly learn structure information in different layers for multi-layer graphs; (**ii**) Our model can also consistently outperform multi-layer GNN models (HAN, mGCN and SSDCM) on most datasets. This verifies our motivation that it's necessary to adaptively preserve shared invariant information and select layers information for each node to facilitate the node classification task; (**iii**) Both X-GOAL and HDMI are state-of-the-art baselines, which adopt contrastive learning on multi-layer graphs. They show greater performance compared with other models. Our model can further outperform them on most datasets. This is because our method can effectively capture inter-layer relations. The potential reason for our method's suboptimal performance on the DBLP dataset is that using the concatenation operation in Eq. 2 can cause an out-of-memory (OOM) issue. Therefore, we replace the concatenation with the addition operation. Furthermore, using embeddings from X-GOAL can further improve the performance.

5.4 Ablation Study

To answer **RQ2**, in this subsection, we conduct an ablation study to evaluate the contribution of each component in AmGNN. Specifically, we consider the following ablations: (**i**) w/o adp, which is a variant by removing the main component that controls the inter-layer aggregation process, i.e., we set γ as 0 for Eq. (12). (**ii**) w/o CL, which

Table 3. Ablation Study on Amazon.

Dataset Metric	Amazon	
	Accuracy	F1-score
w/o adp	88.93 ± 0.70	88.77 ± 0.70
w/o CL	88.97 ± 0.68	88.68 ± 0.72
w/o both	88.75 ± 0.83	88.43 ± 0.88
AmGNN	$\mathbf{89.33 \pm 0.67}$	$\mathbf{89.18 \pm 0.68}$

denotes the variant by removing the key component of multi-layer graph contrastive learning, i.e., we set λ as 0 for Eq. (12). (**iii**) w/o both, which means that we remove these two components. The results are shown in Table 3. We only report the results on Amazon because we can have similar observations on DBLP. All experiments are conducted five times and the means and standard deviations are reported. We can observe that only preserving the shared invariant information across layers (without adp) with multi-layer graph contrastive learning loss can still learn expressive representation for node classification. This observation highlights the effectiveness of our AmGNN in identifying nodes that require the preservation of shared invariant information. Additionally, we have observed that solely adaptively controlling the inter-layer aggregation process (without contrastive learning) can improve the performance of our designed multi-layer graph neural networks. This observation confirms that AmGNN is capable of assisting nodes in selecting trustworthy layers' information for inter-layer aggregation. While we remove these two key components, the performance is worse compared with removing only one component. Our experimental findings confirm the validity of our approach to extract invariant information across layers and selectively aggregate inter-layer information, leading to improved node classification performance. Importantly, the full model (last row) achieves the highest performance, highlighting the complementary nature of the different components in AmGNN.

5.5 Analysis of Confidence Score

To answer **RQ3**, we conduct the visualization experiment in this subsection. Our objective is to investigate the ability of the weights s_i^l in Eq. (10) to identify nodes that require which layers information. For GNNs that focus on aggregating local neighbors' information, nodes that share similar labels with neighbors may have a higher probability to be correctly classified [37]. Hence, nodes opting for layers with higher label matching ratios concerning neighbors in that layer can be deemed as reliable layers, facilitating node classification performance.

To quantify this, we first compute the ratio of neighbors that share the same class as the central nodes. Specifically, we denote the neighbors set of the node v_i as $\mathcal{N}_i^l$ and a set of neighbors having similar labels as v_i as $\mathcal{S}_i^l$. For $v_j \in \mathcal{S}_i^l$, we have $\mathbf{Y}_j = \mathbf{Y}_i$ and $v_j \in \mathcal{S}_i^l$. The label matching ratio for v_i in the layer l is $r_i^l = |\mathcal{S}_i^l|/|\mathcal{N}_i^l|$. And we denote the vector $r_i \in \mathbb{R}^L$, where each dimension is r_i^l. Similarly, we employ the vector $\mathbf{s}_i \in \mathbb{R}^L$ to represent the same concept, where each dimension is denoted as s_i^l. To verify whether $\mathbf{s}_i$ captures relevant layers information to improve the performance of node classification, we calculate the cosine similarity between this local label matching ratio $\mathbf{r}_i$ and $\mathbf{s}_i$. Finally, we visualize the distribution of these cosine similarity values on them, and the results are

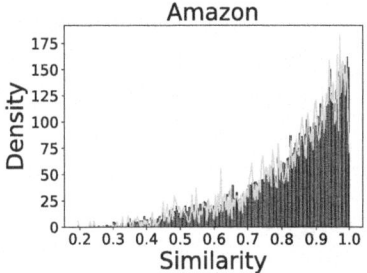

Fig. 4. Visualization Experiments for AmGNN.

shown in Fig. 4. We observe that the majority of cosine similarity values exceed 0.5. The substantial similarity between s_i and r_i shows the effectiveness of the confidence score.

6   Conclusion and Future Work

In conclusion, this paper addresses the problem of node classification in multi-layer graphs by selectively leveraging information from different layers. We propose a novel framework called Adaptive Multi-layer Graph Neural Networks (AmGNN). AmGNN efficiently captures invariant information by introducing a multi-layer graph contrastive learning method that assigns different weights to the contrastive learning loss for each node. We also develop a strategy to determine whether nodes require shared invariant information based on their consistency in predicted results from pretrained GNN models on single layer relations. Additionally, we encourage nodes without consistent prediction results to preserve information from layers with high-confidence predictions. Empirical results on real-world multi-layer graphs demonstrate the effectiveness of AmGNN in node classification tasks.

Several interesting directions need further investigation. First, one limitation is that the concatenation operation in Eq. 3 requires large GPU memory and will cause OOM problems on large-scale graphs. One future direction is how to reduce the GPU memory that our framework needs. Second, for simplicity, we randomly select negative samples for a given node from nodes that are not linked in the original graph. More complicated methods could be used to find negative nodes.

Acknowledgments. This material is based upon work supported by the U.S. Department of Homeland Security under Grant Award Number 17STCIN00001-05-00. The views and conclusions contained in this document are those of the authors and should not be interpreted as necessarily representing the official policies, either expressed or implied, of the U.S. Department of Homeland Security.

Disclosure of Interests. The authors have no competing interests to declare that are relevant to the content of this article.

References

1. Cen, Y., Zou, X., Zhang, J., Yang, H., Zhou, J., Tang, J.: Representation learning for attributed multiplex heterogeneous network. In: Proceedings of SIGKDD (2019)
2. Fan, W., Ma, Y., Li, Q., He, Y., Zhao, E., Tang, J., Yin, D.: Graph neural networks for social recommendation. In: Proceedings of WWW, pp. 417–426 (2019)
3. Fu, D., Xu, Z., Li, B., Tong, H., He, J.: A view-adversarial framework for multi-view network embedding. In: Proceedings of CIKM (2020)
4. Gal, Y., Ghahramani, Z.: Dropout as a bayesian approximation: representing model uncertainty in deep learning. In: Proceedings of ICML, pp. 1050–1059 (2016)

5. Gao, H., Wang, Z., Ji, S.: Large-scale learnable graph convolutional networks. In: Proceedings of SIGKDD, pp. 1416–1424 (2018)
6. Hamilton, W., Ying, Z., Leskovec, J.: Inductive representation learning on large graphs. In: Proceedings of NeurIPS (2017)
7. Hamilton, W.L., Ying, R., Leskovec, J.: Representation learning on graphs: methods and applications. arXiv preprint arXiv:1709.05584 (2017)
8. He, M., Wei, Z., Wen, J.R.: Convolutional neural networks on graphs with chebyshev approximation, revisited. arXiv preprint arXiv:2202.03580 (2022)
9. He, M., Wei, Z., Xu, H., et al.: Bernnet: learning arbitrary graph spectral filters via bernstein approximation. In: Proceedings of NeurIPS (2021)
10. He, R., McAuley, J.: Ups and downs: modeling the visual evolution of fashion trends with one-class collaborative filtering. In: Proceedings of WWW (2016)
11. Jing, B., Feng, S., Xiang, Y., Chen, X., Chen, Y., Tong, H.: X-goal: multiplex heterogeneous graph prototypical contrastive learning. In: CIKM, pp. 894–904 (2022)
12. Jing, B., Park, C., Tong, H.: HDMI: high-order deep multiplex infomax. In: Proceedings of WWW, pp. 2414–2424 (2021)
13. Kendall, A., Gal, Y.: What uncertainties do we need in bayesian deep learning for computer vision? In: Proceedings of NeurIPS (2017)
14. Kipf, T.N., Welling, M.: Semi-supervised classification with graph convolutional networks. arXiv preprint arXiv:1609.02907 (2016)
15. Kipf, T.N., Welling, M.: Variational graph auto-encoders. arXiv preprint arXiv:1611.07308 (2016)
16. Lee, N., Lee, J., Park, C.: Augmentation-free self-supervised learning on graphs. In: Proceedings of AAAI, pp. 7372–7380 (2022)
17. Li, J., Chen, C., Tong, H., Liu, H.: Multi-layered network embedding. In: Proceedings of SDM, pp. 684–692 (2018)
18. Ma, Y., Wang, S., Aggarwal, C.C., Yin, D., Tang, J.: Multi-dimensional graph convolutional networks. In: Proceedings of SDM, pp. 657–665 (2019)
19. Mitra, A., Vijayan, P., Sanasam, R., Goswami, D., Parthasarathy, S., Ravindran, B.: Semi-supervised deep learning for multiplex networks. In: Proceedings of SIGKDD (2021)
20. Mo, Y., Peng, L., Xu, J., Shi, X., Zhu, X.: Simple unsupervised graph representation learning. In: Proceedings of AAAI (2022)
21. Nickel, M., Murphy, K., Tresp, V., Gabrilovich, E.: A review of relational machine learning for knowledge graphs. Proc. IEEE **104**(1), 11–33 (2015)
22. Park, C., Kim, D., Han, J., Yu, H.: Unsupervised attributed multiplex network embedding. In: Proceedings of AAAI, pp. 5371–5378 (2020)
23. Peng, Z., et al.: Graph representation learning via graphical mutual information maximization. In: Proceedings of WWW, pp. 259–270 (2020)
24. Qu, L., Zhu, H., Zheng, R., Shi, Y., Yin, H.: Imgagn: imbalanced network embedding via generative adversarial graph networks. In: Proceedings of SIGKDD (2021)
25. Qu, M., Tang, J., Shang, J., Ren, X., Zhang, M., Han, J.: An attention-based collaboration framework for multi-view network representation learning. In: Proceedings of CIKM, pp. 1767–1776 (2017)
26. Sun, Y., Wang, S., Hsieh, T.Y., Tang, X., Honavar, V.: Megan: a generative adversarial network for multi-view network embedding. In: Proceedings of IJCAI (2019)
27. Tang, J., Zhang, J., Yao, L., Li, J., Zhang, L., Su, Z.: Arnetminer: extraction and mining of academic social networks. In: Proceedings of SIGKDD, pp. 990–998 (2008)

28. Tang, S., Li, B., Yu, H.: Chebnet: efficient and stable constructions of deep neural networks with rectified power units using chebyshev approximations. arXiv preprint arXiv:1911.05467 (2019)
29. Tsitsulin, A., Palowitch, J., Perozzi, B., Müller, E.: Graph clustering with graph neural networks. arXiv preprint arXiv:2006.16904 (2020)
30. Veličković, P., Cucurull, G., Casanova, A., Romero, A., Lio, P., Bengio, Y.: Graph attention networks. arXiv preprint arXiv:1710.10903 (2017)
31. Wang, H., Zhang, J., Zhu, Q., Huang, W.: Augmentation-free graph contrastive learning with performance guarantee. arXiv preprint arXiv:2204.04874 (2022)
32. Wang, X., Ji, H., Shi, C., Wang, B., Ye, Y., Cui, P., Yu, P.S.: Heterogeneous graph attention network. In: Proceedings of WWW (2019)
33. Wang, X., Zhang, M.: How powerful are spectral graph neural networks. In: Proceedings of ICML, pp. 23341–23362 (2022)
34. Xiao, T., Chen, Z., Wang, D., Wang, S.: Learning how to propagate messages in graph neural networks. In: Proceedings of SIGKDD, pp. 1894–1903 (2021)
35. Xu, K., Hu, W., Leskovec, J., Jegelka, S.: How powerful are graph neural networks? In: Proceedings of ICLR (2018)
36. Zhang, M., Chen, Y.: Link prediction based on graph neural networks. In: Proceedings of NeurIPS (2018)
37. Zhu, J., Yan, Y., Zhao, L., Heimann, M., Akoglu, L., Koutra, D.: Beyond homophily in graph neural networks: Current limitations and effective designs. In: Proceedings of NeurIPS (2020)

Detecting Homophobic Speech in Soccer Tweets Using Large Language Models and Explainable AI

Guto Leoni Santos[1]([✉]) [ID], Vitor Gaboardi dos Santos[1] [ID], Colm Kearns[1] [ID],
Gary Sinclair[1] [ID], Jack Black[2] [ID], Mark Doidge[3] [ID], Thomas Fletcher[4] [ID],
Dan Kilvington[4] [ID], Katie Liston[6] [ID], Patricia Takako Endo[5] [ID],
and Theo Lynn[1] [ID]

[1] Dublin City University, Dublin, Ireland
{guto.santos,vitorgaboardidos.santos,colm.g.kearns,
gary.sinclair,theo.lynn}@dcu.ie
[2] Sheffield Hallam University, Sheffield, UK
j.black@shu.ac.uk
[3] Loughborough University, Loughborough, UK
M.Doidge@lboro.ac.uk
[4] Leeds Beckett University, Leeds, UK
{T.E.Fletcher,D.J.Kilvington}@leedsbeckett.ac.uk
[5] Universidade de Pernambuco, Caruaru, Brazil
patricia.endo@upe.br
[6] Ulster University, Belfast, UK
k.liston@ulster.ac.uk

Abstract. Homophobic speech is a form of hate speech. Social media enables hate speech to spread rapidly and widely through the internet, and unlike offline hate speech, can persist indefinitely, thereby prolonging its impact. Due to the adverse impact of hate speech, policymakers have called for greater action from online platforms to moderate and remove hate speech, including homophobic content. While homophobic hate speech is prevalent in online soccer discourses, there are few studies on this empirical context in general and specifically on the use of Large Language Models (LLMs) for detecting such speech. This study addresses this gap by proposing a homophobic speech text classification pipeline. We introduce H-DICT, a new general dictionary for identifying potential homophobic content in documents, and leverage this dictionary to curate and manually label an annotated dataset of homophobic and non-homophobic samples from the UEFA European Football Championships (the Euros) discourse on Twitter. We fine-tune and evaluate five large language models (LLMs) based on the BERT architecture - BERT, DistilBERT, RoBERTa, BERT Hate, and RoBERTa Offensive - and use Integrated Gradients, an explainable AI technique to explain each model's predictions. RoBERTa Offensive, an LLM fine-tuned specifically for detecting offensive language, presented the best performance when compared to the other LLMs.

L. M. Aiello et al. (Eds.): ASONAM 2024, LNCS 15211, pp. 489–504, 2025.
https://doi.org/10.1007/978-3-031-78541-2_30

Keywords: Soccer · Hate speech classification · Homophobic speech · Large language models · Explainable AI

1   Introduction

The Council of Europe defines hate speech as: "all types of expression that incite, promote, spread or justify violence, hatred or discrimination against a person or group of persons, or that denigrates them, by reason of their real or attributed personal characteristics or status such as "race", colour, language, religion, nationality, national or ethnic origin, age, disability, sex, gender identity and sexual orientation" [13]. Such expressions differ in terms of severity, the damage they inflict, and their effects on specific group members in different situations [13]. Homophobia is defined as "the irrational fear of, and aversion to, homosexuality and to lesbian, gay, and bisexual people based on prejudice" [15]. As a gendered concept, homophobic speech is therefore widely accepted as a particular type of hate speech that includes derogatory and threatening language, images, and symbols towards the lesbian, gay, and bisexual community [7]. The adverse impact of hate speech is well documented including emotional and psychological harm [28], social isolation and political radicalisation [4], and in extreme cases, suicide [32]. Similarly, studies suggest that homophobic hate speech has distinct psychological effects on those targeted by such speech, including adverse effects on psychological wellbeing [46] and contributing to depression [35], amongst others [47]. With the increase in hate crimes in general and against the LGBTQ+ community specifically [16], there is increasing pressure by policy-makers on social media platforms to moderate and remove homophobic content [13]. For example, the new European Union Digital Services Act (DSA) mandates online platforms to actively monitor and address issues like hate speech, with financial penalties of up to 6% of their annual global revenue for non-compliance [41].

This paper studies the automatic classification of homophobic hate speech in the soccer discourse on Twitter (now known as X). Social media has provided unprecedented access to sports-related content and opportunities for fans to engage with teams, players, and each other [17]. However, it has also enabled the rapid and widespread propagation of hate speech in the sports discourse [23,37]. While homophobia in soccer has been widely examined in the extant literature [12,19,30], there is a dearth of research on online homophobic hate speech in the soccer context. A recent scoping review by Kearns et al. [23] only identifies six papers on the topic. Similarly, while there has been a rise in the number of studies on homophobic speech detection on social media, particularly using Large Language Models (LLMs) [6,7,9], again there are few studies in the soccer context. We argue that given the linguistic idiosyncrasies of the soccer discourse [5,26], it should be treated as a distinct domain for training language models for homophobic speech detection and classification.

In this paper, we propose a pipeline for automatically detecting and classifying homophobic content in the Twitter discourse in soccer. We use the

UEFA European Football Championships (the Euros) as the empirical context and build a dataset of eight tournaments (four men's and four women's) from 2008 to 2022. After building an annotated dataset, we applied LLMs to classify the tweets and understand how terms and phrases impact performance of the models.

In summary, we make the following contributions:

- A comprehensive general dictionary (H-DICT) of terms, word stems, and phrases for detecting potential homophobic speech. H-DICT is a valuable tool for identifying potential homophobic speech in any document, across various platforms, and is not limited to the soccer domain or the Twitter platform. It can also be used for 'bag of words'-based research or to enhance machine learning and deep learning models.
- A novel manually-annotated dataset focused on homophobic speech on Twitter during the UEFA European Football Championships over a 15-year period. This dataset provides valuable samples of different types of homophobic speech in an international soccer context.
- A performance evaluation of five LLMs fine-tuned using our annotated dataset to classify homophobic speech in textual data in a soccer context. We included models trained in a general context (e.g., Bidirectional Encoder Representations from Transformers (BERT), DistilBERT, and RoBERTa) and models that were fine-tuned to classify hate and offensive speech (BERT Hate and RoBERTa Offensive).
- An analysis of the impact of different input text on the model classification using Explainable Artificial Intelligence (XAI), most specifically Integrated Gradients, to understand why models are classifying the input text as homophobic or not, and then to identify potential improvements for enhancing the models' performance.

Our results suggest that the RoBERTa Offensive model fine-tuned on our annotated dataset achieved the best overall performance.

The rest of this paper is organized as follows: Sect. 2 introduces the background on LLM and XAI. Section 3 summarises related works on detecting homophobic speech on Twitter using Machine Learning (ML) and Deep Learning (DL) models. Section 4 presents the data and methodology, highlighting the dictionary development, dataset collection and labelling, the LLMs employed for homophobic detection, and the XAI method used to understand the model's decision-making. The results for the evaluation of LLMs performance and the outcomes from the XAI analysis are presented in Sect. 5. Section 6 concludes the paper and presents avenues for future research.

2   Background

2.1   Large Language Models

LLMs are advanced language models with a substantial number of parameters and trained on extensive text datasets, leading to notably improved performance across a wide range of applications [10]. The foundational technology behind LLMs is the Transformer [44] architecture, which introduces a new approach to sequence modelling by processing input data concurrently through attention mechanisms and facilitating the capture of extensive contextual dependencies. BERT [14] leverages the Transformer architecture. It has proven to be particularly effective in detecting hate speech across social media platforms, including Twitter (X) [31].

BERT employs a two-step pre-training approach. First, masked language modelling is used to predict a random subset of hidden words in a sentence. Second, sentence prediction predicts whether a given sentence follows another coherently and logically. This pre-training provides BERT with a deep understanding of contextual language representations. Pre-trained on vast corpora such as books and Wikipedia articles, BERT is fine-tuned for specific tasks, such as sentiment analysis, question answering, or, as illustrated in this paper, text classification.

Other variations of the BERT model architecture have been developed to improve performance and address processing constraints. For instance, Distil-BERT [36] is a smaller and faster version of BERT with a distinct pre-training strategy but comparable language understanding capabilities and performance. Robustly optimized BERT approach (RoBERTa) [29] eliminates the next sentence prediction task and extends the model's training to encompass longer text sequences, enriching its contextual comprehension. Additionally, RoBERTa adopts dynamic masking throughout pre-training which leads to the acquisition of more versatile representations.

2.2   Explainable Artificial Intelligence

XAI assists in understanding the decisions generated by LLMs. It is a critical tool in applications where transparency and accountability are essential, such as content moderation on social media platforms. Integrated Gradients [39] is a method used to interpret the predictions made by ML models by attributing the model prediction to its input features. It offers insights into how the input features contribute to the model's decision-making process.

In the context of text classification, the aim is to compute the contribution of each word or token to the prediction of the Natural Language Processing (NLP) model. The Integrated Gradients method involves creating a path consisting of a series of intermediate inputs. These are created by linearly interpolating the baseline (e.g., an empty input) and the actual input. Along this path, the model generates predictions at different steps, with the number of words or tokens changing incrementally at each step. The method then approximates the integral of gradients along this path. The rationale behind this approach lies in systematically assessing the impact of individual features on the model's prediction at each step. This allows for the evaluation of the influence of all features throughout the entire path.

Formally, given a model f with input features x, the Integrated Gradients technique calculates the feature attributions IG_i for each feature x_i as follows:

$$IG_i(x) = (x_i - x_i') \times \int_{\alpha=0}^{1} \frac{\partial f(x' + \alpha(x - x'))}{\partial x_i} d\alpha \qquad (1)$$

where x' is the baseline input, x_i is the i-th feature of the actual input x, $\frac{\partial f}{\partial x_i}$ represents the partial derivative of the model's output with respect to the i-th feature, α denotes the interpolation coefficient between the baseline input x' and the actual input x. By integrating the gradients along this path, this technique provides a better understanding of feature contributions across the input space. This enables stakeholders to interpret the model's decisions and identify influential features.

3   Related Work

While there is a relatively established literature on the use of LLMs for hate speech detection [2,11,21], the research on the use of LLMs for homophobic speech classification on social media is relatively small. Chakravarthi et al. [7] evaluated four LLMs for detecting homophobic and transphobic content on social media in India. All models evaluated were based on the BERT architecture, namely mBERT, XLM-RoBERTa, MuRIL, and Indic-BERT. To address a lack of resources for testing and training models, the authors use a data augmentation via pseudolabeling approach. The models were evaluated using metrics for precision, recall, F1-score, and accuracy. The mBERT model achieved the best performance.

In a related work, Chakravarthi et al. [6] presented a dataset for testing and training models for classifying homophobic and transphobic content on YouTube in English, Tamil, and both English and Tamil. They also present the results of their experiment classifying content using ML and DL models and LLMs, showcasing the results from other researchers using the same dataset. They found that ensemble models worked significantly better than either ML or DL alone. A proposed model using the RoBERTa based model achieved the best results.

Nilsson et al. [34] proposed a pipeline that fine-tunes a multilingual transformer, XLM-RoBERTA, using multitask learning with SBERT as the teacher model to detect homophobic and transphobic content in YouTube comments in English, Tamil, and Malayalam. They compared single-task and multitask models and find that the former outperforms the latter.

Garcia et al. [18] addressed homophobic detection in Mexican Spanish to solve two tasks: (a) detecting whether a given tweet is homophobic and (b) identifying between different types of phobia. They conducted fine-tuning on various monolingual and multilingual LLMs, extracting sentence embeddings for each model. These embeddings were then input into a multi-input neural network. They found out that ensemble approaches lead to the best performance.

As can be seen from above, detecting homophobic speech is not a new ML/DL task, however we make a number of contributions. Firstly, we develop a general dictionary of English language words, word stems, and phrases (H-DICT) that can be used to identify potentially homophobic content, significantly expanding the available dictionaries. Furthermore, it is a general contribution in that it can be used to detect potential homophobic content in any document or platform and independent of the soccer context. This is an important contribution, as language is not static. Many new words, and particularly offensive language, make their way into use through social media. Secondly, while the LLMs we evaluate are present in the literature, such models perform differently in different contexts. We focus on a particular empirical context, soccer, which has been found to have specific linguistic idiosyncrasies that require discrete consideration [5, 26]. Consequently, we make a contribution by both using a unique annotated dataset of soccer tweets featuring homophobic language examples for testing and training ML and DL models for the homophobic speech detection task. Finally, our work applies XAI to identify individual word impact on the models' prediction. This is critical for understanding which content features are influencing the model's classification and, where necessary, fine-tuning training datasets and models to address common misclassifications. XAI is critical in hate speech classification, particularly on social media, to ensure that algorithms transparently balance individuals' rights to freedom of expression in compliance with legislation such as the DSA, thereby minimising instances of unfair censorship or erroneous labelling of content as homophobic or other forms of discrimination.

4   Data and Methods

Figure 1 presents the pipeline used to create the annotated datasets, to fine-tune LLMs to classify homophobic content in the soccer discourse, and evaluate the

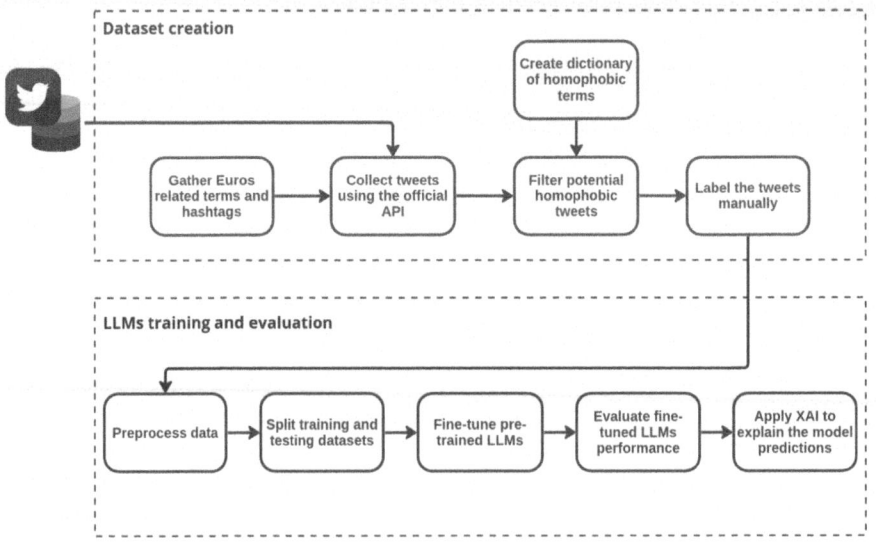

Fig. 1. Homophobic speech text classification pipeline.

models' performance. The pipeline can be divided into two flows- (1) creating an annotated dataset with homophobic content, and (2) training and evaluating LLMs. In the following sections, we will describe the pipeline steps.

4.1   Dataset

For this study, we curated tweets associated with the Euros using the Twitter Enterprise API. Spanning one week before and one week after each tournament, we collected tweets from a total of eight men's and women's tournaments from 2008 to 2022. Our collection criteria comprised tweets featuring the official tournament hashtag, references to the tournament name, mentions of tournament-specific usernames, UEFA, and FIFA. Furthermore, we included hashtags pertaining to all matches in the tournaments and abbreviations for team names (e.g., the match between England and Russia was tagged as #ENGvRUS and #ENGRUS). We also included the official championship hashtags (e.g., #euro2016 and 'euro 2016', etc.), official Twitter accounts (e.g., @euro2016, etc.). We stored all the tweets in a local database for further analysis.

To build our annotated datasets of homophobic speech, we developed a general dictionary of terms, words stems, and phrases. The dictionary (H-DICT) was initially populated with 68 terms sourced from the Hatebase project[1], an online platform designed to aid organisations in moderating online discourse and identifying hate speech. We then expanded this dictionary with additional homophobic terms cited in existing literature and references on homophobic speech. It is important to note that when using such a dictionary, researchers need to also consider plurals, hyphenated forms, misspellings, and word combinations including those found within hashtags or in combination with player's names (e.g., 'Fabregay' instead of 'Fabregas'). For filtering purposes, it is also useful to include specific widely-used terms and phrases even where there is a common word stem e.g., word combinations with 'gay' or 'homo'. We make H-DICT available on a GitHub repository[2] that can be used and extended by another researchers.

Leveraging this dictionary, we employed SQL queries to search through our database for potential homophobic tweets. It is important to note that some terms and word stems may result in false positives due to overlaps of general and domain-specific vocabulary overlapping, as well as semantic ambiguity. For example, 'trans' is a term and word stem commonly used in homophobic speech but is also a common word stem in the language of soccer, e.g. "transferring to" and "transfer window". Terms associated with the LGBTQ+ community, e.g., "pride", may also be used in other domain-specific contexts, e.g., 'English pride' or 'take pride in'. Similarly, some potentially homophobic terms and word stems may overlap with player names, e.g. Lyndon Dyke (Scotland) and Lars Bender (Germany). Consequently, a manual review by three human annotators specializing in hate speech was required to identify a sufficient volume and diversity of

---

[1] https://hatebase.org/.

[2] https://github.com/GutoL/H-DICT.

true homophobic tweets. The final decision on the label is defined where all three agree. For the purpose of this paper, our target dataset size was 1,000 tweets per class featuring homophobic content; the ultimate dataset size was 1,005.

In addition to gathering homophobic tweets, it was necessary to obtain a sample of non-homophobic tweets to allow the training of our models to distinguish between homophobic and non-homophobic content. This involved crafting queries that excluded items from H-DICT when selecting non-homophobic tweets. Human annotators evaluated these tweets to ensure the absence of homophobic connotations and to validate their pertinence to soccer-related discussions. To maintain dataset balance, an equal number of homophobic and non-homophobic tweets were incorporated. Our final dataset contains 2,010 tweets, evenly divided with 1,005 examples of homophobic tweets and an equivalent number of non-homophobic tweets.

It is important to note that there are pre-existing annotated datasets available both for hate speech (e.g. [3]) and homophobic speech (e.g. [8,43]. The former may not contain specific words for homophobic speech, while the latter are in different languages. Additionally and as discussed in Sect. 1, the soccer discourse features linguistic idiosyncrasies [5,26]. In both cases, the existing labelled datasets may not contain specific words and expressions that are related to the soccer context, and we decided to not include them in our study.

While some language models used in this study have the capability to process raw text, we opted to employ text preprocessing techniques to enhance comprehension by eliminating extraneous elements or noise [25]. Consequently, following data collection, we conducted preprocessing procedures involving the conversion of text to lowercase and the removal of stop words, user mentions, URLs, and emojis. Subsequently, the dataset was partitioned into training and testing sets, allocating randomly 80% of the samples for training purposes and reserving the remaining 20% for model evaluation.

4.2   Large Language Models

After building the annotated datasets, we fine-tuned several pre-trained LLMs to classify homophobic content using the text in tweets. This process of fine-tuning LLMs has demonstrated efficacy in attaining state-of-the-art performance across various downstream tasks [27,38].

Five models are used in this study - BERT, [14], DistilBERT [36], RoBERTa [29], BERT Hate [24], and RoBERTa Offensive [3]. All these LLMs are available on the Hugging Face platform. For BERT, we use the uncased version of BERT[3], which does not make a difference between upper and lower case. Similarly, we employ an uncased version of DistilBERT[4]. The main goal of using DistilBERT architecture is to have a lighter version of BERT and check if this model is able to give competitive results against other BERT-based models. We utilize the base model of RoBERTa[5]. Unlike BERT and DistilBERT, RoBERTa is case-sensitive.

---

[3] https://huggingface.co/bert-base-uncased.

[4] https://huggingface.co/distilbert/distilbert-base-uncased.

[5] https://huggingface.co/FacebookAI/roberta-base.

Therefore, we converted all texts to lower case for a fairer comparison. Since the main goal of this paper is to detect homophobic content on Twitter in the English language, we included BERT Hate[6], a monolingual model for hate speech classification of social media content in English language [24]. BERT Hate is a version of the base BERT pre-trained language model fine-tuned with a dataset of 20,227,765 texts, comprising YouTube comments and tweets in English. Finally, we also employed RoBERTa Offensive[7], a version of the RoBERTa model fine-tuned with a dataset composed of 14,100 tweets to identify offensive speech [3].

The fine-tuning process was performed using the following hyperparameters: 10 epochs, 5×10^{-6} learning rate, AdamW optimizer, batch size of 8 samples, and we save the model that provided the lowest loss value during the training. The training and experiments were performed using a computer with Intel(R) Core(TM) i7-12700 CPU at 2.10GHz, 32 GB RAM, and Nvidia GeForce GTX 1660 SUPER. In order to evaluate the LLMs performance, we used traditional metrics: accuracy, precision, recall, and F1-score.

4.3   Explainable AI

After fine-tuning the LLMs, we employed Integrated Gradients to verify the correlation between input data and classification output [39]. This method has gaining popularity since it can be applied to any differentiable model, has strong theoretical foundations, and it is computationally efficient compared to other models. Integrated Gradients assign a score to each input token, indicating the token's influence on the model's prediction. A positive score implies that the token influenced the model's prediction, while a negative score suggests the token had an opposing influence on the model's prediction. By applying Integrated Gradients to some tweets, we discerned the influence of specific terms on the model's classification, assisting us to identify words closely associated with homophobia in a soccer context. The Google wordpiece tokenization mechanism [45] divides the words into tokens, which can generate sub-word units. As the Integrated Gradients method calculates a contribution score for each token, we averaged these scores to obtain a composite score for divided words.

5   Results

Table 1 presents benchmark results for the LLM models. The BERT model was the model with the lowest performance, with all metrics around 93.5%. The DistilBERT model presented slightly better results than BERT, with all metrics around 94.4%. The superiority of DistilBERT over BERT can be explained by our limited dataset (2,010 tweets). As a result, DistilBERT converged faster than the base BERT model since it has fewer parameters to adjust during training [20].

---

[6] https://huggingface.co/IMSyPP/hate_speech_en.
[7] https://huggingface.co/cardiffnlp/twitter-roberta-base-offensive.

The RoBERTa base model presented better results than the base BERT model and DistilBERT, with all the metrics around 96%. This is unsurprising given RoBERTa is trained on a significantly larger corpus than the base BERT model and features additional refinements (e.g., dynamic masking) to the base BERT model. This outperformance of RoBERTa over BERT is consistent with existing literature [33].

Table 1. Summary metrics for each fine-tuned LLM.

Model	Accuracy	Precision	Recall	F1-score
BERT	93.5323	93.5660	93.5323	93.5282
DistilBERT	94.4030	94.4118	94.4030	94.4037
RoBERTa	96.0199	96.0216	96.0199	96.0201
BERT Hate	96.1443	96.1698	96.1443	96.1450
RoBERTa Offensive	**97.0149**	**97.0165**	**97.0149**	**97.0151**

Models already fine-tuned for hate or offensive speech showed the best results for classifying homophobic content in our experiments. The BERT Hate model presented a small improvement when compared to RoBERTa with around 96.1% for all metrics. The RoBERTa Offensive model outperformed all models in the homophobia detection task, presenting all metrics at circa 97.01%. The RoBERTa Offensive model improved all metrics by 3.5%-points over the base BERT model, which performed the worst.

In order to provide a qualitative analysis, Fig. 2 shows the embeddings for the BERT and RoBERTa Offensive models, the worst and best models, respectively. To compute the embeddings, we selected randomly 100 samples of homophobic and non-homophobic tweets from the testing dataset. The tweets were fed into the fine-tuned models, and the embedding representation is the output of the last layer.

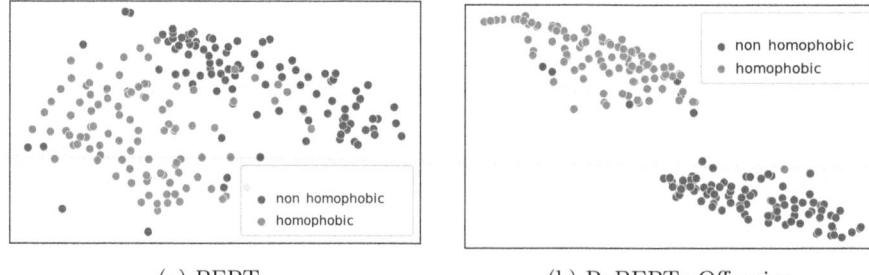

(a) BERT.   (b) RoBERTa Offensive.

Fig. 2. Embeddings for 100 samples from the testing dataset.

Figure 2a illustrates that, even though the BERT model was able to separate the two groups of tweets, there are some points of homophobic tweets that are within the group of non-homophobic tweets. Similarly, in the lower left corner of Fig. 2a, there are some non-homophobic tweets that are within the homophobic group. These finding may indicate misclassifications that adversely affected the model's performance.

On the other hand, the RoBERTa Offensive model created more disjoint groups, as shown in Fig. 2b. The group of non-homophobic tweets is located in the lower right corner, while the homophobic tweets group is located in the upper left corner. There are few non-homophobic tweets that are in the homophobic group, which can be considered false positives. In the context of classifying homophobic content, false positives are less harmful than false negatives, as a tweet that is not homophobic could be classified as homophobic. However, a tweet that is homophobic being classified as non-homophobic is more harmful, since harmful content would go unnoticed by the model. As shown in Fig. 2b, only one tweet that is homophobic is in the non-homophobic group, shown that the RoBERTa Offensive model was able to identify false negatives.

Table 2 shows three examples of homophobic tweets and the predictions made by the BERT and RoBERTa Offensive models. The use of the term "gayest" in Example 1 implies that Mexican Waves, a common stadium activity during sports events, are somehow frivolous, silly, or lacking in masculinity. Using "gay" as a pejorative term in this way is disrespectful and reinforces harmful stereotypes about LGBTQ+ individuals. Both BERT and RoBERTa Offensive models were able to identify the homophobic content in this text.

Table 2. Classification of homophobic tweets with BERT and RoBERTa Offensive models.

#	Tweet Text	BERT Prediction	RoBERTa Prediction
1	Mexican Waves are the gayest thing ever. #Wimbledon #Euro2012	homophobic	homophobic
2	Football now a game for pansies! #SWISPA #EURO2020	non-homophobic	homophobic
3	Since Greizmann grew his hair like a Pansy, he hasn't been the same player #Fra #EURO2020	non-homophobic	non-homophobic

In Example 2, the user uses the term "pansies" which is often used as a derogatory slang term to insult someone's masculinity or imply that they are weak or effeminate, which can perpetuate harmful stereotypes about gender and sexuality. Considering the context of the tweet, it suggests that soccer is becoming less masculine or less tough, implicitly implying that toughness or masculinity is a necessary or desirable quality in sports which in turn can contribute to a culture of toxic masculinity. The BERT model was not able to detect the homophobic content in this tweet, while RoBERTa Offensive model was able to identify that the word "pansies" was used in an offensive way. Example 3 shows an example of homophobic content where both models were not able to classify the text correctly. Similar to Example 2, the term "pansy" is used in an offensive way to insult Griezmann's hairstyle. Again, this language perpetuates harmful stereotypes about masculinity and implies that adopting a certain hairstyle or appearance is undesirable or weak. Therefore, this speech is disrespectful not only to Griezmann, but also to individuals who may choose to express themselves through their appearance in various ways. For this example, both models were not able to classify it correctly, i.e., even the RoBERTa model, which was the model with the best performance, was not able to classify it correctly.

To better understand the misclassification of the RoBERTa Offensive model in Example 3, we used the Integrated Gradients technique. Figure 3 shows the contribution score of each token on the model's prediction. In this case, we are analysing the predictions that say that a homophobic tweet was not homophobic (a false negative example). Figure 3a indicates that the word "Pansy" has the highest negative value, suggesting that it influences the model prediction towards a negative result, in this case it indicates the content is likely homophobic. Despite the expected negative score for this word due to its homophobic connotation, the final model classification was non-homophobic. We believe this misclassification is likely related to the positive weighting RoBERTa gave to the remaining tokens relative to the non-homophobic class. For instance, the token "#EURO2020" had a relatively high score towards the classification, while it should be more neutral.

As a result, we removed the hashtags words ("#EURO2020" and "#FRA") to evaluate the model performance for this specific example, a common pre-processing procedure when dealing the tweets [31]. As shown in Fig. 3b, the model now correctly classified the tweet as homophobic and the sequence of words "like a Pansy" had the highest positives scores, indicating the impact of these tokens on the prediction. These findings are important as it highlights the benefit of using XAI for understanding the relative impact of unexpected content features on the classification task. Using XAI merely for homophobic content would nearly be pointless given the use of the H-DICT dictionary. However, XAI provides a unique insight on the impact of non-homophobic terms, in this case the tournament hashtags. Once corrected, the model's performance is further improved. We hypothesise that this misclassification occurred due to a higher incidence of terms like "#EURO2020" in the non-homophobic samples during the training phase, which lead to a slightly higher influence of these words into the non-homophobic class.

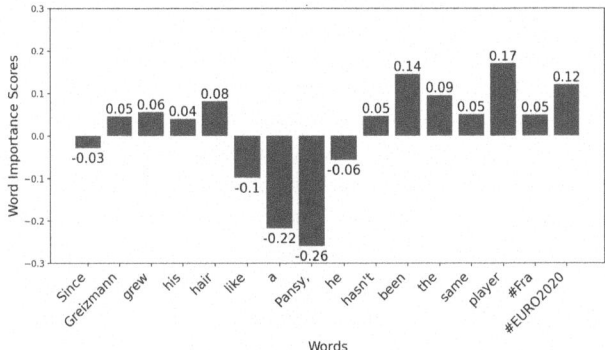

(a) Example of tweet text including hashtags misclassified as non-homophobic.

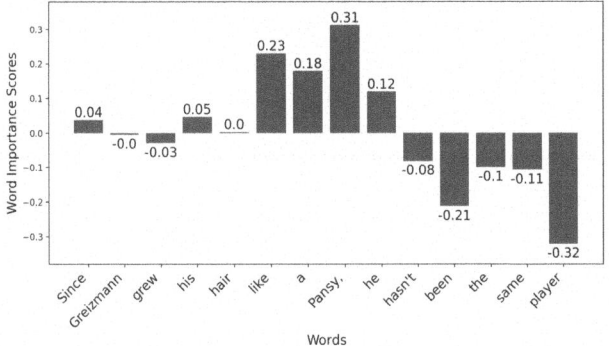

(b) Example of tweet text excluding hashtags classified correctly as homophobic.

Fig. 3. Integrated Gradients results for Example 3 in Table 2 misclassified by the RoBERTa Offensive model.

6   Conclusion

This research explored the critical issue of hate speech and derogatory language on social media platforms, with a specific focus on homophobic content in the online soccer discourse on Twitter. We developed H-DICT, a general dictionary of terms, word stems, and phrases that can be used to identify potentially homophobic content in documents. Using H-DICT, we created an annotated dataset for training and testing models for classifying homophobic speech specifically in soccer. Using this annotated dataset, we then fine-tuned five variations of BERT-based models to classify homophobic speech in English language tweets from the Euros discourse on Twitter. Our performance evaluation of these LLMs, suggested that the RoBERTa Offensive model performed best, with all the metrics circa 97.01%. Furthermore, we used Integrated Gradients to explain how specific tokens contribute to the model prediction. This analysis helped us understand the reason why some tweets were misclassified as non-homophobic, despite the models' ability to identify the key tokens that rendered the tweets homophobic.

Homophobia is not confined to any single language or culture; it manifests in various forms worldwide. Expanding the detection of homophobic speech beyond English is crucial for fostering inclusivity and combatting discrimination on a global scale. In future research, we plan to expand H-DICT with neologisms from sources such as Urban Dictionary and for international languages. Furthermore, we plan to use H-DICT to create datasets from different online platforms (e.g. Facebook, Instagram, and TikTok) and expand for different sports, in order to create a larger and more diverse dataset. This will enable more generalised use, but also greater insights when operationalised on full datasets. As shown Fig. 3, removing the hashtags resulted in an improvement in the model classification. Therefore, we plan to use XAI to identify terms, which can be considered noise, that can be removed during the preprocessing step. In this paper, we used BERT-based models exclusively. In the future, we will evaluate emerging models including Mistral [22], LLaMA [42], and API-based LLM services such as OpenAI GPT models [1] and Google Gemini [40]. Through fine-tuning, these LLMs may be more effective in classifying homophobic hate speech and in particular identifying more nuanced linguistic patterns indicative of homophobic speech.

Acknowledgment. The research in this paper was partially funded by the UK Arts and Humanities Research Council and the Irish Research Council (Grant Number AH/W001624/1) and the Federation Internationale de l'Automobile.

References

1. Achiam, J., et al.: GPT-4 technical report. arXiv preprint arXiv:2303.08774 (2023)
2. Anjum, Katarya, R.: Hate speech, toxicity detection in online social media: a recent survey of state of the art and opportunities. Int. J. Inf. Secur. 1–32 (2023)
3. Barbieri, F., Camacho-Collados, J., Neves, L., Espinosa-Anke, L.T.: Unified benchmark and comparative evaluation for tweet classification. arXiv preprint arXiv:2020.12421 (2020)
4. Bilewicz, M., Soral, W.: Hate speech epidemic. the dynamic effects of derogatory language on intergroup relations and political radicalization. Polit. Psychol. **41**, 3–33 (2020)
5. Billings, A.C.: Defining Sport Communication. Taylor & Francis (2016)
6. Chakravarthi, B.R.: Detection of homophobia and transphobia in youtube comments. Int. J. Data Sci. Anal. 1–20 (2023)
7. Chakravarthi, B.R., Hande, A., Ponnusamy, R., Kumaresan, P.K., Priyadharshini, R.: How can we detect homophobia and transphobia? experiments in a multilingual code-mixed setting for social media governance. Int. J. Inf. Manag. Data Insights **2**(2), 100119 (2022)
8. Chakravarthi, B.R., et al.: Dataset for identification of homophobia and transophobia in multilingual youtube comments. arXiv preprint arXiv:2109.00227 (2021)
9. Chanda, S., Mishra, A., Pal, S.: Sentiment analysis and homophobia detection of code-mixed dravidian languages leveraging pre-trained model and word-level language tag. In: Working Notes of FIRE 2022-Forum for Information Retrieval Evaluation (Hybrid). CEUR (2022)

10. Chang, Y., et al.: A survey on evaluation of large language models. ACM Trans. Intell. Syst. Technol. (2023)
11. Chiu, K.L., Collins, A., Alexander, R.: Detecting hate speech with GPT-3. arXiv preprint arXiv:2103.12407 (2021)
12. Cleland, J., MacDonald, C.: Social media, digital technology, and masculinity in sport. In: Sport, Social Media, and Digital Technology: Sociological Approaches, pp. 49–66. Emerald Publishing Limited (2022)
13. Council of Europe: Combating hate speech. Council of Europe (2022)
14. Devlin, J., Chang, M.W., Lee, K., Toutanova, K.: Bert: pre-training of deep bidirectional transformers for language understanding. arXiv preprint arXiv:1810.04805 (2018)
15. European Union Agency for Fundamental Rights: Homophobia and discrimination on grounds of sexual orientation and gender identity in the EU member states: Part II-The social situation. European Union Agency for Fundamental Rights (2009)
16. FBI: FBI releases 2022 crime in the nation statistics. FBI (2023)
17. Fenton, A., Keegan, B.J., Parry, K.D.: Understanding sporting social media brand communities, place and social capital: a netnography of football fans. Commun. Sport 11(2), 313–333 (2023)
18. García-Díaz, J.A., Jiménez-Zafra, S.M., Valencia-García, R.: Umuteam at homomex 2023: fine-tuning large language models integration for solving hate-speech detection in Mexican Spanish (2023)
19. Glynn, E., Brown, D.H.: Discrimination on football twitter: the role of humour in the othering of minorities. Sport Soc. 26(8), 1432–1454 (2023)
20. Gupta, P., Gandhi, S., Chakravarthi, B.R.: Leveraging transfer learning techniques-bert, roberta, albert and distilbert for fake review detection. In: Proceedings of the 13th Annual Meeting of the Forum for Information Retrieval Evaluation, pp. 75–82 (2021)
21. Jahan, M.S., Oussalah, M.: A systematic review of hate speech automatic detection using natural language processing. Neurocomputing 126232 (2023)
22. Jiang, A.Q., et al.: Mistral 7b. arXiv preprint arXiv:2310.06825 (2023)
23. Kearns, C., et al.: A scoping review of research on online hate and sport. Commun. Sport 11(2), 402–430 (2023)
24. Kralj Novak, P., Scantamburlo, T., Pelicon, A., Cinelli, M., Mozetič, I., Zollo, F.: Handling disagreement in hate speech modelling. In: International Conference on Information Processing and Management of Uncertainty in Knowledge-Based Systems, pp. 681–695. Springer, Cham (2022)
25. Kurniasih, A., Manik, L.P.: On the role of text preprocessing in bert embedding-based DNNs for classifying informal texts. Neuron 1024(512), 927–34 (2022)
26. Lavric, E., Pisek, G., Skinner, A., Stadler, W.: The linguistics of football, vol. 38. Narr Francke Attempto Verlag (2008)
27. Lee, J.S., Hsiang, J.: Patent classification by fine-tuning bert language model. World Patent Inf. 61, 101965 (2020)
28. Leets, L., Giles, H.: Words as weapons–when do they wound? Investigations of harmful speech. Hum. Commun. Res. 24(2), 260–301 (1997)
29. Liu, Y., et al.: Roberta: a robustly optimized bert pretraining approach. arXiv preprint arXiv:1907.11692 (2019)
30. Magrath, R.: 'To try and gain an advantage for my team': homophobic and homosexually themed chanting among english football fans. Sociology 52(4), 709–726 (2018)

31. Mozafari, M., Farahbakhsh, R., Crespi, N.: A bert-based transfer learning approach for hate speech detection in online social media. In: Complex Networks and Their Applications VIII: Volume 1 Proceedings of the Eighth International Conference on Complex Networks and Their Applications COMPLEX NETWORKS 2019 8, pp. 928–940. Springer, Cham (2020)
32. Mullen, B., Smyth, J.M.: Immigrant suicide rates as a function of ethnophaulisms: hate speech predicts death. Psychosom. Med. **66**(3), 343–348 (2004)
33. Murarka, A., Radhakrishnan, B., Ravichandran, S.: Detection and classification of mental illnesses on social media using roberta. arXiv preprint arXiv:2011.11226 (2020)
34. Nilsson, F., Al-Azzawi, S.S.S., Kovács, G.: Leveraging sentiment data for the detection of homophobic/transphobic content in a multi-task, multi-lingual setting using transformers. In: 14th Forum for Information Retrieval Evaluation, FIRE 2022, 9–13 December 2022, Kolkata, India, vol. 3395, pp. 196–207. CEUR-WS (2022)
35. Polders, L.A.: Factors affecting vulnerability to depression among gay men and lesbian women. Ph.D. thesis, University of South Africa (2006)
36. Sanh, V., Debut, L., Chaumond, J., Wolf, T.: Distilbert, a distilled version of bert: smaller, faster, cheaper and lighter. arXiv preprint arXiv:1910.01108 (2019)
37. Santos, G.L., et al.: Kicking prejudice: large language models for racism classification in soccer discourse on social media. In: International Conference on Advanced Information Systems Engineering, pp. 547–562. Springer, Cham (2024)
38. dos Santos, V.G., Santos, G.L., Lynn, T., Benatallah, B.: Identifying citizen-related issues from social media using LLM-based data augmentation. In: International Conference on Advanced Information Systems Engineering, pp. 531–546. Springer, Cham (2024)
39. Sundararajan, M., Taly, A., Yan, Q.: Axiomatic attribution for deep networks. In: International Conference on Machine Learning, pp. 3319–3328. PMLR (2017)
40. Team, G., et al.: Gemini: a family of highly capable multimodal models. arXiv preprint arXiv:2312.11805 (2023)
41. Tourkochoriti, I.: The digital services act and the EU as the global regulator of the internet. Chi. J. Int. L. **24**, 129 (2023)
42. Touvron, H., et al.: Llama 2: open foundation and fine-tuned chat models. arXiv preprint arXiv:2307.09288 (2023)
43. Vásquez, J., Andersen, S., Bel-Enguix, G., Gómez-Adorno, H., Ojeda-Trueba, S.L.: Homo-mex: a Mexican Spanish annotated corpus for LGBT+ phobia detection on twitter. In: The 7th Workshop on Online Abuse and Harms (WOAH), pp. 202–214 (2023)
44. Vaswani, A., et al.: Attention is all you need. In: Advances in Neural Information Processing Systems, vol. 30 (2017)
45. Wu, Y., et al.: Google's neural machine translation system: bridging the gap between human and machine translation. arXiv preprint arXiv:1609.08144 (2016)
46. Zochniak, K., Lewicka, O., Wybrańska, Z., Bilewicz, M.: Homophobic hate speech affects well-being of highly identified LGBT people. J. Lang. Soc. Psychol. 0261927X231174569 (2023)
47. Stefăniță, O., Buf, D.M.: Hate speech in social media and its effects on the LGBT community: a review of the current research. Rom. J. Commun. Public Relations **23**(1), 47–55 (2021)

Author Index

© The Editor(s) (if applicable) and The Author(s), under exclusive license
to Springer Nature Switzerland AG 2025
L. M. Aiello et al. (Eds.): ASONAM 2024, LNCS 15211, pp. 505–506, 2025.
https://doi.org/10.1007/978-3-031-78541-2

The manufacturer's authorised representative in the EU is Springer
Nature Customer Service Centre GmbH, Europaplatz 3, 69115 Heidelberg,
Germany. If you have any concerns regarding our products, please
contact ProductSafety@springernature.com

Printed and bound by CPI Group (UK) Ltd, Croydon, CR0 4YY

29/04/2026

02099544-0017